MATHEMATICS OF FINANCE

WITH CANADIAN APPLICATIONS

S. A. HUMMELBRUNNER

K. SUZANNE COOMBS
Kwantlen University College

Contributing Author
BRUCE COOMBS
Kwantlen University College

FIFTH EDITION

Toronto

National Library of Canada Cataloguing in Publication

Hummelbrunner, S. A. (Siegfried August)
Mathematics of finance with Canadian applications / S.A. Hummelbrunner, Suzanne Coombs.

Includes index.
ISBN 0-13-129028-2

1. Business mathematics—Textbooks. I. Coombs, Suzanne II. Title.

HF5691.H86 2005 650'.01'513 C2004-902627-5

ISBN 0-13-129028-2

Vice President, Editorial Director: Michael Young
Senior Acquisitions Editor: Gary Bennett
Director of Marketing: Bill Todd
Senior Developmental Editor: Madhu Ranadive
Production Editor: Joel Gladstone
Copy Editor: Rodney Rawlings
Production Coordinator: Janette Lush
Page Layout: Carolyn Sebestyen
Permissions Research: Nicola Winstanley
Art Director: Julia Hall
Interior and Cover Design: Anthony Leung
Cover Image: Getty Images/Photodisc Collection

2 3 4 5 09 08 07 06 05

Printed and bound in the USA.

Brief Contents

Contents

Preface

INTRODUCTION

Mathematics of Finance is intended for use in introductory mathematics of finance courses in business administration programs. In a more general application it also provides a comprehensive basis for those who wish to review and extend their understanding of business mathematics.

The primary objective of the text is to increase the student's knowledge and skill in the solution of practical financial and mathematical problems encountered in the business community. It also provides a supportive base for mathematical topics in finance, accounting, and marketing.

ORGANIZATION

Mathematics of Finance is a teaching text using the objectives approach. The systematic and sequential development of the material is supported by carefully selected and worked examples. These detailed step-by-step solutions presented in a clear and colourful layout are particularly helpful in allowing students, in either independent studies or in the traditional classroom setting, to carefully monitor their own progress.

Each topic in each chapter is followed by an Exercise containing numerous drill questions and application problems. The Review Exercise, Self-Test, and the Case Studies at the end of each chapter integrate the material studied.

The text is based on Canadian practice, and reflects current trends using available technology—specifically the availability of reasonably priced electronic pocket calculators. Students using this book should have access to calculating equipment having a power function and a natural logarithm function. The use of such calculators eliminates the arithmetic constraints often associated with financial problems and frees the student from reliance on financial tables.

The power function and the natural logarithm function are often needed to determine values that will be used for further computation. Such values should not be rounded and all available digits should be retained. The student is encouraged to use the memory to retain such values.

When using the memory the student needs to be aware that the number of digits retained in the registers of the calculator is greater than the number of digits displayed. Depending on whether the memory or the displayed digits are used, slight differences may occur. Such differences will undoubtedly be encountered when working through the examples presented in the text. However, they are insignificant and should not be of concern. In most cases the final answers will agree, whichever method is used.

Students are encouraged to use preprogrammed financial calculators, though this is not essential. The use of preprogrammed calculators facilitates the solving of most financial problems and is demonstrated extensively in chapters 3 to 10.

NEW TO THIS EDITION

In this fifth edition major revisions have been made to the text. To reflect current practices in Canada and to better suit the needs of users of this book, a number of important changes have been made in content and organization.

Appendix I, entitled Trade Discount, Cash Discount, Markup, and Markdown, is new to the fifth edition. Merchandising terminology has been clarified and standardized. Appendix I also includes all the elements of a regular chapter.

In Chapter 1 (Simple Interest), objectives have been simplified with a focus on outcomes. The manual calculation of days between dates has been placed in a chapter appendix. Financial calculator instructions have been increased and updated and the mathematical steps in solutions have been simplified. Consistent terminology has been used.

In Chapter 2 (Simple Interest Applications), additional coverage of credit card loans has been included. Coverage of demand loans has been separated from coverage of lines of credit. The discussion of discounting promissory notes has been reduced and the computer application to develop a loan repayment schedule has been updated.

In Chapter 3 (Compound Interest—Future Value and Present Value), computations involving whole number and fractional exponents within future value and present value sections have been combined. The use of the terms FV and PV within the formulas has been introduced and used from the beginning of the chapter.

In Chapter 4 (Compound Interest—Further Topics), calculator solutions have been added, along with many new problems. The sections on finding interest rates, and effective and equivalent interest rates, have been expanded as well.

The topics for Chapters 5, 6, and 7 have been reorganized to emphasize the type of annuity; Chapter 11 focuses on Ordinary Simple Annuities while Chapter 6 focuses on Ordinary General Annuities. In each of these chapters, the sections have a similar order. The learner finds the future value, the present value, the periodic payment, the number of payments, and then the interest rate.

Chapter 7 covers Simple Annuities Due, General Annuities Due, Deferred Annuities, and Perpetuities. In these chapters, an identification of the nominal interest rate per year and the number of compounding periods per year have been added. A new explanation of Down Payment has been added in Chapter 5. Many new exercise questions have been added in all three chapters, as well as additional calculator solutions.

In general, interest rates used reflect the current economic climate in Canada. Calculator tips and solutions have been updated or added. Spreadsheet instructions and Internet site references have been updated. More *Pitfalls and Pointers* have been added to assist in performing tasks and interpreting word problems, and sections have been rewritten to clarify the explanations. Many more word problems have been added and the problems have been ranked by difficulty, with an icon indicating the more difficult problems. New *Business Math News Boxes* and case studies have been added. Examples involving both business and personal situations are included. Where appropriate, the use of six decimal places has been standardized.

The pedagogical elements of the previous edition have been retained. In response to requests and suggestions by users of the book, a number of new features for this edition have been included. They are described below.

FEATURES

- A new colourful and student-friendly design has been created for the book, making it more accessible and less intimidating to learners of all levels.
- Instructions for using preprogrammed financial calculators have been further expanded in this seventh edition. Although any preprogrammed financial calculator may be used, this edition includes extensive instructions for using the Texas Instruments BA II Plus financial calculator. Equivalent instructions are given in Appendix II for the Texas Instruments BA-35 Solar, the Sharp EL-733A, and the Hewlett Packard 10B financial calculators.
- To reduce the amount of "translation" required to go from the formulas in the text to the keystrokes on the preprogrammed financial calculator, the compounding and annuity formulas in Chapters 3 to 10 have been restated in this edition. From the beginning of Chapter 3, the P has been replaced with PV, S has been replaced with FV, and A has been replaced with PMT in these formulas. In addition, the compounding interval, C/Y, has been identified within the calculator solutions.

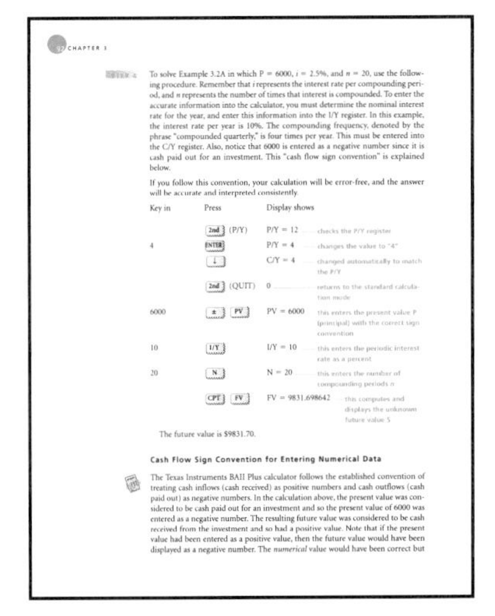

92 CHAPTER 3

To solve Example 3.2A in which P = 6000, i = 2.5%, and n = 20, use the following procedure. Remember that i represents the interest rate per compounding period, and n represents the number of times that interest is compounded. To enter the accurate information into the calculator, you must determine the nominal interest rate for the year, and enter this information into the I/Y register. In this example, the interest rate per year is 10%. The compounding frequency, denoted by the phrase "compounded quarterly," is four times per year. This must be entered into the C/Y register. Also, notice that 6000 is entered as a negative number since it is cash paid out for an investment. This "cash flow sign convention" is explained below.

If you follow this convention, your calculation will be error-free, and the answer will be accurate and interpreted consistently.

Key in	Press	Display shows	
	2nd (P/Y)	P/Y = 12	checks the P/Y register
4	ENTER	P/Y = 4	changes the value to "4"
	↓	C/Y = 4	changed automatically to match the P/Y
	2nd (QUIT)	0	returns to the standard calculation mode
6000	± PV	PV = 6000	this enters the present value P (principal) with the correct sign convention
10	I/Y	I/Y = 10	this enters the periodic interest rate as a percent
20	N	N = 20	this enters the number of compounding periods n
	CPT FV	FV = 9831.698642	this computes and displays the unknown future value S

The future value is $9831.70.

Cash Flow Sign Convention for Entering Numerical Data

The Texas Instruments BAII Plus calculator follows the established convention of treating cash inflows (cash received) as positive numbers and cash outflows (cash paid out) as negative numbers. In the calculation above, the present value was considered to be cash paid out for an investment and so the present value of 6000 was entered as a negative number. The resulting future value was considered to be cash received from the investment and so had a positive value. Note that if the present value had been entered as a positive value, then the future value would have been displayed as a negative number. The *numerical* value would have been correct but

A new Spreadsheet Template Disk was developed by Bruce Coombs. It is a CD-ROM that accompanies the text and contains Word files and Excel spreadsheet files. It requires Windows 95 or newer versions of Windows to run. The disk contains 26 self-contained tutorials that show how to use one of Excel's special spreadsheet functions to solve business problems. Each tutorial gives a brief description of the function, a general example showing how the function might be used, directions for creating a spreadsheet template to use the function, and a list of questions

from the text that can be answered using the spreadsheet function. In addition, an Excel file containing ready-to-use examples of each function is included. The student can use these spreadsheets to enter values and obtain immediate results. The Spreadsheet Template Disk also contains spreadsheet templates for the spreadsheet applications described in the text: loan repayment schedules (Chapter 2), accumulation of principal schedules (Chapter 3), amortization schedules (Chapter 8), and sinking fund schedules (Chapter 9). An Excel icon in the text highlights information on the use of Excel to solve problems and the Excel spreadsheet applications, directing students to the spreadsheet template disk.

Excel

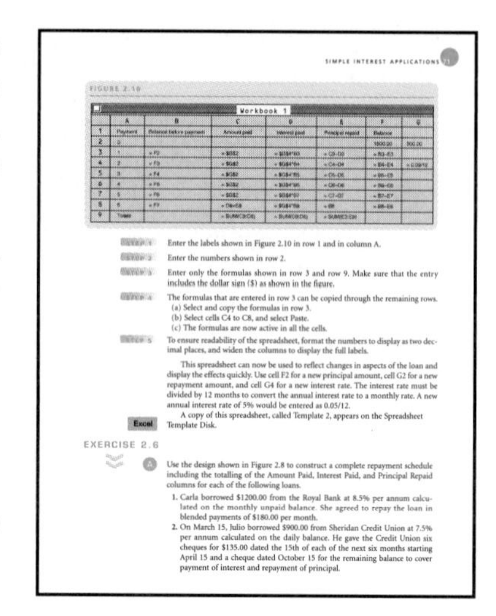
SIMPLE INTEREST APPLICATIONS

FIGURE 2.10

Workbook 1

STEP 1 Enter the labels shown in Figure 2.10 in row 1 and in column A.

STEP 2 Enter the numbers shown in row 2.

STEP 3 Enter only the formulas shown in row 3 and row 9. Make sure that the entry includes the dollar sign ($) as shown in the figure.

STEP 4 The formulas that are entered in row 3 can be copied through the remaining rows.
(a) Select and copy the formulas in row 3.
(b) Select cells C4 to C8, and select Paste.
(c) The formulas are now active in all the cells.

STEP 5 To ensure readability of the spreadsheet, format the numbers to display as two decimal places, and widen the columns to display the full labels.

This spreadsheet can now be used to reflect changes in aspects of the loan and display the effects quickly. Use cell F2 for a new principal amount, cell G2 for a new repayment amount, and cell G4 for a new interest rate. The interest rate must be divided by 12 months to convert the annual interest rate to a monthly rate. A new annual interest rate of 5% would be entered as 0.05/12.

Excel A copy of this spreadsheet, called Template 2, appears on the Spreadsheet Template Disk.

EXERCISE 2.6

A Use the design shown in Figure 2.8 to construct a complete repayment schedule including the totalling of the Amount Paid, Interest Paid, and Principal Repaid columns for each of the following loans.

1. Carla borrowed $1200.00 from the Royal Bank at 8.5% per annum calculated on the monthly unpaid balance. She agreed to repay the loan in blended payments of $180.00 per month.
2. On March 15, Julio borrowed $900.00 from Sheridan Credit Union at 7.5% per annum calculated on the daily balance. He gave the Credit Union six cheques for $135.00 dated the 15th of each of the next six months starting April 15 and a cheque dated October 15 for the remaining balance to cover payment of interest and repayment of principal.

A spreadsheet icon in the text highlights both the questions in the text that can be solved using an Excel spreadsheet function. Appendix A and B are also included on the Spreadsheet Template Disk.

- A set of learning objectives is listed at the beginning of each chapter.
- Each chapter opens with a description of a situation familiar to students to emphasize the practical applications of the material to follow.

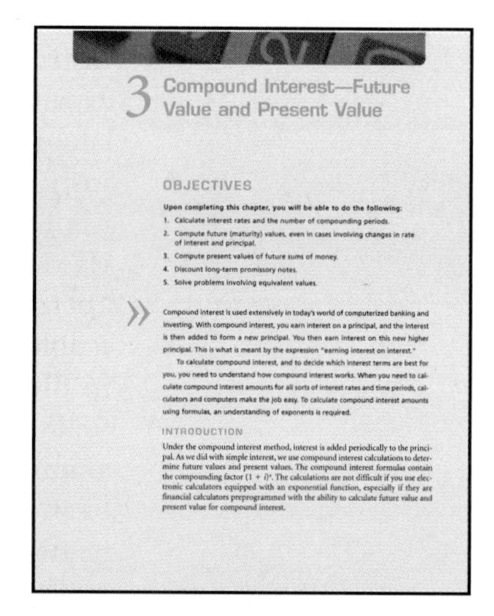
3 Compound Interest—Future Value and Present Value

OBJECTIVES

Upon completing this chapter, you will be able to do the following:

1. Calculate interest rates and the number of compounding periods.
2. Compute future (maturity) values, even in cases involving changes in rate of interest and principal.
3. Compute present values of future sums of money.
4. Discount long-term promissory notes.
5. Solve problems involving equivalent values.

Compound interest is used extensively in today's world of computerized banking and investing. With compound interest, you earn interest on a principal, and the interest is then added to form a new principal. You then earn interest on this new higher principal. This is what is meant by the expression "earning interest on interest."

To calculate compound interest, and to decide which interest terms are best for you, you need to understand how compound interest works. When you need to calculate compound interest amounts for all sorts of interest rates and time periods, calculators and computers make the job easy. To calculate compound interest amounts using formulas, an understanding of exponents is required.

INTRODUCTION

Under the compound interest method, interest is added periodically to the principal. As we did with simple interest, we use compound interest calculations to determine future values and present values. The compound interest formulas contain the compounding factor $(1 + i)^n$. The calculations are not difficult if you use electronic calculators equipped with an exponential function, especially if they are financial calculators preprogrammed with the ability to calculate future value and present value for compound interest.

- A Business Math News Box is presented in most chapters. This element consists of short excerpts based on material appearing in newspapers, magazines, or websites, followed by a set of questions. These boxes demonstrate how widespread business math applications are in the real world.

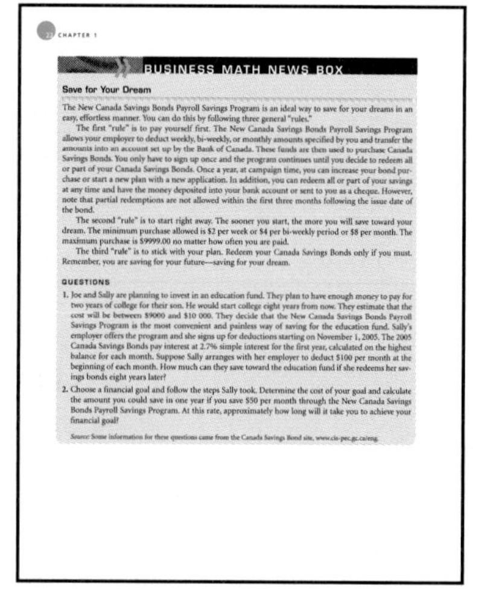
CHAPTER 1

BUSINESS MATH NEWS BOX

Save for Your Dream

The New Canada Savings Bonds Payroll Savings Program is an ideal way to save for your dreams in an easy, effortless manner. You can do this by following three general "rules."

The first "rule" is to pay yourself first. The New Canada Savings Bonds Payroll Savings Program allows your employer to deduct weekly, bi-weekly, or monthly amounts specified by you and transfer the amounts into an account set up by the Bank of Canada. These funds are then used to purchase Canada Savings Bonds. You only have to sign up once and the program continues until you decide to redeem all or part of your Canada Savings Bonds. Once a year, at campaign time, you can increase your bond purchase or start a new plan with a new application. In addition, you can redeem all or part of your savings at any time and have the money deposited into your bank account or sent to you as a cheque. However, note that partial redemptions are not allowed within the first three months following the issue date of the bond.

The second "rule" is to start right away. The sooner you start, the more you will save toward your dream. The minimum purchase allowed is $2 per week or $4 per bi-weekly period or $8 per month. The maximum purchase is $9999.00 no matter how often you are paid.

The third "rule" is to stick with your plan. Redeem your Canada Savings Bonds only if you must. Remember, you are saving for your future—saving for your dream.

QUESTIONS

1. Joe and Sally are planning to invest in an education fund. They plan to have enough money to pay for two years of college for their son. He would start college eight years from now. They estimate that the cost will be between $9000 and $10 000. They decide that the New Canada Savings Bonds Payroll Savings Program is the most convenient and painless way of saving for the education fund. Sally's employer offers the program and she signs up for deductions starting on November 1, 2005. The 2005 Canada Savings Bonds pay interest at 2.7% simple interest for the first year, calculated on the highest balance for each month. Suppose Sally arranges with her employer to deduct $100 per month at the beginning of each month. How much can they save toward the education fund if she redeems her savings bonds eight years later?
2. Choose a financial goal and follow the steps Sally took. Determine the cost of your goal and calculate the amount you could save in one year if you save $50 per month through the New Canada Savings Bonds Payroll Savings Program. At this rate, approximately how long will it take you to achieve your financial goal?

Source: Some information for these questions came from the Canada Savings Bond site, www.cis-pec.gc.ca/eng.

- At least one new Pointers and Pitfalls box in each chapter emphasizes good practices, highlights ways to avoid common errors, shows how to use a financial calculator efficiently, or gives hints for tackling business math situations to reduce math anxiety.

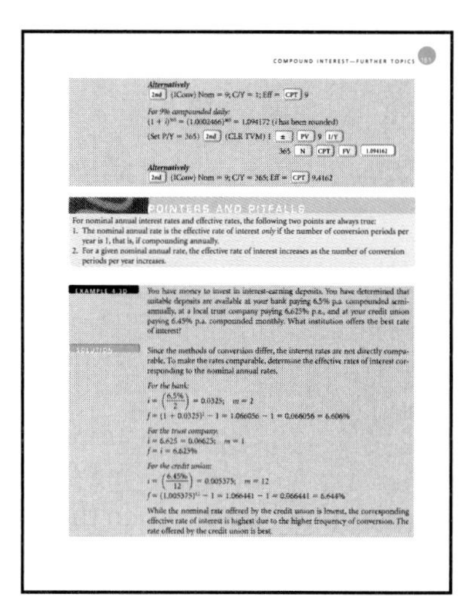

COMPOUND INTEREST—FURTHER TOPICS

Alternatively
2nd (ICONV) Nom = 9; C/Y = 1; Eff = CPT 9

For 9% compounded daily:
$(1 + i)^{365} = (1.0002466)^{365} = 1.094172$ (i has been rounded)
(Set P/Y = 365) 2nd (CLR TVM) 1 ± PV 9 I/Y
365 N CPT FV 1.094162

Alternatively
2nd (ICONV) Nom = 9; C/Y = 365; Eff = CPT 9.4162

POINTERS AND PITFALLS

For nominal annual interest rates and effective rates, the following two points are always true:
1. The nominal annual rate is the effective rate of interest *only* if the number of conversion periods per year is 1, that is, if compounding annually.
2. For a given nominal annual rate, the effective rate of interest increases as the number of conversion periods per year increases.

EXAMPLE 4.3D You have money to invest in interest-earning deposits. You have determined that suitable deposits are available at your bank paying 6.5% p.a. compounded semi-annually, at a local trust company paying 6.625% p.a., and at your credit union paying 6.45% p.a. compounded monthly. What institution offers the best rate of interest?

SOLUTION Since the methods of conversion differ, the interest rates are not directly comparable. To make the rates comparable, determine the effective rates of interest corresponding to the nominal annual rates.

For the bank:
$i = \left(\frac{6.5\%}{2}\right) = 0.0325; \quad m = 2$
$f = (1 + 0.0325)^2 - 1 = 1.066056 - 1 = 0.066056 = 6.606\%$

For the trust company:
$i = 6.625 = 0.06625; \quad m = 1$
$f = i = 6.625\%$

For the credit union:
$i = \left(\frac{6.45\%}{12}\right) = 0.005375; \quad m = 12$
$f = (1.005375)^{12} - 1 = 1.066441 - 1 = 0.066441 = 6.644\%$

While the nominal rate offered by the credit union is lowest, the corresponding effective rate of interest is highest due to the higher frequency of conversion. The rate offered by the credit union is best.

- A Did You Know? box in each chapter offers interesting practical and mathematical facts.

- Numerous Examples with worked-out Solutions are provided throughout the book, offering easy-to-follow, step-by-step instructions.

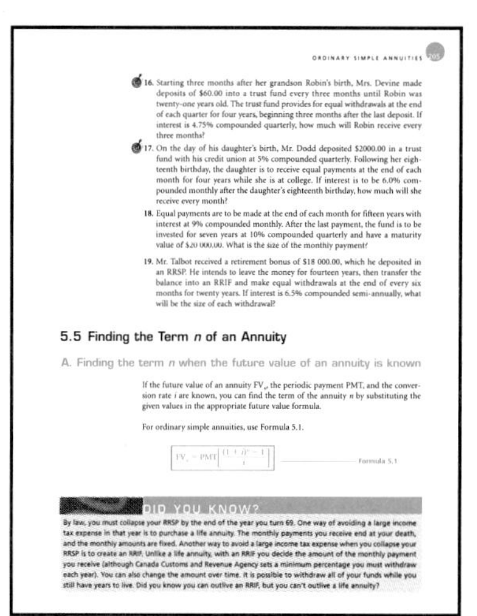

ORDINARY SIMPLE ANNUITIES

16. Starting three months after her grandson Robin's birth, Mrs. Devine made deposits of $60.00 into a trust fund every three months until Robin was twenty-one years old. The trust fund provides for equal withdrawals at the end of each quarter for four years, beginning three months after the last deposit. If interest is 4.75% compounded quarterly, how much will Robin receive every three months?

17. On the day of his daughter's birth, Mr. Dodd deposited $2000.00 in a trust fund with his credit union at 5% compounded quarterly. Following her eighteenth birthday, the daughter is to receive equal payments at the end of each month for four years while she is at college. If interest is to be 6.0% compounded monthly after the daughter's eighteenth birthday, how much will she receive every month?

18. Equal payments are to be made at the end of each month for fifteen years with interest at 9% compounded monthly. After the last payment, the fund is to be invested for seven years at 10% compounded quarterly and have a maturity value of $20 000.00. What is the size of the monthly payment?

19. Mr. Talbot received a retirement bonus of $18 000.00, which he deposited in an RRSP. He intends to leave the money for fourteen years, then transfer the balance into an RRIF and make equal withdrawals at the end of every six months for twenty years. If interest is 6.5% compounded semi-annually, what will be the size of each withdrawal?

5.5 Finding the Term n of an Annuity

A. Finding the term n when the future value of an annuity is known

If the future value of an annuity FV_n, the periodic payment PMT, and the conversion rate i are known, you can find the term of the annuity n by substituting the given values in the appropriate future value formula.

For ordinary simple annuities, use Formula 5.1.

$$FV_n = PMT\left[\frac{(1+i)^n - 1}{i}\right]$$ Formula 5.1

DID YOU KNOW?

By law, you must collapse your RRSP by the end of the year you turn 69. One way of avoiding a large income tax expense in that year is to purchase a life annuity. The monthly payments you receive end at your death, and the monthly amounts are fixed. Another way to avoid a large income tax expense when you collapse your RRSP is to create an RRIF. Unlike a life annuity, with an RRIF you decide the amount of the monthly payment you receive (although Canada Customs and Revenue Agency sets a minimum percentage you must withdraw each year). You can also change the amount over time. It is possible to withdraw all of your funds while you still have years to live. Did you know you can outlive an RRIF, but you can't outlive a life annuity?

- Programmed solutions using the Texas Instruments BA II Plus calculator are offered for all examples in Chapters 3 to 10. Since this calculator display can be pre-set, it is suggested that the learner set the display to show six decimal places to match the mathematical calculations in the body of the text. Both mathematical and calculator solutions for all Exercises, Review Exercises, and Self-Tests are included in the Instructor's Solutions Manual. An icon highlights information on the use of the BA II Plus Calculator.

- Key Terms are introduced in the text in boldface type. A Glossary at the end of each chapter lists each term with its definition and a page reference to where the term was first defined in the chapter.

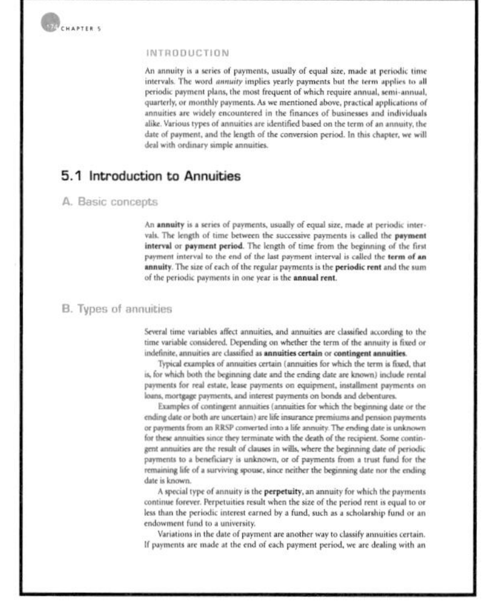

CHAPTER 5

INTRODUCTION

An annuity is a series of payments, usually of equal size, made at periodic time intervals. The word *annuity* implies yearly payments but the term applies to all periodic payment plans, the most frequent of which require annual, semi-annual, quarterly, or monthly payments. As we mentioned above, practical applications of annuities are widely encountered in the finances of businesses and individuals alike. Various types of annuities are identified based on the term of an annuity, the date of payment, and the length of the conversion period. In this chapter, we will deal with ordinary simple annuities.

5.1 Introduction to Annuities

A. Basic concepts

An **annuity** is a series of payments, usually of equal size, made at periodic intervals. The length of time between the successive payments is called the **payment interval** or **payment period**. The length of time from the beginning of the first payment interval to the end of the last payment interval is called the **term of an annuity**. The size of each of the regular payments is the **periodic rent** and the sum of the periodic payments in one year is the **annual rent**.

B. Types of annuities

Several time variables affect annuities, and annuities are classified according to the time variable considered. Depending on whether the term of the annuity is fixed or indefinite, annuities are classified as **annuities certain** or **contingent annuities**.

Typical examples of annuities certain (annuities for which the term is fixed, that is, for which both the beginning date and the ending date are known) include rental payments for real estate, lease payments on equipment, installment payments on loans, mortgage payments, and interest payments on bonds and debentures.

Examples of contingent annuities (annuities for which the beginning date or the ending date or both are uncertain) are life insurance premiums and pension payments or payments from an RRSP converted into a life annuity. The ending date is unknown for these annuities since they terminate with the death of the recipient. Some contingent annuities are the result of clauses in wills, where the beginning date of periodic payments to a beneficiary is unknown, or of payments from a trust fund for the remaining life of a surviving spouse, since neither the beginning date nor the ending date is known.

A special type of annuity is the **perpetuity**, an annuity for which the payments continue forever. Perpetuities result when the size of the period rent is equal to or less than the periodic interest earned by a fund, such as a scholarship fund or an endowment fund to a university.

Variations in the date of payment are another way to classify annuities certain. If payments are made at the end of each payment period, we are dealing with an

- Main Equations are highlighted in the chapters and repeated in a Summary of Formulas at the ends of the chapters. Each main formula is presented in colour and labelled numerically (with the letter A suffix if equivalent forms of the formula are presented later). By contrast, equivalent formulae are presented in black and labelled with the number of the related main formula followed by the letter B or C.

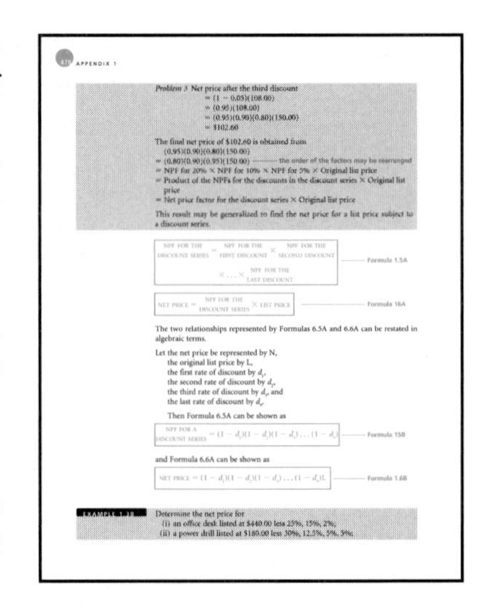

APPENDIX 1

Problem 3 Net price after the third discount
= (1 − 0.05)(108.00)
= (0.95)(108.00)
= (0.95)(0.90)(0.80)(150.00)
= $102.60

The final net price of $102.60 is obtained from
(0.95)(0.90)(0.80)(150.00)
= (0.80)(0.90)(0.95)(150.00) ——— the order of the factors may be rearranged
= NPF for 20% × NPF for 10% × NPF for 5% × Original list price
= Product of the NPFs for the discounts in the discount series × Original list price
= Net price factor for the discount series × Original list price

This result may be generalized to find the net price for a list price subject to a discount series.

NPF FOR THE DISCOUNT SERIES = NPF FOR THE FIRST DISCOUNT × NPF FOR THE SECOND DISCOUNT × . . . × NPF FOR THE LAST DISCOUNT ——— Formula 1.5A

NET PRICE = NPF FOR THE DISCOUNT SERIES × LIST PRICE ——— Formula 16A

The two relationships represented by Formulas 6.5A and 6.6A can be restated in algebraic terms.

Let the net price be represented by N,
the original list price by L,
the first rate of discount by d_1,
the second rate of discount by d_2,
the third rate of discount by d_3, and
the last rate of discount by d_n.

Then Formula 6.5A can be shown as

NPF FOR A DISCOUNT SERIES $= (1 - d_1)(1 - d_2)(1 - d_3) \ldots (1 - d_n)$ ——— Formula 15B

and Formula 6.6A can be shown as

NET PRICE $= (1 - d_1)(1 - d_2)(1 - d_3) \ldots (1 - d_n)L$ ——— Formula 1.6B

EXAMPLE 1.3B Determine the net price for
(i) an office desk listed at $440.00 less 25%, 15%, 2%;
(ii) a power drill listed at $180.00 less 30%, 12.5%, 5%, 5%;

- A list of the Main Formulas begins on the inside of the front cover and continues at the end of the text for easy reference.

- An Exercise set is provided at the end of each section in every chapter. If students choose, they can use the suggested Excel spreadsheet functions to answer those questions marked with the spreadsheet logo. In addition, each chapter contains a Review Exercise set and a Self-Test. If students choose, they can use an Excel spreadsheet function to answer those questions marked with a spreadsheet logo, but they must decide which function to use. Answers to all the odd-numbered Exercises, Review Exercises, and Self-Tests are given at the back of the book. Solutions for questions answered using Excel spreadsheet functions are given in the Instructor's area of the Hummelbrunner Companion Website.

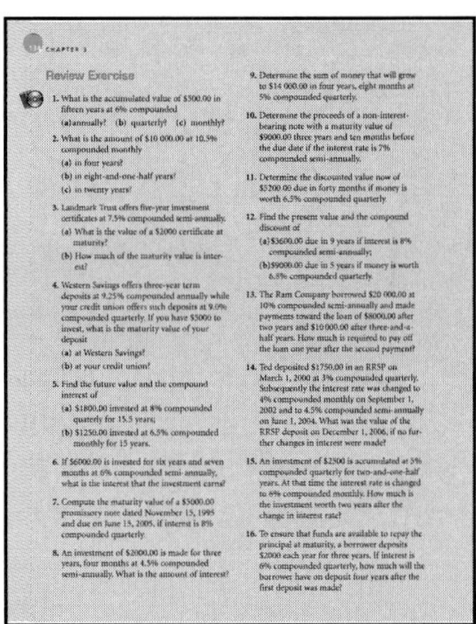

CHAPTER 3

Review Exercise

1. What is the accumulated value of $500.00 in fifteen years at 6% compounded
(a) annually? (b) quarterly? (c) monthly?
2. What is the amount of $10 000.00 at 10.5% compounded monthly
(a) in four years?
(b) in eight-and-one-half years?
(c) in twenty years?
3. Landmark Trust offers five-year investment certificates at 7.5% compounded semi-annually.
(a) What is the value of a $2000 certificate at maturity?
(b) How much of the maturity value is interest?
4. Western Savings offers three-year term deposits at 9.25% compounded annually while your credit union offers such deposits at 9.0% compounded quarterly. If you have $5000 to invest, what is the maturity value of your deposit
(a) at Western Savings?
(b) at your credit union?
5. Find the future value and the compound interest of
(a) $1800.00 invested at 8% compounded quaterly for 15.5 years;
(b) $1250.00 invested at 6.5% compounded monthly for 15 years.
6. If $6000.00 is invested for six years and seven months at 6% compounded semi-annually, what is the interest that the investment earns?
7. Compute the maturity value of a $5000.00 promissory note dated November 15, 1995 and due on June 15, 2005, if interest is 8% compounded quarterly.
8. An investment of $2000.00 is made for three years, four months at 4.5% compounded semi-annually. What is the amount of interest?
9. Determine the sum of money that will grow to $14 000.00 in four years, eight months at 5% compounded quarterly.
10. Determine the proceeds of a non-interest-bearing note with a maturity value of $9000.00 three years and ten months before the due date if the interest rate is 7% compounded semi-annually.
11. Determine the discounted value now of $5200.00 due in forty months if money is worth 6.5% compounded quarterly.
12. Find the present value and the compound discount of
(a) $3600.00 due in 9 years if interest is 8% compounded semi-annually;
(b) $9000.00 due in 5 years if money is worth 6.8% compounded quarterly.
13. The Ram Company borrowed $20 000.00 at 10% compounded semi-annually and made payments toward the loan of $8000.00 after two years and $10 000.00 after three-and-a-half years. How much is required to pay off the loan one year after the second payment?
14. Ted deposited $1750.00 in an RRSP on March 1, 2000 at 3% compounded quarterly. Subsequently the interest rate was changed to 4% compounded monthly on September 1, 2002 and to 4.5% compounded semi-annually on June 1, 2004. What was the value of the RRSP deposit on December 1, 2006, if no further changes in interest were made?
15. An investment of $2500 is accumulated at 5% compounded quarterly for two-and-one-half years. At that time the interest rate is changed to 6% compounded monthly. How much is the investment worth two years after the change in interest rate?
16. To ensure that funds are available to repay the principal at maturity, a borrower deposits $2000 each year for three years. If interest is 6% compounded quarterly, how much will the borrower have on deposit four years after the first deposit was made?

- A set of Challenge Problems is provided in each chapter. These problems give users the opportunity to apply the skills learned in the chapter to questions that are pitched at a higher level than the Exercises.

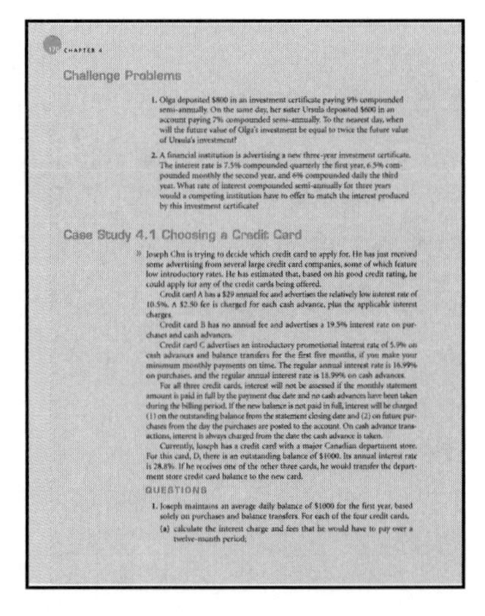

CHAPTER 4

Challenge Problems

1. Olga deposited $800 in an investment certificate paying 9% compounded semi-annually. On the same day, her sister Ursula deposited $600 in an account paying 7% compounded semi-annually. To the nearest day, when will the future value of Olga's investment be equal to twice the future value of Ursula's investment?
2. A financial institution is advertising a new three-year investment certificate. The interest rate is 7.5% compounded quarterly the first year, 6.5% compounded monthly the second year, and 6% compounded daily the third year. What rate of interest compounded semi-annually for three years would a competing institution have to offer to match the interest produced by this investment certificate?

Case Study 4.1 Choosing a Credit Card

Joseph Chu is trying to decide which credit card to apply for. He has just received some advertising from several large credit card companies, some of which feature low introductory rates. He has estimated that, based on his good credit rating, he could apply for any of the credit cards being offered.

Credit card A has a $29 annual fee and advertises the relatively low interest rate of 10.5%. A $2.50 fee is charged for each cash advance, plus the applicable interest charges.

Credit card B has no annual fee and advertises a 19.5% interest rate on purchases and cash advances.

Credit card C advertises an introductory promotional interest rate of 5.9% on cash advances and balance transfers for the first five months, if you make your minimum monthly payments on time. The regular annual interest rate is 16.99% on purchases, and the regular annual interest rate is 18.99% on cash advances.

For all three credit cards, interest will not be assessed if the monthly statement amount is paid in full by the payment due date and no cash advances have been taken during the billing period. If the new balance is not paid in full, interest will be charged (1) on the outstanding balance from the statement closing date and (2) on future purchases from the day the purchases are posted to the account. On cash advance transactions, interest is always charged from the date the cash advance is taken.

Currently, Joseph has a credit card with a major Canadian department store. For this card, D, there is an outstanding balance of $1000. Its annual interest rate is 28.8%. If he receives one of the other three cards, he would transfer the department store credit card balance to the new card.

QUESTIONS

1. Joseph maintains an average daily balance of $1000 for the first year, based solely on purchases and balance transfers. For each of the four credit cards,
(a) calculate the interest charge and fees that he would have to pay over a twelve-month period;

- Twenty-two Case Studies are included in the book, two near the end of each chapter. They present comprehensive realistic scenarios followed by a set of questions, and illustrate some of the important types of practical applications of the chapter material. At least one half of the Case Studies represent personal-finance applications (indicated by a credit-card logo) while the rest are business applications.

- A new set of Useful Internet Sites is provided at the end of each chapter, with the URL and a brief description provided for each site. These sites are related to the chapter topic or to companies mentioned in the chapter, or they show how business math is important to the day-to-day operations of companies and industries.

Maturity value see **Future value**

Nominal rate of interest the stated rate at which the compounding is done one or more times per year; usually stated as an annual rate *(p. 82)*

Periodic rate of interest value of interest is obtained by dividing the nominal annual rate by the number of compounding periods per year *(p. 82)*

Present value the principal at any time that will grow at compound interest to a given future value over a given number of compounding periods at a given rate of interest *(p. 102)*

Proceeds see **Present value**

USEFUL INTERNET SITES

www.globefund.com/
Globefund.com This popular site for mutual fund information and analysis is hosted by the *Globe and Mail* Website

www.fin.gc.ca/fin-eng.html
Department of Finance This site has information on the preparation of the federal government's budget that shows what is happening to taxes, as well as information related to interest rates and the economy.

www.cannex.com
CANNEX This site has a large list of comparative interest rates for GICs and term deposits offered by financial institutions in Canada, the United States, and Australia.

SUPPLEMENTS

The following supplements have been carefully prepared to accompany this seventh edition.

- **Instructor's Solutions Manual** provides complete mathematical and calculator solutions to all the Exercises, Review Exercises, Self-Tests, Business Math News Box questions, Challenge Problems, and Case Studies in the textbook.
- A **Companion Website** provides students with self-test questions that include Case Studies, Multiple Choice, Fill-in-the-Blank, True/False, and Internet-based Exercises. Hints are included for most of the questions, and immediate feedback is available to assist students to study and test themselves. Links to other websites of interest and an Internet search of key business math terms have been updated for the Fifth Edition. New to this edition is information about Web-based calculators and the use of interactive charts and graphs that accompany them. Visit the Hummelbrunner Website at **www.pearsoned.ca/hummelbrunner** and then click on the *Mathematics of Finance with Canadian Applications*, Fifth Edition cover.
- On-line **Student Self-Tests** is an exciting new Web-based student tutorial created for this Fifth Edition. Students will now be able to create multiple iterations of chapter quizzes online, using algorithmically generated exercises derived from their text. Created in Pearson's Course Compass online course management shell, using TestGen software, the questions come with thorough tutorial feedback, linked back in to the textbook, so students can more readily master the concepts of business math.

The Instructor's Solutions Manual is available through the Companion Website.

ACKNOWLEDGEMENTS

Special thanks must be given to Bruce M. Coombs of Kwantlen University College and Langara College for his help in creating and updating the Spreadsheet Template Disk and the Companion Website. He listened, suggested, proofed, and agonized over this edition throughout the entire process of updating. His contribution was invaluable. Also great thanks to Jim Roberts of Red River College, who checked and revised all material pertaining to the TI BA II Plus calculator.

We would like to express our thanks to the many people who offered thoughtful suggestions and recommendations for updating and improving the book. We would particularly like to thank the following instructors for providing formal reviews.

Millie Atkinson (Mohawk College)
Wayne Bemister (Red River College)
Sam Boutilier (University College of Cape Breton)
Ross Bryant (Conestoga College)
Helen Catania (Centennial College)
Michael R. Conte (Durham College)
Laurel Donaldson (Douglas College)
Tom Fraser (Niagara College)
Brune Fullone (George Brown College)
Frank Gruen (British Columbia Institute of Technology)
Amoel Lisecki (Southern Alberta Institute of Technology)
Judy Palm (Malaspina College)
Harry Matsugu (Humber College)
Colleen Quinn (Seneca College)
David Roberts (Devry Institute, Calgary)
Jim Roberts (Red River College)
Aruna Sarkar (Algonquin College)
Susan Vallery (Fleming College)
Carol Ann Waite (Sheridan College)
Donald J. Webber (College of the North Atlantic)
Allen Zhu (Capilano College)

We would also like to thank the many people at Pearson Education Canada who helped with the development and production of this book, especially to the production editor, Joel Gladstone; the production coordinator, Janette Lush; the copy editor, Rodney Rawlings; the proofreader, Lesley Mann; and to Ross Meacher, who completed a technical check of the entire manuscript. Special thanks to editors Kelly Torrance and Madhu Ranadive for their supportive directions.

Sieg Hummelbrunner
Suzanne Coombs

The Pearson Education Canada **COMPANION WEBSITE**

A Great Way to Learn and Instruct Online

The Pearson Education Canada Companion Website is easy to navigate and is organized to correspond to the chapters in this textbook. Whether you are a student in the classroom or a distance learner you will discover helpful resources for in-depth study and research that empower you in your quest for greater knowledge and maximize your potential for success in the course.

[www.pearsoned.ca/hummelbrunner]

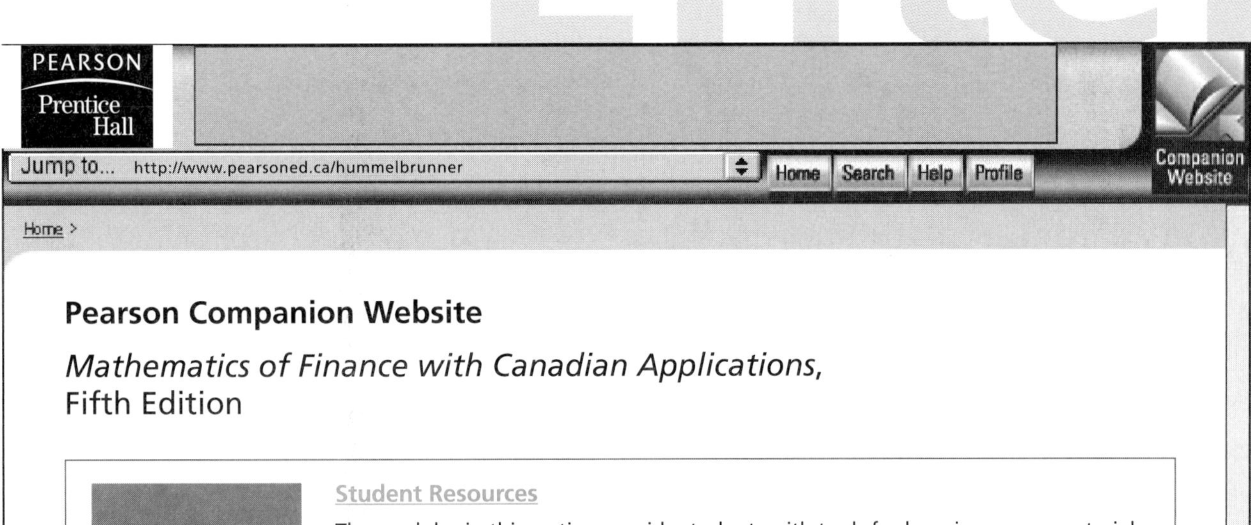

Pearson Companion Website

Mathematics of Finance with Canadian Applications,
Fifth Edition

Student Resources

The modules in this section provide students with tools for learning course material. These modules include:

- Chapter Objectives
- Multiple Choice Questions
- True and False Questions
- Links to Useful Internet Sites
- Cases
- Information on Web-based calculators

In the quiz modules students can send answers to the grader and receive instant feedback on their progress through the Results Reporter. Coaching comments and references to the textbook may be available to ensure that students take advantage of all available resources to enhance their learning experience.

Instructor Resources

The modules in this section provide instructors with additional teaching tools. Downloadable PowerPoint Presentations, Electronic Transparencies, and an Instructor's Manual are just some of the materials that may be available in this section. Where appropriate, this section will be password protected.

1 Simple Interest

OBJECTIVES

Upon completing this chapter, you will be able to do the following:

1. Determine the number of days between two dates.
2. Compute the amount of simple interest using the formula $I = Prt$.
3. Compute the principal, interest rate, or time using variations of the formula $I = Prt$.
4. Compute the maturity value (future value) using the formula $S = P(1 + rt)$.
5. Compute the principal (present value) using the formula $P = \frac{S}{1+rt}$.
6. Compute equivalent or dated values for specified focal dates.

Transactions that involve borrowing or lending money for short periods of time take place every day in business or in our personal lives. Businesses lend money when they extend credit to customers or clients, or when they make short-term investments. Businesses borrow money when they purchase on credit from vendors, use a line of credit from a financial institution, or utilize credit cards to make a purchase. Individuals borrow money for many reasons. We borrow to pay for an education, to buy a vehicle, or to pay for a large purchase by using a credit card. Interest needs to be considered on those loans.

INTRODUCTION

In business transactions, money is lent or borrowed daily. To compensate the lenders for the use of their money, interest is paid. The amount of interest paid is based on three factors: the amount of money borrowed, the rate of interest at which it is borrowed, and the time period for which it is borrowed.

1.1 Determining the Number of Days Between Two Dates

A. Counting exact time

The time period for which interest is charged is the **interest period**. The exact number of days must be determined when two dates are involved.

The number of days between two dates can be determined by using one of the manual techniques or by using a financial calculator or computer. This text will use the method of counting the starting date but not the ending date.

Counting days requires knowing the order of the 12 months in the year and the number of days in each month. The number of days in each of the 12 months in order (ignoring leap years) is listed below.

1. January	31	2. February	28	3. March	31
4. April	30	5. May	31	6. June	30
7. July	31	8. August	31	9. September	30
10. October	31	11. November	30	12. December	31

Leap years are ones that are evenly divisible by 4, such as 2008, 2012, and 2016. An extra day is added to February if a year is a leap year.

A detailed discussion of further manual techniques is contained in the appendix to this chapter.

POINTERS AND PITFALLS

To use the TI BA II Plus financial calculator, choose the DATE worksheet by pressing 2nd DATE. The first date, DT1, is usually the earlier date. The date format can be set to show the U.S. format, mm/dd/yy, or the European format, dd/mm/yy. When the U.S. format is used, enter the date by choosing the one- or two-digit number representing the month, followed by a period, then enter two digits for the day and the last two digits for the year. Only one period is entered between the month and the day. Press Enter to save the data. To move to the next label, press the down arrow. The second date, DT2, is usually the later date. Enter the second date in the same manner as the first date, with one or two digits for the number of the month, a period, two digits for the day, and two digits for the year. Following the second date is the "days between dates" calculation. Press the down arrow to access this part of the worksheet. To show the exact days, press CPT (Compute) to instruct the calculator to perform the calculation. When the first date precedes the second date, the days between dates will appear as a positive number. If the second date is entered as the earlier date, the days between dates will appear as a negative number. The fourth label within the worksheet sets the calculator to show the actual number of days between the dates indicated, including adjustments for leap years. Set this label to ACT by pressing 2nd SET.

EXAMPLE 1.1A Compute the exact number of days between April 23, 2005 and July 21, 2005.

SOLUTION

Manual: The starting date is April 23, 2005 —— to be counted
The ending date is July 21, 2005 —— do *not* count

Number of days

April	8
May	31
June	30
July	20
TOTAL	89

POINTERS AND PITFALLS

Using the DATE function on the calculator, it is possible to determine the number of days between two dates. Press

2nd	DATE	DT1	04.2305	Enter	↓
		DT2	07.2105	Enter	↓
CPT		DBD			Result is 89

The month is entered first, with one or two digits, followed by a period; then the day is entered using two digits; and then the year is entered using two digits.

If one of the dates and the desired days between the dates are entered, it is possible to determine the second date.

EXAMPLE 1.1B Compute the exact number of days between November 10, 2007 and March 30, 2008.

Manual: The starting date is November 10, 2007 —— to be counted
The ending date is March 30, 2008 —— do *not* count
Number of days

November	21
December	31
January	31
February	29 (2008 is a leap year)
March	29
TOTAL	141

Excel If you choose, you can use Excel's *Coupon Days*, or days between dates (**COUPDAYSNC**) function, to find the number of days between two dates. Refer to **COUPDAYSNC** on the Spreadsheet Template Disk to learn how to use this Excel function.

EXERCISE 1.1

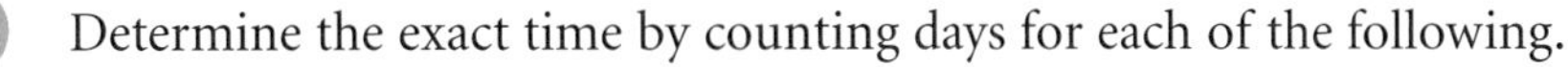

A. Determine the exact time by counting days for each of the following.

1. January 18, 2005 to May 10, 2005
2. August 30, 2005 to April 1, 2006
3. November 1, 2007 to April 15, 2008
4. December 24, 2006 to February 5, 2008

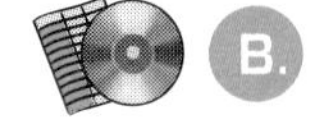

B. Determine the date represented.

1. 244 days after April 1, 2007
2. 619 days before March 30, 2008
3. 341 days before March 11, 2008
4. 634 days after August 25, 2006

1.2 Finding the Amount of Simple Interest

A. Basic concepts and formula

Interest is the rent charged for the use of money. The amount of **simple interest** is determined by the relationship
Interest = Principal × Rate × Time

$$I = Prt$$ —— Formula 1.1A

where I is the amount of interest earned, measured in dollars and cents;
P is the principal sum of money earning the interest, measured in dollars and cents;
r is the simple annual (yearly or **nominal**) **rate** of interest, expressed as a percent, which can be converted into a decimal;
t is the **time period** in years.

Simple interest is often used in business, through short-term loans to and from financial institutions, vendors, and customers.

B. Matching *r* and *t*

While the time may be stated in days, months, or years, the rate of interest is generally stated as a yearly charge, often followed by "per annum" or "p.a." In using the simple interest formula, it is imperative that the time *t* correspond to the interest rate *r*. Time expressed in months or days often needs to be converted into years.

EXAMPLE 1.2A

State r and t for each of the following.
(i) Rate 4%; time 3 years
(ii) Rate 6.5% p.a. (per annum); time 18 months
(iii) Rate 5.25% p.a.; time 243 days

SOLUTION

(i) The annual rate $r = 4\% = 0.04$
The time in years $t = 3$

(ii) The annual rate $r = 6.5\% = 0.065$
The time in years $t = \dfrac{18}{12}$

Note: To convert months into years, divide by 12.

(iii) The annual rate $r = 5.25\% = 0.0525$
The time in years $t = \dfrac{243}{365}$

Note: To convert days into years, divide by 365.

C. Computing the amount of interest

When the principal, rate, and time are known, the amount of interest can be determined by the formula $I = Prt$.

EXAMPLE 1.2B

Compute the amount of interest for
(i) $3600.00 at 6.25% p.a. (per annum) for 3 years;
(ii) $5240.00 at 4.5% p.a. for 16 months;
(iii) $1923.60 at 3% p.a. for 215 days.

SOLUTION

(i) $P = \$3600.00;\ r = 6.25\% = 0.0625;\ t = 3$
$I = Prt = (3600.00)(0.0625)(3) = \$675.00$

(ii) $P = \$5240.00;\ r = 4.5\% = 0.045;\ t = 16 \text{ months} = \dfrac{16}{12}$
$I = Prt = (5240.00)(0.045)\left(\dfrac{16}{12}\right) = \$314.40$

(iii) $P = \$1923.60;\ r = 3\% = 0.03;\ t = 215 \text{ days} = \dfrac{215}{365}$
$I = Prt = (1923.60)(0.03)\left(\dfrac{215}{365}\right) = \$33.99$

EXAMPLE 1.2C

Compute the amount of interest on \$785.95 borrowed at 18% p.a. from January 30, 2005, until March 21, 2005.

SOLUTION

Number of days			
	January	2	(30 and 31)
	February	28	
	March	20	
	TOTAL	50	

$$P = 785.95; \ r = 18\% = 0.018; \ t = \frac{50}{365}$$

$$I = (785.95)(0.18)\left(\frac{50}{365}\right) = \$19.38$$

EXAMPLE 1.2D

Compute the amount of interest on \$1240.00 earning 3.75% p.a. from September 30, 2007, to May 15, 2008.

SOLUTION

The starting date is September 30, 2003 (DT1). The ending date is May 15, 2008 (DT2). Note that 2008 is a leap year.

Days between dates (DBD) = 228

$$P = 1240.00; \ r = 3.75\% = 0.0375; \ t = \frac{228}{365}$$

$$I = (1240.00)(0.0375)\left(\frac{228}{365}\right) = \$29.05$$

Note: 365 is used to convert the number of days into years even if the year involved is a leap year.

DID YOU KNOW?

During a regular year, which is a year with 365 days, when the period of an investment or loan is given in days, the Canadian approach is to express the time period t in years by dividing the number of days by 365. By comparison, Americans often approximate the yearly time period by dividing by 360 rather than 365. During a leap year, which is a year with 366 days, there is some inconsistency in the approach taken to calculate t. While the Canada Customs and Revenue Agency, as well as numerous financial institutions, divides by 366 to calculate t for a leap year, the more general approach is to divide by 365.

In this textbook, 365 is used as a divisor for t calculations in both regular and leap years.

Excel

If you choose, you can use Excel's *Accrued Interest (ACCRINT)* function to compute the amount of interest when the time is given in days. Refer to **ACCRINT** on the Spreadsheet Template Disk to learn how to use this Excel function.

EXERCISE 1.2

State r and t for each of the following:

1. Rate is 3½%; time is 1¼ years
2. Rate is 9¾%; time is 21 months
3. Rate is 8.25%; time is 183 days
4. Rate is 5½%; time is 332 days

Compute the amount of interest for each of the following.

1. $5000.00 at 9¾% for 2¼ years
2. $645.00 at 6¼% for 1 ¾ years
3. $1755.00 at 4.65% for 14 months
4. $1651.43 at 4.9% for 9 months
5. $980.00 at 11.5% for 244 days
6. $1697.23 at 3.4% for 163 days

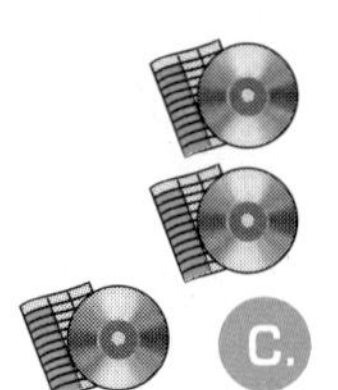

C. Compute the amount of interest for each of the following.

1. $275.00 at 9.25% from November 30, 2005 to May 5, 2006
2. $1090.60 at 7.8% from October 12, 2007 to April 24, 2008
3. $424.23 at 8¾% from April 4, 2006 to November 4, 2006
4. $1713.09 at 4.4% from August 30, 2006 to March 30, 2007

1.3 Finding the Principal, Rate, or Time

A. Formulas derived from the simple interest formula

The simple interest formula $I = Prt$ contains the four variables I, P, r, and t. If any three of the four are given, the value of the unknown variable can be computed by substituting the known values in the formula or by solving for the unknown variable first and then substituting in the resulting derived formula.

The three derived formulas are

(i) To find the principal P,

$$P = \frac{I}{rt}$$ **Formula 1.1B**

(ii) To find the rate of interest r,

$$r = \frac{I}{Pt}$$ **Formula 1.1C**

(iii) To find the time period t,

$$t = \frac{I}{Pr}$$ ——— **Formula 1.1D**

Note:

(a) In Formula 1.1C, if the time period t is expressed in years, the value of r represents an annual rate of interest in decimal form.

(b) In Formula 1.1D, if the rate of interest r is an annual rate, the value of t represents years in decimal form.

POINTERS AND PITFALLS

This diagram is a useful aid in remembering the various forms of the simple interest formula I = Prt.

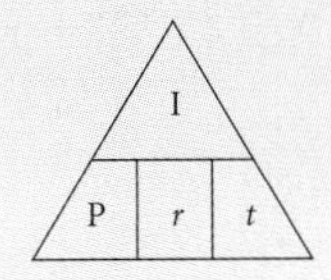

B. Finding the principal

When the amount of interest, the rate of interest, and the time period are known, the principal can be determined.

EXAMPLE 1.3A What principal will earn interest of \$18.20 at 3.25% in 8 months?

SOLUTION $I = 18.20; \quad r = 3.25\%; \quad t = \frac{8}{12}$

(i) Using the formula I = Prt,

$$18.20 = (P)(0.0325)\left(\frac{8}{12}\right)$$ ——— by substitution

$$18.20 = (P)(0.0216667)$$ ——— $(0.0325)\left(\frac{8}{12}\right)$

$$P = \frac{18.20}{0.0216667}$$ ——— divide 18.20 by the coefficient of P

$$= \$840.00$$

(ii) Using the derived formula $P = \frac{I}{rt}$,

$$P = \frac{18.20}{(0.0325)\left(\frac{8}{12}\right)}$$ ——— by substitution

$$= \frac{18.20}{0.021667}$$

$$= \$840.00$$

EXAMPLE 1.3B Determine the amount of money that must be invested for 245 days at 9.75% to earn \$71.99.

SOLUTION $I = 71.99; \quad r = 9.75\% = 0.0975; \quad t = \frac{245}{365}$

(i) Using the formula $I = Prt$,

$$71.99 = (P)(0.0975)\left(\frac{245}{365}\right)$$
$$71.99 = (P)(0.065445)$$
$$P = \frac{71.99}{0.065445}$$
$$= \$1100.00$$

(ii) Using the derived formula $P = \frac{I}{rt}$,

$$P = \frac{71.99}{(0.0975)\left(\frac{245}{365}\right)}$$
$$= \frac{71.99}{0.065445}$$
$$= \$1100.00$$

C. Finding the rate

When the amount of interest, the principal, and the time period are known, the rate of interest can be determined.

EXAMPLE 1.3C Find the annual rate of interest required for \$744.00 to earn \$75.95 in 14 months.

SOLUTION $I = 75.95; \quad P = 744.00; \quad t = \frac{14}{12}$

(i) Using the formula $I = Prt$,

$$75.95 = (744)(r)\left(\frac{14}{12}\right)$$
$$75.95 = (868)(r)$$
$$r = \frac{75.95}{868}$$
$$= 0.0875$$
$$= 8.75\%$$ ——— convert to a percent

(ii) Using the derived formula $r = \frac{I}{Pt}$

$$r = \frac{75.95}{(744.00)\left(\frac{14}{12}\right)}$$
$$= \frac{75.95}{868.00} = 0.0875 = 8.75\%$$

EXAMPLE 1.3D Find the yearly rate of interest on a principal of \$1600.00 earning interest of \$22.36 in 120 days.

SOLUTION $I = 22.36; \quad P = 1600.00; \quad t = \frac{120}{365}$

(i) Using the formula $I = Prt$,

$$22.36 = (1600.00)(r)\left(\frac{120}{365}\right)$$

$$22.36 = (526.0274)(r)$$

$$r = \frac{22.36}{526.0274} = 0.0425 = 4.25\%$$

(ii) Using the derived formula $r = \frac{I}{Pt}$,

$$r = \frac{22.36}{(1600.00)\left(\frac{120}{365}\right)}$$

$$= \frac{22.36}{526.027397}$$

$$= 0.0425$$

$$= 4.25\%$$

D. Finding the time

When the amount of interest, the principal, and the rate of interest are known, the time period can be determined.

EXAMPLE 1.3E Find the number of years required for \$745.00 to earn \$178.80 simple interest at 8% p.a.

SOLUTION $I = 178.80; \quad P = 745.00; \quad r = 8\% = 0.08$

(i) Using the formula $I = Prt$,

$$178.80 = (745.00)(0.08)(t)$$

$$178.80 = (59.60)(t)$$

$$t = \frac{178.80}{59.60}$$

$$= 3.00 \text{ (years)}$$

(ii) Using the derived formula $t = \frac{I}{Pr}$,

$$t = \frac{178.80}{(745.00)(0.08)}$$

$$= 3.00 \text{ (years)}$$

Note: The value of t in the formula $I = Prt$ will be in years. If the time period is to be stated in months or in days, it is necessary to multiply the initial value of t by 12 for months or 365 for days.

EXAMPLE 1.3F Determine the number of months required for a deposit of \$1320.00 to earn \$16.50 interest at 3.75%.

SOLUTION $I = 16.50; \quad P = 1320.00; \quad r = 3.75\% = 0.0375$

(i) Using the formula $I = Prt$,

$$16.50 = (1320.00)(0.0375)(t)$$
$$16.50 = (49.50)(t)$$
$$t = \frac{16.50}{49.50}$$
$$= 0.333333 \text{ years}$$
$$= (0.333333)(12) \text{ months}$$
$$= 4 \text{ months}$$

(ii) Using the derived formula $t = \frac{I}{Pr}$,

$$t = \frac{16.50}{(1320.00)(0.0375)} \text{ years}$$
$$= \frac{16.50}{49.50} \text{ years} = \frac{1}{3} \text{ years}$$
$$= 4 \text{ months}$$

EXAMPLE 1.3G For how many days would a loan of \$1500.00 be outstanding to earn interest of \$36.16 at 5.5% p.a.?

SOLUTION $I = 36.16; \quad P = 1500.00; \quad r = 5.5\% = 0.055$

(i) Using the formula $I = Prt$,

$$36.16 = (1500.00)(0.055)(t)$$
$$36.16 = (82.50)(t)$$
$$t = \frac{36.16}{82.50}$$
$$= 0.438303 \text{ years}$$
$$= (0.438303)(365) \text{ days}$$
$$= 160 \text{ days}$$

(ii) Using the derived formula $t = \frac{I}{Pr}$,

$$t = \frac{36.16}{(1500.00)(0.055)} \text{ years}$$
$$= \frac{36.16}{82.50} \text{ years}$$
$$= 0.438303 \text{ years}$$
$$= (0.438303)(365) \text{ days}$$
$$= 160 \text{ days}$$

EXERCISE 1.3

A. Determine the missing value for each of the following.

	Interest	Principal	Rate	Time
1.	\$67.83	?	9.5%	7 months
2.	\$106.25	?	4.25%	250 days
3.	\$215.00	\$2400.00	?	10 months
4.	\$53.40	\$750.00	?	315 days
5.	\$36.17	\$954.00	3.25%	? (months)
6.	\$52.64	\$1295.80	9.75%	? (months)
7.	\$7.14	\$344.75	5.25%	? (days)
8.	\$68.96	\$830.30	10.75%	? (days)

B. Find the value indicated for each of the following.

1. Find the principal that will earn \$148.32 at 6.75% in 8 months.
2. Determine the deposit that must be made to earn \$39.27 in 225 days at 2.75%.
3. A loan of \$880.00 can be repaid in 15 months by paying the principal sum borrowed plus \$104.50 interest. What was the rate of interest charged?

4. At what rate of interest will \$1387.00 earn \$63.84 in 200 days?
5. In how many months will \$1290.00 earn \$100.51 interest at 8½%?
6. Determine the number of days it will take \$564.00 to earn \$15.09 at 7¾%.

7. What principal will earn \$39.96 from June 18, 2005 to December 15, 2005 at 9.25%?

8. What rate of interest is required for \$740.48 to earn \$42.49 interest from September 10, 2007 to March 4, 2008?
9. Philip wants to supplement his pension by \$2000 per month with income from his investments. His investments pay him monthly and earn 6% p.a. What value of investments must Philip have in his portfolio to generate enough interest to give him his desired income?

10. Bunny's Antiques received \$88.47 interest on a 120-day term deposit of \$7800.00. At what rate of interest was the term deposit invested?

11. Anne's Dress Shop borrowed \$3200.00 to buy material. The loan was paid off seven months later by a lump-sum payment of \$3368.00. What was the simple rate of interest at which the money was borrowed?

12. Bill filed his income tax return with the Canada Customs and Revenue Agency (CCRA) after the April 30 deadline. He calculated that he owed the CCRA $3448.00, but did not include a payment for this amount when he sent in his tax return. The CCRA's Notice of Assessment indicated agreement with Bill's tax calculation. It also showed that the balance due was $3827.66, which included a 10% late-filing penalty and interest at 9% p.a. For how many days was Bill charged interest?

13. On August 15, 2006, Low Rider Automotive established a line of credit at its bank, with interest at 8.75% p.a. This line of credit was used to purchase $5000.000 of inventory and supplies. Low Rider paid $5113.87 to satisfy the incurred debt. On what date did Low Rider Automotive honour the line of credit?

14. On April 1, Faircloud Variety Company deposited $24 000.00 into a daily interest savings account earning simple interest of 1.5%. Interest is paid to the account at the end of every calendar quarter.

(**a**) How much interest was paid to Faircloud's account on June 30?
(**b**) How much interest was paid to Faircloud's account on September 30?
(**c**) Give two reasons why the answers to (a) and (b) are not the same.

15. Mishu wants to invest an inheritance of $50 000 for one year. His credit union offers 3.95% for a one-year term or 3.85% for a six-month term.

(**a**) How much will Mishu receive after one year if he invests at the one-year rate?
(**b**) How much will Mishu receive after one year if he invested for six months at a time at 3.85% each time?
(**c**) What would the one-year rate have to be to yield the same amount of interest as the investment described in part (b)?

16. Prairie Grains Cooperative wants to invest $45 000.00 in a short-term deposit. The bank offers 1.3% interest for a one-year term and 1.1% for a six-month term.

(**a**) How much would Prairie Grains receive if the $45 000.00 is invested for one year?
(**b**) How much would Prairie Grains receive at the end of one year if the $45 000.00 is invested for six months and then the principal and interest earned is reinvested for another six months?
(**c**) What would the one-year rate have to be to yield the same amount of interest as the investment described in part (b)?

1.4 Computing Future Value (Maturity Value)

A. Basic concept

When you borrow money, you are obligated to repay both the sum borrowed (the principal) and any interest due. Therefore, the **future value of a sum of money** (or **maturity value**) is the value obtained by adding the original principal and the interest due.

FUTURE VALUE (OR MATURITY VALUE) = PRINCIPAL + INTEREST

$$S = P + I$$ — Formula 1.2

EXAMPLE 1.4A Determine the future value (maturity value), principal, or interest as indicated.

(i) The principal is $2200.00 and the interest is $240.00. Find the future value (maturity value).

SOLUTION

$P = 2200.00; \quad I = 240.00$

$$S = P + I$$
$$= 2200.00 + 240.00$$
$$= \$2440.00$$

The future value is $2440.00.

(ii) The principal is $850.00 and the future value (maturity value) is $920.00. Compute the amount of interest.

SOLUTION

$P = 850.00; \quad S = 920.00$

$$S = P + I$$
$$I = S - P$$
$$I = 920.00 - 850.00$$
$$I = \$70.00$$

The amount of interest is $70.00.

(iii) The future value (maturity value) is $430.00 and the interest is $40.00. Compute the principal.

SOLUTION

$S = 430.00; \quad I = 40.00$

$$S = P + I$$
$$P = S - I$$
$$P = 430.00 - 40.00$$
$$P = \$390.00$$

The principal is $390.00.

B. The future value formula $S = P(1 + rt)$

To obtain the future value (maturity value) formula for simple interest, the formulas $I = Prt$ and $S = P + I$ are combined .

$S = P + I$
$S = P + Prt$ ——— substitute Prt for I
$S = P(1 + rt)$ ——— take out the common factor P

$$S = P(1 + rt)$$ ——— Formula 1.3A

EXAMPLE 1.4B Find the future value (maturity value) of an investment of \$720.00 earning 4% p.a. for 146 days.

SOLUTION $P = 720.00; \quad r = 4\% = 0.04; \quad t = \frac{146}{365}$

$$\begin{aligned} S &= P(1 + rt) \\ &= (720.00)\left[1 + (0.04)\left(\frac{146}{365}\right)\right] \\ &= (720.00)(1 + 0.016) \\ &= (720.00)(1.016) \\ &= \$731.52 \end{aligned}$$

The future value of the investment is \$731.52.

EXAMPLE 1.4C Find the maturity value of a deposit of \$1250.00 invested at 2.75% p.a. from October 15, 2005 to May 1, 2006.

SOLUTION The time period in days = 17 + 30 + 31 + 31 + 28 + 31 + 30 + 0 = 198.

$P = 1250.00; \quad r = 2.75\% = 0.0275; \quad t = \frac{198}{365}$

$$\begin{aligned} S &= P(1 + rt) \\ &= (1250.00)\left[1 + (0.0275)\left(\frac{198}{365}\right)\right] \\ &= (1250.00)(1 + 0.014918) \\ &= (1250.00)(1.014918) \\ &= \$1268.65 \end{aligned}$$

The maturity value of the deposit is \$1268.65.

EXERCISE 1.4

A. Use the future value (maturity value) formula to answer each of the following.

1. Find the future value of \$480.00 at $3\frac{1}{2}\%$ for 220 days.
2. Find the future value of \$1100.00 invested at 6.75% for 360 days.

3. Find the maturity value of \$732.00 invested at 9.8% from May 20, 2006 to November 23, 2006.

4. Find the maturity value of \$775.00 invested at 6.25% from March 1, 2005 to October 20, 2005.
5. Compute the future value of \$820.00 over nine months at $4\frac{3}{4}\%$.
6. Compute the future value of \$570.00 over seven months at $5\frac{1}{2}\%$.
7. What payment is required to pay off a loan of \$1200.00 at 7% fourteen months later?
8. What payment is required to pay off a loan of \$2000.00 at 6% eighteen months later?

1. Paul invested \$2500.00 in a 180-day term deposit at 3.45% p.a. What is the maturity value of the deposit?
2. Suzette invested \$800.00 in a 210-day term deposit at 2.75% p.a. What is the maturity value of the deposit?
3. Jack signed a two-year loan of \$12 000.00 at 7.5% p.a. How much will he owe at maturity?
4. Hao Lin borrowed \$8000.00 at 8.55% p.a. for fifteen months. How much will he owe at maturity?
5. On September 30, 2006, Red Flag Inn invested \$26 750.00 in a short-term investment of 215 days. An investment of this length earns 1.3% p.a. How much will the investment be worth at maturity?
6. Jo-Mar Consulting invested \$17 200.00 on July 18, 2005 into a short-term investment of 150 days. The investment would earn 1.85% p.a. How much will the investment be worth at maturity?
7. Speedy Courier invested \$13 500.00 into a 270-day term deposit. What is the maturity value if the rate of interest is 3.65%?
8. Diamond Drilling invested \$40 000.00 into a 240-day term deposit earning interest at 2.43%. What is the maturity value of the investment?

1.5 Finding the Principal (Present Value)

A. Finding the principal when the maturity value (future value) is known

When the maturity value, the rate, and the time are given, calculation of the principal utilizes the formula $S = P(1 + rt)$.

When interest is paid for the use of money, the value of any sum of money subject to interest changes with time. This change is called the **time value of money.** The **present value** of an amount at any given time is the principal needed to grow to that amount at a given rate of interest over a given period of time.

Since the problem of finding the present value is equivalent to finding the principal when the future value, rate, and time are given, the future value formula $S = P(1 + rt)$ applies. However, as the problem of finding the present value of an amount is one of the frequently recurring problems in financial analysis, it is useful to solve the future value formula for P to obtain the present value formula.

$$S = P(1 + rt)$$ — starting with the future value formula

$$\frac{S}{(1+rt)} = \frac{P(1+rt)}{(1+rt)}$$ — divide both sides by $(1 + rt)$

$$\frac{S}{(1+rt)} = P$$ — reduce the fraction $\frac{(1+rt)}{(1+rt)}$ to 1

This is the present value formula for simple interest.

$$P = \frac{S}{(1 + rt)}$$ — **Formula 1.3B**

EXAMPLE 1.5A Find the principal that will amount to \$1249.50 in nine months at 5.5% per annum.

SOLUTION $S = 1249.50;\ r = 5.5\% = 0.055;\ t = \frac{9}{12}$

$$P = \frac{S}{(1 + rt)}$$ — use the present value formula when S is known

$$P = \frac{1249.50}{1 + (0.055)\frac{(9)}{(12)}}$$

$$P = \frac{1249.50}{1.04125}$$

$$P = \$1200.00$$

The principal is \$1200.00.

EXAMPLE 1.5B What sum of money must be invested on January 31, 2008 to amount to $7700.00 on August 18, 2008 at 10% p.a.?

SOLUTION The time period in days = 1 + 29 (leap year) + 31 + 30 + 31 + 30 + 31 + 17 = 200.

$$P = \frac{S}{(1 + rt)}$$ ——— use the present value formula when S is known

$S = 7700.00;\ r = 10\% = 0.10;\ t = \frac{200}{365}$

$$P = \frac{7700.00}{1 + (0.10)\left(\frac{200}{365}\right)}$$

$$P = \frac{7700.00}{1.054795}$$

$$P = \$7300.00$$

Note:
1. The total interest earned is $7700.00 − $7300.00 = $400.00.
2. The daily amount of interest is $\frac{\$400.00}{200} = \$2.00$.

B. The present value concept and formula

When interest is paid for the use of money, the value of any sum of money subject to interest changes with time. This change is called the ***time value of money.***

To illustrate the concept of time value of money, Example 1.5B is represented on the time graph shown in Figure 1.1.

FIGURE 1.1 **Time Graph for Example 1.5B**

Jan. 31, 2008	$r = 10\%$	Aug. 18, 2008
Original Principal $7300		Maturity Value $7700

The original principal of $7300.00 will grow to $7700.00 at 10% in 200 days. Interest of $400.00 will be earned in those 200 days, indicating that the original principal will grow by $2.00 each day. The value of the investment changes day by day. On January 31, 2008, the $7300.00 principal is known as the present value of the August 18, 2008 $7700.00.

EXAMPLE 1.5C Compute the value of an investment eight months before the maturity date that earns interest at 6% p.a. and has a maturity value of $884.00.

FIGURE 1.2 Time Graph for Example 1.5C

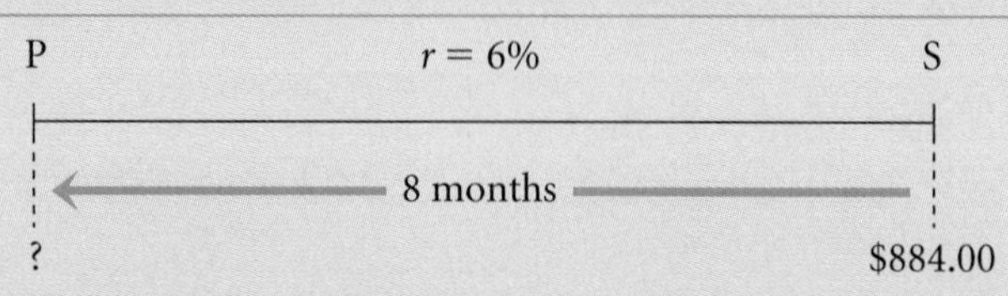

SOLUTION

$S = 884.00;\ r = 6\% = 0.06;\ t = \frac{8}{12}$

$$P = \frac{S}{1 + rt}$$

$$P = \frac{884.00}{1 + (0.06)\left(\frac{8}{12}\right)} \quad \text{using Formula 1.3B}$$

$$= \frac{884.00}{1.04}$$

$$= \$\,850.00$$

The present value of the investment is $850.00.

EXAMPLE 1.5D Compute the amount of money that, deposited in an account on April 1, 2005, will grow to $657.58 by September 10, 2005 at 4.75% p.a.

SOLUTION

The time in days = 30 + 31 + 30 + 31 + 31 + 9 = 162

$S = 657.58;\ r = 4.75\% = 0.0475;\ t = \frac{162}{365}$

$$P = \frac{S}{1 + rt}$$

$$P = \frac{657.58}{1 + (0.0475)\left(\frac{162}{365}\right)}$$

$$P = \frac{657.58}{1.021082}$$

$$P = \$644.00$$

The amount of money is $644.00.

EXERCISE 1.5

A. Find the principal and the missing value in each of the following.

	Future Value (Maturity Value)	Interest Amount	Interest Rate	Time
1.	$279.30	?	4%	15 months
2.	$729.30	$117.30	?	20 months
3.	?	$29.67	8.6%	8 months
4.	?	$27.11	9.5%	240 days
5.	$2109.24	$84.24	5.2%	?
6.	$1035.38	?	7.5%	275 days

B. Solve each of the following.

1. What principal will have a future value of $1241.86 at 3.9% in five months?

2. What amount of money will accumulate to $480.57 in 93 days at 4.6%?

3. Determine the present value of a debt of $1760.00 due in four months if interest at 9¾% is allowed.

4. Compute the present value of a debt of $708.13 eighty days before it is due if money is worth 5.3%.

5. The annual Deerfield Golf Club membership fees of $1750.00 are due on March 1. Club management offers a reduction of membership fees of 18.9% p.a. to members who pay the dues by September 1 of the previous year. How much must a member pay on September 1 if she chooses to take advantage of the club management's offer?

6. You are the accountant for Peel Credit Union. The lawyer for a member has sent a cheque for $7345.64 in full settlement of the member's loan balance including interest at 6.25% for 11 months. How much of the payment is interest?

7. On March 15, Bozana bought a government-guaranteed short-term investment maturing on September 12. How much did Bozana pay for the investment if she will receive $10 000.00 on September 12 and interest is 5.06%?

8. On October 29, 2005, Toddlers' Toys borrowed money with a promise to pay $23 520.18 on March 5, 2006. This loan included interest at 6.5%. How much money did Pixley borrow on October 29?

9. On March 15, 2006, Ling purchased a government-guaranteed short-term investment maturing on September 12, 2006. How much did Ling pay for the investment if $10 000.00 will be received on September 12, 2006 and interest is 1.35% p.a.?

BUSINESS MATH NEWS BOX

Save for Your Dream

The New Canada Savings Bonds Payroll Savings Program is an ideal way to save for your dreams in an easy, effortless manner. You can do this by following three general "rules."

The first "rule" is to pay yourself first. The New Canada Savings Bonds Payroll Savings Program allows your employer to deduct weekly, bi-weekly, or monthly amounts specified by you and transfer the amounts into an account set up by the Bank of Canada. These funds are then used to purchase Canada Savings Bonds. You only have to sign up once and the program continues until you decide to redeem all or part of your Canada Savings Bonds. Once a year, at campaign time, you can increase your bond purchase or start a new plan with a new application. In addition, you can redeem all or part of your savings at any time and have the money deposited into your bank account or sent to you as a cheque. However, note that partial redemptions are not allowed within the first three months following the issue date of the bond.

The second "rule" is to start right away. The sooner you start, the more you will save toward your dream. The minimum purchase allowed is $2 per week or $4 per bi-weekly period or $8 per month. The maximum purchase is $9999.00 no matter how often you are paid.

The third "rule" is to stick with your plan. Redeem your Canada Savings Bonds only if you must. Remember, you are saving for your future—saving for your dream.

QUESTIONS

1. Joe and Sally are planning to invest in an education fund. They plan to have enough money to pay for two years of college for their son. He would start college eight years from now. They estimate that the cost will be between $9000 and $10 000. They decide that the New Canada Savings Bonds Payroll Savings Program is the most convenient and painless way of saving for the education fund. Sally's employer offers the program and she signs up for deductions starting on November 1, 2005. The 2005 Canada Savings Bonds pay interest at 2.7% simple interest for the first year, calculated on the highest balance for each month. Suppose Sally arranges with her employer to deduct $100 per month at the beginning of each month. How much can they save toward the education fund if she redeems her savings bonds eight years later?
2. Choose a financial goal and follow the steps Sally took. Determine the cost of your goal and calculate the amount you could save in one year if you save $50 per month through the New Canada Savings Bonds Payroll Savings Program. At this rate, approximately how long will it take you to achieve your financial goal?

Source: Some information for these questions came from the Canada Savings Bond site, www.cis-pec.gc.ca/eng.

1.6 Computing Equivalent Values

A. Dated values

If an amount of money is subject to a rate of interest, it will grow over time. Thus, the value of the amount of money changes with time. This change is known as the ***time value of money***. For example, if you invested $1000.00 today at 4% p.a. simple interest, your investment has a value of $1000.00 today, $1010.00 in three months, $1020.00 in six months, and $1040.00 in one year.

The value of the original amount at any particular time is a **dated value**, or **equivalent value**, of that amount. The dated value combines the original sum with the interest earned up to the dated value date. Each dated value at a different time is equivalent to the original amount of money. The table below shows four dated values for $1000.00 invested at 4% p.a. The longer the time is from today, the greater is the dated value. This is so because interest has been earned on the principal over a longer time period.

Time	Dated Value
Today	$1000.00
3 months from today	$1010.00
6 months from today	$1020.00
1 year from today	$1040.00

Timberwest Company owes Abco Inc. $500.00 and payment is due today. Timberwest asks for an extension of four months to pay off the obligation. How much should they expect to pay in four months' time if money is worth 6%?

Since Abco could invest the $500.00 at 6% p.a., Timberwest should be prepared to pay the dated value. This dated value includes interest for the additional four-month time period. It represents the amount to which the $500.00 will grow in four months (the future value) and is found using Formula 1.3A.

$$\begin{aligned} S &= P(1 + rt) \\ &= 500.00\left[1 + (0.06)\left(\frac{4}{12}\right)\right] \\ &= 500.00(1 + 0.02) \\ &= \$510.00 \end{aligned}$$

In addition, Red Rock Construction owes Abco Inc. $824.00, due to be paid six months from now. Suppose Red Rock Construction offers to pay the debt today. How much should Red Rock Construction pay Abco Inc. if money is worth 6%?

Since Abco Inc. could invest the payment at 6%, the payment should be the sum of money that will grow to $824.00 in six months at 6%. By definition, this amount of money is the present value of the $824.00. The present value represents today's dated value of the $824.00 and is found using Formula 1.3B.

$$P = \frac{S}{1 + rt}$$

$$= \frac{824.00}{1 + (0.06)\left(\frac{6}{12}\right)}$$

$$= \frac{824.00}{1 + 0.03}$$

$$= \$800.00$$

Because of the time value of money, sums of money given at different times are not directly comparable. For example, imagine you are given a choice between \$2000.00 today and \$2200.00 one year from now. It does not automatically follow, from the point of view of investing money, that either the larger amount of money or the chronologically earlier amount of money is preferable.

To make a rational choice, we must allow for the rate of interest money can earn and choose a comparison date or **focal date** to obtain the dated values of the amounts of money at a specific time.

Equivalent values on the same date are directly comparable and may be obtained for simple interest by using either the maturity value (future value) formula, Formula 1.3A, $S = P(1 + rt)$, or the present value formula, Formula 1.3B,

$$P = \frac{S}{1 + rt}.$$

B. Choosing the appropriate formula

The choice of which formula to use for computing dated values depends on the due date of the sum of money relative to the selected focal (or comparison) date.

(a) If the due date falls before the focal date, use the future value (maturity value) formula.

FIGURE 1.3 **When to Use the Future Value (or Maturity Value) Formula**

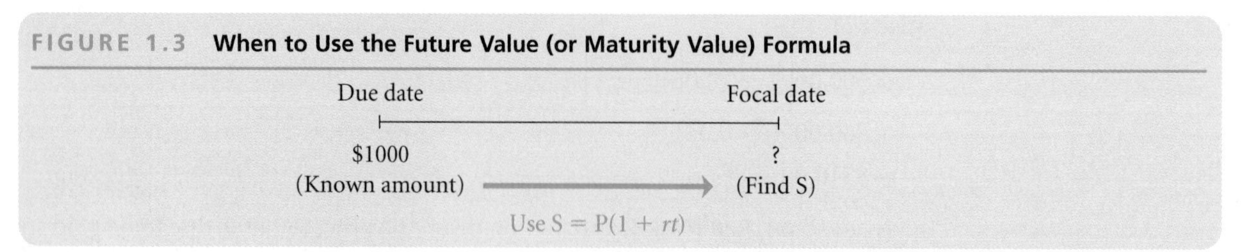

Explanation of Figure 1.3: We are looking for a future value relative to the given value. This future value will be higher than the known value by the interest that accumulates on the known value from the due date to the focal date. Because this is a future value problem (note that the arrow points to the right), the future value (or maturity value) formula $S = P(1 + rt)$ applies.

(b) If the due date falls after the focal date, use the present value formula.

FIGURE 1.4 **When to Use the Present Value Formula**

Focal date — Due date

? (Find P) ← $1000 (Known amount)

Use $P = \frac{S}{1 + rt}$

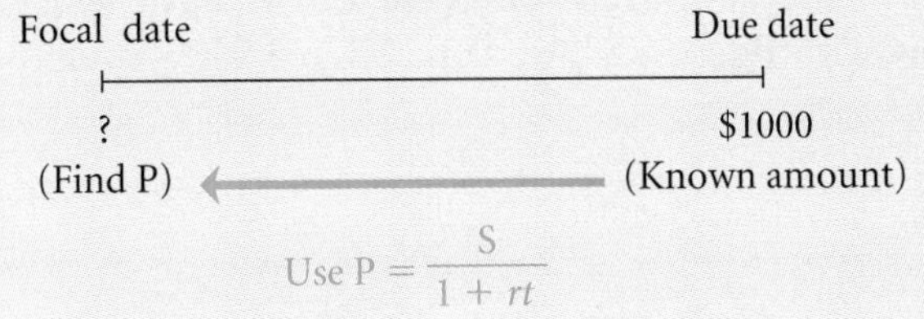

Explanation of Figure 1.4: We are looking for an earlier value relative to the given value. This earlier value will be less than the given value by the interest that would accumulate on the unknown earlier value from the focal date to the due date. We are, in fact, looking for the principal that will grow to the given value. Because this is a present value problem (note that the arrow points to the left), the present value formula $P = \frac{S}{1 + rt}$ is appropriate.

C. Finding the equivalent single payment

EXAMPLE 1.6A A debt can be paid off by payments of $872.00 one year from now and $1180.00 two years from now. Determine the single payment now that would fully repay the debt. Allow for simple interest at 9% p.a.

SOLUTION See Figure 1.5 for the graphic representation of the dated values. Refer to Figures 1.3 and 1.4 to determine which formula is appropriate.

FIGURE 1.5 **Graphical Representation of the Dated Values**

Now — 1 Year — 2 Years

? (Focal date) ← $872.00 ← $1180.00

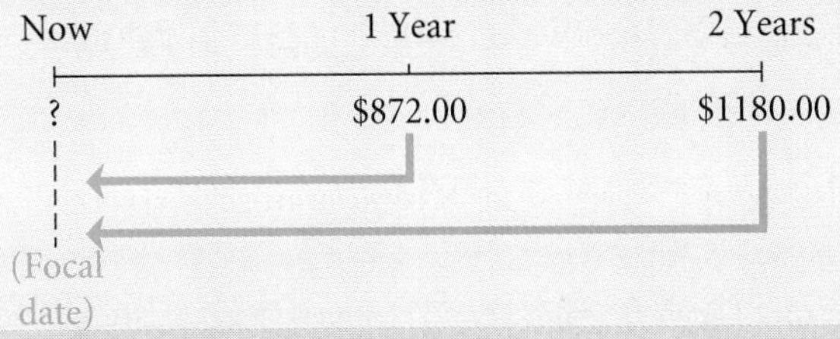

Since the focal date is *earlier* relative to the dates for the given sums of money (the arrows point to the left), the present value formula $P = \frac{S}{1 + rt}$ is appropriate.

(i) The dated (present) value of the $872.00 at the focal date

$$P = \frac{872.00}{1 + (0.09)(1)} = \frac{872.00}{1.09} = 800.00$$

(ii) The dated (present) value of the $1180.00 at the focal date

$$P = \frac{1180.00}{1 + (0.09)(2)} = \frac{1180.00}{1.18} = 1000.00$$

(iii) Single payment required now = 800.00 + 1000.00 = $1800.00

EXAMPLE 1.6B You are owed payments of \$400 due today, \$500 due in five months, and \$618 due in one year. You have been approached to accept a single payment nine months from now with interest allowed at 12% p.a. How much will the single payment be?

SOLUTION

FIGURE 1.6 Graphical Representation of the Dated Values

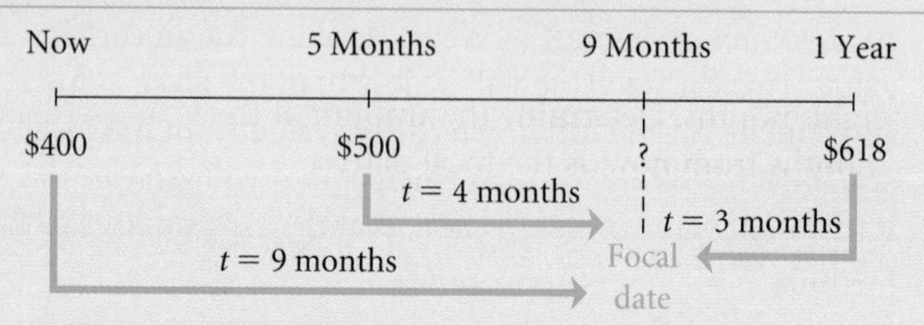

Since the focal date is in the *future* relative to the \$400 now and the \$500 five months from now (the arrows point to the right), the future value formula $S = P(1 + rt)$ is appropriate for these two amounts. However, since the focal date is *earlier* relative to the \$618 one year from now (the arrow points to the left), the present value formula $P = \frac{S}{1 + rt}$ is appropriate for this amount.

(i) The dated (future) value of \$400 at the focal date

$$P = 400.00; \quad r = 12\% = 0.12; \quad t = \frac{9}{12}$$

$$S = 400\left[1 + (0.12)\left(\frac{9}{12}\right)\right] = 400(1 + 0.09) = 400(1.09) = \$436.00$$

(ii) The dated (future) value of \$500 at the focal date

$$P = 500.00; \quad r = 12\% = 0.12; \quad t = \frac{4}{12}$$

$$S = 500\left[1 + (0.12)\left(\frac{4}{12}\right)\right] = 500(1 + 0.04) = 500(1.04) = \$520.00$$

(iii) The dated (present) value of \$618 at the focal date

$$S = 618.00; \quad r = 12\% = 0.12; \quad t = \frac{3}{12}$$

$$P = \frac{618.00}{1 + (0.12)\left(\frac{3}{12}\right)} = \frac{618.00}{1 + 0.03} = \frac{618.00}{1.03} = \$600.00$$

(iv) The single payment to be made nine months from now will be = 436.00 + 520.00 + 600.00 = \$1556.00.

D. Finding the value of two or more equivalent payments

The equivalent values (the dated value of the original scheduled payments) obtained when using simple interest formulas are influenced by the choice of focal date. Choose a focal date on the basis of the date when an equivalent value must be calculated. When two or more equivalent values must be calculated, choose one of the dates for which an equivalent value must be calculated.

EXAMPLE 1.6C Scheduled payments of \$400.00 due now and \$700.00 due in five months are to be settled by a payment of \$500.00 in three months and a final payment in eight months. Determine the amount of the final payment at 6% p.a., using eight months from now as the focal date.

SOLUTION Let the value of the final payment be $\$x$.

(i) Use a time diagram to represent the given data.

FIGURE 1.7 Graphical Representation of Data

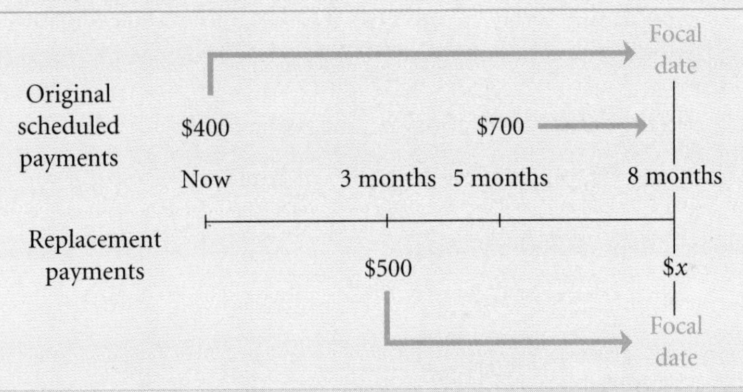

(ii) Dated value of the original scheduled payments

(a) The value at the focal date of the \$400 payment due eight months before the focal date is found by using the future value formula.

$$S = 400\left[1 + (0.06)\left(\frac{8}{12}\right)\right] = 400(1 + 0.04) = 400(1.04) = \$416.00$$

(b) The value at the focal date of the \$700 payment due three months before the focal date is found by using the future value formula.

$$S = 700\left[1 + (0.06)\left(\frac{3}{12}\right)\right] = 700(1 + 0.015) = 700(1.015) = \$710.50$$

(iii) Dated value of the replacement payments

(a) The value at the focal date of the \$500 payment made five months before the focal date is found by using the future value formula.

$$S = 500\left[1 + (0.06)\left(\frac{5}{12}\right)\right] = 500(1 + 0.025) = 500(1.025) = \$512.50$$

(b) The value at the focal date of the final payment is $\$x$ (no adjustment for interest is necessary for an amount of money located at the focal date).

(iv) The **equation of values** at the focal date is now set up by matching the dated values of the original debts to the dated values of the replacement payments.

THE SUM OF THE DATED VALUES OF THE REPLACEMENT PAYMENTS = THE SUM OF THE DATED VALUES OF THE ORIGINAL SCHEDULED PAYMENTS

$$500\left[1 + (0.06)\left(\frac{5}{12}\right)\right] + x = 400\left[1 + (0.06)\left(\frac{8}{12}\right)\right] + 700\left[1 + (0.06)\left(\frac{3}{12}\right)\right]$$

$$512.50 + x = 416.00 + 710.50$$

$$512.50 + x = 1126.50$$

$$x = 1126.50 - 512.50$$

$$x = 614.00$$

The final payment to be made in eight months is \$614.00.

EXAMPLE 1.6D

Clarkson Developments was supposed to pay Majestic Flooring \$2000.00 60 days ago and \$1300.00 in 30 days. Majestic Flooring agreed to accept three equal payments due today, 60 days from today, and 120 days from today. Compute the size of the equal payments at 10% p.a. Use today as the focal date.

SOLUTION

Let the size of the equal payments be $\$x$.

(i) Graphical representation of data

FIGURE 1.8 **Graphical Representation of Data**

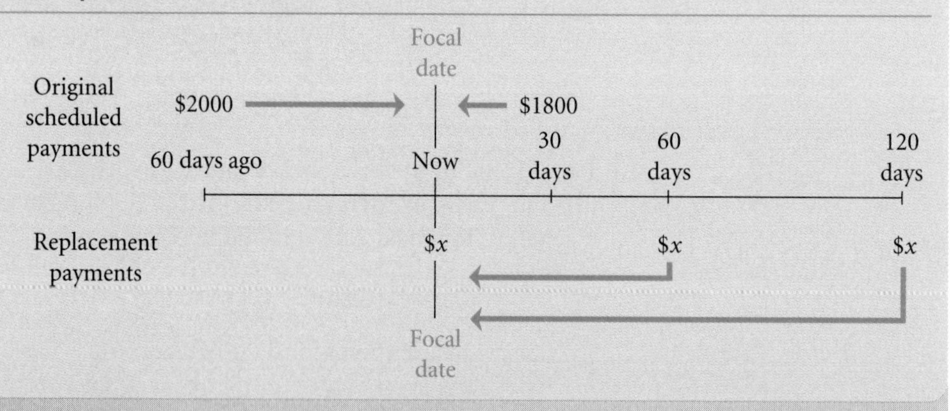

(ii) Dated value of the original scheduled payments at the focal date

(a) Because the \$2000 payment is due 60 days before the focal date, the future value formula is appropriate.

$$S = 2000\left[1 + (0.10)\left(\frac{60}{365}\right)\right] = 2000(1 + 0.016438) = \$2032.88$$

(b) Because the $1800 payment is due 30 days after the focal date, the present value formula is appropriate.

$$P = \frac{1800}{1 + (0.10)\left(\frac{30}{365}\right)} = \frac{1800}{1 + 0.008219} = \$1785.33$$

(iii) Dated value of the replacement payments at the focal date

(a) Since the first payment is to be made at the focal date, its value is $\$x$.

(b) Because the second payment is to be made 60 days after the focal date, the present value formula is appropriate.

$$P = \frac{x}{1 + (0.10)\left(\frac{60}{365}\right)} = \frac{x}{1 + 0.016438}$$

$$= \frac{1}{1.016438}(x) = \$0.983828x$$

(c) Because the third payment is to be made 120 days after the focal date, the present value formula is appropriate.

$$P = \frac{x}{1 + (0.10)\left(\frac{120}{365}\right)} = \frac{x}{1 + 0.032877}$$

$$= \frac{1}{1.032877}(x) = \$0.968170x$$

(iv) The equation of values (dated value of the replacement payments = dated value of the original scheduled payments)

$$x + 0.983828x + 0.968170x = 2032.88 + 1785.33$$
$$2.951998x = 3818.21$$
$$x = \frac{3818.21}{2.951998}$$
$$x = 1293.43$$

The size of each of the three equal payments is $1293.43.

EXAMPLE 1.6E

Two debts, one of $4000 due in three months with interest at 8% and the other of $3000 due in eighteen months with interest at 10%, are to be discharged by making two equal payments. What is the size of the equal payments if the first is due one year from now, the second two years from now, money is worth 12%, and the chosen focal date is one year from now?

SOLUTION

Let the size of the equal payments be $\$x$.

(i) Graphical representation of data

FIGURE 1.9 Graphical Representation of Data

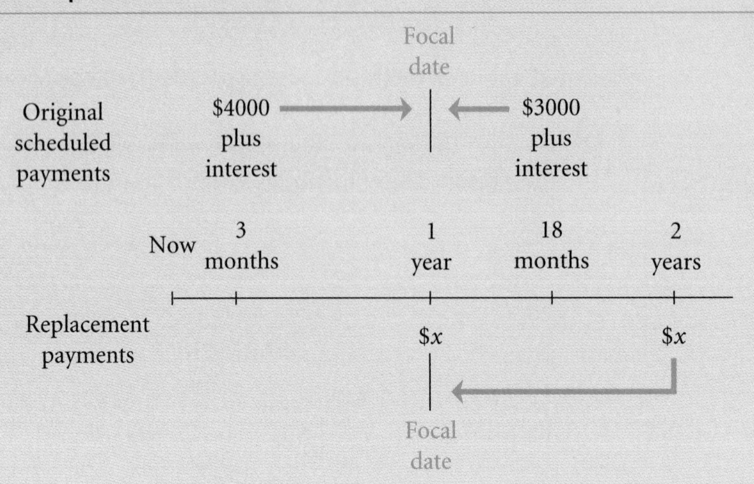

(ii) Maturity value of the original debts

Since the two original debts are interest-bearing, their maturity values need to be determined first so that the principal *and* interest of these original debts are discharged by the two equal payments.

(a) For the $4000 debt: $P = 4000$; $r = 8\% = 0.08$; $t = \frac{3}{12}$

$$\text{Maturity value, } S = 4000\left[1 + (0.08)\left(\frac{3}{12}\right)\right] = 4000(1.02) = \$4080$$

(b) For the $3000 debt: $P = 3000$; $r = 10\% = 0.10$; $t = \frac{18}{12}$

$$\text{Maturity value, } S = 3000\left[1 + (0.10)\left(\frac{18}{12}\right)\right] = 3000(1.15) = \$3450$$

(iii) Dated value of the money value of the original payments at 12%

(a) Because the maturity value of $4080 is due nine months before the focal date, the future value formula is appropriate.

$$S = 4080\left[1 + (0.12)\left(\frac{9}{12}\right)\right] = 4080(1.09) = \$4447.20$$

(b) Because the maturity value of $3450 is due six months after the focal date, the present value formula is appropriate.

$$P = \frac{3450}{1 + (0.12)\left(\frac{6}{12}\right)} = \frac{3450}{1.06} = \$3254.72$$

(iv) Dated value of the replacement payments

(a) Since the first replacement payment is made at the focal date, its value is $\$x$.

(b) Because the second replacement payment is made 12 months after the focal date, the present value formula is appropriate.

$$P = \frac{x}{1 + (0.12)\left(\frac{12}{12}\right)} = \frac{x}{1.12} = 0.892857x$$

(v) The equation of values

$$x + 0.892857x = 4447.20 + 3254.72$$
$$1.892857x = 7701.92$$
$$x = \frac{7701.92}{1.892857}$$
$$x = 4068.94$$

The size of the equal payments is $4068.94.

E. Loan repayments

Loans by financial institutions to individuals are usually repaid by **blended payments,** which are equal periodic payments that include payment of interest and repayment of principal. To repay the loan, the sum of the present values of the periodic payments must equal the original principal. The concept of equivalent values is used to determine the size of the blended payments.

EXAMPLE 1.6F

A loan of $2000 made at 8.5% p.a. is to be repaid in four equal payments due at the end of the next four quarters respectively. Determine the size of the quarterly payments if the agreed focal date is the date of the loan.

SOLUTION

Let the size of the equal quarterly payments be represented by $\$x$.

(i) Graphical representation of data

FIGURE 1.10 **Graphical Representation of Data**

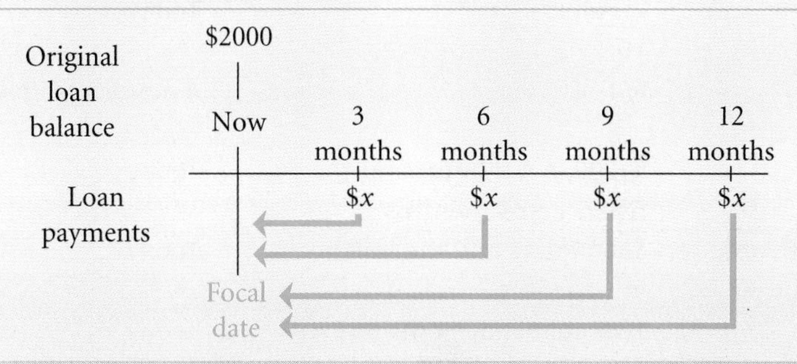

(ii) Dated value of the loan balance at the focal date is $2000.00.

(iii) Dated value of the loan payments at the focal date

The payments are due three, six, nine, and twelve months after the focal date respectively. Their values are

(a) $P_1 = \dfrac{x}{1 + (0.085)\left(\frac{3}{12}\right)} = \dfrac{x}{1 + 0.02125} = \dfrac{1}{1.02125}(x) = 0.979192x$

$$\text{(b) } P_2 = \frac{x}{1 + (0.085)\left(\frac{6}{12}\right)} = \frac{x}{1 + 0.0425} = \frac{1}{1.0425}(x) = 0.959233x$$

$$\text{(c) } P_3 = \frac{x}{1 + (0.085)\left(\frac{9}{12}\right)} = \frac{x}{1 + 0.06375} = \frac{1}{1.06375}(x) = 0.940071x$$

$$\text{(d) } P_4 = \frac{x}{1 + (0.085)\left(\frac{12}{12}\right)} = \frac{x}{1 + 0.085} = \frac{1}{1.085}(x) = 0.921659x$$

$$\text{(iv) } 0.979192x + 0.959233x + 0.940071x + 0.921659x = 2000.00$$

$$3.800155x = 2000.00$$

$$x = 526.29$$

The size of the quarterly payment is \$526.29.

EXERCISE 1.6

Find the equivalent replacement payments indicated for each of the following scheduled payments.

	Original Scheduled Payments	Replacement Payments	Focal Date	Rate
1.	\$800 due today	In full	4 months from today	11%
2.	\$1200 due 4 months ago	In full	Today	6%
3.	\$600 due in 2 months	In full	7 months from today	7%
4.	\$1000 due in 8 months	In full	2 months from today	10%
5.	\$500 due 4 months ago, \$600 due in 2 months	In full	Today	5%
6.	\$800 due today, \$700 due in 2 months	In full	4 months from today	9%
7.	\$2000 due today	\$1200 in 4 months and the balance in 8 months	Today	6%
8.	\$400 due 1 month ago, \$600 due in 3 months	\$500 today and the balance in 6 months	Today	8%
9.	\$1500 due today	Two equal payments due in 2 and 7 months	Today	10%
10.	\$1800 due 30 days ago	Three equal payments due today, in 30 days, and in 60 days	Today	9%
11.	\$1500 due 4 months ago, \$1200 due in 8 months with 10% interest	\$700 due now and two equal payments due in 6 months and in 12 months	Today	5%
12.	\$2000 due in 6 months with interest at 8%, \$1600 due in 2 years with interest at 7%	Three equal payments due in 6, 12, and 18 months respectively	1 year from today	11%

B. Solve each of the following problems.

1. Scheduled debt payments of $600 each are due three months and six months from now respectively. If interest at 10% is allowed, what single payment is required to settle the two scheduled payments today?

2. A loan payment of $1000 was due 60 days ago and another payment of $1200 is due 30 days from now. What single payment 90 days from now will pay off the two obligations if interest is to be 8% and the agreed focal date is 90 days from now?

3. Loan payments of $400 due three months ago and $700 due today are to be repaid by a payment of $600 one month from today and the balance four months from today. If money is worth 6% and the agreed focal date is one month from today, what is the size of the final payment?

4. Jessica should have made two payments of $800 each. The first was due 60 days ago and the second payment was due 30 days ago. The two original scheduled payments are to be settled by two equal payments to be made today and 60 days from now respectively. If interest allowed is 7.25% and the agreed focal date is today, what is the size of the equal payments?

5. Jerry borrowed $4000.00. The loan of $4000 is to be repaid by three equal payments due in four, eight, and twelve months from now respectively. Determine the size of the equal payments at 8.5% with a focal date of today.

6. On March 1, 2005, Bear Mountain Tours borrowed $1500.00. Three equal payments are required, on April 30, June 20, and August 10, and a final payment of $400.00 on September 30 of the same year. If the focal date is September 30, what is the amount of the equal payments at 6.75%?

7. Two debt payments, of $1900.00 due today and $1200.00 due with interest at 9% nine months from now, are to be settled by two equal payments due in three and six months respectively. Determine the amount of the equal payments if money is worth 7% and focal date is today.

8. Wayne's Windows should have made a payment of $1200.00 to Shane's Woodcraft one year ago. They are also scheduled to pay $1500.00 due with interest of 8% in nine months. Shane's Woodcraft agreed to accept three equal payments due today, six months from now, and one year from now at 9.5%. Determine the amount of the equal payments if the focal date is one year from today.

Appendix: Determining the Number of Days Using Manual Techniques

A. Counting exact time

The interest period is the time period for which interest is charged. When the interest period involves two dates, the exact number of days between the two dates must be determined. In counting the number of days, the traditional practice was to count the ending date but not the starting date. However, the current practice, associated with the computerization of interest calculation, is to count the starting date but *not* the ending date. This text uses the current method of counting the starting date.

APPENDIX 1.1

Find the number of days between January 30, 2005 and June 1, 2005.

SOLUTION

The starting date is January 30			to be counted
The ending date is June 1			do not count
Number of days	January	2	(31 − 29)
	February	28	
	March	31	
	April	30	
	May	31	
	June	0	
	TOTAL	122	

Alternatively, in determining the number of days between dates, count the ending date but *not* the starting date. The results of using either method should be the same.

The starting date is January 30			do *not* count
The ending date is June 1			to be counted
Number of days	January	1	(31 − 30)
	February	28	
	March	31	
	April	30	
	May	31	
	June	1	
	TOTAL	122	

APPENDIX 1.2 Find the number of days between November 12, 2005 and May 5, 2006.

SOLUTION

The starting date is November 12 —— to be counted
The ending date is May 5 —— do *not* count

Number of days November 2005	19	—— (30 − 11)
December	31	
January 2006	31	
February	28	
March	31	
April	30	
May	4	
TOTAL	174	

B. Using a table

Exact time can also be obtained from a table listing the number of each day of the year. (See Table 1.1 on page 36.) When using such a table, take care to distinguish between two cases:

(i) the starting date and the ending date are in the same year;

(ii) the ending date is in the year following the starting date or in a later year.

APPENDIX 1.3 Find the number of days between May 29, 2005 and August 3, 2005.

SOLUTION

The starting date is May 29 →	Day 149
The ending date is August 3 →	Day 215
The difference in days is (215 − 149)	66

APPENDIX 1.4 Find the number of days between September 1, 2005 and April 1, 2006.

SOLUTION

The starting date is September 1, 2005 →	Day 244
Number of days in 2001 is 365	
Number of days remaining in 2005 is (365 − 244) =	121
The ending date is April 1, 2006 →	Day 91
The total number of days is (121 + 91) =	212

TABLE 1.1 The Number of Each Day of the Year

Day of Month	Jan.	Feb.	Mar.	Apr.	May	June	July	Aug.	Sept.	Oct.	Nov.	Dec.	Day of Month
1	1	32	60	91	121	152	182	213	244	274	305	335	1
2	2	33	61	92	122	153	183	214	245	275	306	336	2
3	3	34	62	93	123	154	184	215	246	276	307	337	3
4	4	35	63	94	124	155	185	216	247	277	308	338	4
5	5	36	64	95	125	156	186	217	248	278	309	339	5
6	6	37	65	96	126	157	187	218	249	279	310	340	6
7	7	38	66	97	127	158	188	219	250	280	311	341	7
8	8	39	67	98	128	159	189	220	251	281	312	342	8
9	9	40	68	99	129	160	190	221	252	282	313	343	9
10	10	41	69	100	130	161	191	222	253	283	314	344	10
11	11	42	70	101	131	162	192	223	254	284	315	345	11
12	12	43	71	102	132	163	193	224	255	285	316	346	12
13	13	44	72	103	133	164	194	225	256	286	317	347	13
14	14	45	73	104	134	165	195	226	257	287	318	348	14
15	15	46	74	105	135	166	196	227	258	288	319	349	15
16	16	47	75	106	136	167	197	228	259	289	320	350	16
17	17	48	76	107	137	168	198	229	260	290	321	351	17
18	18	49	77	108	138	169	199	230	261	291	322	352	18
19	19	50	78	109	139	170	200	231	262	292	323	353	19
20	20	51	79	110	140	171	201	232	263	293	324	354	20
21	21	52	80	111	141	172	202	233	264	294	325	355	21
22	22	53	81	112	142	173	203	234	265	295	326	356	22
23	23	54	82	113	143	174	204	235	266	296	327	357	23
24	24	55	83	114	144	175	205	236	267	297	328	358	24
25	25	56	84	115	145	176	206	237	268	298	329	359	25
26	26	57	85	116	146	177	207	238	269	299	330	360	26
27	27	58	86	117	147	178	208	239	270	300	331	361	27
28	28	59	87	118	148	179	209	240	271	301	332	362	28
29	29		88	119	149	180	210	241	272	302	333	363	29
30	30		89	120	150	181	211	242	273	303	334	364	30
31	31		90		151		212	243		304		365	31

Note: In leap years, February 29 becomes day 60 and the numbers in the table increase by 1 for all following days.

C. Leap years

Leap years are years that are evenly divisible by 4, such as 2008, 2012, and 2016. An extra day is added to February if a year is a leap year. However, a centennial year is not a leap year unless the number is *evenly* divisible by 400. This means that the year 1900 was not a leap year but the year 2000 was. It did have the extra day.

APPENDIX 1.5

Find the number of days between December 12, 2007 and April 1, 2008
(i) by counting; (ii) by using Table 1.1.

SOLUTION

(i) The starting date is December 12, 2007 — to be counted
The ending date is April 1, 2008 — do *not* count

Number of days			
	December 2007	20	(31 − 11)
	January 2008	31	
	February	29	leap year
	March	31	
	April	0	
	TOTAL	111	

(ii) The starting date is December 12, 2007 → Day 346
The number of days in 2007 is 365
The number of days remaining in 2007 is $(365 - 346) =$ 19
The ending date is April 1, 2008 → Day 92
(for a leap year, increase the table number 91 by 1 to 92)

TOTAL $= 19 + 92 = 111$

Review Exercise

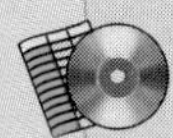

1. Determine the number of days for

(a) April 25, 2005 to October 14, 2005;

(b) July 30, 2007 to February 29, 2008.

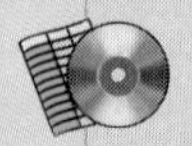

2. Compute the amount of interest for

(a) $1975.00 at 5.5% for 215 days;

(b) $844.65 at 8.25% from May 30, 2005 to January 4, 2006.

3. What principal will earn

(a) $34.80 interest at 5% in 219 days?

(b) $34.40 interest at 9.75% from October 30, 2005 to June 1, 2006?

4. Answer each of the following.

(a) What was the rate of interest if the interest on a loan of $675.00 for 284 days was $39.39?

(b) How long will it take for $2075.00 to earn $124.29 interest at 8¼% p.a.?

(c) If $680 is worth $698.70 after three months, what interest rate was charged?

(d) How many months will it take $750.00 to grow to $795.00 at 7.2% p.a.?

5. Solve each of the following.

(a) What principal will have a maturity value of $785.96 at 10% in 175 days?

(b) What is the present value of $6300.00 due in 16 months at 7.75%?

6. What principal will earn $24.87 at 4.75% in 156 days?

7. What sum of money will earn $148.57 from September 1, 2005 to April 30, 2006 at 7.5%?

8. At what rate of interest must a principal of $1545.00 be invested to earn interest of $58.93 in 150 days?

9. At what rate of interest will $1500.00 grow to $1562.04 from June 1, 2005 to December 1, 2005?

10. In how many months will $2500.00 earn $51.04 interest at 3.5%?

11. In how many days will $3100.00 grow to $3195.72 at 5.75%?

12. Compute the accumulated value of $4200.00 at 4.5% after 11 months.

13. What is the amount to which $1550.00 will grow from June 10, 2005 to December 15, 2005 at 6.5%?

14. What amount of money will accumulate to $1516.80 in eight months at 8%?

15. What principal will amount to $3367.28 if invested at 9% from November 1, 2007 to May 31, 2008?

16. What is the present value of $3780.00 due in nine months if interest is 5%?

17. Compute the present value on June 1, 2005 of $1785.00 due on October 15, 2005 if interest is 7.5%.

18. Payments of $1750.00 and $1600.00 are due four months from now and nine months from now, respectively. What single payment is required to pay off the two scheduled payments today if interest is 9%?

19. A loan payment of $1450.00 was due 45 days ago and a payment of $1200.00 is due in 60 days. What single payment made 30 days from now is required to settle the two payments if interest is 7% and the focal date is 30 days from now?

20. Scheduled payments of $800.00 due two months ago and $1200.00 due in one month are to be repaid by a payment of $1000.00 today and the balance in three months. What is the amount of the final payment if interest is 7.75% and the focal date is one month from now?

21. An obligation of $10 000.00 is to be repaid by equal payments due in 60 days, 120 days, and 180 days. What is the amount of the equal payments if money is worth 6.5% and the focal date is today?

22. Payments of $4000.00 each due in four, nine, and eleven months from now are to be settled by three equal payments due today, six months from now, and twelve months from now. What is the amount of the equal payments if interest is 7.35% and the agreed focal date is today?

23. A loan of $5000.00 due in one year is to be repaid by three equal payments due today, six months from now, and one year from now. What is the amount of the equal payments if interest is 6.5% and the focal date is today?

24. Three debts, the first for $1000.00 due two months ago, the second for $1200.00 due in two months, and the third for $1400.00 due in four months, are to be paid by a single payment today. How much is the single payment if money is worth 8.25% p.a. and the focal date is today?

25. Loan payments of $700.00 due three months ago and of $1000.00 due today are to be paid by a payment of $800.00 in two months and a final payment in five months. If 9% interest is allowed, and the focal date is five months from now, what is the amount of the final payment?

26. A loan of $5000.00 is to be repaid in three equal installments due 60, 120, and 180 days after the date of the loan. If the focal date is the date of the loan and interest is 6.9% p.a., compute the amount of the installments.

27. Three loan payments, the first for $2000.00 due three months ago, the second for $1400.00 with interest of 7.5% due in nine months, and the third for $1200.00 with interest of 5.75% due in sixteen months, are to be paid in three equal installments due today, four months from now, and one year from now. If money is worth 8.5% and the focal date is today, determine the amount of the equal payments. (Compute the maturity values of the original debts first.)

28. A loan of $7000.00 is to be settled by four equal payments due today, and one year, two years, and three years from now. Determine the amount of the equal payments if money is worth 5.75% and the focal date is today.

Self-Test

1. Compute the amount of interest earned by $1290.00 at 3.5% p.a. in 173 days.

2. In how many months will $8500.00 grow to $8818.75 at 5% p.a.?

3. What interest rate is paid if the interest on a loan of $2500.00 for six months is $81.25?

4. What principal will have a maturity value of $10 000.00 at 8.25% p.a. in three months?

5. What is the amount to which $6000.00 will grow at 3.75% p.a. in ten months?

6. What principal will earn $67.14 interest at 6.25% for 82 days?

7. What is the present value of $4400.00 due at 3.25% p.a. in 243 days?

8. What rate of interest is paid if the interest on a loan of $2500.00 is $96.06 from November 14, 2005 to May 20, 2006?

9. How many days will it take for $8500.00 to earn $689.72 at 8.25% p.a.?

10. What principal will earn $55.99 interest at 9.75% p.a. from February 4, 2008 to July 6, 2008?

11. What amount invested will accumulate to $7500.00 at 3.75% p.a. in 88 days?

12. Compute the amount of interest on $835.00 at 7.5% p.a. from October 8, 2007 to August 4, 2008.

13. A loan of $3320.00 is to be repaid by three equal payments due in 92 days, 235 days, and 326 days. Determine the amount of the equal payments at 8.75% p.a. with a focal date of today.

14. Loan payments of $1725.00 due today, $510.00 due in 75 days, and $655.00 due in 323 days are to be combined into a single payment to be made 115 days from now. What is that single payment if money is worth 8.5% p.a. and the focal date is 115 days from now?

15. Scheduled payments of $1010.00 due five months ago and $1280.00 due today are to be repaid by a payment of $615.00 in four months and the balance in seven months. If money is worth 7.75% p.a. and the focal date is in seven months, what is the amount of the final payment?

16. A payment of $1310.00 due five months ago and a second payment of $1225.00 with interest at 12% p.a. due in three months are to be settled by two equal payments due now and seven months from now. Compute the amount of the equal payments at 6.5%, with the focal date now.

Challenge Problems

1. Nora borrows $37 500.00 on September 28, 2007 at 7% p.a. simple interest, to be repaid on October 31, 2008. She has the option of making payments toward the loan before the due date. Nora pays $6350.00 on February 17, 2008, $8250.00 on July 2, 2008, and $7500.00 on October 1, 2008. Compute the payment required to pay off the debt on the focal date of October 31, 2008.

2. A supplier will give Shark Unibase Company a discount of 2% if an invoice is paid 60 days before its due date. Suppose Shark wants to take advantage of this discount but needs to borrow the money. It plans to pay back the loan in 60 days. What is the highest annual simple interest rate at which Shark Unibase can borrow the money and still save by paying the invoice 60 days before its due date?

Case Study 1.1 Loans, Loans, Loans

Radical Sports stocks and sells seasonal sporting equipment for hiking and skiing. In December, due to a spell of warm weather, sales of snow skiing and snowboarding equipment had not been as high as expected. Since Radical Sports had already purchased the equipment from their suppliers, payment was due. Jay Caccione, the manager, was trying to borrow the cash to make the payment of $18 000.00. At the company's bank, he would be able to obtain a one-month loan for the full amount, at 6.5% per annum simple interest. At a second bank, he would have three months to pay, with interest at 7%.

Later that same day, in talking with his suppliers, Jay learned that they were willing to finance the purchases at 8.05% per annum if he signed a promissory note. Two equal payments would then be due, two months and four months from now.

QUESTIONS

1. What are the maturity values of the two different bank loans?
2. What equal payments would have to be made to the supplier?
3. As of today's date, compute the equivalent values of each of the three options. Which option should Jay choose?

Case Study 1.2 Pay Now or Pay Later?

» Purchasing insurance protects against potential financial loss in the future. Along with licensing, the purchase of insurance for vehicles is generally required in Canada. To pay for vehicle insurance, some options are available.

Taylor's Delivery Service is about to purchase insurance on its new delivery van. Taylor's insurance agent has outlined three payment plans for them to provide coverage for the next twelve months.

Plan One requires that the full year's premium of $3200.00 be paid at the beginning of the year.

Plan Two allows payment of the annual premium in two instalments. The first installment would have to be paid immediately and would amount to one-half of the annual premium, plus a $50.00 service charge. The second installment would be paid in six months' time, and would amount to the remaining half of the premium.

Plan Three allows for twelve equal monthly payments of $285.00 each. These payments would be made on the same date within each month, starting immediately.

Taylor's Delivery Service pays its insurance using Plan One. However, the company realizes that it is missing out on interest it could have earned on this money when it pays the full year's premium at the beginning of the policy term. Taylor's considers this to be the "cost" of paying its bill in one lump sum. Taylor's is interested in knowing the cost of its insurance payment options.

QUESTIONS

1. Suppose Taylor's Delivery Service could earn 3.5% p.a. simple interest on its money over the next year. Assume the first premium payment is due today (the first day of the insurance policy term) and the focal date is today. Ignoring all taxes, compute the cost to Taylor's of paying the insurance using each of the three payment plans.
2. Suppose Taylor's Delivery Service expects to earn 2% p.a. simple interest on any money it invests in the first two months of this year, and 1.5% p.a. simple interest on any money it invests during the rest of this year. Assume the focal date is today. Which option—Plan One, Plan Two, or Plan Three—will have the least cost for the company?
3. Examine a vehicle insurance policy of your own or of a family member. Calculate what you could earn on your money at today's rates of interest. What is the cost of this insurance policy if you pay the annual premium at the beginning of the policy term? Make the focal date the first day of the policy term.

SUMMARY OF FORMULAS

Formula 1.1A

$I = Prt$

Finding the amount of interest when the principal, the rate, and the time are known

Formula 1.1B

$P = \frac{I}{rt}$

Finding the principal directly when the amount of interest, the rate of interest, and the time are known

Formula 1.1C

$r = \frac{I}{Pt}$

Finding the rate of interest directly when the amount of interest, the principal, and the time are known

Formula 1.1D

$t = \frac{I}{Pr}$

Finding the time directly when the amount of interest, the principal, and the rate of interest are known

Formula 1.2

$S = P + I$

Finding the future value (maturity value) when the principal and the amount of interest are known

Formula 1.3A

$S = P(1 + rt)$

Finding the future value (maturity value) at simple interest directly when the principal, rate of interest, and time are known

Formula 1.3B

$P = \frac{S}{1 + rt}$

Finding the present value at simple interest when the future value (maturity value), the rate of interest, and the time are known

GLOSSARY

Blended payments equal periodic payments that include payment of interest and repayment of principal, usually paid by individuals to financial institutions *(p. 31)*

Dated value the value of a sum of money at a specific time relative to its due date, including interest *(p. 23)*

Equation of values the equation obtained when matching the dated values of the original payments at an agreed focal date to the dated values of the replacement payments at the same focal date *(p. 28)*

Equivalent value see **Dated value**

Focal date a specific time chosen to compare the time value of one or more dated sums of money *(p. 24)*

Future value of a sum of money the value obtained when the amount of interest is added to the original principal *(p. 15)*

Interest rent paid for the use of money *(p. 5)*

Interest period the time period for which interest is charged *(p. 5)*

Leap year a year with an extra day in February *(p. 3)*

Maturity value see **Future value of a sum of money**

Nominal rate the yearly or annual rate of interest charged on the principal of a loan *(p. 5)*

Present value the principal that grows to a given future value (maturity value) over a given period of time at a given rate of interest *(p. 18)*

Simple interest interest calculated on the original principal by the formula $I = Prt$, and paid only when the principal is repaid *(p. 5)*

Time value of money a concept of money value that allows for a change in the value of a sum of money over time if the sum of money is subject to a rate of interest *(p. 18)*

USEFUL INTERNET SITES

www.cis-pec.gc.ca

Canada Investment and Savings (CIS) CIS is a special operating agency of the Department of Finance that markets and manages savings and investment products for Canadians. Their site provides information on products, including Canada Savings Bonds, as well as rates, savings programs, and publications.

www.tse.com

Toronto Stock Exchange (TSX) Visit the TSX site to see the latest market trends, get quotes, and read about the market data services offered.

www.m-x.ca

Montreal Stock Exchange Visit this site to see companies traded on the exchange, its history, mission, operations, and FAQs.

www.standardandpoors.com

Standard and Poor's Standard and Poor's research services provide data, analysis, and economic forecasts, and analyze economic events and trends in business and government around the globe to help people make informed business decisions.

2 Simple Interest Applications

OBJECTIVES

Upon completing this chapter, you will be able to do the following:

1. Identify and define promissory notes and their related terms.
2. Compute the maturity value of interest-bearing promissory notes.
3. Compute the present value of promissory notes and treasury bills.
4. Compute interest and balances for demand loans.
5. Compute interest and balances for lines of credit and credit card loans.
6. Construct repayment schedules for loans using blended payments.

A business often needs to borrow money for periods of less than one year. It may need to pay for products purchased or for work done by employees now, but will not collect the money owed to it until some time in the future. A business may need to replace higher-interest-rate loans, such as credit card debts, with lower-interest-rate loans, such as a bank loan or line of credit. Financial institutions are making it easier for businesses and individuals to borrow money, and are making it more convenient to repay what is borrowed. In addition to the traditional promissory notes, demand loans and lines of credit are being made available for use.

INTRODUCTION

Businesses often encounter situations that involve the application of simple interest. Simple interest calculation is usually restricted to financial instruments subject to time periods of less than one year. In this chapter we apply simple interest to short-term promissory notes, bills, lines of credit, credit card loans, and demand loans.

2.1 Promissory Notes—Basic Concepts and Computations

A. Nature of promissory notes and illustration

A **promissory note** is a written promise by one party to pay a certain sum of money, with or without interest, at a specific date or on demand, to another party. (See illustration Figure 2.1.)

FIGURE 2.1 **Promissory Note**

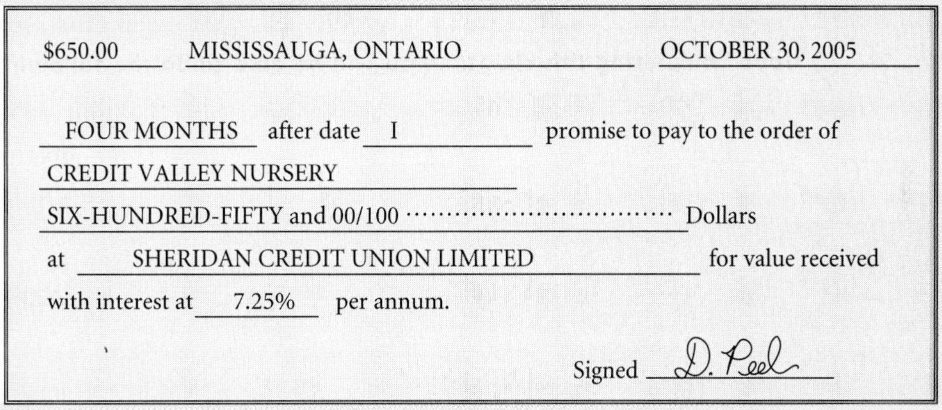

$650.00 MISSISSAUGA, ONTARIO OCTOBER 30, 2005

FOUR MONTHS after date I promise to pay to the order of

CREDIT VALLEY NURSERY

SIX-HUNDRED-FIFTY and 00/100 ································ Dollars

at SHERIDAN CREDIT UNION LIMITED for value received

with interest at 7.25% per annum.

Signed D. Peel

This is an **interest-bearing promissory note** because it is a note subject to the rate of interest stated on the face of the note.

B. Related terms explained

The following information is directly available in the promissory note (see items a, b, c, d, e, f below) or can be determined (see items g, h, i, j).

(a) The **maker** of the note is the party making the promise to pay. —— (D. Peel)

(b) The **payee** of the note is the party to whom the promise to pay is made. —— (Credit Valley Nursery)

(c) The **face value** of the note is the sum of money (principal) specified. —— ($650.00)

(d) The **rate of interest** is stated as a simple annual rate based on the face value. —— (7.25%)

(e) The **date of issue** or **issue date** is the date on which the note was made. —— (October 30, 2005)

(f) The **term of the promissory note** is the length of time before the note matures (becomes payable). —— (four months)

(g) The **due date** or **date of maturity** is the date on which the note is to be paid. —— (See Subsection C)

(h) The **interest period** is the time period from the date of issue to the legal due date. —— (See Subsection C)

(i) The **amount of interest** is payable together with the face value on the legal due date. ——— (See Subsection C)

(j) The **maturity value** is the amount payable on the due date (face value plus interest). ——— (See Subsection C)

The Canadian law relating to promissory notes adds **three days of grace** to the term of the note to obtain the **legal due date** (Bills of Exchange Act, Section 41). This is to allow for the situation of the repayment date falling on a statutory holiday. In this case, without three days of grace, you would either have to pay the note early, or take a penalty for paying three days late. Therefore, three days are added to the due date of a promissory note. Interest must be paid for those three days of grace, but there is no late payment penalty and your credit rating remains good. Today, with electronic banking, you can arrange to pay your note at any time, even on the weekend, so you may not need the three days of grace. If you decide not to include the three days of grace, write "No Grace Days" on the note when you negotiate the loan.

DID YOU KNOW?

Promissory notes are often used by retailers as a means of delaying payment of goods delivered from suppliers. For example, the owner of an outdoor centre may wish to receive an order of new lawnmowers from a particular manufacturer in March, but may not be able to pay for the lawnmowers until July. In this case, the supplier might agree to deliver the goods in March in exchange for a four-month promissory note. If the supplier needs cash prior to the due date, the promissory note can be sold to another party, such as a bank, at a discounted value.

C. Computed values

EXAMPLE 2.1A

For the promissory note illustrated in Figure 8.1, determine
(i) the due date;
(ii) the interest period;
(iii) the amount of interest;
(iv) the maturity value.

SOLUTION

(i) *Finding the due date*
Add three days of grace to the term of the note to obtain the legal due date. Since calendar months vary in length, the month in which the term ends does not necessarily have a date that corresponds to the date of issue. In such cases, the last day of the month is used as the end of the term of the note. Three days of grace are added to that date to determine the legal due date. (Throughout this chapter, we have included three days of grace in the exercises. However, as we pointed out earlier, electronic banking has reduced the need for three days of grace.)

With reference to the promissory note in Figure 2.1,

- the date of issue is October 30, 2005;
- the term of the note is four months;
- the month in which the term ends is February 2006;
- the end of the term is February 28 (since February has no day corresponding to day 30, the last day of the month is used to establish the end of the term of the note);
- the legal due date (adding 3 days) is March 3.

(ii) *Determining the interest period*
If the note bears interest, the interest period covers the number of days from the date of issue of the note to the legal due date. (Remember to count the first day but not the last day of the interest period.)

October 30 to March 3 $(2 + 30 + 31 + 31 + 28 + 2) = 124$ days

(iii) *Computing the amount of interest*
The interest payable on the note is the simple interest based on the face value of the note for the interest period at the stated rate. It is found using the simple interest formula

$$I = Prt$$ Formula 1.1A

$$I = (650.00)(0.0725)\left(\frac{124}{365}\right) = \$16.01$$

(iv) *Finding the maturity value of the note*
The maturity value of the promissory note is the total amount payable at the legal due date.

Face value + Interest = 650.00 + 16.01 = $666.01

Excel

In Excel, use the following functions to do these calculations:

COUPDAYSNC	Days between dates
ACCRINTM	Accrued interest
YIELDDISC	Simple interest yields
ACCRINTM	Accrued interest to maturity

EXERCISE 2.1

Determine each of the items listed from the information provided in the promissory note below.

$530.00 OAKVILLE, ONTARIO DECEMBER 30, 2005

FIVE MONTHS after date I promise to pay

to the order of JANE WELTON

FIVE-HUNDRED-THIRTY and 00/100 .... Dollars

at SHERIDAN CENTRAL BANK for value received

with interest at 6.5% per annum.

Signed E. Salt

1. Date issued

2. Legal due date

3. Face value

4. Interest rate

5. Interest period (days)

6. Amount of interest

7. Maturity value

B. For each of the following notes, determine

(a) the legal due date;
(b) the interest period (in days);
(c) the amount of interest;
(d) the maturity value.

1. The face value of a five-month, 6% note dated September 30, 2007 is $840.

2. A note for $760 dated March 20, 2006, with interest at 5% per annum, is issued for 120 days.

3. A sixty-day, 6.5% note for $1250 is issued January 31, 2004.

4. A four-month, 5.25% note for $2000 is issued July 31, 2005.

2.2 Maturity Value of Interest-Bearing Promissory Notes

A. Using the formula $S = P(1 + rt)$

Since the maturity value of a promissory note is the principal (face value) plus the interest accumulated to the legal due date, the future value formula for simple interest will determine the maturity value directly.

$$S = P(1 + rt)$$ ——— **Formula 1.3A**

S = maturity value of the promissory note;
P = the face value of the note;
r = the rate of interest on the note;
t = the interest period (the number of days between the *date of issue* and the *legal due date*).

B. Worked examples

EXAMPLE 2.2A For the promissory note illustrated in Figure 2.1, determine the maturity value using the future value formula $S = P(1 + rt)$.

SOLUTION $P = 650.00;\quad r = 0.0725;\quad t = \frac{124}{365}$

$$S = 650.00\left[1 + (0.0725)\left(\frac{124}{365}\right)\right] = 650.00(1 + 0.024630) = \$666.01$$

EXAMPLE 2.2B Find the maturity value of an \$800, six-month note with interest at 7.5% dated May 31, 2006.

SOLUTION The date of issue is May 31, 2006;
the term of the note is six months;
the term ends November 30, 2006;
the legal due date is December 3, 2006;
the interest period (May 31 to December 3) has 186 days.

$P = 800.00;\quad r = 0.075;\quad t = \frac{186}{365}$

$$S = 800.00\left[1 + (0.075)\left(\frac{186}{365}\right)\right] = 800.00(1 + 0.038219) = \$830.58$$

EXAMPLE 2.2C Determine the maturity value of a 90-day, \$750 note dated December 15, 2007, with interest at 8%.

SOLUTION The date of issue is December 15, 2007;
the term is 90 days;
the term ends March 14, 2008 (from 90 days take away 17 days remaining in December, 31 days for January, 29 days for February, 2008 being a leap year, which leaves 13 days for March);
the legal due date (adding the three days of grace) is March 17;
the interest period (December 15 to March 17) has 93 days.

$P = 750.00;\quad r = 0.08;\quad t = \frac{93}{365}$

$$S = 750.00\left[1 + (0.08)\left(\frac{93}{365}\right)\right] = 750.00(1 + 0.020384) = \$765.29$$

EXERCISE 2.2

A. Use the future value formula to compute the maturity value of each of the following promissory notes.

1. A four-month, 5.25% note for \$620 is issued May 25, 2005.
2. A \$350 note is issued on October 30, 2007 at 4.5% for 90 days.
3. A 150-day note for \$820 with interest at 5% is dated June 28, 2006.
4. A seven-month, \$575 note dated November 1, 2007 earns interest at 7.5%.

2.3 Present Value of Promissory Notes

A. Finding the face value

The face value (or principal) of promissory notes can be obtained by solving the future value formula $S = P(1 + rt)$ for P, that is, by using the present value formula

$$P = \frac{S}{1 + rt}$$ —— **Formula 1.3B**

P = the face value (or present value) of the note at the date of issue;
S = the maturity value;
r = the rate of interest;
t = the interest period.

EXAMPLE 2.3A A five-month note dated January 31, 2007 and bearing interest at 8% p.a.(per annum, that is, per year) has a maturity value of $558.11. Find the face value of the note.

SOLUTION

FIGURE 2.2 **Graphical Representation of Data**

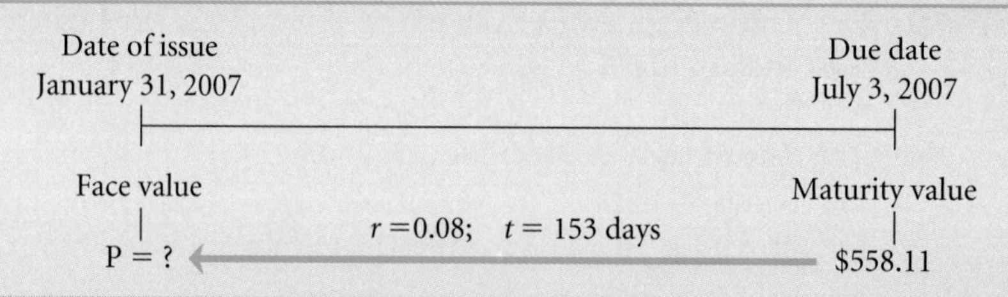

The term of the note ends June 30;
the legal due date is July 3;
the interest period (January 31 to July 3) has 153 days.

$S = 558.11; \quad r = 0.08; \quad t = \frac{153}{365}$

$$P = \frac{558.11}{1 + (0.08)\left(\frac{153}{365}\right)} = \frac{558.11}{1 + 0.033534} = \$540.00$$

B. Present value of promissory notes

The present value of a promissory note is its value any time before the due date, allowing for the rate money is worth, the time between the present date and its date of maturity, and its maturity value.

The first step in determining the present value is to ascertain whether the note is an interest-bearing or a non-interest-bearing note.

If the rate of interest is stated, the promissory note is an interest-bearing note. The face value of the note is the principal amount borrowed. To determine the present value of an interest-bearing note, two steps are required:

1. First, the maturity value of the note must be calculated, using the stated interest rate.
2. Next, the present value of the note must be calculated, using the **rate money is worth.**

If there is no rate of interest stated, the promissory note is a non-interest-bearing note. Interest on money borrowed may be implied, but is not stated. The face value of the note is the amount to be repaid at maturity. To determine the present value of a non-interest-bearing note, use the rate at which money is worth. The present value can be computed in one step.

As the two rates are likely to be *different*, take care in using them.

EXAMPLE 2.3B

A seven-month note for $1500 is issued on March 31, 2006, bearing interest at 9%. Compute the present value of the note on the date of issue if money is worth 7%.

SOLUTION

(i) First, determine the *maturity value* of the note. Use the date of issue as the focal date.
The term of the note ends October 31;
the legal due date is November 3;
the interest period (March 31 to November 3) has 217 days;
the interest rate to be used is 9%.

FIGURE 2.3 **Graphical Representation of Data**

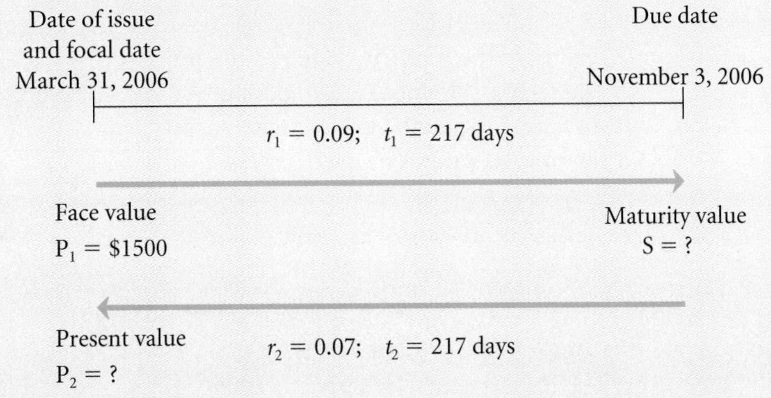

Date of issue
and focal date
March 31, 2006

Due date
November 3, 2006

$r_1 = 0.09$; $t_1 = 217$ days

Face value
$P_1 = \$1500$

Maturity value
$S = ?$

Present value
$P_2 = ?$

$r_2 = 0.07$; $t_2 = 217$ days

$$P_1 = 1500.00; \quad r_1 = 0.09; \quad t_1 = \frac{217}{365}$$

$$S = 1500\left[1 + (0.09)\left(\frac{217}{365}\right)\right] = 1500(1 + 0.053507) = \$1580.26$$

(ii) Second, use the maturity value found in part (i) to determine the *present value* at the specified date.
The focal date (date of issue) is March 31;
the legal due date is November 3;
the interest period (March 31 to November 3) has 217 days;
the interest rate to be used is 7%.

$$S = 1580.26; \quad r_2 = 0.07; \quad t_2 = \frac{217}{365}$$

$$P_2 = \frac{1580.26}{1 + (0.07)\left(\frac{217}{365}\right)} = \frac{1580.26}{1 + 0.041616} = \$1517.12$$

Note: The present value at the date of issue is more than the face value of the note because the interest rate on the note (9.0%) is more than the rate money is worth (7.0%).

EXAMPLE 2.3C A 180-day note for $2000 with interest at 7% is dated September 18, 2006. Compute the value of the note on December 1, 2006 if money is worth 5%.

SOLUTION

FIGURE 2.4 Graphical Representation of Data

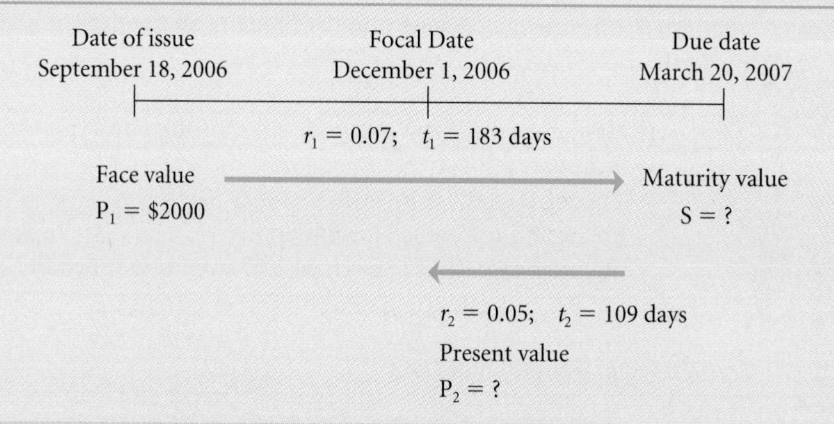

i) Compute the maturity value of the note.
The term of the note is 180 days plus 3 days of grace;
therefore, the term ends March 20, 2007;
the interest rate to be used is 7%.

$$P_1 = 2000; \; r_1 = 0.07; \; t_1 = \frac{183}{365}$$

$$S = 2000\left[1 + (0.07)\frac{(183)}{(365)}\right] = 2000(1 + 0.035096) = \$2070.19$$

ii) Compute the present value.
The focal date is December 1, 2006;
the interest period (March 20, 2007 to December 1, 2006) has 109 days;
the interest rate to be used is 5%.

$$S = 2070.19; \; r_2 = 0.05; \; t_2 = \frac{109}{365}$$

$$P_2 = \frac{2070.19}{\left[1 + (0.05)\left(\frac{109}{365}\right)\right]} = \frac{2070.19}{(1 + 0.014932)} = \$2039.73$$

EXAMPLE 2.3D Compute the present value on the date of issue of a non-interest-bearing $950, three-month promissory note dated April 30, 2006 if money is worth 6.5%.

SOLUTION

FIGURE 2.5 **Graphical Representation of Data**

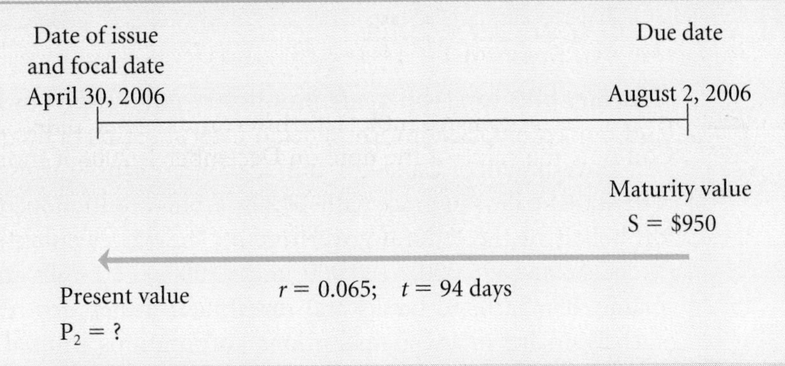

(i) Since this is a non-interest-bearing note, the maturity value of the note is the face value, $950.
The term of the note ends July 30, 2006, plus 3 days of grace;
therefore, the legal due date is August 2, 2006;
the interest period (April 30 to August 2) has 94 days;
the interest rate to be used is 6.5%.

$S = 950;\ r = 0.065;\ t = \frac{94}{365}$

$$P = \frac{950}{\left[1 + (0.065)\left(\frac{94}{365}\right)\right]} = \frac{950}{(1 + 0.016740)} = \$934.36$$

EXAMPLE 2.3E Compute the value of a non-interest-bearing $400, 120-day note dated March 2, 2005 on May 15, 2005 if money is worth 9%.

SOLUTION The 120-day term ends June 30;
the legal due date is July 3;
the focal date is May 15;
the interest period (May 15 to July 3) has 49 days;
the maturity value of the note is its face value, $400.00.

FIGURE 2.6 **Graphical Representation of Data**

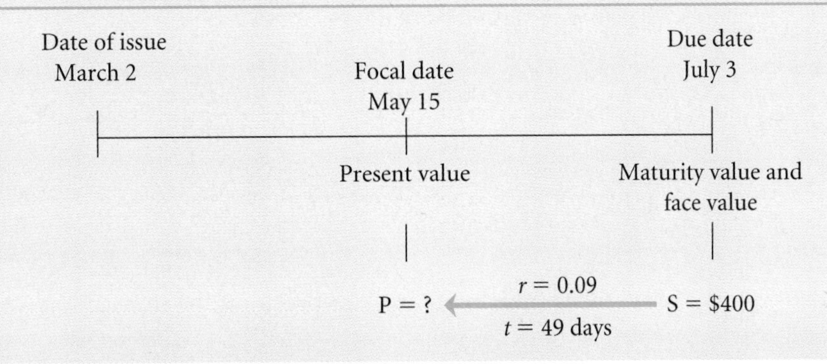

$$S = 400.00; \quad r = 0.09; \quad t = \frac{49}{365}$$

$$P = \frac{400.00}{1 + (0.09)\left(\frac{49}{365}\right)} = \frac{400.00}{1 + 0.012082} = \$395.22$$

C. Present value of treasury bills

Treasury bills (or **T-bills**) are promissory notes issued by the federal government and most provincial governments to meet short-term financing requirements.

Government of Canada T-bills are for terms of 91 days, 182 days, and 364 days. There are no days of grace with T-bills. T-bills are auctioned by the Bank of Canada on behalf of the federal government. They are available in denominations of \$1000, \$5000, \$25 000, \$100 000, and \$1 000 000. T-bills are bought at the auction mainly by chartered banks and investment dealers for resale to other investors, such as smaller financial institutions, corporations, mutual funds, and individuals.

T-bills are promissory notes that do not carry an interest rate. The issuing government guarantees payment of the face value at maturity. The investor purchases T-bills at a discounted price reflecting a rate of return that is determined by current market conditions. The discounted price is determined by computing the present value of the T-bills.

EXAMPLE 2.3F

An investment dealer bought a 91-day Canada T-bill to yield an annual rate of return of 2.66%.

(i) What was the price paid by the investment dealer for a T-bill with a face value of \$100 000?

(ii) The investment dealer resold the \$100 000 T-bill the same day to an investor to yield 2.55%. What was the investment dealer's profit on the transaction?

SOLUTION

(i) *Find the purchase price,* P_1

The maturity value is the face value of the T-bill, \$100 000.00; the discount period has 91 days (no days of grace are allowed on T-bills).

$$S = 100\ 000.00; \quad r_1 = 0.0266; \quad t_1 = \frac{91}{365}$$

$$P_1 = \frac{S}{1 + r_1 t_1} = \frac{100\ 000.00}{1 + 0.0266\left(\frac{91}{365}\right)} = \frac{100\ 000.00}{1 + 0.006632} = 99\ 341.17$$

The investment dealer paid \$99 341.17 for a \$100 000 T-bill.

(ii) *Find the resale price,* P_2

$$S = 100\ 000.00; \quad r_2 = 0.0255; \quad t_2 = \frac{91}{365}$$

$$P_2 = \frac{100\ 000.00}{1 + 0.0255\left(\frac{91}{365}\right)} = \frac{100\ 000.00}{1 + 0.006358} = 99\ 368.22$$

Investment dealer's profit = Resale price − Price paid by dealer

$= P_2 - P_1$

$= 99\ 368.22 - 99\ 341.17 = 27.05$

The investment dealer's profit on the transaction was \$27.05.

EXAMPLE 2.3G

An investor purchased \$250 000 in 364-day T-bills 315 days before maturity to yield 2.86%. He sold the T-bills 120 days later to yield 2.45%.

(i) How much did the investor pay for the T-bills?
(ii) For how much did the investor sell the T-bills?
(iii) What rate of return did the investor realize on the investment?

SOLUTION

(i) *Find the purchase price of the T-bills,* P_1

$$S = 250\,000.00; \quad r_1 = 0.0286; \quad t_1 = \frac{315}{365}$$

$$P_1 = \frac{250\,000.00}{1 + 0.0286\left(\frac{315}{365}\right)} = \frac{250\,000.00}{1 + 0.024682} = 243\,978.13$$

The investor paid \$243 978.13.

(ii) *Find the selling price of the T-bills,* P_2

The time to maturity at the date of sale is 195 days (315 − 120).

$$S = 250\,000.00; \quad r_2 = 0.0245; \quad t_2 = \frac{195}{365}$$

$$P_2 = \frac{250\,000.00}{1 + 0.0245\left(\frac{195}{365}\right)} = \frac{250\,000.00}{1 + 0.013089} = 246\,770.03$$

The investor sold the T-bills for \$246 770.03.

(iii) The investment of \$243 978.13 grew to \$246 770.03 in 120 days.
To compute the rate of return, use the future value formula (Formula 2.3A).

$$P = 243\,978.13; \quad S = 246\,770.03; \quad t = \frac{120}{365}$$

$$S = P(1 + rt)$$

$$246\,770.03 = 243\,978.13\left[1 + r\left(\frac{120}{365}\right)\right]$$

$$\frac{246\,770.03}{243\,978.13} = 1 + \left(\frac{120}{365}\right)r$$

$$1.011443 = 1 + 0.328767r$$

$$0.011443 = 0.328767r$$

$$\frac{0.011443}{0.328767} = r$$

$$r = 0.034806 = 3.48\%$$

The investor realized a rate of return of about 3.48%.

Alternatively

The gain realized represents the interest.

$$I = 246\,770.03 - 243\,978.13 = 2791.90$$

Using the formula $r = \frac{I}{Pt}$,

$$r = \frac{2791.90}{243\,978.13\left(\frac{120}{365}\right)} = \frac{2791.90}{80211.988} = 0.034807 = 3.48\%$$

EXERCISE 2.3

A. Compute the face value of each of the following promissory notes.

1. A six-month note dated April 9, 2005 with interest at 4% has a maturity value of $510.19.

2. The maturity value of a 150-day, 5% note dated March 25, 2006 is $1531.44.

B. Find the present value, on the date indicated, of each of the following promissory notes:

1. A non-interest-bearing note for $1500 issued August 10, 2006 for three months if money is worth 7.5%, on the date of issue;

2. A non-interest-bearing note for $750 issued February 2, 2008 for 180 days if money is worth 6.25%, on June 1, 2008;

3. A sixty-day, 9% note for $1600 issued October 28 if money is worth 7%, on November 30;

4. A four-month note for $930 dated April 1 with interest at 6.5% if money is worth 8%, on June 20.

C. Answer each of the following questions.

1. What is the price of a 182-day, $100 000 Government of Canada treasury bill that yields 2.79% per annum?

2. An investment dealer bought a 182-day Government of Canada treasury bill at the price required to yield an annual rate of return of 2.79%.

(a) What was the price paid by the investment dealer if the T-bill has a face value of $1 000 000?

(b) Later the same day, the investment dealer sold this T-bill to a large corporation to yield 2.65%. What was the investment dealer's profit on this transaction?

3. An investment dealer acquired a $5000, 91-day Province of Alberta treasury bill on its date of issue at a price of $4966.20. What was the annual rate of return?

4. On June 2 you bought for $24 767.75 a $25 000 Province of New Brunswick treasury bill maturing October 7. What was the discount rate at which you bought the treasury bill?

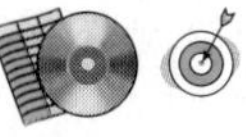
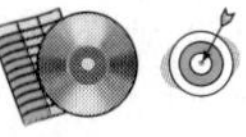

5. An investor purchased a 91-day, $100 000 T-bill on its issue date for $99 326.85. After holding it for 42 days, she sold the T-bill for a yield of 2.72%.

(a) What was the original yield of the T-bill?

(b) For what price was the T-bill sold?

(c) What rate of return (per annum) did the investor realize while holding this T-bill?

6. On April 1, $25 000 364-day treasury bills were auctioned off to yield 2.92%.

(a) What is the price of each $25 000 T-bill on April 1?

(b) What is the yield rate on August 15 if the market price is $24 377.64?

(c) Calculate the market value of each $25 000 T-bill on October 1 if the rate of return on that date is 4.545%.

(d) What is the rate of return realized if a $25 000 T-bill purchased on April 1 is sold on November 20 at a market rate of 4.625%?

2.4 Demand Loans

A. Nature of demand loans

A **demand loan** is a loan for which repayment, in full or in part, may be required at any time, or made at any time. The financial instrument representing a demand loan is called a **demand note.**

When borrowing on a demand note, the borrower receives the full face value of the note. The lender may demand payment of the loan in full or in part at any time. Conversely, the borrower may repay all of the loan or any part at any time without notice and without interest penalty. Interest, based on the unpaid balance, is usually payable monthly. The interest rate on such loans is normally not fixed for the duration of the loan but fluctuates with market conditions. Thus the total interest cost cannot be predicted with certainty. Note that the method of counting days is to count the first day but not the last.

B. Examples

EXAMPLE 2.4A

Rose 'n Blooms borrowed $1200.00 from the Royal Bank on a demand note. It agreed to repay the loan in six equal monthly installments (each payment is made at the end of the month) and also authorized the bank to collect interest monthly from its bank account at 6% p.a. calculated on the unpaid balance. What will the loan cost?

SOLUTION

$$\text{Monthly payment of principal} = \frac{1200.00}{6} = \$200.00$$

$$\text{Monthly rate of interest} = \frac{6\%}{12} = 0.5\% = 0.005$$

Month	Loan Amount Owing During Month		Interest Collected for Month	
1	$1200.00	Original	$6.00	(1200)(0.005)
2	$1000.00	1200 − 200	$5.00	(1000)(0.005)
3	$ 800.00	1000 − 200	$4.00	(800)(0.005)
4	$ 600.00	800 − 200	$3.00	(600)(0.005)
5	$ 400.00	600 − 200	$2.00	(400)(0.005)
6	$ 200.00	400 − 200	$1.00	(200)(0.005)
	Total interest cost		$21.00	

EXAMPLE 2.4B

On August 17, 2006 Sheridan Toy Company borrowed $30 000.00 from Peel Credit Union on a demand note to finance its inventory. Interest on the loan, calculated on the daily balance, is charged against the borrower's current account on the 17th of each month while the loan is in force. The company makes a payment of $5000.00 on September 24, makes a further payment of $10 000.00 on October 20, and pays the balance on December 10. The interest on demand loans on

August 17 was 6% p.a. The rate was changed to 7% effective October 1, to 8.5% effective November 1, and to 8% effective December 1. Determine the cost of financing the loan.

SOLUTION

Date	Interest Period	Principal	Rate	Interest Due	
Sept. 17	Aug. 17–Sept. 17	$30 000.00	6.0%	$152.88	— $(30\,000)(0.06)\left(\frac{31}{365}\right)$
Oct. 17	Sept. 17–Sept. 24	$30 000.00	6.0%	34.52	— $(30\,000)(0.06)\left(\frac{7}{365}\right)$
	Sept. 24–Sept. 30*	$25 000.00	6.0%	28.77	— $(25\,000)(0.06)\left(\frac{7}{365}\right)$
	Oct. 1–Oct. 17	$25 000.00	7.0%	76.71	— $(25\,000)(0.07)\left(\frac{16}{365}\right)$
				$292.88	
Nov. 17	Oct. 17–Oct. 20	$25 000.00	7.0%	$ 14.38	— $(25\,000)(0.07)\left(\frac{3}{365}\right)$
	Oct. 20–Oct. 31*	$15 000.00	7.0%	34.52	— $(15\,000)(0.07)\left(\frac{12}{365}\right)$
	Nov. 1–Nov. 17	$15 000.00	8.5%	55.89	— $(15\,000)(0.85)\left(\frac{16}{365}\right)$
				$104.79	
Dec. 10	Nov. 17–Nov. 30*	$15 000.00	8.5%	$ 48.90	— $(15\,000)(0.85)\left(\frac{14}{365}\right)$
	Dec. 1–Dec. 10	$15 000.00	8.0%	29.59	— $(15\,000)(0.8)\left(\frac{9}{365}\right)$
				$ 78.49	
	Total cost of financing ⟶			$476.16	

*Inclusive

C. Partial payments

Demand loans and similar debts are sometimes paid off by a series of **partial payments.** The commonly used approach to dealing with this type of loan repayment is the **declining balance approach**, requiring that each partial payment is applied first to the accumulated interest. Any remainder is then used to reduce the outstanding principal. Thus, interest is always calculated on the unpaid balance and the new unpaid balance is determined after each partial payment.

The following step-by-step procedure is useful in dealing with such problems.

(a) Compute the interest due to the date of the partial payment.

(b) Compare the interest due computed in part (a) with the partial payment received and do part (c) if the partial payment is *greater than* the interest due or do part (d) if the partial payment is *less than* the interest due.

(c) *Partial payment greater than interest due*

  (i) Deduct the interest due from the partial payment.

  (ii) Deduct the remainder in part (i) from the principal balance to obtain the new unpaid balance.

(d) *Partial payment less than interest due*
In this case, the partial payment is not large enough to cover the interest due.
(i) Deduct the partial payment from the interest due to determine the unpaid interest due at the date of the principal payment.
(ii) Keep a record of this balance and apply any future partial payments to this unpaid interest first.

POINTERS AND PITFALLS

When determining the number of days in any given time period between interest rate changes or between two partial payments, remember to always count the first day and omit the last day. As well, keep in mind that

(a) The date on which there is a change in the interest rate is counted as the first day at the new interest rate.
(b) The date on which a partial payment is made is counted as the first day at the new outstanding principal balance.

EXAMPLE 2.4C

On April 20, 2007 Bruce borrowed $4000.00 at 5% on a note requiring payment of principal and interest on demand. Bruce paid $600.00 on May 10 and $1200.00 on July 15. What payment is required on September 30 to pay the note in full?

SOLUTION

April 20			
Original loan balance →		$4000.00	
May 10			
Deduct			
First partial payment →	$ 600.00		
Less interest			
April 20–May 10 →	10.96		$(4000)(0.05)\left(\frac{20}{365}\right)$
		589.04	
Unpaid balance →		$3410.96	
July 15			
Deduct			
Second partial payment →	$1200.00		
Less interest			
May 10–July 15 →	30.84		$(3410.96)(0.05)\left(\frac{66}{365}\right)$
		1169.16	
Unpaid balance →		$2241.80	
September 30			
Add			
Interest			
July 15–Sept. 30 →		23.65	$(2241.80)(0.05)\left(\frac{77}{365}\right)$
Payment required to pay the note in full →		$2265.45	

EXAMPLE 2.4D

The Provincial Bank lent $20 000.00 to the owner of the Purple Pelican on April 1, 2006 for commercial improvements. The loan was secured by a demand note subject to a variable rate of interest. This rate was 7% on April 1. The rate of interest was raised to 9% effective August 1 and reduced to 8% effective November 1. Partial payments, applied to the loan by the declining balance method, were made as follows: June 10, $1000.00; September 20, $400.00; November 15, $1200.00. How much interest is due to the Provincial Bank on December 31?

SOLUTION

April 1				
Original loan balance →			$20 000.00	
June 10				
Deduct				
First partial payment →		$1000.00		
Less interest				
April 1–June 10 →		268.49		$(20\,000)(0.07)\left(\frac{70}{365}\right)$
			731.51	
Unpaid loan balance →			$19 268.49	
September 20				
Deduct				
Second partial payment		$ 400.00		
Less interest				
June 10–Sept. 20:				
June 10–July 31 (inclusive)	$192.16			$(19\,268.49)(0.07)\left(\frac{52}{365}\right)$
Aug. 1–Sept. 20	$237.56	429.72		$(19\,268.49)(0.09)\left(\frac{50}{365}\right)$
Unpaid interest to Sept. 20 →		$ 29.72		
Unpaid loan balance →			$19 268.49	
November 15				
Deduct				
Third partial payment →		$1200.00		
Less interest				
Unpaid interest to Sept. 20 →	$ 29.72			(see above)
Sept. 20–Oct. 31 (inclusive) →	199.55			$(19\,268.49)(0.09)\left(\frac{42}{365}\right)$
Nov. 1–Nov. 15 →	59.13			$(19\,268.49)(0.08)\left(\frac{14}{365}\right)$
		288.40		
			911.60	
Unpaid loan balance →			$18 356.89	
December 31				
Interest due				
Nov. 15–Dec. 31 →			$ 185.08	$(18\,356.89)(0.08)\left(\frac{46}{365}\right)$

EXERCISE 2.4

Determine the total interest cost for each of the following loans.

1. On June 15, 2005, Jean-Luc borrowed $2500.00 from his bank secured by a demand note. He agreed to repay the loan in five equal monthly installments and authorized the bank to collect the interest monthly from his bank account at 6.0% per annum calculated on the unpaid balance.
2. James TV & Stereo borrowed $1800.00 from the Teachers' Credit Union. The loan was to be repaid in six equal monthly payments, plus interest of 7.5% per annum calculated on the unpaid balance.
3. Jamie borrowed $900.00 from the Essex District Credit Union. The loan agreement provided for repayment of the loan in four equal monthly payments plus interest at 12% per annum calculated on the unpaid balance.
4. Erindale Automotive borrowed $8000.00 from the Bank of Montreal on a demand note on May 10. Interest on the loan, calculated on the daily balance, is charged to Erindale's current account on the 10th of each month. Erindale made a payment of $2000.00 on July 20, a payment of $3000.00 on October 1, and repaid the balance on December 1. The rate of interest on the loan on May 10 was 8% per annum. The rate was changed to 9.5% on August 1 and to 8.5% on October 1.

5. The Tomac Swim Club arranged short-term financing of $12 500.00 on July 20 with the Bank of Commerce and secured the loan with a demand note. The club repaid the loan by payments of $6000.00 on September 15, $3000.00 on November 10, and the balance on December 30. Interest, calculated on the daily balance and charged to the club's current account on the last day of each month, was at 9.5% per annum on July 20. The rate was changed to 8.5% effective September 1 and to 9% effective December 1.

B. Answer each of the following.

1. On March 10, Fat Tires Ltd. borrowed $10 000.00 with an interest rate of 5.5%. The loan was repaid in full on November 15, with payments of $2500.00 on June 30, and $4000.00 on September 4. What was the final payment?
2. Automotive Excellence Inc. borrowed $20 000.00 on August 12 with an interest rate of 6.75% per annum. On November 1 $7500.00 was repaid, and on December 15 $9000.000 was repaid. Automotive Excellence paid the balance of the loan on February 20. What was the final payment?
3. A loan of $6000.00 made at 11% per annum on March 10 is repaid in full on November 15. Payments were made of $2000.00 on June 30 and $2500.00 on September 5. What was the final payment?
4. D. Slipp borrowed $15 000.00 on August 12. She paid $6000.00 on November 1, $5000.00 on December 15, and the balance on February 20. The rate of interest on the loan was 10.5%. How much did she pay on February 20?
5. The Continental Bank made a loan of $20 000.00 on March 25 to Dr. Hirsch to purchase equipment for her office. The loan was secured by a demand loan subject to a variable rate of interest that was 7% on March 25.

The rate of interest was raised to 8.5% effective July 1 and to 9.5% effective September 1. Dr. Hirsch made partial payments on the loan as follows: $600.00 on May 15; $800.00 on June 30; and $400.00 on October 10. The terms of the note require payment of any accrued interest on October 31. How much must Dr. Hirsch pay on October 31?

6. Dirk Ward borrowed $12 000.00 for investment purposes on May 10 on a demand note providing for a variable rate of interest and payment of any accrued interest on December 31. He paid $300.00 on June 25, $150 on September 20, and $200.00 on November 5. How much is the accrued interest on December 31 if the rate of interest was 7.5% on May 10, 6% effective August 1, and 5% effective November 1?

2.5 Lines of Credit and Credit Card Loans

A **line of credit** is a pre-approved loan agreement between a financial institution and a borrower. The borrower may withdraw money, up to an agreed maximum, at any time. Interest is charged only on the amount withdrawn from the line of credit. A minimum repayment may be required each month. The borrower may repay any additional amount at any time without further penalty. The rate of interest charged for money borrowed on a line of credit is lower than the rate of interest charged on most credit cards. The interest rate can change over time.

A **credit card** is a plastic card entitling the bearer to a revolving line of credit with a pre-established credit limit. Interest rates are set by the credit card issuers, vary considerably, and can be changed at any time. Generally, interest rates charged on credit cards are higher than rates charged on loans made by financial institutions. Some credit card issuers require an annual fee to be paid by the user. When the bearer withdraws cash by presenting the credit card, a cash advance fee is usually charged. Depending on the issuer of the credit card, the cash advance fee may be deducted directly from the cash advance at the time the money is received or it may be posted to the account on the day the cash is received. Interest on the cash advance accrues starting from the day the money is withdrawn. The bearer of the credit card, with an authorizing signature, may make purchases instead of using cash to pay.

An **unsecured line of credit** is a line of credit with no assets promised to the lender to cover non-payment of the loan. Since no security is offered to the lender, the limit of an unsecured line of credit depends on the individual's credit rating and past relationship with the lender.

A **secured line of credit** is a line of credit with assets promised to the lender to cover non-payment of the loan. For example, homeowners might pledge the value of their home, that is, their home equity, to secure a line of credit. In general, the limit of a secured line of credit is higher than the limit of an unsecured one. Furthermore, the interest rate of a secured line of credit is lower than the interest rate of an unsecured one.

Lines of credit secured by home equity are becoming quite popular and are used by some borrowers as an alternative to mortgages. Home equity lines of credit provide access to larger credit limits.

EXAMPLE 2.5A

Suppose your business has secured a line of credit and receives the following statement of account for the month of February.

Date	Transaction Description	Deposit	Withdrawal	Balance
Feb. 01	Balance			−600.00
04	Cheque 262		500.00	−1100.00
10	Deposit	2050.00		950.00
16	Cheque 263		240.00	710.00
20	Cheque 264		1000.00	−290.00
22	Cheque 265		80.00	−370.00
27	Cheque 266		150.00	−520.00
28	Interest earned	?		
	Line of credit interest		?	
	Overdraft interest		?	
	Service charge		?	

Note: "−" indicates a negative balance.

The limit on your line of credit is \$1000. You receive daily interest of 1.5% p.a. on positive balances and pay daily interest of 7% p.a. on *negative (line of credit) balances.* Overdraft interest is 18% p.a. on the daily amount exceeding your line of credit limit. There is a service charge of \$5.00 for each transaction causing an overdraft or adding to an overdraft.

Determine

(i) the amount of interest earned;
(ii) the amount of interest charged on the line of credit;
(iii) the amount of interest charged on overdrafts;
(iv) the amount of the service charge;
(v) the account balance on February 28.

SOLUTION

(i) Interest earned (on positive balances)

February 10 to February 15 inclusive: 6 days at 1.5% on \$950.00

$$I = 950.00(0.015)\left(\frac{6}{365}\right) = \$0.23$$

February 16 to February 19 inclusive: 4 days at 1.5% on \$710.00

$$I = 710.00(0.015)\left(\frac{4}{365}\right) = \$0.12$$

Total interest earned $= 0.23 + 0.12 = \$0.35$

(ii) Line of credit interest charged (on negative balances up to \$1000.00)

February 1 to February 3 inclusive: 3 days at 7% on \$600.00

$$I = 600.00(0.07)\left(\frac{3}{365}\right) = \$0.35$$

February 4 to February 9 inclusive: 6 days at 7% on \$1000.00

$$I = 1000.00(0.07)\left(\frac{6}{365}\right) = \$1.15$$

February 20 to February 21 inclusive: 2 days at 7% on \$290.00

$$I = 290.00(0.07)\left(\frac{2}{365}\right) = \$0.11$$

February 22 to February 26 inclusive: 5 days at 7% on $370.00

$I = 370.00(0.07)\left(\frac{5}{365}\right) = \$0.35$

February 27 to February 28 inclusive: 2 days at 7% on $520.00

$I = 520.00(0.07)\left(\frac{2}{365}\right) = \$0.20$

Total line of credit interest charged
$= 0.35 + 1.15 + 0.11 + 0.35 + 0.20 = \$2.16$

(iii) Since your line of credit limit is $1000.00, overdraft interest is charged on the amount in excess of a negative balance of $1000.00. You were in overdraft from February 4 to February 9 inclusive in the amount of $100.00.

Overdraft interest $= 100.00(0.18)\left(\frac{6}{365}\right) = \$0.30$

(iv) You had one transaction causing an overdraft or adding to an overdraft.
Service charge $= 1(5.00) = \$5.00$

(v) The account balance on February 28
$= -520.00 + 0.35 - 2.16 - 0.30 - 5.00 = -\$527.11$

EXAMPLE 2.5B

You have applied for and received a credit card. The interest rate charged is 18.9% per annum. You note the following transactions for the month of September.

September 6 Purchased textbooks and supplies for a total of $250.00.
September 10 Withdrew $100 as a cash advance through your credit card.
September 30 Received the credit card statement, showing a minimum balance owing of $25.00. A payment date of October 10 is stated on the statement.

Answer each of the following.

(i) Compute the amount of interest charged on the cash advance from September 10 until September 30.
(ii) You decide to pay the amount owing, in full, on October 1. How much must you pay?
(iii) Instead of paying the full amount, you decide to pay the minimum, $25.00, on October 1. What would be the balance owing after the payment?
(iv) If there are no further transactions during October, how much is owing at the end of October?

SOLUTION

(i) Interest charged (on cash advance):
September 10 to September 30 inclusive: $100 at 18.9% for 21 days.

$I = 100.00(0.189)\left(\frac{21}{365}\right) = \$1.09$

(ii) $\$250.00 + 100.00 + 1.09 = \$351.09$

(iii) $\$351.09 - 25.00 = \$326.09$

(iv) $I = 326.09(0.189)\left(\frac{31}{365}\right) = \$5.23$

$\$326.09 + 5.23 = \$331.32$

DID YOU KNOW?

Credit Card Calculator: What Does It Cost You to Pay the Minimum?

Will that debt ever be paid off? That's what you may ask if you have been paying only the minimum amount due stated on your credit card statement. The issuer of your credit card determines the minimum amount due.

The minimum payment on credit card debt is generally calculated as a percent of your current balance. The minimum payment drops as your balance is paid, but thanks to the magic of compounding you'll end up paying for a long, long time.

Let's assume that your credit card balance is $1000, the interest rate on your credit card is 18%, and the minimum payment is calculated based on 2.5% of the balance. Your minimum payment would be $25.00 for the first month.

If you continue to make the minimum payment required, your payment will decrease, and your balance will decrease, but it will take you 153 months to pay off your debt. In addition to repaying the original $1000, you will have paid $1115.41 in interest!

If, however, you make a payment of $25.00 for each month, not just paying the minimum payment required, you would eliminate the debt in 62 months. You will have repaid the original $1000, plus interest of $538.62.

Now, if you can resist that extra donut each day, and instead pay $50.00 each month toward your debt, the debt will be repaid in 24 months, and the additional interest will be only $197.83.

Source: Bankrate Inc. website, www.bankrate.com.

EXERCISE 2.5

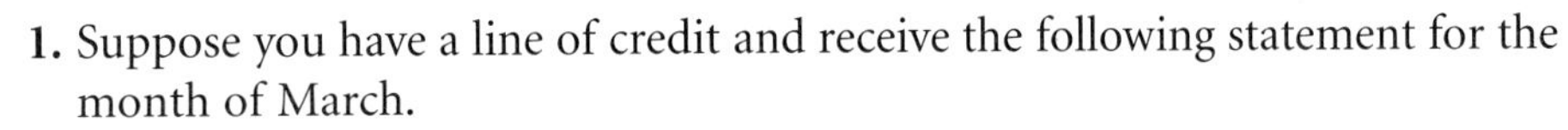

A. Determine the missing information for each of the following lines of credit.

1. Suppose you have a line of credit and receive the following statement for the month of March.

Date	Transaction Description	Deposit	Withdrawal	Balance
Feb. 28	Balance			−527.71
Mar. 02	Cheque 264		600.00	−1127.71
05	Cheque 265		300.00	−1427.71
10	Deposit	2000.00		572.29
16	Cheque 266		265.00	307.29
20	Cheque 267		1000.00	−692.71
22	Cheque 268		83.50	−776.21
27	Cheque 269		165.00	−941.21
31	Interest earned	?		
	Line of credit interest		?	
	Overdraft interest		?	
	Service charge		?	?

Note: "−" indicates a negative balance.

The limit on a line of credit is $1000. Daily interest of 1.25% p.a. is received on positive balances and daily interest of 8% p.a. is paid on negative (line of credit) balances. Overdraft interest is 18% p.a. on the daily amount exceeding the line of credit limit. There is a service charge of $5.00 for each transaction causing an overdraft or adding to an overdraft.

(a) Calculate the amount of interest earned.
(b) Calculate the amount of interest charged on the line of credit.
(c) Calculate the amount of interest charged on overdrafts.
(d) Calculate the amount of the service charge.
(e) What is the account balance on March 31?

2. Exotic Furnishings Ltd. has a line of credit secured by the equity in the business. The limit on the line of credit is $45 000. Transactions for the period April 1 to September 30 are shown below. Exotic owed $25 960.06 on its line of credit on April 1.

Date	Principal Withdrawal	Principal Payment	Interest Payment	Balance
Apr. 01				−25 960.06
30		200.00	?	
May 23	5 000.00			
31		200.00	?	
June 30		200.00	?	
July 19	5 000.00			
31		200.00	?	
Aug. 05	10 500.00			
31		200.00	?	
Sept. 30		200.00	?	

Note: "−" indicates a negative balance.

The line of credit agreement requires regular payment of $200.00 on the principal plus interest (including overdraft interest) by electronic transfer after closing on the last day of each month. Overdraft interest is 17% p.a. The line of credit interest is variable. It was 6.00% on April 1, 5.50% effective June 20, and 5.00% effective September 10.

(a) Calculate the interest payments on April 30, May 31, June 30, July 31, August 31, and September 30.
(b) What is the account balance on September 30?

BUSINESS MATH NEWS BOX

Exploring Personal Lines of Credit

Your local bank branch recently distributed the information shown below.

Consolidate high-interest credit cards with our convenient Line of Credit

A line of credit is a great way to consolidate balances from your credit cards and loans to create one easy-to-manage monthly payment. These examples show you how much your unsecured line of credit can help you save in just six months:

Your Balance	Some Department Store Cards 28.8%	Other Credit Cards 18.9%	Unsecured Line of Credit	In Six Months You Can Save Up To
$8000	$1152	$756	$360	$792
$5000	$720	$472	$224	$496
$2500	$360	$236	$112	$248

(The savings above are based on the rates remaining the same for six months. Calculations have been rounded down to the nearest dollar. Unsecured line of credit rates are subject to change without notice.)

QUESTIONS

1. In the table above, calculations have been rounded down to the nearest dollar. Recalculate the figures in the last row of the table for the balance of $2500, rounding to two decimal places. What effect does rounding have on the six-month savings (the balance in column 5)?
2. Suppose you had a balance of $5000 on a credit card charging interest at a rate of 17.25%. Suppose you consolidated this balance onto the unsecured line of credit, which charges interest at a rate of 9%. How much interest would you save in six months?
3. Suppose you used money from your unsecured line of credit to purchase equipment costing $3000 for your business. For the first two months, the unsecured line of credit had an interest rate of 9%. The rate then increased to 9.5% for the next four months. How much interest would you have paid on the $3000 balance after six months?

2.6 Loan Repayment Schedules

A. Purpose

In the case of loans repaid in fixed installments (often called a **blended payment**), the constant periodic payment is first applied to pay the accumulated interest. The remainder of the payment is then used to reduce the unpaid balance of the principal.

FIGURE 2.7 Statement of Disclosure

STATEMENT OF DISCLOSURE

(COST OF LOAN AND ANNUAL INTEREST RATE) PURSUANT TO THE CONSUMER PROTECTION ACT

Name of Credit Union ___SHERIDAN___ Account No. ___13465–274___

1) Balance of existing loan (if any) $ ___2000.00___

2) Add new amount loaned $ ___4000.00___

3) Full amount of loan $ ___6000.00___

4) Cost of Borrowing expressed in dollars and cents (Interest calculated on full amount of loan (Item 3)) $ ___1473.00___

5) Annual Interest Rate charged (calculated in accordance with the Consumer Protection Act) ___9___ %

6) If any charge is made to the borrower in addition to interest herein noted, it must be disclosed here. $ ______
(Description) ....
Frequency of installments ___60 MONTHS___ Amount of installments $ ___124.55___ First instalment due ______ 20 ___

I, the undersigned, acknowledge receipt of this statement of cost of loan and annual interest rate, prior to the advance of the credit.

DATE ______ X ....
SIGNATURE OF BORROWER

COMPLETE IN DUPLICATE
Original to Borrower

NOTE: Where more than one maker or co-maker, separate Disclosure Forms should be signed for individually.

While lenders are obliged to disclose to the borrower the total cost of borrowing as well as the interest rate (see Figure 2.7), a detailed statement of the cost of borrowing as well as the effect of the periodic payments on the principal may be obtained by constructing a **loan repayment schedule**, or an **amortization schedule**.

The information usually contained in such a schedule includes

(a) the payment number or payment date;

(b) the amount paid at each payment date;

(c) the interest paid by each payment;

(d) the principal repaid by each payment;

(e) the unpaid loan balance after each payment.

Figure 2.8 provides a possible design for such schedules and the same design is used in the solution to Example 2.6A.

FIGURE 2.8 Basic Design of a Loan Repayment Schedule

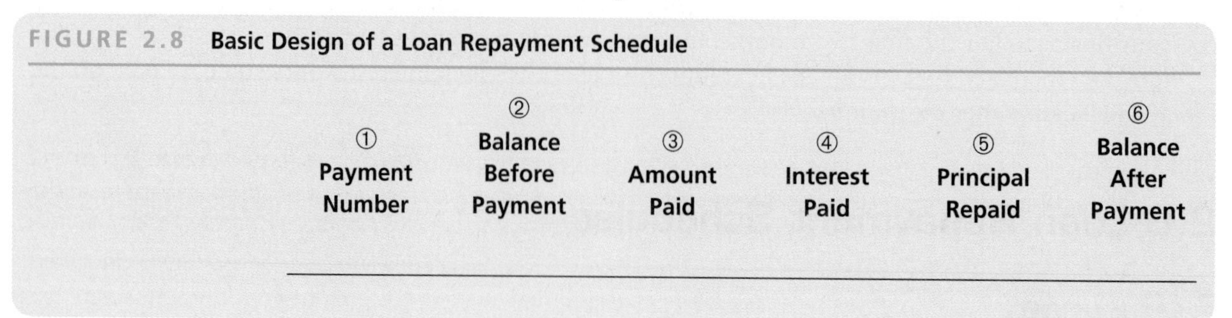

① Payment Number	② Balance Before Payment	③ Amount Paid	④ Interest Paid	⑤ Principal Repaid	⑥ Balance After Payment

B. Construction of loan repayment schedules illustrated

EXAMPLE 2.6A

Great Lakes Marina borrowed $1600.00 from Sheridan Credit Union at 9% p.a. and agreed to repay the loan in monthly installments of $300.00 each, such payments to cover interest due and repayment of principal. Use the design shown in Figure 2.8 to construct a complete repayment schedule including the totalling of columns ③, ④, and ⑤ ("Amount Paid," "Interest Paid," and "Principal Repaid").

SOLUTION

See Figure 2.9 and the explanatory notes that follow.

FIGURE 2.9 **Loan Repayment Schedule for Example 2.6A**

① Payment Number	② Balance Before Payment	③ Amount Paid (1)	④ Interest Paid (2)	⑤ Principal Repaid (3)	⑥ Balance After Payment (4)
0					1600.00 (5)
1	1600.00	300.00	12.00 (6)	288.00 (7)	1312.00 (8)
2	1312.00	300.00	9.84 (9)	290.16 (10)	1021.84 (11)
3	1021.84	300.00	7.66	292.34	729.50
4	729.50	300.00	5.47	294.53	434.97
5	434.97	300.00	3.26	296.74	138.23 (12)
6	138.23	139.27 (15)	1.04 (14)	138.23 (13)	0.00
Totals (16)		1639.27 (18)	39.27 (19)	1600.00 (17)	

Explanatory notes:

(1) The Amount Paid shown in column ③ is the agreed-upon monthly payment of $300.00.

(2) The Interest Paid shown in column ④ is at 9% per annum. This figure is converted into a periodic (monthly) rate of 9%/12 (0.75% per month) to facilitate the computation of the monthly amount of interest paid. (See notes (6) and (9).)

(3) The amount of Principal Repaid each month shown in column ⑤ is found by subtracting the Interest Paid for the month (column ④) from the Amount Paid for the month (column ③). (See notes (7) and (10).)

(4) The Balance After Payment for a month shown in column ⑥ is found by subtracting the Principal Repaid for the month (column ⑤) from the Balance Before Payment for the month (column ②) OR from the previous Balance After Payment figure (column ⑥). (See notes (8) and (11).)

(5) The original loan balance of $1600.00 is introduced as the starting amount for the schedule and is the only amount shown in Line 0.

(6) Interest paid in Payment Number 1
= 0.75% of 1600.00 = (0.0075)(1600.00) = $12.00

(7) Principal Repaid by Payment Number 1
= 300.00 − 12.00 = $288.00
(8) Balance After Payment for Payment Number 1
= 1600.00 − 288.00 = $1312.00
(9) Interest Paid in Payment Number 2
= 0.75% of 1312.00 = (0.0075)(1312.00) = $9.84
(10) Principal Repaid by Payment Number 2
= 300.00 − 9.84 = $290.16
(11) Balance After Payment for Payment Number 2
= 1312.00 − 290.16 = $1021.84
(12) The Balance After Payment for Payment Number 5 of $138.23 is smaller than the regular monthly payment of $300.00. The next payment need only be sufficient to pay the outstanding balance of $138.23 plus the interest due. (See notes (13), (14), and (15).)
(13) Principal Repaid in Payment Number 6 must be $138.23 to pay off the outstanding loan balance.
(14) Interest Paid in Payment Number 6 is the interest due on $138.23 = 0.75% of 138.23 = (0.0075)(138.23) = $1.04.
(15) Amount paid in Payment Number 6
= 138.23 + 1.04 = $139.27.
(16) The Totals of columns ③, ④, and ⑤ serve as a check of the arithmetic accuracy of the payment schedule. (See notes (17), (18), and (19).)
(17) Principal Repaid, the total of column ⑤, must equal the original loan balance of $1600.00.
(18) Amount Paid, the total of column ③, must equal the total of all the payments made (five payments of $300.00 each plus the final payment of $139.27).
(19) Interest paid, the total of column ④, must be the difference between the totals of columns ③ and ⑤ = 1639.27 − 1600.00 = $39.27.

C. Computer application—loan repayment schedule

This exercise assumes a basic understanding of spreadsheet applications; however, an individual who has no previous experience with spreadsheets could complete this task.

Excel

Microsoft Excel and other spreadsheets allow the user to easily create a flexible loan repayment schedule. The finished repayment schedule will reflect any changes made to the loan amount, the interest rate, or the schedule of payments. (The following example was created in Excel, but most spreadsheets work in a similar manner.)

The following Excel application is based on Example 2.6A, which uses six months to repay the loan. An additional row of formulas similar to row 3 would be added for each additional month required to repay the loan.

FIGURE 2.10

Workbook 1

	A	B	C	D	E	F	G
1	Payment	Balance before payment	Amount paid	Interest paid	Principal repaid	Balance	
2	0					1600.00	300.00
3	1	= F2	= G2	= G4*B3	= C3–D3	= B3–E3	
4	2	= F3	= G2	= G4*B4	= C4–D4	= B4–E4	= 0.09/12
5	3	= F4	= G2	= G4*B5	= C5–D5	= B5–E5	
6	4	= F5	= G2	= G4*B6	= C6–D6	= B6–E6	
7	5	= F6	= G2	= G4*B7	= C7–D7	= B7–E7	
8	6	= F7	= D8+E8	= G4*B8	= B8	= B8–E8	
9	Totals		= SUM(C3:C8)	= SUM(D3:D8)	= SUM(E3:E8)		

STEP 1 Enter the labels shown in Figure 2.10 in row 1 and in column A.

STEP 2 Enter the numbers shown in row 2.

STEP 3 Enter only the formulas shown in row 3 and row 9. Make sure that the entry includes the dollar sign ($) as shown in the figure.

STEP 4 The formulas that are entered in row 3 can be copied through the remaining rows.
(a) Select and copy the formulas in row 3.
(b) Select cells C4 to C8, and select Paste.
(c) The formulas are now active in all the cells.

STEP 5 To ensure readability of the spreadsheet, format the numbers to display as two decimal places, and widen the columns to display the full labels.

This spreadsheet can now be used to reflect changes in aspects of the loan and display the effects quickly. Use cell F2 for a new principal amount, cell G2 for a new repayment amount, and cell G4 for a new interest rate. The interest rate must be divided by 12 months to convert the annual interest rate to a monthly rate. A new annual interest rate of 5% would be entered as 0.05/12.

Excel A copy of this spreadsheet, called Template 2, appears on the Spreadsheet Template Disk.

EXERCISE 2.6

A. Use the design shown in Figure 2.8 to construct a complete repayment schedule including the totalling of the Amount Paid, Interest Paid, and Principal Repaid columns for each of the following loans.

1. Carla borrowed $1200.00 from the Royal Bank at 8.5% per annum calculated on the monthly unpaid balance. She agreed to repay the loan in blended payments of $180.00 per month.
2. On March 15, Julio borrowed $900.00 from Sheridan Credit Union at 7.5% per annum calculated on the daily balance. He gave the Credit Union six cheques for $135.00 dated the 15th of each of the next six months starting April 15 and a cheque dated October 15 for the remaining balance to cover payment of interest and repayment of principal.

Review Exercise

1. A four-month promissory note for $1600.00 dated June 30 bears interest at 6.5%.
   (a) What is the due date of the note?
   (b) What is the amount of interest payable at the due date?
   (c) What is the maturity value of the note?
2. Determine the maturity value of a 120-day note for $1250.00 dated May 23 and bearing interest at 5.75%.
3. Compute the face value of a 120-day note dated September 10 bearing interest at 6.75% whose maturity value is $1534.12.
4. The maturity value of a seven-month promissory note issued July 31, 2005 is $3275.00. What is the present value of the note on the date of issue if interest is 7.75%?
5. Compute the maturity value of a 150-day, 6% promissory note with a face value of $5000.00 dated August 5.
6. What is the face value of a three-month promissory note dated November 30, 2007, with interest at 4.5 percent if its maturity value is $950.89?
7. A 90-day, $800 promissory note was issued July 31 with interest at 8%. What is the present value of the note on October 20?
8. On June 1, 2006, a four-month promissory note for $1850 with interest at 5% was issued. Compute the proceeds of the note on August 28, 2006, if money is worth 6.5%.
9. Determine the value of a $1300 non-interest-bearing note four months before its maturity date of July 13, 2006 if money is worth 7%.
10. Compute the proceeds of a five-month, $7000 promissory note dated September 6, 2005 with interest at 5.5% if the note is paid on November 28, 2005 when money is worth 6.5%.
11. An investment dealer paid $24 256.25 to acquire a $25 000.00, 182-day Government of Canada treasury bill at the weekly auction. What was the rate of return on this T-bill?
12. Government of Alberta 364-day T-bills with a face value of $1 000 000 were purchased on April 7 for $971 578. The T-bills were sold on May 16 for $983 500.
    (a) What was the market yield rate on April 7?
    (b) What was the yield rate on May 16?
    (c) What was the rate of return realized?
13. Mel's Photography borrowed $15 000 on March 10 on a demand note. The loan was repaid by payments of $4000 on June 20, $3000 on September 1, and the balance on November 15. Interest, calculated on the daily balance and charged to Mel's Photography current account on the last day of each month, was at 5.5% on March 10 but was changed to 6.25% effective June 1 and to 6% effective October 1. How much did the loan cost?
14. Quick Print Press borrowed $20 000 from the Provincial Bank on May 25 at 7.5% and secured the loan by signing a promissory note subject to a variable rate of interest. Quick Print made partial payments of $5000 on July 10 and $8000 on September 15. The rate of interest was increased to 8% effective August 1 and to 8.5% effective October 1. What payment must Quick Print make on October 31 if, under the terms of the loan agreement, any interest accrued as of October 31 is to be paid on October 31?
15. Muriel has a line of credit with a limit of $10 000.00. She owed $8195.00 on July 1. Principal withdrawals for the period July 1 to November 30 were $3000.00 on August 20 and $600.00 on October 25. The line of credit agreement requires regular payments of $300.00 on the 15th day of each month. Muriel has made all required payments. Interest (including overdraft interest) is charged to the account on the last day of each month. The interest rate was 8% on July 1, but was changed to 7.5% effective September 15.

Overdraft interest is 16% for any balance in excess of $10 000.00.

(a) Calculate the interest charges on July 31, August 31, September 30, October 31, and November 30.

(b) Calculate the account balance on November 30.

16. You borrowed $3000 at 9% per annum calculated on the unpaid monthly balance and agreed to repay the principal together with interest in monthly payments of $500 each. Construct a complete repayment schedule.

Self-Test

1. For the following promissory note, determine the amount of interest due at maturity.

$565.00	TORONTO, ONTARIO	JANUARY 10, 2008
FIVE MONTHS after date we promise to pay to the order of		
WILSON LUMBER COMPANY		
EXACTLY FIVE-HUNDRED-SIXTY-FIVE and 00/100 -------------- Dollars		
at WILSON LUMBER COMPANY		for value received
with interest at 8.25% per annum.		
Due ______________		(seal) ______________
		(seal) ______________

2. Find the maturity value of a $1140.00, 7.75%, 120-day note dated February 19, 2006.

3. Determine the face value of a four-month promissory note dated May 20, 2007 with interest at 7.5% p.a. if the maturity value of the note is $1190.03.

4. Find the present value of a non-interest-bearing seven-month promissory note for $1800 dated August 7, 2007, on December 20, 2007, if money is then worth 6%.

5. A 180-day note dated September 14, 2006 is made at 5.25% for $1665.00. What is the present value of the note on October 18, 2006, if money is worth 6.5%?

6. What is the price of a 91-day, $25 000 Government of Canada treasury bill that yields 3.28% per annum?

7. An investor purchased a 182-day, $100 000 T-bill on its issue date. It yielded 3.85%. The investor held the T-bill for 67 days, then sold it for $98 853.84.

   (a) What was the original price of the T-bill?

   (b) When the T-bill was sold, what was its yield?

8. The owner of Jane's Boutique borrowed $6000.00 from Halton Community Credit Union on June 5, 2006. The loan was secured by a demand note with interest calculated on the daily balance and charged to the store's account on

the 5th day of each month. The loan was repaid by payments of $1500.00 on July 15, $2000.00 on October 10, and $2500.00 on December 30. The rate of interest charged by the credit union was 8.5% on June 5. The rate was changed to 9.5% effective July 1 and to 10% effective October 1. Determine the total interest cost on the loan.

9. Herb's Restaurant borrowed $24 000.00 on March 1, 2005 on a demand note providing for a variable rate of interest. While repayment of principal is open, any accrued interest is to be paid on November 30. Payments on the loan were made as follows: $600.00 on April 15, $400.00 on July 20, and $400.00 on October 10. The rate of interest was 7% on March 1 but was changed to 8.5% effective August 1 and to 7.5% effective November 1. Using the declining balance method to record the partial payments, determine the accrued interest on November 30.

10. Jingyi has a line of credit from her local bank with a limit of $10 000.00. On March 1, 2006, she owed $7265.00. From March 1 to June 30, she withdrew principal amounts of $3000.00 on April 10 and $500.00 on June 20. According to the line of credit agreement, Jingyi must make a regular payment of $200.00 on the 15th of each month. She has made these payments. Interest (including overdraft interest) is charged to the account on the last day of each month. On March 1, the interest rate was 9%, but it was changed to 8.5% effective May 15. Overdraft interest is 18% for any balance in excess of $10 000.00.

(a) Calculate the interest charges on March 31, April 30, May 31, and June 30.

(b) What is the account balance on June 30?

11. Use the design shown in Figure 2.8 to construct a complete repayment schedule, including the totalling of the Amount Paid, Interest Paid, and Principal Repaid columns, for a loan of $4000 repaid in monthly installments of $750.00 each including interest of 6.5% per annum calculated on the unpaid balance.

Challenge Problems

1. Mike Kornas signed a 12-month, 11% p.a. simple interest promissory note for $12 000 with MacDonald's Furniture. After 100 days, MacDonald's Furniture sold the note to the Royal Bank at a rate of 13% p.a. Royal Bank resold the note to Friendly Finance Company 25 days later at a rate of 9% p.a. Find the gain or loss on this note for each company and bank involved.

2. A father wanted to show his son what it might be like to borrow money from a financial institution. When his son asked if he could borrow $120, the father lent him the money and set up the following arrangements. He charged his son $6 for the loan of $120. The son therefore received $114 and agreed to pay his father 12 installments of $10 a month, beginning one month from today, until the loan was repaid. Find the approximate rate of simple interest the father charged on this loan.

Case Study 2.1 The Business of Borrowing

» Magnusson's Computer Store agrees to purchase some new computer monitors costing $5000 plus 7% GST. Doug Magnusson, the store's owner, was informed that if he paid cash on receipt of the goods he could take a cash discount of 4% of the invoice price before GST. GST would then be added to the new invoice price. Doug would like to take advantage of this discount, but the store is short of cash right now. A number of customers are expected to pay their invoices in the next 30 to 60 days.

Doug went to his bank manager to negotiate a short-term loan to pay for the monitors when they arrive and take advantage of the cash discount. The bank manager suggested a 90-day promissory note bearing interest at 7%. Doug agreed to the note and suggested that the three days of grace should be added to give him more repayment flexibility.

QUESTIONS

1. What is the maturity value of the 90-day promissory note using three days of grace for the goods including GST?
2. Suppose Doug decides he does not need the three days of grace. What effect would this have on the maturity value of the note?
3. Doug later discussed his situation with a friend, who suggested that Doug could have negotiated a short-term loan for 90 days instead of using a promissory note. What is the highest annual simple interest rate at which Doug could have borrowed the money and still saved by taking the cash discount?

Case Study 2.2 Dealing with Debt

» Don and Rosemary Schaus were concerned about their high debts. They borrowed money from their bank to purchase their house, car, and computer. They must make regular monthly payments for these three loans. For some of their other expenses, they also owe $4000 to MasterCard and $3500 to Visa. Don and Rosemary decided to meet with a consumer credit counsellor to gain control of their debts.

The consumer credit counsellor explained to Don and Rosemary the details of their loans and credit card debts. Don and Rosemary were shocked to discover that whereas their computer and car loans had an interest rate of 9.0% p.a., their credit cards had an interest rate of 17.5% p.a. The counsellor pointed out that the interest rate on their three loans was reasonable. However, because the interest rate on the credit cards was so high, she advised Don and Rosemary to borrow money at a lower interest rate and pay off the credit card debts.

The credit counsellor suggested that they should consider obtaining a line of credit. She explained that the rate of interest on the line of credit would likely be a few percentage points higher than the prime rate, but much lower than the rate of interest charged on credit card balances. Don and Rosemary would have to make a minimum payment every month, similar to a credit card, that would be applied to pay all the interest and a portion of the principal balance owing on the line of credit. The line of credit would allow them to make monthly payments higher than the minimum so that they could pay as much toward the principal balance as they could afford. Due to the much lower interest rate on a line of credit compared to a typical credit card, the money they would save on interest each month could be paid toward the principal. A line of credit appealed to Don and Rosemary. It helped them feel more in control of their finances and gave them the resolve to pay off their credit card debts.

The next day, Don and Rosemary met with their bank manager and were approved for a $10 000.00 line of credit. Immediately, they paid off the $4000.00 owed to MasterCard and the $3500.00 owed to Visa with money from the line of credit. They then decided to pay off the line of credit over the next ten months by making monthly payments equal to one-tenth of the original line of credit balance plus the simple interest owed on the remaining line of credit balance. The simple interest rate on the line of credit is expected to be 8.25% over the next ten months. Don and Rosemary agreed to cut up their credit cards and not charge any more purchases until they had paid off their line of credit.

QUESTIONS

1. Suppose Don and Rosemary pay off their credit cards with their line of credit on April 20. They will make their monthly payments on the 20th of each month, beginning in May. Create a schedule showing their monthly payments for the next ten months. How much interest will they pay using this repayment plan?
2. Suppose Don and Rosemary had not gotten a line of credit but kept their credit cards. They decided not to make any more credit card purchases. Instead, they made monthly payments equal to one-tenth of the original credit card balance plus the simple interest owed on the remaining credit card balance. They will make their monthly payments on the 20th of each month, beginning in May. Create a schedule showing their monthly payments for the next ten months. How much interest would they have paid using this repayment plan?
3. How much money did Don and Rosemary save on interest by getting the line of credit?
4. What are the requirements for obtaining a line of credit from your financial institution?

SUMMARY OF FORMULAS

Formula 1.1A

$I = Prt$ — Finding the amount of interest on promissory notes

Formula 1.3A

$S = P(1 + rt)$ — Finding the maturity value of promissory notes directly

Formula 1.3B

$P = \dfrac{S}{1 + rt}$ — Finding the present value of promissory notes or treasury bills given the maturity value

GLOSSARY

Amortization schedule see **Loan repayment schedule**

Amount of interest interest, in dollars and cents, payable to the payee on the legal due date *(p. 46)*

Blended payment the usual method of repaying a personal consumer loan by a fixed periodic (monthly) payment that covers payment of interest and repayment of principal *(p. 67)*

Credit card a card entitling the bearer to a revolving line of credit with a pre-approved credit limit *(p. 62)*

Date of issue the date on which a promissory note is made *(p. 45)*

Date of maturity *see* **Legal due date**

Declining balance approach the commonly used approach to applying partial payments to demand loans whereby each partial payment is first applied to pay the interest due and then applied to the outstanding principal *(p. 58)*

Demand loan a loan for which repayment in full or in part may be required at any time or made at any time *(p. 57)*

Demand note the financial instrument representing a demand loan *(p. 57)*

Due date *see* **Legal due date**

Face value the sum of money specified on the promissory note *(p. 45)*

Interest-bearing promissory note a note subject to the rate of interest stated on the note *(p. 45)*

Interest period the time, in days, from the date of issue to the legal due date for promissory notes *(p. 46)*

Issue date *see* **Date of issue**

Legal due date the date on which the promissory note is to be paid; it includes three days of grace unless "No Grace Days" is written on the promissory note *(p. 46)*

Line of credit a pre-approved loan amount issued by a financial institution for use by an individual or business at any time for any purpose; interest is charged only for the time money is borrowed on the line of credit; a minimum monthly payment is required (similar to a credit card); the interest rate can change over time *(p. 62)*

Loan repayment schedule a detailed statement of installment payments, interest cost, repayment of principal, and outstanding balance of principal for an installment plan *(p. 68)*

Maker the party making the promise to pay by signing the promissory note *(p. 45)*

Maturity value the amount (face value plus interest) that must be paid on the legal due date to honour the note *(p. 46)*

Partial payments a series of payments on a debt *(p. 58)*

Payee the party to whom the promise to pay is made *(p. 45)*

Promissory note a written promise to pay a specified sum of money after a specified period of time or on demand, with or without interest as specified *(p. 45)*

Rate money is worth the prevailing rate of interest *(p. 51)*

Rate of interest the simple annual rate of interest based on the face value *(p. 45)*

Secured line of credit a line of credit with assets pledged as security *(p. 62)*

T-bills *see* **Treasury bills**

Term of a promissory note the time period for which the note was written (in days or months) *(p. 45)*

Three days of grace the number of days added to the term of a note in Canada to determine its legal due date *(p. 46)*

Treasury bills promissory notes issued at a discount from their face values by the federal government and most provincial governments to meet short-term financing requirements (the maturity value of treasury bills is the same as their face value) *(p. 54)*

Unsecured line of credit a line of credit where no assets are pledged by the borrower to the lender to cover non-payment of the line of credit *(p. 62)*

USEFUL INTERNET SITES

www.royalbank.com

Royal Bank Visit the Daily Numbers area of this site for current interest rates on investment products. This site also provides links to detailed rate schedules for mortgages, T-bills, personal accounts, and RRSPs.

www.bankofcanada.ca/en/monmrt.htm

Bank of Canada Current T-bill rates are posted on this section of the Bank of Canada site. A link to selected historical interest rates is also provided.

www.bankrate.com

Bankrate.com Go to "Calculators" and select "Credit Cards" to determine what it really costs to pay the minimum payment on your credit card. (This is a U.S. site. Some of the interest rates displayed will not be applicable in Canada.)

3 Compound Interest—Future Value and Present Value

OBJECTIVES

Upon completing this chapter, you will be able to do the following:

1. Calculate interest rates and the number of compounding periods.
2. Compute future (maturity) values, even in cases involving changes in rate of interest and principal.
3. Compute present values of future sums of money.
4. Discount long-term promissory notes.
5. Solve problems involving equivalent values.

Compound interest is used extensively in today's world of computerized banking and investing. With compound interest, you earn interest on a principal, and the interest is then added to form a new principal. You then earn interest on this new higher principal. This is what is meant by the expression "earning interest on interest."

To calculate compound interest, and to decide which interest terms are best for you, you need to understand how compound interest works. When you need to calculate compound interest amounts for all sorts of interest rates and time periods, calculators and computers make the job easy. To calculate compound interest amounts using formulas, an understanding of exponents is required.

INTRODUCTION

Under the compound interest method, interest is added periodically to the principal. As we did with simple interest, we use compound interest calculations to determine future values and present values. The compound interest formulas contain the compounding factor $(1 + i)^n$. The calculations are not difficult if you use electronic calculators equipped with an exponential function, especially if they are financial calculators preprogrammed with the ability to calculate future value and present value for compound interest.

3.1 Basic Concepts and Computations

A. Basic procedure for computing compound interest

The term **compound interest** refers to a procedure for computing interest whereby the interest for a specified time period is added to the original principal. The resulting amount becomes the new principal for the next time period. The interest earned in earlier periods earns interest in future periods.

The compound interest method is generally used to calculate interest for long-term investments. The amount of compound interest for the first interest period is the same as the amount of simple interest, but for further interest periods the amount of compound interest becomes increasingly greater than the amount of simple interest.

The basic procedure for computing compound interest and the effect of compounding is illustrated in Table 3.1. The table also provides a comparison of compound interest and simple interest for an original principal of $10 000.00 invested at 10% per annum for six years.

TABLE 3.1 **Compound Interest versus Simple Interest for a Principal of $10 000.00 Invested at 10% per Annum for 6 Years**

		At Compound Interest		At Simple Interest	
Year		Interest Computation	Amount	Interest Computation	Amount
	Original Principal		10 000.00		10 000.00
1	Add Interest	(0.10)(10 000.00)	1 000.00	(0.10)(10 000.00)	1 000.00
	Amount End Year 1		11 000.00		11 000.00
2	Add Interest	(0.10)(11 000.00)	1 100.00	(0.10)(10 000.00)	1 000.00
	Amount End Year 2		12 100.00		12 000.00
3	Add Interest	(0.10)(12 100.00)	1 210.00	(0.10)(10 000.00)	1 000.00
	Amount End Year 3		13 310.00		13 000.00
4	Add Interest	(0.10)(13 310.00)	1 331.00	(0.10)(10 000.00)	1 000.00
	Amount End Year 4		14 641.00		14 000.00
5	Add Interest	(0.10)(14 641.00)	1 464.10	(0.10)(10 000.00)	1 000.00
	Amount End Year 5		16 105.10		15 000.00
6	Add Interest	(0.10)(16 105.10)	1 610.51	(0.10)(10 000.00)	1 000.00
	Amount End Year 6		17 715.61		16 000.00

The method of computation used in Table 3.1 represents the step-by-step approach used in maintaining a compound interest record, such as a savings account. Note that the amount of interest is determined for each interest period on the basis of the previous balance and is then added to that balance.

Note the following about the end results after six years:

	At compound interest	*At simple interest*
Amount after six years	$17 715.61	$16 000.00
Less original principal	10 000.00	$10 000.00
Amount of interest	$7 715.61	$6 000.00

In this case, the compound interest exceeds the simple interest by $1715.61. This difference represents the amount of interest earned by interest added to the principal at the end of each compounding period.

B. Computer application—accumulation of principal using a spreadsheet

Table 3.1 displays the manual calculations for the accumulation of a principal of $10 000.00 earning compound interest at 10% per annum for six years. Microsoft Excel and other spreadsheet programs can be used to create a file that will immediately display the results of a change in the principal, interest rate, or number of years of accumulation.

The following steps are general instructions for creating a file to calculate the compound interest data in Table 3.1. The formulas in the spreadsheet file were created using Excel; however, most spreadsheets work similarly.

Excel

Though this exercise assumes a basic understanding of spreadsheet applications, someone without previous spreadsheet experience will be able to complete this task.

STEP 1 Enter the labels shown in Figure 3.1 in row 1 and in column A.

STEP 2 Enter the principal in cell B2. Do not type in the dollar sign or a comma.

STEP 3 Enter only the formulas shown in cells C2, D2, and B3. Make sure that the formula entry includes the dollar ($) sign as shown in Figure 3.1. Use Copy and Paste to enter the remainder of the formulas shown.

STEP 4 The formulas that were entered in Step 3 can be copied through the remaining cells.

(a) Select and Copy the formulas in cells C2 and D2.

(b) Select cells C3 to C7, and Paste.

(c) Select and Copy the formula in cell B3.

(d) Select cells B4 to B7, and Paste.

(e) The formulas are now active in all the cells.

STEP 5 To ensure readability of the spreadsheet, format the numbers to display as two decimal places, and widen the columns to display the full labels.

This spreadsheet can now be used to reflect changes in aspects of the investment. You will be able to input the changes and see the effects quickly. Use cell B2 for a new principal amount and cell E1 for a new interest rate.

FIGURE 3.1

Workbook 1

	A	B	C	D	E
1	Year	Amount, beginning of year	Interest	Amount, end of year	0.10
2	1	10000	= B2*E1	= B2+C2	
3	2	= D2	= B3*E1	= B3+C3	
4	3	= D3	= B4*E1	= B4+C4	
5	4	= D4	= B5*E1	= B5+C5	
6	5	= D5	= B6*E1	= B6+C6	
7	6	= D6	= B7*E1	= B7+C7	

C. The future value formula for compound interest

While the compound interest record method is useful in maintaining a savings account record, it is impractical for computational purposes. As in the case of simple interest, the **future value**, **maturity value**, or **amount** of a loan or investment can be found by using the future value formula.

For *simple interest*, the future value formula is $S = P(1 + rt)$, Formula 1.3A.

For *compound interest*, the formula for the future value is

$$S = P(1 + i)^n$$ restated as:

$$FV = PV(1 + i)^n \qquad \text{Formula 3.1A}$$

S or FV = the future or maturity value;
P or PV = the original principal;
i = the periodic rate of interest;
n = the number of compounding periods for the term of the loan or investment.

The results of Table 3.1 could have been obtained by using the two future value formulas.

For simple interest: $P = 10\,000.00$; $r = 0.10$; $t = 6$

$$\begin{aligned} S = P(1 + rt) &= 10\,000.00[1 + (0.10)(6)] \\ &= 10\,000.00(1 + 0.60) \\ &= 10\,000.00(1.60) \\ &= \$16\,000.00 \end{aligned}$$

For compound interest: $PV = 10\,000.00$; $i = 0.10$; $n = 6$

$$\begin{aligned} FV = PV(1 + i)^n &= 10\,000.00(1 + 0.10)^6 \\ &= 10\,000.00(1.10)^6 \\ &= 10\,000.00(1.10)(1.10)(1.10)(1.10)(1.10)(1.10) \\ &= 10\,000.00(1.771561) \\ &= \$17\,715.61 \end{aligned}$$

When using the compound interest formula, determining the factor $(1 + i)^n$ is the main computational problem. The value of this factor, called the **compounding factor** or **accumulation factor**, depends on the values of i and n.

D. Determining the periodic rate of interest

The value of i, the **periodic rate of interest,** is determined from the stated rate of interest to be used in the compounding situation. The stated rate is called the **nominal rate of interest**. Since the nominal rate of interest is usually stated as an annual rate, the value of i depends on the **compounding** (or **conversion**) **frequency** per year. The value of i is obtained by dividing the nominal annual rate by the number of **compounding** (or **conversion**) **periods** per year.

The compounding (conversion) periods commonly used in business and finance cover a number of months, usually an exact divisor of twelve, and are listed in Table 3.2.

TABLE 3.2 **Commonly Used Compounding Frequencies and Conversion Periods**

Compounding (Conversion) Frequency	Length of Compounding (Conversion) Period	Number of Compounding (Conversion) Periods per Year
Annual	12 months (1 year)	1
Semi-annual	6 months	2
Quarterly	3 months	4
Monthly	1 month	12

The relationship between the periodic rate of interest and the nominal annual rate of interest can be stated in the form of a formula.

$$\text{PERIODIC RATE OF INTEREST, } i = \frac{\text{NOMINAL (ANNUAL) RATE}}{\text{NUMBER OF COMPOUNDING (CONVERSION) PERIODS PER YEAR}}$$

Therefore,

$$i = \frac{j}{m}$$ Formula 3.2

where i = periodic rate of interest
j = nominal annual rate of interest
m = number of compounding (conversion) periods per year

EXAMPLE 3.1A

Determine the periodic rate of interest i for
(i) 5% p.a. compounded annually;
(ii) 7% p.a. compounded semi-annually;
(iii) 12% p.a. compounded quarterly;
(iv) 10.5% p.a. compounded monthly.

SOLUTION

	(i)	(ii)	(iii)	(iv)
The nominal annual rate j	5%	7%	12%	10.5%
The compounding (conversion) frequency	annually	semi-annually	quarterly	monthly
The length of the compounding (conversion) period	12 months	6 months	3 months	1 month
The number of compounding (conversion) periods per year m	1	2	4	12
The periodic rate of interest $i = \frac{j}{m}$	$\frac{5\%}{1}$	$\frac{7\%}{2}$	$\frac{12\%}{4}$	$\frac{10.5\%}{12}$
	$= 5.0\%$	$= 3.5\%$	$= 3.0\%$	$= 0.875\%$

DID YOU KNOW?

An investment will grow in value much more rapidly at compound interest than at simple interest. Furthermore, if money is invested at compound interest, it will grow in value more quickly if the compounding frequency is greater. The following chart and graph show the comparative growth in value of a $10 000 investment over a 30-year period, based on a 10% simple interest rate versus a 10% compound interest rate (at various compounding frequencies), to underscore these concepts.

Interest Rate	Future Value (S) of $10 000 After 30 Years
10% p.a. simple interest	$40 000.00
10% p.a. compounded annually	$174 494.02
10% p.a. compounded quarterly	$193 581.50
10% p.a. compounded daily	$200 771.53

You can use the appropriate formulas from Chapters 1 and 3 to confirm these figures.

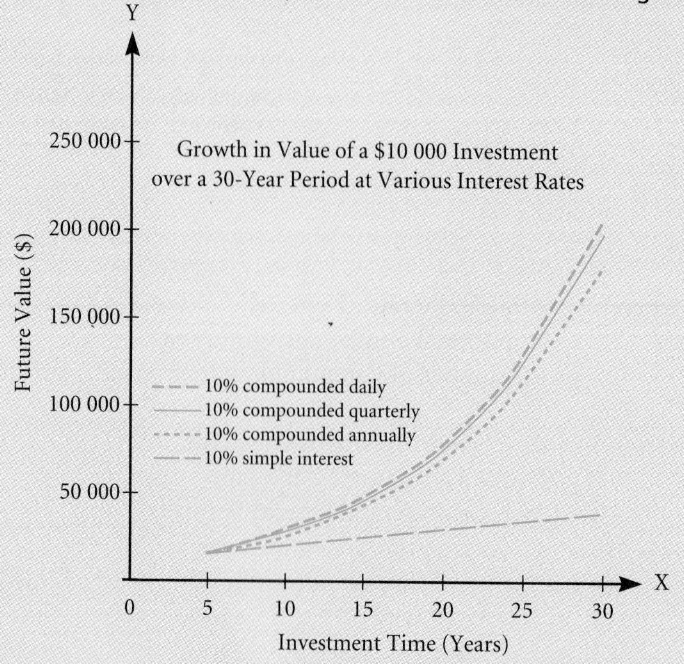

E. Determining the number of compounding (conversion) periods in the term of an investment or loan

To find the number of compounding (conversion) periods in the term of an investment or a loan, multiply the number of years in the term by the number of compounding periods per year.

EXAMPLE 3.1B

Determine the number of compounding periods when
(i) compounding annually for 14 years;
(ii) compounding semi-annually for 15 years;
(iii) compounding quarterly for 12.5 years;
(iv) compounding monthly for 10.75 years;
(v) compounding quarterly for 30 months;
(vi) compounding semi-annually for 42 months.

SOLUTION

	Term (in years)	Compounding Frequency	Number of Compounding Periods per Year, m	Number of Compounding Periods in Term, n
(i)	14	annually	1	14(1) = 14
(ii)	15	semi-annually	2	15(2) = 30
(iii)	12.5	quarterly	4	12.5(4) = 50
(iv)	10.75	monthly	12	10.75(12) = 129
(v)	$\frac{30}{12} = 2.5$	quarterly	4	2.5(4) = 10
(vi)	$\frac{42}{12} = 3.5$	semi-annually	2	3.5(2) = 7

POINTERS AND PITFALLS

Many students have difficulty determining the value of m when compounding is stated as "quarterly." This term means that interest is compounded every quarter of a year. Since there are four quarters in a year, m then becomes 4, with each quarter of a year being 3 months in length. The value of m is never 3.

F. Setting up the compounding factor $(1 + i)^n$

The compounding (accumulation) factor $(1 + i)^n$ can be set up by first determining i and n and then substituting i and n in the general form of the factor, $(1 + i)^n$.

EXAMPLE 3.1C Set up the compounding factor $(1 + i)^n$ for

(i) 5% p.a. compounded annually for 14 years;
(ii) 7% p.a. compounded semi-annually for 15 years;
(iii) 12% p.a. compounded quarterly for 12.5 years;
(iv) 10.5% p.a. compounded monthly for 10.75 years;
(v) 8% p.a. compounded quarterly for 30 months;
(vi) 9.5% p.a. compounded semi-annually for 42 months.

SOLUTION

	i	m	n	$(1 + i)^n$
(i)	5% = 0.05	1	14(1) = 14	$(1 + 0.05)^{14} = 1.05^{14}$
(ii)	3.5% = 0.035	2	15(2) = 30	$(1 + 0.035)^{30} = 1.035^{30}$
(iii)	3.0% = 0.03	4	12.5(4) = 50	$(1 + 0.03)^{50} = 1.03^{50}$
(iv)	0.875% = 0.00875	12	10.75(12) = 129	$(1 + 0.00875)^{129} = 1.00875^{129}$
(v)	2% = 0.02	4	2.5(4) = 10	$(1 + 0.02)^{10} = 1.02^{10}$
(vi)	4.75% = 0.0475	2	3.5(2) = 7	$(1 + 0.0475)^{7} = 1.0475^{7}$

G. Computing the numerical value of the compounding factor $(1 + i)^n$

The numerical value of the compounding factor can now be computed using an electronic calculator. For calculators equipped with the exponential function feature [y^x], the numerical value of the compounding factor can be computed directly.

STEP 1 Enter the numerical value of $(1 + i)$ in the keyboard.

STEP 2 Press the function key [y^x].

STEP 3 Enter the numerical value of n in the keyboard.

STEP 4 Press [=].

STEP 5 Read the answer in the display.

The numerical value of the compounding factors in Example 3.1C are obtained as follows.

		(i)	(ii)	(iii)	(iv)	(v)	(vi)
STEP 1	Enter	1.05	1.035	1.03	1.00875	1.02	1.0475
STEP 2	Press	y^x	y^x	y^x	y^x	y^x	y^x
STEP 3	Enter	14	30	50	129	10	7
STEP 4	Press	=	=	=	=	=	=
STEP 5	Read	1.979932	2.806794	4.383906	3.076647	1.218994	1.383816

Note: Do not be concerned if your calculator shows a difference in the last decimal. There is no error. It reflects the precision of the calculator and the number of decimal places formatted to show on the display of the calculator.

With the increasing availability of inexpensive electronic calculators, the two traditional methods of determining the compounding factor $(1 + i)^n$— logarithms and tables—are rapidly falling into disuse. Neither method is used in this text.

EXERCISE 3.1

A. Determine m, i, and n for each of the following.

1. 12% compounded annually for 5 years

2. 7.4% compounded semi-annually for 8 years

3. 5.5% compounded quarterly for 9 years

4. 7% compounded monthly for 4 years

5. 11.5% compounded semi-annually for 13.5 years

6. 4.8% compounded quarterly for $5\frac{3}{4}$ years

7. 8% compounded monthly for 12.5 years

8. 10.75% compounded quarterly for 3 years, 9 months

9. 12.25% compounded semi-annually for 54 months

10. 8.1% compounded monthly for 15.5 years

B. Set up and compute the compounding factor $(1 + i)^n$ for each of the questions in part A.

C. Answer each of the following questions.

1. For a sum of money invested at 10% compounded quarterly for 12 years, state

(a) the number of compounding periods;
(b) the periodic rate of interest;
(c) the compounding factor $(1 + i)^n$;
(d) the numerical value of the compounding factor.

2. For each of the following periodic rates of interest, determine the nominal annual compounding rate.

(a) i = 2%; compounding is quarterly
(b) i = 0.75%; compounding is monthly
(c) i = 5.5%; compounding is semi-annual
(d) i = 9.75%; compounding is annual

DID YOU KNOW?

A loonie invested for 100 years at 2% p.a. compounded annually accumulates to about \$7.24. If the 2% p.a. is compounded daily, the result is about \$7.39.

A loonie invested for 100 years at 10% p.a. compounded annually accumulates to about \$13 780.61. If the 10% p.a. is compounded daily, the result is about \$21 996.26.

A loonie invested for 100 years at 20% p.a. compounded annually accumulates to about \$82 817 974.00. If the 20% p.a. is compounded daily, the result is about \$482 510 000.00.

The rate of compound interest and the compounding frequency make all the difference!

3.2 Using the Formula for the Future Value of a Compound Amount $FV = PV(1 + i)^n$

A. Finding the future value (maturity value) of an investment

EXAMPLE 3.2A Find the amount to which \$6000.00 will grow if invested at 10% per annum compounded quarterly for five years.

SOLUTION The original principal PV = 6000.00;
the nominal annual rate $j = 10\%$;
the number of compounding periods per year $m = 4$;
the quarterly rate of interest $i = \frac{10\%}{4} = 2.5\% = 0.025$;
the number of compounding periods (quarters) $n = (5)(4) = 20$.

$$
\begin{aligned}
FV &= PV(1 + i)^n && \text{using Formula 3.1A}\\
&= 6000.00(1 + 0.025)^{20} && \text{substituting for P, } i, n\\
&= 6000.00(1.025)^{20} && \text{exponential form of factor}\\
&= 6000.00(1.6386164) && \text{using a calculator}\\
&= \$9831.70
\end{aligned}
$$

EXAMPLE 3.2B What is the future value after 78 months of \$2500 invested at 5.25% p.a. compounded semi-annually?

SOLUTION The original principal PV = 2500.00;
the nominal annual rate $j = 5.25$;
the number of compounding periods per year $m = 2$;
the semi-annual rate of interest $i = \frac{5.25\%}{2} = 2.625\% = 0.02625$;
the number of compounding periods (each period is six months)
$n = \left(\frac{78}{12}\right)(2) = (6.5)(2) = 13$.

$$
\begin{aligned}
FV &= PV(1 + i)^n\\
&= 2500.00(1 + 0.02625)^{13}\\
&= 2500.00(1.02625)^{13}\\
&= 2500.00(1.400526)\\
&= \$3501.32
\end{aligned}
$$

EXAMPLE 3.2C Accumulate a deposit of \$1750.00 made into a registered retirement savings plan from March 1, 1985 to December 1, 2005 at 6.5% p.a. compounded quarterly.

SOLUTION The original principal PV = 1750.00; $j = 6.5\%$; $m = 4$;

the quarterly rate of interest $i = \dfrac{6.5\%}{4} = 1.625\% = 0.01625$;

the time period from March 1, 1985 to December 1, 2005 contains 20 years and 9 months, or 20.75 years: $n = (20.75)(4) = 83$.

$$\begin{aligned} FV &= PV(1 + i)^n \\ &= 1750.00(1 + 0.01625)^{83} \\ &= 1750.00(1.01625)^{83} \\ &= 1750.00(3.811065) \\ &= \$6669.36 \end{aligned}$$

B. Using preprogrammed financial calculators

Compound interest calculations, which can become complex, are performed frequently and repeatedly. Doing the calculations algebraically can enhance your understanding and appreciation of the theory but it can also be time-consuming, laborious, and subject to mechanical errors. Using preprogrammed financial calculators can save time and reduce or eliminate mechanical errors, assuming they are set up properly and numerical sign conventions are observed when entering data and interpreting results.

Different models of financial calculators vary in their operation and labelling of the function keys and faceplate. Appendix II, entitled "Instructions and Tips for Four Preprogrammed Financial Calculator Models," highlights the relevant variations for students using Texas Instruments' BAII Plus and BA-35 Solar, Sharp's EL-733A, and Hewlett-Packard's 10B calculators. Appendix II is intended to help you use one of these four calculators and *supplements* the instruction booklet that came with your calculator. Refer to the instruction booklet for your particular model to become familiar with your calculator.

Specific function keys on preprogrammed financial calculators correspond to the five variables used in compound interest calculations. Function keys used for the calculator models presented in Appendix II are shown in Table 3.3.

TABLE 3.3 Financial Calculator Function Keys that Correspond to Variables Used in Compound Interest Calculations

Variable	Algebraic Symbol	Function Key: TI BAII+	TI BA-35S	Sharp EL-733A	HP 10B
The number of compounding periods	n	N	N	n	N
The rate of interest[1]	i	I/Y C/Y	%i	i	I/YR
The periodic annuity payment[2]	R	PMT	PMT	PMT	PMT
The present value or principal	P	PV	PV	PV	PV
The future value or maturity value	S	FV	FV	FV	FV

Notes: 1. The periodic rate of interest is entered as a percent and not as a decimal equivalent (as it is when using the algebraic method to solve compound interest problems.) For example, 8% is entered as "8" not ".08". With some calculators, the rate of interest is the periodic rate. In the case of the BAII Plus, the rate of interest entered is the rate per year.

2. The periodic annuity payment function key PMT is used only for annuity calculations, which are introduced in Chapter 5.

The function keys are used to enter the numerical values of the known variables into the appropriate preprogrammed calculator registers. The data may be entered in any order. The answer is then displayed by using a computation key or by depressing the key representing the unknown variable, depending on the calculator model.

Before entering the numerical data to complete compound interest calculations, it is important to verify that your calculator has been set up correctly to ensure error-free operation. There are a number of items to check during this "pre-calculation" phase. Specifically, does the calculator require a mode change within a register to match the text presentation? Does the calculator have to be in the financial mode? Are the decimal places set to the correct number to ensure the required accuracy?

Further checks must be made when entering data during the "calculation" phase. For example, have the function key registers been cleared? What numerical data require a minus sign to avoid errors in operation and incorrect answers? How can the data entered be confirmed? Responses to these queries in the "pre-calculation" and "calculation" phases for four preprogrammed financial calculators are given in Appendix II, along with general information.

Instructions in this text are given for the Texas Instruments BAII Plus calculator. Refer to Appendix II for instructions for setting up and using the Texas Instruments BA-35 Solar, Sharp EL-733A, and Hewlett-Packard 10B calculators.

Using the Texas Instruments BAII Plus to Solve Compound Interest Problems

Follow the steps below to compute the future value of a sum of money using the formula $FV = PV(1 + i)^n$ and a Texas Instruments BAII Plus calculator. Compare your result with Example 3.2A.

Pre-calculation Phase (Initial Set-up)

STEP 1 The P/Y register, and behind it the C/Y register, must be set to match the calculator's performance to the text presentation. The P/Y register is used to represent the number of regular payments per year. If the text of the question does not discuss regular payments per year, this should be set to equal the C/Y in the calculator. The C/Y register is used to represent the number of compounding periods per year, that is, the compounding frequency. The description of the compounding frequency is usually contained within the phrase that describes the nominal interest rate. An example would be "8% p.a. compounded quarterly." This means that the nominal, or annual, interest rate of 8% is compounded four times each year at 8%/4, or 2%, each period. The compounding frequency of 4 is entered into the C/Y register within the calculator.

Key in	Press	Display shows	
	2nd (P/Y)	P/Y = 12	— checks the P/Y register
4	ENTER	P/Y = 4	— changes the value to "4"
	↓	C/Y = 4	— changed automatically to match the P/Y
	2nd (QUIT)	0	— returns to the standard calculation mode

STEP 2 Verify that the decimal format is set to the number you require. A setting of "9" represents a floating decimal point format. The default setting is "2."

Key in	*Press*	*Display shows*	
	2nd (Format)	DEC = 2	— checks the decimal format
6	ENTER	DEC = 6	— changes to "6"
	2nd (QUIT)	0	— returns to the standard calculation mode

This calculator is ready for financial calculations in its standard mode.

Calculation Phase

STEP 3 Always clear the function key registers before beginning compound interest calculations.

Key in	*Press*	*Display shows*	
	2nd (CLR TVM)	0	— clears the function key registers

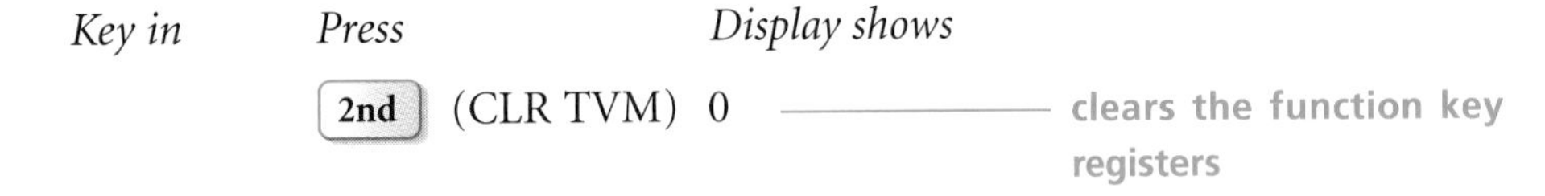

STEP 4 To solve Example 3.2A in which P = 6000, i = 2.5%, and n = 20, use the following procedure. Remember that i represents the interest rate per compounding period, and n represents the number of times that interest is compounded. To enter the accurate information into the calculator, you must determine the nominal interest rate for the year, and enter this information into the I/Y register. In this example, the interest rate per year is 10%. The compounding frequency, denoted by the phrase "compounded quarterly," is four times per year. This must be entered into the C/Y register. Also, notice that 6000 is entered as a negative number since it is cash paid out for an investment. This "cash flow sign convention" is explained below.

If you follow this convention, your calculation will be error-free, and the answer will be accurate and interpreted consistently.

Key in	Press	Display shows	
	2nd (P/Y)	P/Y = 12	checks the P/Y register
4	ENTER	P/Y = 4	changes the value to "4"
	↓	C/Y = 4	changed automatically to match the P/Y
	2nd (QUIT)	0	returns to the standard calculation mode
6000	± PV	PV = 6000	this enters the present value P (principal) with the correct sign convention
10	I/Y	I/Y = 10	this enters the periodic interest rate as a percent
20	N	N = 20	this enters the number of compounding periods n
	CPT FV	FV = 9831.698642	this computes and displays the unknown future value S

The future value is $9831.70.

Cash Flow Sign Convention for Entering Numerical Data

The Texas Instruments BAII Plus calculator follows the established convention of treating cash inflows (cash received) as positive numbers and cash outflows (cash paid out) as negative numbers. In the calculation above, the present value was considered to be cash paid out for an investment and so the present value of 6000 was entered as a negative number. The resulting future value was considered to be cash received from the investment and so had a positive value. Note that if the present value had been entered as a positive value, then the future value would have been displayed as a negative number. The *numerical* value would have been correct but the result would have been a negative number. "Error 5" is displayed when calculating i or n if both the present value and future value are entered using the same sign. Therefore, to avoid errors, always enter the present value as a negative number for compound interest calculations. Enter all other values as positive numbers. This topic is discussed further in Appendix II and throughout this text as required.

Excel

Excel has a ***Future Value (FV)*** function you can use to calculate the future value of an investment subject to compound interest. Refer to **FV** on the Spreadsheet Template Disk to learn how to use this Excel function.

C. Finding the future value when n is a fractional value

The value of n in the compounding factor $(1 + i)^n$ is not restricted to integral values; n may take any fractional value. The future value can be determined by means of the formula $FV = PV(1 + i)^n$ whether the time period contains an integral number of conversion periods or not.

EXAMPLE 3.2D

Find the accumulated value of $1000.00 invested for two years and nine months at 10% p.a. compounded annually.

SOLUTION

The entire time period is 2 years and 9 months; the number of whole conversion periods is 2; the fractional conversion period is $^9/_{12}$ of a year.

$PV = 1000.00;\ I/Y = 10;\ P/Y = 1;\ i = 10\% = 0.10;\ n = 2\frac{9}{12} = 2.75$
$FV = 1000.00(1.10)^{2.75} = 1000.00(1.299660) = \$1299.66$

Programmed Solution

(Set P/Y = 1) 2nd (CLR TVM) 1000 ± PV 10 I/Y
2.75 N CPT FV 1299.660393

EXAMPLE 3.2E

Determine the compound amount of $400.00 invested at 6% p.a. compounded quarterly for three years and five months.

SOLUTION

$PV = 400.00;\quad i = 1.5\% = 0.015$

$n = \left(3\frac{5}{12}\right)(4) = \left(\frac{41}{12}\right)(4) = \frac{41}{3} = 13\frac{2}{3} = 13.666667$

$FV = 400.00(1.015)^{13.666667} = 400.00(1.225658) = \$490.26$

Programmed Solution

(Set P/Y = 4) 2nd (CLR TVM) 400 ± PV 6 I/Y
13.666667 N CPT FV 490.2631345

EXAMPLE 3.2F Find the maturity value of a promissory note for $2000.00 dated February 1, 2000, and due on October 1, 2006, if interest is 7.5% p.a. compounded semi-annually.

SOLUTION PV = 2000.00; I/Y = 7.5; P/Y = 2; $i = \frac{7.5\%}{2} = 3.75\% = 0.0375$;

The time period February 1, 2000 to October 1, 2006 contains 6 years and 8 months: $n = \left(6\frac{8}{12}\right)(2) = \left(\frac{80}{12}\right)(2) = 13.333333.$

$FV = 2000.00(1.0375)^{13.333333} = 2000.00(1.633709) = \$3267.42$

Programmed Solution

(Set P/Y = 2) [2nd] (CLR TVM) 2000 [±] [PV] 7.5 [I/Y]

13.333333 [N] [CPT] [FV] [3267.418274]

EXAMPLE 3.2G A debt of $3500.00 dated August 31, 2005 is payable together with interest at 9% p.a. compounded quarterly on June 30, 2008. Determine the amount to be paid.

SOLUTION PV = 3500.00; I/Y = 9; P/Y = 4; $i = \frac{9\%}{4} = 2.25\% = 0.0225$; the time period August 31, 2005 to June 30, 2008, contains 2 years and 10 months; the number of quarters $n = 11.333333$.

$$\begin{aligned} FV &= 3500.00(1.0225)^{11.333333} \\ &= 3500.00(1.286819) \\ &= \$4503.87 \end{aligned}$$

Programmed Solution

(Set P/Y = 4) [2nd] (CLR TVM) 3500 [±] [PV] 9 [I/Y]

11.333333 [N] [CPT] [FV] [4503.867756]

D. Applications involving changes in interest rate or principal

EXAMPLE 3.2H A deposit of \$2000.00 earns interest at 6% p.a. compounded monthly for four years. At that time, the interest rate changes to 7% p.a. compounded quarterly. What is the value of the deposit three years after the change in the rate of interest?

SOLUTION The data given can be represented on a time diagram as shown in Figure 3.2.

FIGURE 3.2 Graphical Representation of Data

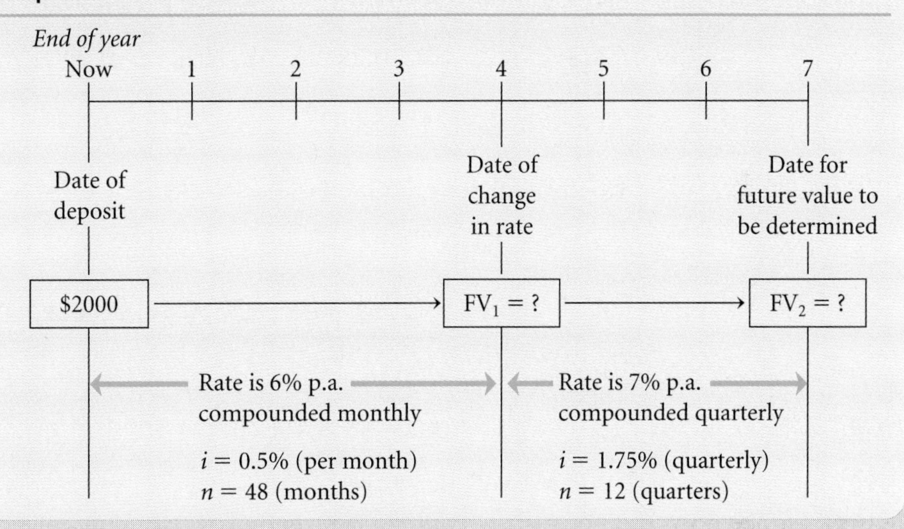

STEP 1 Determine the accumulated value of the original deposit at the time the interest rate changes, that is, after four years.

$$\text{PV} = 2000.00;\ \text{I/Y} = 6;\ \text{P/Y} = 12;\ i = \frac{6\%}{12} = 0.5\% = 0.005;\ n = 48$$

$$\text{FV}_1 = 2000.00(1 + 0.005)^{48} = 2000.00(1.270489) = \$2540.98$$

STEP 2 Use the accumulated value after four years as new principal and calculate its accumulated value three years later using the new rate of interest.

$$\text{PV} = 2540.98;\ \text{I/Y} = 7;\ \text{P/Y} = 4;\ i = \frac{7\%}{4} = 1.75\% = 0.0175;\quad n = 12$$

$$\text{FV}_2 = 2540.98(1 + 0.0175)^{12} = 2540.98(1.231439) = \$3129.06$$

Solution by Preprogrammed Calculator

Keys are applicable to Texas Instruments BAII Plus calculator (this is the case throughout the rest of the chapter).

STEP 1

Key in	Press	Display shows	
	2nd (P/Y)		
12	ENTER	P/Y = 12	
	↓	C/Y = 12	
	2nd (QUIT)	0	
	2nd (CLR TVM)	0	
2000	± PV	− 2000	
6	I/Y	6	
48	N	48	
	CPT FV	2540.978322	answer to Step 1 ($FV_1$2540.978322)

Do *not* clear your display. Proceed to Step 2.

STEP 2

Key in	Press	Display shows	
	± PV	−2540.978322	this step enters the new principal and the proper sign convention, since this amount is reinvested (a cash outflow)
	2nd (P/Y)	P/Y = 12	
4	ENTER	P/Y = 4	
	↓	C/Y = 4	
	2nd (QUIT)	0	
	RCL FV		
7	I/Y	7	
12	N	12	
	CPT FV	3129.060604	final answer (FV_2=$3129.06)

EXAMPLE 3.2I

A debt of $500 accumulates interest at 8% p.a. compounded quarterly from April 1, 2004 to July 1, 2005, and 9% p.a. compounded monthly thereafter. Determine the accumulated value of the debt on December 1, 2006.

SOLUTION

STEP 1

Determine the accumulated value of the debt on July 1, 2005.

$PV = 500.00$; $I/Y = 8$; $P/Y = 4$; $i = \frac{8\%}{4} = 2\% = 0.02$;

the period April 1, 2004 to July 1, 2005 contains 15 months: $n = 5$

$FV_1 = 500.00(1.02)^5 = 500.00(1.104081) = \$552.04$

STEP 2 Use the result of Step 1 as new principal and find its accumulated value on December 1, 2006.

$PV = 552.04$; $I/Y = 9$; $P/Y = 12$; $i = \frac{9\%}{12} = 0.75\% = 0.0075$;

the period July 1, 2005 to December 1, 2006 contains 17 months: $n = 17$

$FV_2 = 552.04(1.0075)^{17} = 552.04(1.135445) = \$626.81$

Programmed Solution

STEP 1 (Set P/Y = 4) [2nd] (CLR TVM) 500 [±] [PV] 8 [I/Y] 5 [N] [CPT] [FV]

Result: [552.040402]

STEP 2 552.040402 [±] [PV] (Set P/Y = 12) 9 [I/Y] 17 [N] [CPT] [FV]

Result: [626.811268]

EXAMPLE 3.2J Jay opened a registered retirement savings plan account with his credit union on February 1, 1997 with a deposit of \$2000.00. He added \$1900.00 on February 1, 1998 and another \$1700.00 on February 1, 2001. What will his account amount to on August 1, 2007 if the plan earns a fixed rate of interest of 7% p.a. compounded semi-annually?

SOLUTION

FIGURE 3.3 **Graphical Representation of Data**

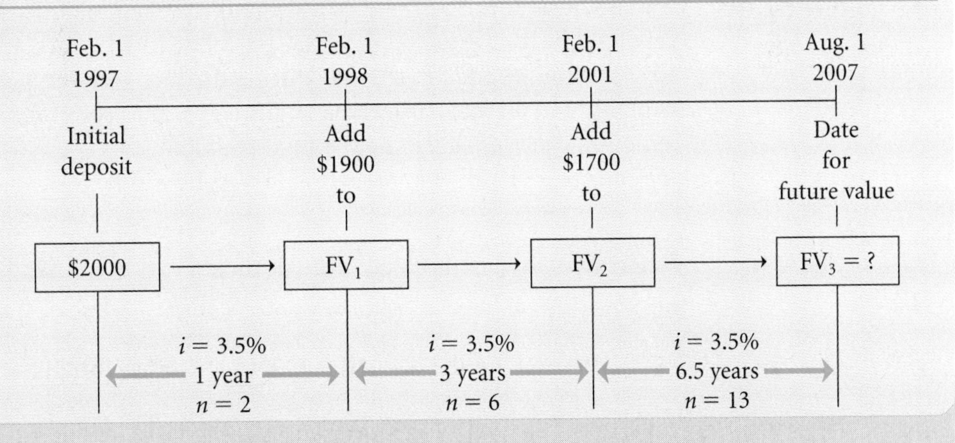

STEP 1 Determine the future value, FV_1, of the initial deposit on February 1, 1998.

$PV = 2000.00$; $I/Y = 7$; $P/Y = 2$; $i = \frac{7\%}{2} = 3.5\% = 0.035$;

the period February 1, 1997 to February 1, 1998 contains 1 year: $n = 2$

$FV_1 = 2000.00(1.035)^2 = 2000.00(1.071225) = \$2142.45$

STEP 2 Add the deposit of \$1900.00 to the amount of \$2142.45 to obtain the new principal as of February 1, 1998, and determine its future value, FV_2, on February 1, 2001.

$PV = 2142.45 + 1900.00 = 4042.45; \quad i = 0.035;$
the period February 1, 1998 to February 1, 2001 contains 3 years: $n = 6$
$FV_2 = 4042.45(1.035)^6 = 4042.45(1.229255) = \$4969.20$

STEP 3 Add the deposit of $1700.00 to the amount of $4969.20 to obtain the new principal as of February 1, 2001, and determine its future value, FV_3, on August 1, 2007.

$PV = 4969.20 + 1700.00 = 6669.20; \quad i = 0.035;$
the period February 1, 2001 to August 1, 2007 contains 6.5 years: $n = 13$
$FV_3 = 6669.20(1.035)^{13} = 6669.20(1.563956) = \$10\,430.34$

Programmed Solution

STEP 1 (Set P/Y = 2) 2nd (CLR TVM) 2000 ± PV 7 I/Y 2 N CPT FV

Result: 2142.45

STEP 2 + 1900 = 4042.45 ± PV 6 N CPT FV

Result: 4969.203194

STEP 3 + 1700 = 6669.203194 ± PV 13 N CPT FV

Result: 10 430.34075

Note: There is no need to key in the interest rate in Steps 2 and 3—it is already programmed from Step 1.

EXAMPLE 3.2K

A demand loan of $10 000.00 is repaid by payments of $5000.00 in one year, $6000.00 in four years, and a final payment in six years. Interest on the loan is 10% p.a. compounded quarterly during the first year, 8% p.a. compounded semi-annually for the next three years, and 7.5% p.a. compounded annually for the remaining years. Determine the final payment.

SOLUTION

STEP 1 Determine the accumulated value of the debt at the time of the first payment.

$PV = 10\,000.00; \; I/Y = 10; \; P/Y = 4; \; i = \frac{10\%}{4} = 2.5\% = 0.025; \; n = 4$

$FV_1 = 10\,000.00(1.025)^4 = 10\,000.00(1.10381289) = \$11\,038.13$

STEP 2 Subtract the payment of $5000.00 from the accumulated value of $11 038.13 to obtain the debt balance. Now determine its accumulated value at the time of the second payment three years later.

FIGURE 3.4 Graphical Representation of Data

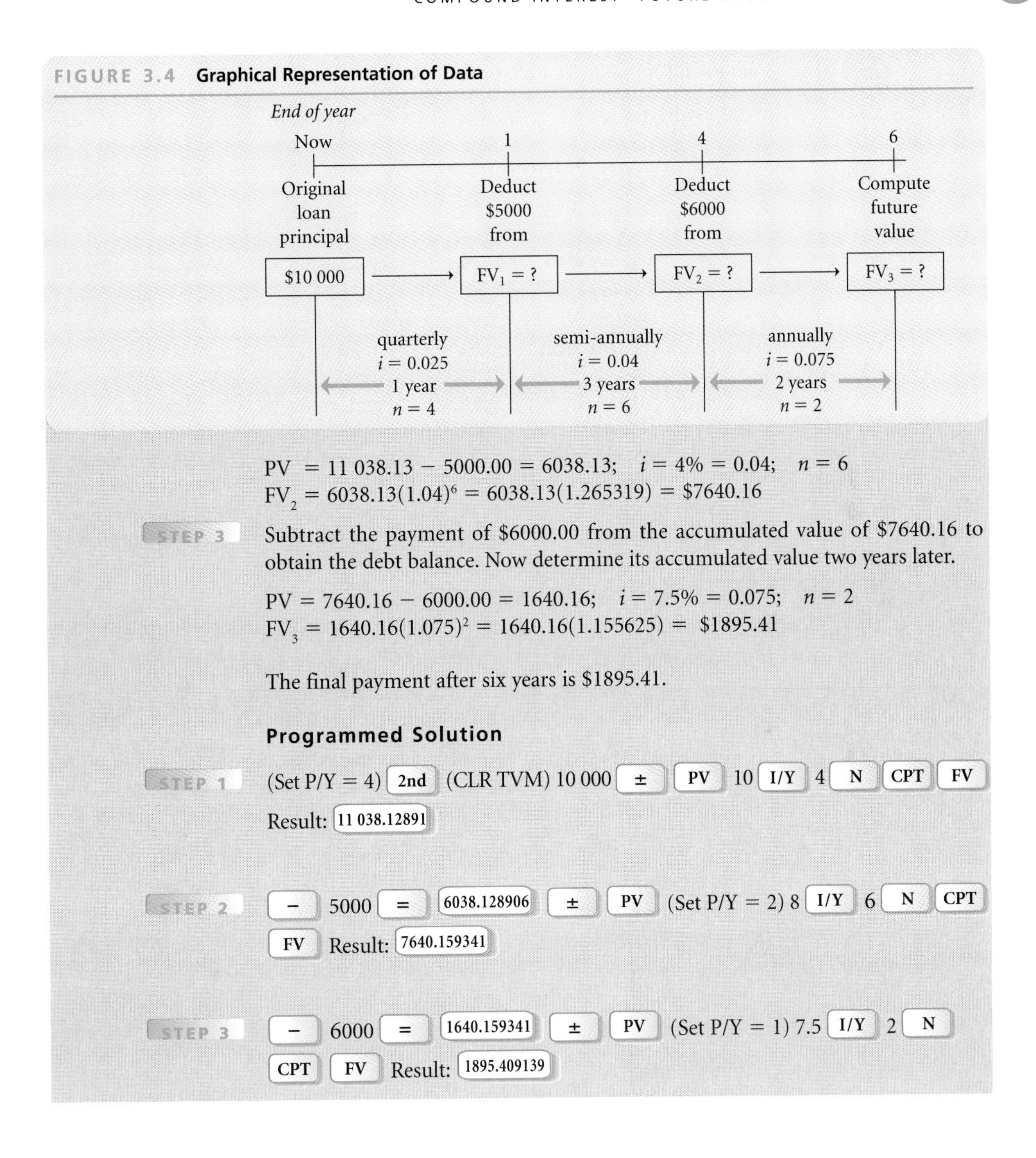

$$PV = 11\,038.13 - 5000.00 = 6038.13; \quad i = 4\% = 0.04; \quad n = 6$$
$$FV_2 = 6038.13(1.04)^6 = 6038.13(1.265319) = \$7640.16$$

STEP 3 Subtract the payment of $6000.00 from the accumulated value of $7640.16 to obtain the debt balance. Now determine its accumulated value two years later.

$$PV = 7640.16 - 6000.00 = 1640.16; \quad i = 7.5\% = 0.075; \quad n = 2$$
$$FV_3 = 1640.16(1.075)^2 = 1640.16(1.155625) = \$1895.41$$

The final payment after six years is $1895.41.

Programmed Solution

STEP 1 (Set P/Y = 4) 2nd (CLR TVM) 10 000 ± PV 10 I/Y 4 N CPT FV

Result: 11 038.12891

STEP 2 − 5000 = 6038.128906 ± PV (Set P/Y = 2) 8 I/Y 6 N CPT FV Result: 7640.159341

STEP 3 − 6000 = 1640.159341 ± PV (Set P/Y = 1) 7.5 I/Y 2 N CPT FV Result: 1895.409139

EXERCISE 3.2

Excel

If you choose, you can use Excel's ***Future Value (FV)*** function to calculate the future value of an investment subject to compound interest. Refer to **FV** on the Spreadsheet Template Disk to learn how to use this Excel function.

  A. Find the future value for each of the investments in the table below.

	Principal	Nominal Rate	Frequency of Conversion	Time
1.	$ 400.00	7.5%	annually	8 years
2.	1000.00	3.5%	semi-annually	12 years
3.	1250.00	6.5%	quarterly	9 years
4.	500.00	12%	monthly	3 years
5.	1700.00	8%	quarterly	14.75 years
6.	840.00	5.5%	semi-annually	8.5 years
7.	2500.00	8%	monthly	12.25 years
8.	150.00	10.8%	quarterly	27 months
9.	480.00	9.4%	semi-annually	42 months
10.	1400.00	4.8%	monthly	18.75 years
11.	2500.00	7%	annually	7 years, 6 months
12.	400.00	9%	quarterly	3 years, 8 months
13.	1300.00	5%	semi-annually	9 years, 3 months
14.	4500.00	3.5%	monthly	7.5 months

 B. Answer each of the following questions.

1. What is the maturity value of a five-year term deposit of $5000.00 at 6.5% compounded semi-annually? How much interest did the deposit earn?

2. How much will a registered retirement savings deposit of $1500.00 be worth in 15 years at 8% compounded quarterly? How much of the amount is interest?

3. You made a registered retirement savings plan deposit of $1000.00 on December 1, 2000 at a fixed rate of 5.5% compounded monthly. If you withdraw the deposit on August 1, 2007, how much will you receive?

4. Roy's parents made a trust deposit of $500.00 on October 31, 1986 to be withdrawn on Roy's eighteenth birthday on July 31, 2004. To what will the deposit amount on that date at 7% compounded quarterly?

5. What is the accumulated value of $100.00 invested for eight years at 9% p.a. compounded
   (a) annually? **(b)** semi-annually? **(c)** quarterly? **(d)** monthly?

6. To what future value will a principal of $500.00 amount in five years at 7.5% p.a. compounded
   (a) annually? **(b)** semi-annually? **(c)** quarterly? **(d)** monthly?

7. What is the future value of and the amount of compound interest for $100.00 invested at 8% compounded quarterly for
   (a) 5 years? **(b)** 10 years? **(c)** 20 years?

8. Find the future value of and the compound interest on $500.00 invested at 4.5% compounded monthly for
   (a) 3.5 years; **(b)** 6 years; **(c)** 11.5 years.

9. A demand loan for $5000.00 with interest at 9.75% compounded semi-annually is repaid after five years, ten months. What is the amount of interest paid?

10. Suppose $4000.00 is invested for four years, eight months at 8.5% compounded annually. What is the compounded amount?

11. Determine the maturity value of a $600.00 promissory note dated August 1, 2000 and due on June 1, 2005, if interest is 5% p.a. compounded semi-annually.

12. Find the maturity value of a promissory note for $3200.00 dated March 31, 1998 and due on August 31, 2004, if interest is 7% compounded quarterly.

13. A debt of $8000.00 is payable in seven years and five months. Determine the accumulated value of the debt at 10.8% p.a. compounded annually.

14. A $6000.00 investment matures in three years, eleven months. Find the maturity value if interest is 9% p.a. compounded quarterly.

15. The Canadian Consumer Price Index was approximately 98.5 (base year 1992) at the beginning of 1991. If inflation continued at an average annual rate of 3%, what would the index have been at the beginning of 2004?

16. Peel Credit Union expects an average annual growth rate of 20% for the next five years. If the assets of the credit union currently amount to $2.5 million, what will the forecasted assets be in five years?

17. A local bank offers $5000.00 five-year certificates at 6.75% compounded semi-annually. Your credit union makes the same type of deposit available at 6.5% compounded monthly.

(**a**) Which investment gives more interest over the five years?
(**b**) What is the difference in the amount of interest?

18. The Continental Bank advertises capital savings at 7.25% compounded semi-annually while TD Canada Trust offers premium savings at 7% compounded monthly. Suppose you have $1000.00 to invest for two years.

(**a**) Which deposit will earn more interest?
(**b**) What is the difference in the amount of interest?

Answer each of the following questions.

1. A deposit of $2000.00 earns interest at 7% p.a. compounded quarterly. After two and a half years, the interest rate is changed to 6.75% compounded monthly. How much is the account worth after six years?

2. An investment of $2500.00 earns interest at 4.5% p.a. compounded monthly for three years. At that time the interest rate is changed to 5% compounded quarterly. How much will the accumulated value be one-and-a-half years after the change?

3. A debt of $800.00 accumulates interest at 10% compounded semi-annually from February 1, 2002 to August 1, 2004, and 11% compounded quarterly thereafter. Determine the accumulated value of the debt on November 1, 2007.

4. Accumulate $1300.00 at 8.5% p.a. compounded monthly from March 1, 2002 to July 1, 2004, and thereafter at 8% p.a. compounded quarterly. What is the amount on April 1, 2007?

5. Pat opened an RRSP deposit account on December 1, 2002 with a deposit of $1000.00. He added $1000.00 on July 1, 2003 and $1000.00 on November 1, 2004. How much is in his account on January 1, 2006, if the deposit earns 6% p.a. compounded monthly?

6. Terri started an RRSP on March 1, 2000 with a deposit of $2000.00. She added $1800.00 on December 1, 2002 and $1700.00 on September 1, 2004. What is the accumulated value of her account on December 1, 2007, if interest is 7.5% compounded quarterly?

7. A debt of $4000.00 is repaid by payments of $1500.00 in nine months, $2000.00 in 18 months, and a final payment in 27 months. If interest was 10% compounded quarterly, what was the amount of the final payment?

8. Sheridan Service has a line of credit loan with the bank. The initial loan balance was $6000.00. Payments of $2000.00 and $3000.00 were made after four months and nine months respectively. At the end of one year, Sheridan Service borrowed an additional $4000.00. Six months later, the line of credit loan was converted into a collateral mortgage loan. What was the amount of the mortgage if the line of credit interest was 9% compounded monthly?

9. A demand loan of $3000.00 is repaid by payments of $1500.00 after two years, $1500.00 after four years, and a final payment after seven years. Interest is 9% compounded quarterly for the first year, 10% compounded semi-annually for the next three years, and 10% compounded monthly thereafter. What is the size of the final payment?

10. A variable rate demand loan showed an initial balance of $12 000.00, payments of $5000.00 after eighteen months, $4000.00 after thirty months, and a final payment after five years. Interest was 11% compounded semi-annually for the first two years and 12% compounded monthly for the remaining time. What was the size of the final payment?

3.3 Present Value and Compound Discount

A. The present value concept and related terms

EXAMPLE 3.3A Find the principal that will amount in six years to $17 715.61 at 10% p.a. compounded annually.

SOLUTION The problem may be graphically represented as shown in Figure 3.5.

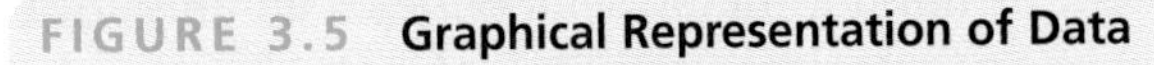

FIGURE 3.5 **Graphical Representation of Data**

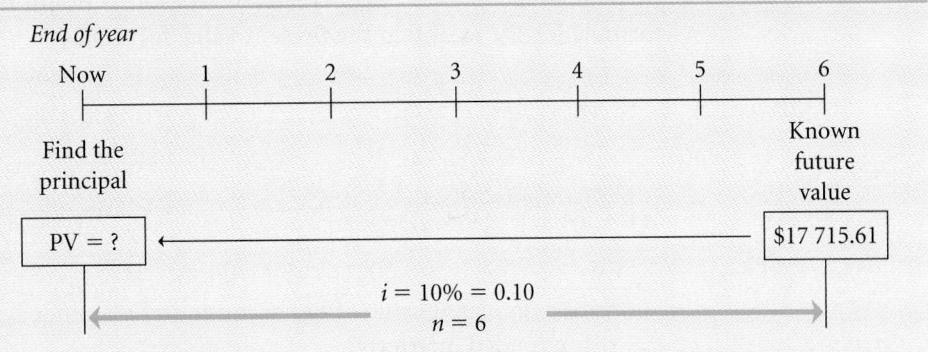

This problem is the inverse of the problem used to illustrate the meaning of compound interest. Instead of knowing the value of the principal and finding its future value, we know that the future value is $17 715.61. What we want to determine is the value of the principal.

To solve the problem, we use the future value formula $FV = PV(1 + i)^n$ and substitute the known values.

$FV = 17\,715.61$; $I/Y = 10$; $P/Y = 1$; $i = \frac{10\%}{1} = 10\% = 0.10$; $n = 6$

$17\,715.61 = PV(1.10)^6$ ——— by substituting in $FV = PV(1 + i)^n$

$17\,715.61 = PV(1.771561)$ ——— computing $(1.10)^6$

$PV = \frac{17\,715.61}{1.771561}$ ——— solve for PV by dividing both sides by 1.771561

$PV = \$10\,000.00$

The principal that will grow to $17715.61 in six years at 10% p.a. compounded annually is $10 000.00.

This principal is called the **present value** or **discounted value** or **proceeds** of the known future amount.

The difference between the known future amount of $17 715.61 and the computed present value (principal) of $10 000.00 is the **compound discount** and represents the compound interest accumulating on the computed present value.

The process of computing the present value or discounted value or proceeds is called **discounting.**

B. The present value formula

The present value of an amount at a given time at compound interest is defined as the principal that will grow to the given amount if compounded at a given periodic rate of interest over a given number of conversion periods.

Since the problem of finding the present value is equivalent to finding the principal when the future value, the periodic rate of interest, and the number of conversion periods are given, the formula for the future value formula, $FV = PV(1 + i)^n$, applies.

However, because the problem of finding the present value of an amount is frequently encountered in financial analysis, it is useful to solve the future value formula for PV to obtain the present value formula.

$$\text{FV} = \text{PV}(1+i)^n$$ ——— start with the future value formula, Formula 3.1A

$$\frac{\text{FV}}{(1+i)^n} = \frac{\text{PV}(1+i)^n}{(1+i)^n}$$ ——— divide both sides by the compounding factor $(1+i)^n$

$$\frac{\text{FV}}{(1+i)^n} = \text{PV}$$ ——— reduce the fraction $\frac{(1+i)^n}{(1+i)^n}$ to 1

The present value formula for compound interest is:

$$\boxed{\text{PV} = \frac{\text{FV}}{(1+i)^n}}$$ ——— **Formula 3.1B**

EXAMPLE 3.3B

Find the present value of \$6836.56 due in nine years at 6% p.a. compounded quarterly.

SOLUTION

FV = 6836.56; I/Y = 6; P/Y = 4; $i = \frac{6\%}{4} = 1.5\% = 0.015$; $n = 36$

$$\text{PV} = \frac{\text{FV}}{(1+i)^n}$$ ——— using the present value formula

$$= \frac{6836.56}{(1+0.015)^{36}}$$ ——— by substitution

$$= \frac{6836.56}{1.709140}$$

$$= \$4000.00$$

Note: The division of 6836.56 by 1.7091395, like any division, may be changed to a multiplication by using the reciprocal of the divisor.

$$\frac{6836.56}{1.709140}$$ ——— the division to be changed into a multiplication

$$= 6836.56\left(\frac{1}{1.709140}\right)$$ ——— the reciprocal of the divisor 1.709140 is found by dividing 1 by 1.709140

$$= 6836.56(0.585090)$$ ——— computed value of the reciprocal

$$= \$4000.00$$

For calculators equipped with the reciprocal function key [1/x], converting the division into a multiplication is easily accomplished by first computing the compounding factor and then using the [1/x] key to obtain the reciprocal.

EXAMPLE 3.3C

What principal will amount to \$5000.00 seven years from today if interest is 9% p.a. compounded monthly?

SOLUTION

Finding the principal that amounts to a future sum of money is equivalent to finding the present value.

$$FV = 5000.00;\ I/Y = 9;\ P/Y = 12;\ i = \frac{9\%}{12} = 0.75\% = 0.0075;\quad n = 84$$

$$PV = \frac{5000.00}{(1.0075)^{84}} \text{ ——— using Formula 3.1B}$$

$$= \frac{5000.00}{1.873202} \text{ ——— computing the factor } (1.0075)^{84}$$

$$= 5000.00(0.533845) \text{ ——— using the reciprocal function key}$$

$$= \$2669.23$$

Using the reciprocal of the divisor to change division into multiplication is reflected in the practice of stating the present value formula with a negative exponent.

$$\frac{1}{a^n} = a^{-n} \text{ ——— negative exponent rule}$$

$$\frac{1}{(1+i)^n} = (1+i)^{-n}$$

$$\frac{FV}{(1+i)^n} = FV(1+i)^{-n}$$

Formula 9.1B, the present value formula, can be restated in multiplication form using a negative exponent.

$$\boxed{PV = FV(1+i)^{-n}} \text{ ——— Formula 3.1C}$$

The factor $(1 + i)^{-n}$ is called the **discount factor** and is the reciprocal of the compounding factor $(1 + i)^n$.

C. Using preprogrammed financial calculators to find present value

As explained in Section 3.2B, preprogrammed calculators provide quick solutions to compound interest calculations. Three of the four variables are entered and the value of the fourth variable is retrieved.

To solve Example 3.3C, in which FV = 5000, i = 0.75%, n = 84, and PV is to be determined, use the following procedure. (Remember that the interest rate for the year, 9% must be entered into the calculator as I/Y.)

Key in	Press	Display shows	
	2nd (CLR TVM)	0	clears the function key registers
	2nd (P/Y)	0	checks the P/Y register

12	ENTER	P/Y = 12	changes the value to "12"
	↓	C/Y = 12	changed automatically to match the P/Y
	2nd (QUIT)	0	returns to the standard calculation mode
5000	FV	5000	enters the future value amount FV
9	I/Y	9	enters the conversion rate i
84	N	84	enters the number of compounding periods n
	CPT PV	−2669.226329	retrieves the unknown principal (present value) PV, an investment or cash outflow as indicated by the negative sign

The principal is $2669.23.

Excel

Excel has a ***Present Value (PV)*** function you can use to calculate the present value of an investment subject to compound interest. Refer to **PV** on the Spreadsheet Template Disk to learn how to use this Excel function.

D. Finding the present value when *n* is a fractional value

Use Formula 3.1A, $FV = PV(1+i)^n$, or Formula 3.1B, $PV = \frac{FV}{(1+i)^n}$, or Formula 3.1C, $PV = FV(1+i)^{-n}$ where n is a fractional value representing the *entire* time period, FV is a known value, and PV is to be determined.

EXAMPLE 3.3D

Find the present value of $2000.00 due in three years and eight months if money is worth 8% p.a. compounded quarterly.

SOLUTION

$FV = 2000.00$; $I/Y = 8$; $P/Y = 4$; $i = \frac{8\%}{4} = 2\% = 0.02$;

$n = \left(3\frac{8}{12}\right)(4) = 14\frac{2}{3} = 14.666667$

$PV = \frac{FV}{(1+i)^n}$ ——— using Formula 3.1B

$= \frac{2000.00}{(1+0.02)^{14.666667}}$ ——— use as many decimals as are available in your calculator

$= 2000.00(0.747936)$ ——— multiply by the reciprocal

$= \$1495.87$

Programmed Solution

(Set P/Y = 4) 2nd (CLR TVM) 2000 FV 8 I/Y

14.666667 N CPT PV -1495.870992

EXAMPLE 3.3E

Determine the principal that will accumulate to \$2387.18 from September 1, 2000 to April 1, 2004 at 5% p.a. compounded semi-annually.

SOLUTION

Finding the principal that will grow to the given amount of \$2387.18 is equivalent to finding the present value or discounted value of this amount.

The time period September 1, 2000 to April 1, 2004 contains three years and seven months; that is, it consists of seven whole conversion periods of six months each and a fractional conversion period of one month.

Use $PV = \frac{FV}{(1+i)^n}$

$FV = 2387.18$; $I/Y = 5$; $P/Y = 2$; $i = \frac{5\%}{2} = 2.5\% = 0.025$;

$n = \left(3\frac{7}{12}\right)(2) = 7\frac{1}{6} = 7.166667$

$$PV = \frac{2387.18}{(1.025)^{7.166667}}$$
$$= \frac{2387.18}{1.193588}$$
$$= 2387.18(0.837810)$$
$$= \$2000.00$$

Programmed Solution

(Set P/Y = 2) [2nd] (CLR TVM) 2387.18 [FV] 5 [I/Y]

7.166667 [N] [CPT] [PV] [-2000.003696]

EXERCISE 3.3

If you choose, you can use Excel's ***Present Value (PV)*** function to calculate the present value of an investment subject to compound interest. Refer to **PV** on the Spreadsheet Template Disk to learn how to use this Excel function.

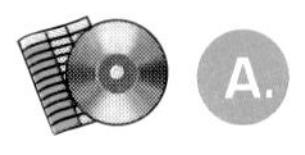

A. Find the present value of each of the following amounts.

	Amount	Nominal Rate	Frequency of Conversion	Time
1.	\$1000.00	8%	quarterly	7 years
2.	1500.00	6.5%	semi-annually	10 years
3.	600.00	8%	monthly	6 years
4.	350.00	7.5%	annually	8 years
5.	1200.00	9%	monthly	12 years
6.	3000.00	12.25%	semi-annually	5 years, 6 months
7.	900.00	6.4%	quarterly	9 years, 3 months
8.	500.00	8.4%	monthly	15 years
9.	1500.00	4.5%	annually	15 years, 9 months
10.	900.00	5.5%	semi-annually	8 years, 10 months
11.	6400.00	7%	quarterly	5 years, 7 months
12.	7200.00	6%	monthly	21.5 months

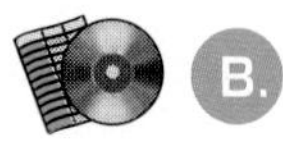

B. Answer each of the following questions.

1. Find the present value and the compound discount of $1600.00 due four-and-a-half years from now if money is worth 4% compounded semi-annually.
2. Find the present value and the compound discount of $2500.00 due in six years, three months, if interest is 6% compounded quarterly.
3. Find the principal that will amount to $1250.00 in five years at 10% p.a. compounded quarterly.
4. What sum of money will grow to $2000.00 in seven years at 9% compounded monthly?
5. A debt of $5000.00 is due November 1, 2011. What is the value of the obligation on February 1, 2005, if money is worth 7% compounded quarterly?
6. How much would you have to deposit in an account today to have $3000.00 in a five-year term deposit at maturity if interest is 7.75% compounded annually?
7. What is the principal that will grow to $3000.00 in eight years, eight months at 9% compounded semi-annually?
8. Find the sum of money that accumulates to $1600.00 at 5% compounded quarterly in six years, four months.

3.4 Application—Discounting Negotiable Financial Instruments at Compound Interest

A. Discounting long-term promissory notes

Long-term promissory notes (written for a term longer than one year) are usually subject to compound interest. As with short-term promissory notes, long-term promissory notes are negotiable and can be bought and sold (*discounted*) at any time before maturity.

The principles involved in discounting long-term promissory notes are similar to those used in discounting short-term promissory notes by the simple discount method *except* that no requirement exists to add three days of grace in determining the legal due date of a long-term promissory note.

The discounted value (or proceeds) of a long-term promissory note is the present value at the date of discount of the maturity value of the note. It is found using the present value formula $PV = \frac{FV}{(1 + i)^n}$ or $PV = FV(1 + i)^{-n}$.

For non-interest-bearing notes, the maturity value is the face value. However, for interest-bearing promissory notes, the maturity value must be determined first by using the future value formula $FV = PV(1 + i)^n$.

Like promissory notes, long-term bonds promise to pay a specific face value at a specified future point in time. In addition, there is a promise to periodically pay a specified amount of interest. Long-term bonds will be covered in detail in Chapter 9.

B. Discounting non-interest-bearing promissory notes

Since the face value of a non-interest-bearing note is also its maturity value, the proceeds of a non-interest-bearing note are the present value of its face value at the date of discount.

EXAMPLE 3.4A Determine the proceeds of a non-interest-bearing note for $1500.00 discounted two-and-a-quarter years before its due date at 9% p.a. compounded monthly.

SOLUTION The maturity value FV = 1500.00;

the rate of discount I/Y = 9; P/Y = 12; $i = \frac{9\%}{12} = 0.75\% = 0.0075$;

the number of conversion periods $n = (2.25)(12) = 27$.

$$\begin{aligned} PV &= FV(1+i)^{-n} \quad \text{——— using restated Formula 3.1C} \\ &= 1500.00(1+0.0075)^{-27} \\ &= 1500.00\left(\frac{1}{1.223535}\right) \\ &= 1500.00(0.817304) \\ &= \$1225.96 \end{aligned}$$

Programmed Solution

(Set P/Y = 12) [2nd] (CLR TVM) 1500 [FV] 9 [I/Y] 27 [N] [CPT] [PV] [-1225.955705]

EXAMPLE 3.4B A three-year, non-interest-bearing promissory note for $6000.00 dated August 31, 2004 was discounted on October 31, 2005 at 6% p.a. compounded quarterly. Determine the proceeds of the note.

SOLUTION The due date of the note is August 31, 2008; the discount period October 31, 2005 to August 31, 2008 contains 2 years and 10 months.

FV = 6000.00; I/Y = 6; P/Y = 4; $i = \frac{6\%}{4} = 1.5\% = 0.015$;

$$n = \left(2\frac{10}{12}\right)(4) = 11\tfrac{1}{3} = 11.333333$$

$$\begin{aligned} PV &= FV(1+i)^{-n} \\ &= 6000.00(1+0.015)^{-11.333333} \\ &= 6000.00(0.844731) \\ &= \$5068.38 \end{aligned}$$

Programmed Solution

(Set P/Y = 4) [2nd] (CLR TVM) 6000 [FV] 6 [I/Y] 11.333333 [N] [CPT] [PV] [-5068.383148]

EXAMPLE 3.4C

You signed a promissory note at the Continental Bank for \$3000.00 due in 27 months. If the bank charges interest at 8% p.a. compounded semi-annually, determine the proceeds of the note.

SOLUTION

The amount shown on the note is the sum of money due in 27 months, that is, the maturity value of the note.

FV = 3000.00; I/Y = 8; P/Y = 2; $i = \frac{8\%}{2} = 4\% = 0.04$; $n = \left(\frac{27}{12}\right)(2) = 4.5$

$$
\begin{aligned}
PV &= FV(1 + i)^{-n} \\
&= 3000.00(1 + 0.04)^{-4.5} \\
&= 3000.00(0.838205) \\
&= \$2514.61
\end{aligned}
$$

Programmed Solution

(Set P/Y = 2) [2nd] (CLR TVM) 3000 [FV] 8 [I/Y] 4.5 [N] [CPT] [PV] [-2514.613414]

C. Discounting interest-bearing promissory notes

The proceeds of an interest-bearing note are equal to the present value at the date of discount of the value of the note at maturity. Therefore, the maturity value of an interest-bearing promissory note must be determined before finding the discounted value.

EXAMPLE 3.4D

Determine the proceeds of a promissory note for \$3600.00 with interest at 6% p.a. compounded quarterly, issued September 1, 2004, due on June 1, 2010, and discounted on December 1, 2006 at 8% p.a. compounded semi-annually.

SOLUTION

STEP 1

Find the maturity value of the note using Formula 9.1A, $FV = PV(1 + i)^n$.

PV = 3600.00; I/Y = 6; P/Y = 4; $i = \frac{6\%}{4} = 1.5\% = 0.015$; the interest period, September 1, 2004 to June 1, 2010, contains 5 years and 9 months: $n = (5\frac{9}{12})(4) = 23$.

$$
\begin{aligned}
FV &= 3600.00(1 + 0.015)^{23} \\
&= 3600.00(1.408377) \\
&= \$5070.16
\end{aligned}
$$

STEP 2

Find the present value at the date of discount of the maturity value found in Step 1 using $PV = FV(1 + i)^{-n}$.

FV= 5070.16; I/Y = 8; P/Y = 2; $i = \frac{8\%}{2} = 4\% = 0.04$;

the discount period, December 1, 2006 to June 1, 2010, contains 3 years and 6 months: $n = (3\frac{6}{12})(2) = 7$.

$$\begin{aligned} PV &= 5070.16(1 + 0.04)^{-7} \\ &= 5070.16(0.759918) \\ &= \$3852.90 \end{aligned}$$

The proceeds of the note on December 1, 2006 are \$3852.90. The method and the data are represented graphically in Figure 3.6.

FIGURE 3.6 Graphical Representation of Method and Data

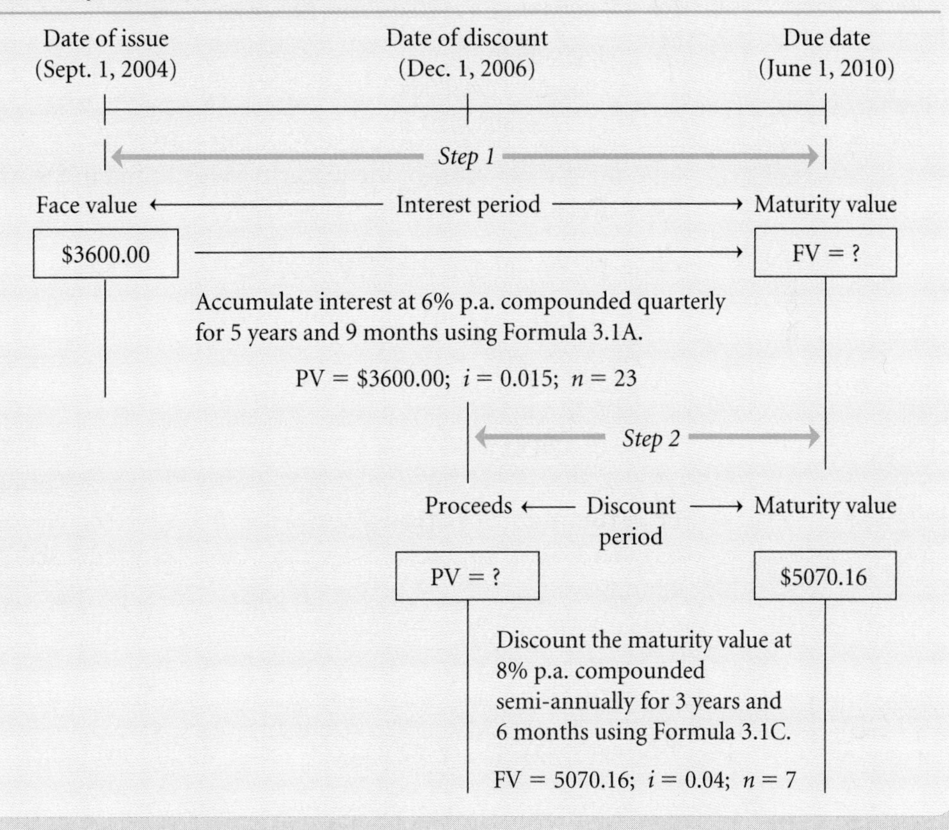

Programmed Solution

STEP 1

Result

(Set P/Y = 4) [2nd] (CLR TVM) 3600 [±] [PV] 6 [I/Y] 23 [N] [CPT] [FV] [5070.157757]

the answer to Step 1 is now programmed as [FV] for Step 2

STEP 2

Result

(Set P/Y = 2) 8 [I/Y] 7 [N] [CPT] [PV] [-3852.903195]

EXAMPLE 3.4E

A five-year note for \$8000.00 bearing interest at 6% p.a. compounded monthly is discounted two years and five months before the due date at 5% p.a. compounded semi-annually. Determine the proceeds of the note.

SOLUTION

STEP 1 Find the maturity value using $FV = PV(1 + i)^n$.

$PV = 8000.00;\ I/Y = 6;\ P/Y = 12;\ i = \frac{6\%}{12} = 0.5\% = 0.005;\ n = 60$

$$\begin{aligned} FV &= 8000.00(1.005)^{60} \\ &= 8000.00(1.348850) \\ &= \$10\,790.80 \end{aligned}$$

STEP 2 Find the present value of the maturity value found in Step 1 using $PV = FV(1 + i)^{-n}$.

$FV = 10\,790.80;\ I/Y = 5;\ P/Y = 2;\ i = \frac{5\%}{2} = 2.5\% = 0.025;$

$n = \left(\frac{29}{12}\right)(2) = 4.833333$

$$\begin{aligned} PV &= 10\,790.80(1.025)^{-4.833333} \\ &= 10\,790.80(0.887499) \\ &= \$9576.83 \end{aligned}$$

Programmed Solution

STEP 1

(Set P/Y = 12) 2nd (CLR TVM) 8000 ± PV 6 I/Y 60 N CPT FV 10 790.80122

STEP 2

(Set P/Y = 2) 5 I/Y 4.833333 N CPT PV -9576.82776

EXERCISE 3.4

Excel If you choose, you can use Excel's Future Value ***Future Value (FV)*** or ***Present Value (PV)*** functions to answer the questions below. Refer to **FV** and **PV** on the Spreadsheet Template Disk to learn how to use these Excel functions.

A. Find the proceeds and the compound discount for each of the long-term promissory notes shown in the table below. Note that the first six are non-interest-bearing promissory notes.

	Face Value	Date of Issue	Term	Int. Rate	Frequency of Conversion	Date of Discount	Discount Rate	Frequency of Conversion
1.	$2000.00	30-06-2000	5 years	—	—	31-12-2002	5%	semi-annually
2.	700.00	01-04-1998	10 years	—	—	01-07-2003	10%	quarterly
3.	5000.00	01-04-1998	10 years	—	—	01-08-2003	5%	annually
4.	900.00	31-08-1999	8 years	—	—	30-06-2004	4%	quarterly
5.	3200.00	31-03-2000	6 years	—	—	31-10-2003	8%	quarterly
6.	1450.00	01-10-1997	9 years	—	—	01-12-2002	6%	semi-annually
7.	780.00	30-09-1997	10 years	8%	annually	30-04-2001	8%	quarterly
8.	2100.00	01-02-1996	12 years	6%	monthly	01-07-2003	7%	semi-annually
9.	1850.00	01-11-2002	5 years	10%	quarterly	01-10-2004	9%	semi-annually
10.	3400.00	31-01-1999	7 years	9%	monthly	31-12-2003	7.5%	quarterly
11.	1500.00	31-05-1999	8 years	7%	annually	31-05-2004	8%	semi-annually
12.	4000.00	30-09-2001	4 years	5%	semi-annually	31-03-2003	4%	quarterly
13.	800.00	01-02-2000	7 yr., 9 mo.	9%	quarterly	01-11-2005	9%	monthly
14.	2200.00	31-10-1999	8.25 years	6%	monthly	31-01-2002	7%	quarterly

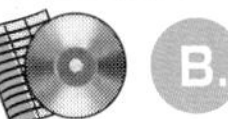

B. Find the proceeds of each of the following promissory notes.

1. A non-interest-bearing promissory note for $6000.00, discounted 54 months before its due date at 6% compounded quarterly.
2. A $4200.00, non-interest-bearing note due August 1, 2010 discounted on March 1, 2006 at 7.5% compounded monthly.
3. A promissory note with a maturity value of $1800.00 due on September 30, 2009 discounted at 8.5% compounded semi-annually on March 31, 2006.
4. A fifteen-year promissory note discounted after six years at 9% compounded quarterly with a maturity value of $7500.00.
5. A five-year promissory note for $3000.00 with interest at 8% compounded semi-annually, discounted 21 months before maturity at 9% compounded quarterly.
6. A $5000.00, seven-year note bearing interest at 8.0% compounded quarterly, discounted two-and-a-half years after the date of issue at 6.0% compounded monthly.
7. A six-year, $900.00 note bearing interest at 10% compounded quarterly, issued June 1, 2003, discounted on December 1, 2008 to yield 8.5% compounded semi-annually.
8. A ten-year promissory note dated April 1, 2000 with a face value of $1300.00 bearing interest at 7% compounded semi-annually, discounted seven years later when money was worth 9% compounded quarterly.

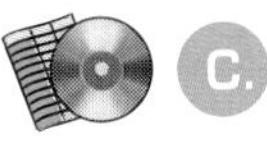

C. Solve each of the following problems:

1. Determine the proceeds of an investment with a maturity value of $10 000.00 if discounted at 9% compounded monthly 22.5 months before the date of maturity.

2. Compute the discounted value of $7000.00 due in three years, five months if money is worth 8% compounded quarterly.

3. Find the discounted value of $3800.00 due in six years, eight months if interest is 7.5% compounded annually.

4. Calculate the proceeds of $5500.00 due in seven years, eight months discounted at 4.5% compounded semi-annually.

5. A four-year non-interest-bearing promissory note for $3750.00 is discounted 32 months after the date of issue at 5.5% compounded semi-annually. Find the proceeds of the note.

6. A seven-year non-interest-bearing note for $5200.00 is discounted three years, eight months before its due date at 9% compounded quarterly. Find the proceeds of the note.

7. A non-interest-bearing eight-year note for $4500.00 issued August 1, 2002 is discounted April 1, 2006 at 6.5% compounded annually. Find the compound discount.

8. A $2800.00 promissory note issued without interest for five years on September 30, 2004 is discounted on July 31, 2007 at 8% compounded quarterly. Find the compound discount.

9. A six-year note for $1750.00 issued on December 1, 2001 with interest at 6.5% compounded annually is discounted on March 1, 2004 at 7% compounded semi-annually. What are the proceeds of the note?

10. A ten-year note for $1200.00 bearing interest at 6% compounded monthly is discounted at 8% compounded quarterly three years and ten months after the date of issue. Find the proceeds of the note.

11. Four years, seven months before its due date, a seven-year note for $2650.00 bearing interest at 9% compounded quarterly is discounted at 8% compounded semi-annually. Find the compound discount.

12. On April 15, 2004, a ten-year note dated June 15, 1999 is discounted at 10% compounded quarterly. If the face value of the note is $4000.00 and interest is 8% compounded quarterly, find the compound discount.

BUSINESS MATH NEWS BOX

Starting Early—The Power of Compound Growth

One of the most costly mistakes Canadians make with their RRSPs is not starting them early enough. According to a recent Angus Reid survey, only about 22% of eligible Canadians make their first RRSP contribution between the ages of 25 and 29. Fully 40% don't make that first contribution until they're between 30 and 44.

The longer you wait, the more you miss out on one of the key benefits of your registered plan: the incredible power of long-term tax-deferred compound growth.

The more years you have to invest, and the higher your investment return, the more dramatic this growth potential will be.

Watch Your Savings Grow

This table illustrates the effect of compound growth on a single investment of $1000. The higher the rate of return, and the longer you invest, the more powerful the effect.

Rate of Return	Years Invested		
	20	25	30
4%	$2 191*	$ 2 666	$ 3 243
6%	$3 207	$ 4 292	$ 5 743
8%	$4 661	$ 6 848	$10 063
10%	$6 727	$10 835	$17 449

*The effect of taxation is not taken into account in these examples.

QUESTIONS

1. What principal is used to calculate the figures in the table? Show all calculations
2. Suppose you were 25 years old when you made your first $1000 RRSP contribution. What would be the value of your investment when you reach age 60, given each of the following rates of return?

   (a) 4% (b) 6% (c) 8% (d) 10%

3. Compare the values of the investment you calculated in question 2 with the values of the investment made at age 30 given in the table. For each interest rate, what is the difference in the values of the investment at age 60?
4. Suppose you were 30 years old when you made your first $1000 RRSP contribution. Suppose your investment would earn 6% for the first ten years and 8% for the next twenty years. What would be the value of your contribution when you turned 60?

Source: Investor's Group. Used with permission.

3.5 Equivalent Values

A. Equations of value

Because of the time value of money, amounts of money have different values at different times, as explained in Chapter 1. When sums of money fall due or are payable at different times, they are not directly comparable. To make such sums of money comparable, a point in time—the **comparison date** or **focal date**—must be chosen. Allowance must be made for interest from the due dates of the sums of money to the selected focal date; that is, the dated values of the sums of money must be determined.

Any point in time may be chosen as the focal date; the choice does not affect the final answers. It is advisable, however, to choose a date on which an amount is unknown. The choice of date determines which formula to be used. For compound interest, equations of value need to be set up.

Which formula is appropriate depends on the position of the due dates relative to the focal date. The following rules apply:

(a) If an amount in the future of, or after, the focal date is to be determined, use the future value formula, $FV = PV(1 + i)^n$ (see Figure 3.7).

(b) If an amount in the past of, or before, the focal date is to be determined,use the present value formula, $PV = FV(1 + i)^{-n}$ (see Figure 3.8).

B. Finding the equivalent single payment

Equivalent values are the dated values of an original sum of money.

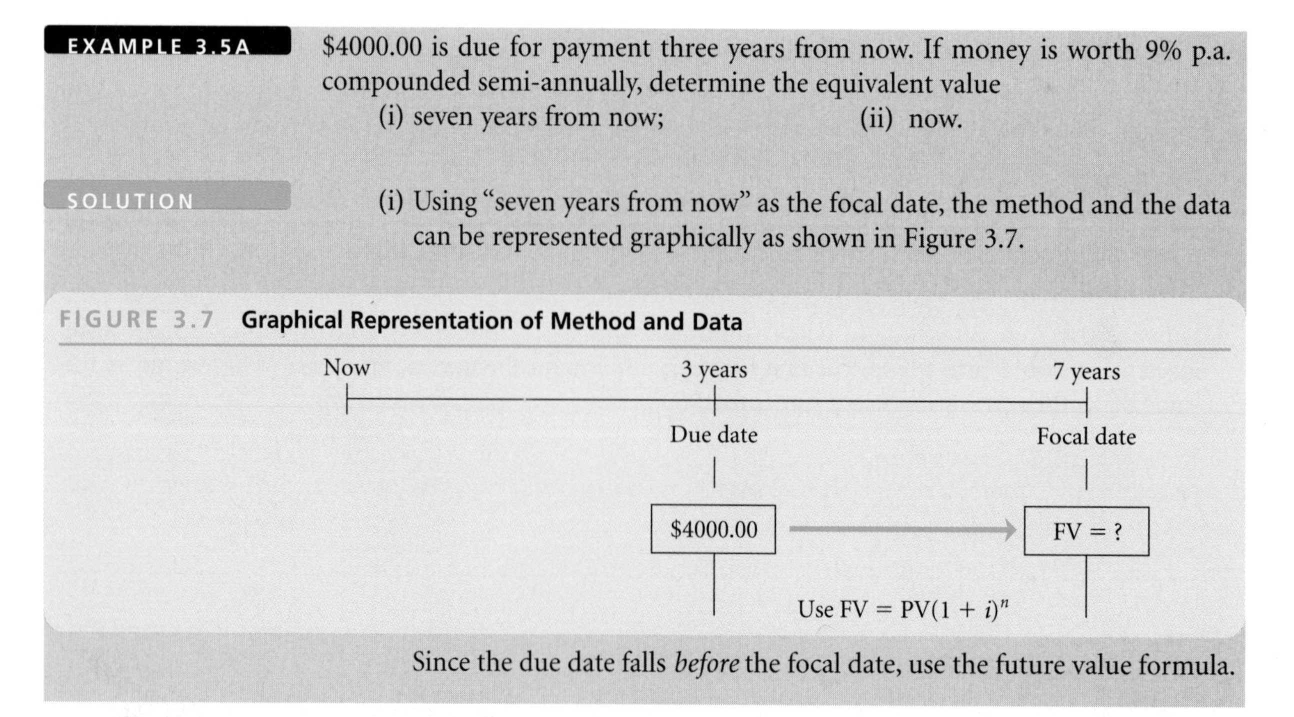

EXAMPLE 3.5A

$4000.00 is due for payment three years from now. If money is worth 9% p.a. compounded semi-annually, determine the equivalent value

(i) seven years from now; (ii) now.

SOLUTION

(i) Using "seven years from now" as the focal date, the method and the data can be represented graphically as shown in Figure 3.7.

FIGURE 3.7 **Graphical Representation of Method and Data**

Since the due date falls *before* the focal date, use the future value formula.

$PV = 4000.00$; $I/Y = 9$; $P/Y = 2$; $i = \frac{9\%}{2} = 0.045$; $n = 4(2) = 8$

$FV = 4000.00(1 + 0.045)^8 = 4000.00(1.4221006) = \$5688.40$

The equivalent value of the $4000.00 seven years from now is $5688.40.

(ii) Using "now" as the focal date, the method and the data can be represented graphically as shown in Figure 3.8.

FIGURE 3.8 **Graphical Representation of Method and Data**

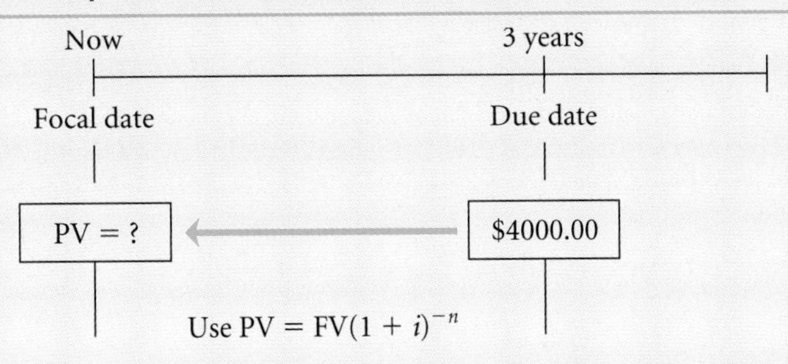

Since the due date falls *after* the focal date, use the present value formula.

$FV = 4000.00$; $i = \frac{9\%}{2} = 0.045$; $n = 3(2) = 6$

$PV = 4000.00(1 + 0.045)^{-6} = 4000.00(0.767896) = \$3071.58$

The equivalent value of the $4000.00 now is $3071.58.

Programmed Solution

(i) (Set P/Y = 2) [2nd] (CLR TVM) 4000 [±] [PV] 9 [I/Y] 8 [N] [CPT] [FV] [5688.402451]

(ii) [2nd] (CLR TVM) 4000 [FV] 9 [I/Y] 6 [N] [CPT] [PV] [-3071.582953]

EXAMPLE 3.5B

Joanna plans to pay off a debt by payments of $1600.00 one year from now, $1800.00 eighteen months from now, and $2000.00 thirty months from now. Determine the single payment now that would settle the debt if money is worth 8% p.a. compounded quarterly.

SOLUTION

While any date may be selected as the focal date, a logical choice for the focal date is the time designated "now," since the single payment "now" is wanted. As is shown in Figure 3.9, the due dates of the three scheduled payments are after the focal date. Therefore, the present value formula $PV = FV(1 + i)^{-n}$ is appropriate for finding the equivalent values of each of the three scheduled payments.

The equivalents of the three scheduled payments at the selected focal date are

$$PV_1 = 1600.00(1 + 0.02)^{-4} = 1600.00(0.923845) = \$1478.15$$
$$PV_2 = 1800.00(1 + 0.02)^{-6} = 1800.00(0.887971) = \$1598.35$$
$$PV_3 = 2000.00(1 + 0.02)^{-10} = 2000.00(0.820348) = \$1640.70$$

The equivalent single payment to settle the debt now is $4717.20.

FIGURE 3.9 **Graphical Representation of Method and Data**

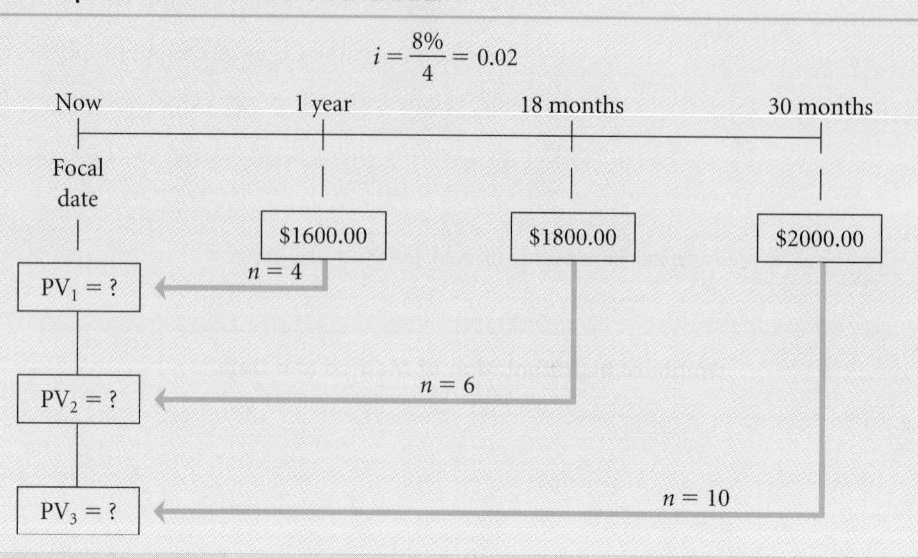

Programmed Solution

PV_1 (Set P/Y = 4) 2nd (CLR TVM) 1600 FV 8 I/Y 4 N CPT PV −1478.152682 +/− STO 1

PV_2 2nd (CLR TVM) 1800 FV 8 I/Y 6 N CPT PV −1598.348488 +/− STO 2

PV_3 2nd (CLR TVM) 2000 FV 8 I/Y 10 N CPT PV −1640.6966 +/− STO 3

Note: Remember that the negative sign indicates a cash payment or outflow. Disregard the negative signs when finding equivalent single payments.

Since you know you will need these numbers later, store them in memory. Use your RCL key to bring them back when needed as follows:

RCL 1 + RCL 2 + RCL 3

1478.152682 + 1598.348488 + 1640.6966 = 4717.19777= $4717.20

EXAMPLE 3.5C

Debt payments of $400.00 due five months ago, $600.00 due today, and $800.00 due in nine months are to be combined into one payment due three months from today at 12% p.a. compounded monthly.

SOLUTION

The logical choice for the focal date is "3 months from now," the date when the equivalent single payment is to be made.

As shown in Figure 3.10, the first two scheduled payments are due before the focal date; the future value formula $FV = PV(1 + i)^n$ should be used. However, the third scheduled payment is due *after* the focal date which means that, for it, the present value formula $PV = FV(1 + i)^{-n}$ applies.

The equivalent values (designated E_1, E_2, E_3) of the scheduled debt payments at the selected focal date are

$E_1 = 400.00(1 + 0.01)^8 = 400.00(1.082857) = \$\ 433.14$

$E_2 = 600.00(1 + 0.01)^3 = 600.00(1.030301) = \$\ 618.18$

$E_3 = 800.00(1 + 0.01)^{-6} = 800.00(0.942045) = \$\ 753.64$

The equivalent single payment to settle the debt three months from now is $1804.96

FIGURE 3.10 **Graphical Representation of Method and Data**

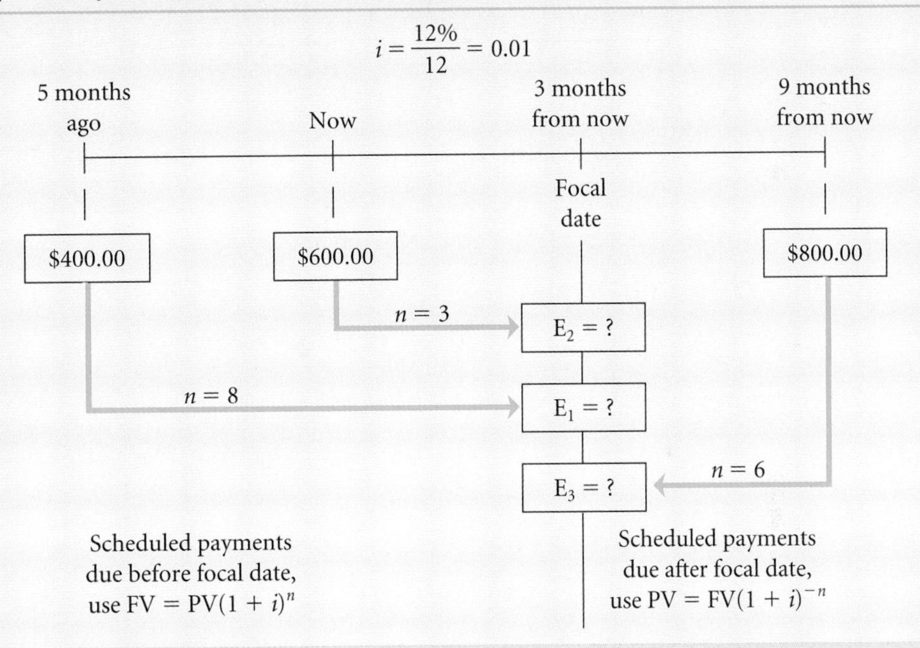

Programmed Solution

E_1 (Set P/Y = 12) 2nd (CLR TVM) 400 ± PV 12 I/Y 8 N CPT FV 433.1426823 STO 1

E_2 2nd (CLR TVM) 600 ± PV 12 I/Y 3 N CPT FV 618.1806 STO 2

E_3 2nd (CLR TVM) 800 FV 12 I/Y 6 N CPT PV −753.636188 +/− STO 3

Remember to treat this payment as a positive amount.

RCL 1 + RCL 2 + RCL 3

433.142682+ 618.1806 + 753.636188 = 1804.959470 = $1804.96

EXAMPLE 3.5D Payments of $500.00 are due at the end of each of the next five years. Determine the equivalent single payment five years from now (just after the last scheduled payment is due) if money is worth 10% p.a. compounded annually.

SOLUTION Select as the focal date "five years from now."

Let the equivalent single payment be represented by E and the dated values of the first four scheduled payments be represented by E_1, E_2, E_3, E_4 as indicated in Figure 3.11. Then the following equation of values can be set up.

$$\begin{aligned} E &= 500.00 + E_4 + E_3 + E_2 + E_1 \\ &= 500.00 + 500.00(1.1) + 500.00(1.1)^2 + 500.00(1.1)^3 + 500.00(1.1)^4 \\ &= 500.00[1 + (1.1) + (1.1)^2 + (1.1)^3 + (1.1)^4] \\ &= 500.00(1 + 1.1 + 1.21 + 1.331 + 1.4641) \\ &= 500.00(6.1051) \\ &= 3052.55 \end{aligned}$$

The equivalent single payment after five years is $3052.55.

FIGURE 3.11 **Graphical Representation of Method and Data**

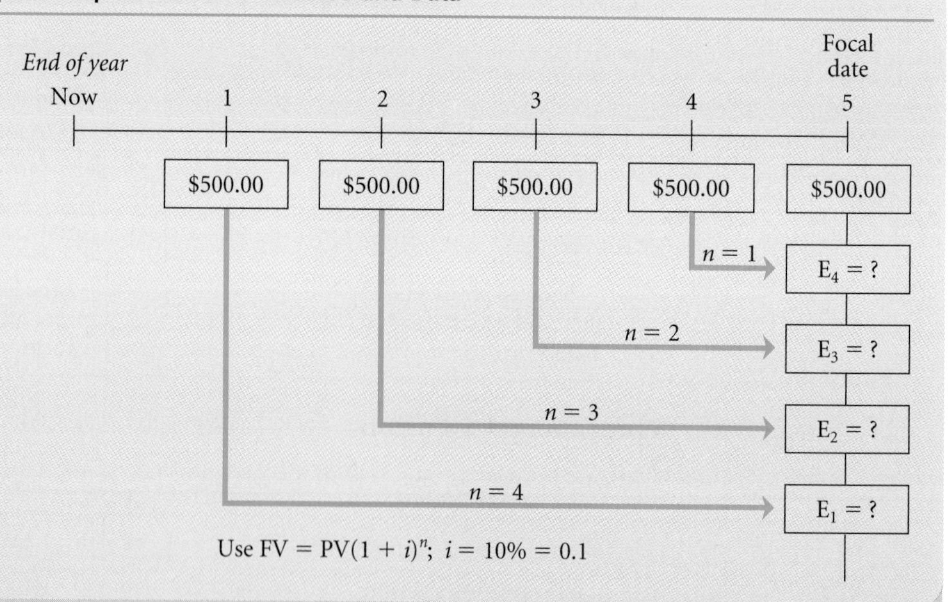

Programmed solution

(Set P/Y =1) E_1 [2nd] (CLR TVM) 500 [±] [PV] 10 [I/Y] 4 [N]
[CPT] [FV] [732.05] [STO] 1

E_2 [2nd] (CLR TVM) 500 [±] [PV] 10 [I/Y] 3 [N]
[CPT] [FV] [665.50] [STO] 2

E_3 [2nd] (CLR TVM) 500 [±] [PV] 10 [I/Y] 2 [N]
[CPT] [FV] [605.00] [STO] 3

E_4 [2nd] (CLR TVM) 500 [±] [PV] 10 [I/Y] 1 [N]

[CPT] [FV] [550.00] [STO] 4

[RCL] 1 + [RCL] 2 + [RCL] 3 + [RCL] 4 + 500 =

or

$732.05 + 665.50 + 605.00 + 550.00 + 500.00 = \$3052.55$

EXAMPLE 3.5E

Payments of $200.00 are due at the end of each of the next five quarters. Determine the equivalent single payment that will settle the debt now if interest is 9% p.a. compounded quarterly.

SOLUTION

Select as the focal date "now."

Let the equivalent single payment be represented by E and the dated values of the five scheduled payments by E_1, E_2, E_3, E_4, E_5 respectively as shown in Figure 3.12. Then the following equation of values can be set up.

$$
\begin{aligned}
E &= E_1 + E_2 + E_3 + E_4 + E_5 \\
&= 200.00(1.0225)^{-1} + 200.00(1.0225)^{-2} + 200.00(1.0225)^{-3} \\
&\quad + 200.00(1.0225)^{-4} + 200.00(1.0225)^{-5} \\
&= 200.00[(1.0225)^{-1} + (1.0225)^{-2} + (1.0225)^{-3} + (1.0225)^{-4} + (1.0225)^{-5}] \\
&= 200.00(0.977995 + 0.956474 + 0.935427 + 0.914843 + 0.894712) \\
&= 200.00(4.679452) \\
&= 935.89
\end{aligned}
$$

The equivalent single payment now is $935.89.

FIGURE 3.12 **Graphical Representation of Method and Data**

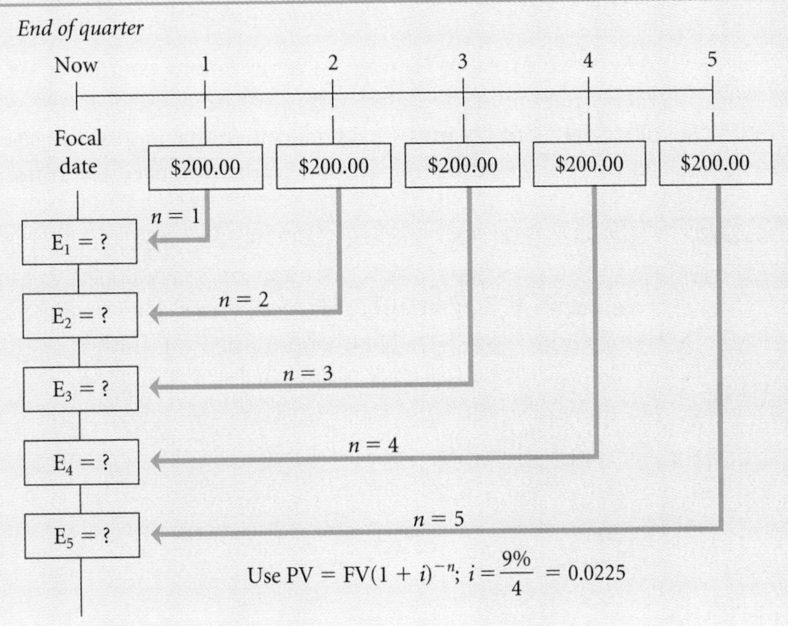

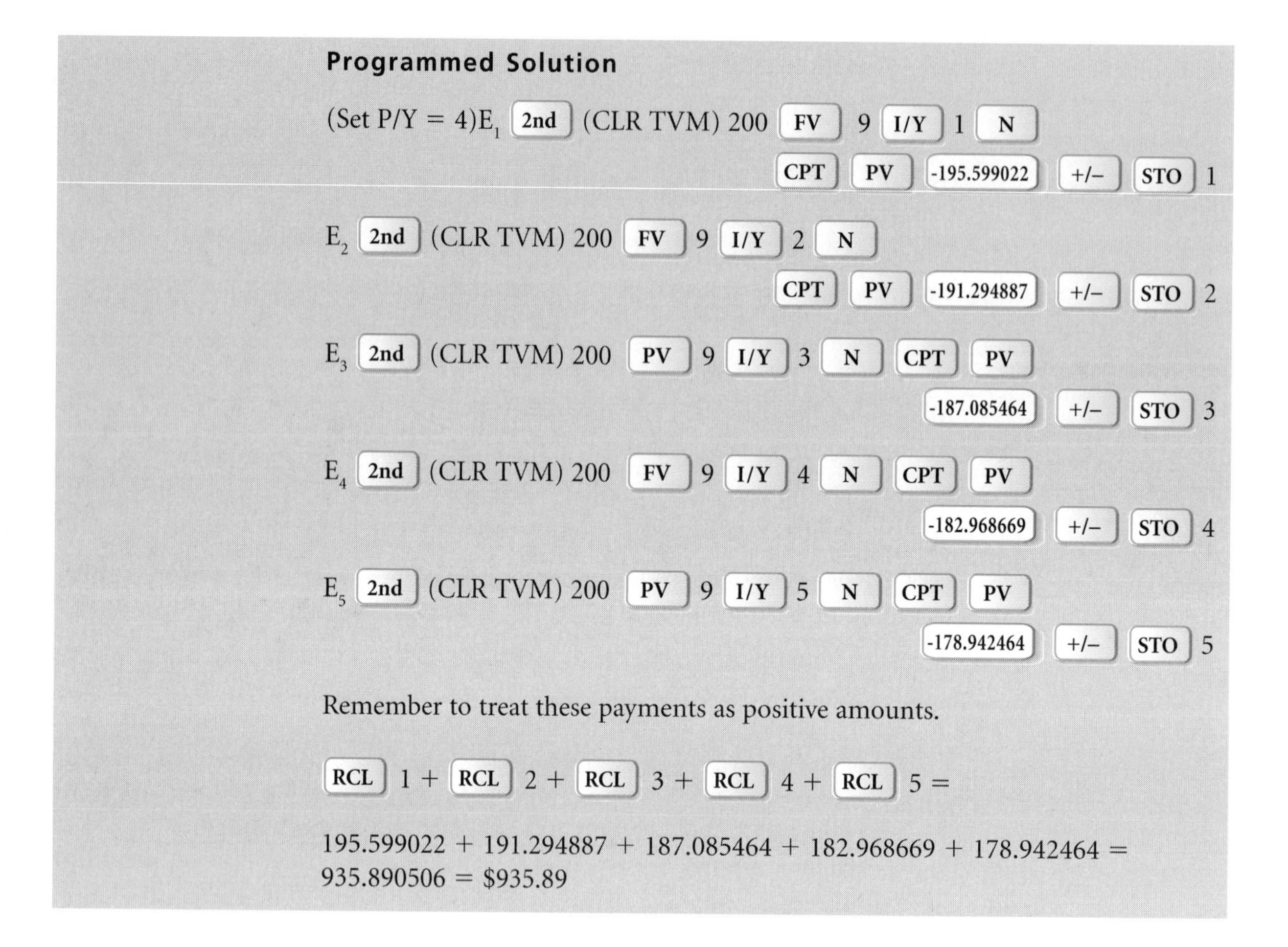

Programmed Solution

(Set P/Y = 4)E_1 [2nd] (CLR TVM) 200 [FV] 9 [I/Y] 1 [N] [CPT] [PV] [-195.599022] [+/−] [STO] 1

E_2 [2nd] (CLR TVM) 200 [FV] 9 [I/Y] 2 [N] [CPT] [PV] [-191.294887] [+/−] [STO] 2

E_3 [2nd] (CLR TVM) 200 [PV] 9 [I/Y] 3 [N] [CPT] [PV] [-187.085464] [+/−] [STO] 3

E_4 [2nd] (CLR TVM) 200 [FV] 9 [I/Y] 4 [N] [CPT] [PV] [-182.968669] [+/−] [STO] 4

E_5 [2nd] (CLR TVM) 200 [PV] 9 [I/Y] 5 [N] [CPT] [PV] [-178.942464] [+/−] [STO] 5

Remember to treat these payments as positive amounts.

[RCL] 1 + [RCL] 2 + [RCL] 3 + [RCL] 4 + [RCL] 5 =

195.599022 + 191.294887 + 187.085464 + 182.968669 + 178.942464 = 935.890506 = \$935.89

C. Finding the value of two or more equivalent replacement payments

When two or more equivalent replacement payments are needed, an equation of values matching the dated values of the original scheduled payments against the dated values of the proposed replacement payments on a selected focal date should be set up. This procedure is similar to the one used for simple interest in Chapter 1.

EXAMPLE 3.5F

Scheduled debt payments of \$1000.00 due today and \$2000.00 due one year from now are to be settled by a payment of \$1500.00 three months from now and a final payment eighteen months from now. Determine the size of the final payment if interest is 10% p.a. compounded quarterly.

SOLUTION

Let the size of the final payment be \$$x$. The logical focal date is the date of the final payment.

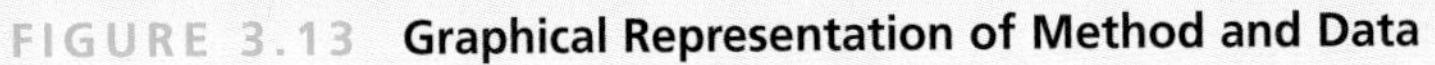

FIGURE 3.13 **Graphical Representation of Method and Data**

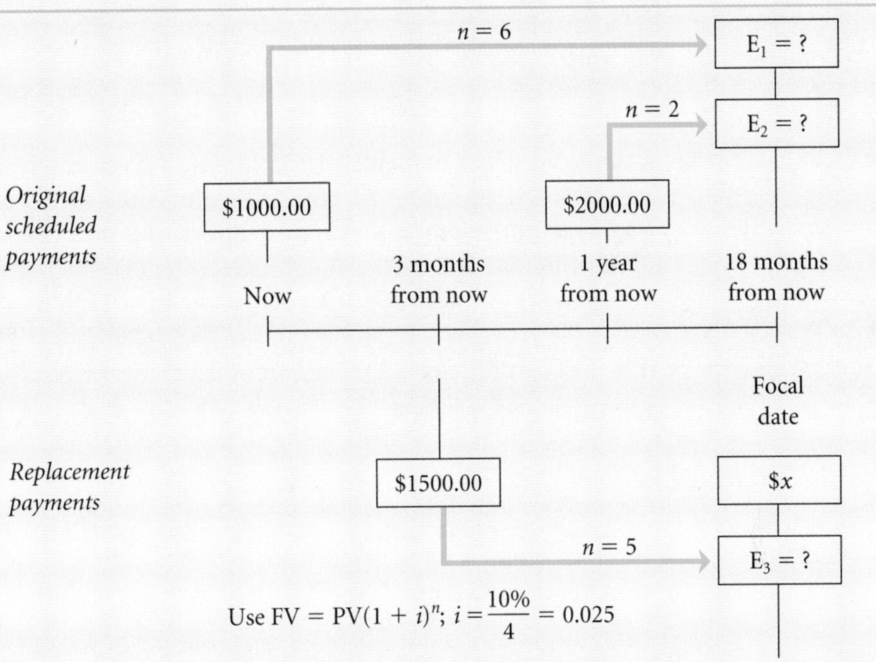

As shown in Figure 3.13, the two original scheduled debt payments and the first replacement payment are due before the selected focal date. The future value formula $FV = PV(1 + i)^n$ applies. Because the final payment is dated on the focal date, its dated value is $\$x$.

The equivalent values of the original scheduled debt payments at the selected focal date, designated E_1 and E_2, are matched against the equivalent values of the replacement payments, designated E_3 and x, giving rise to the equation of values.

$$\begin{aligned} E_1 + E_2 &= x + E_3 \\ 1000.00(1.025)^6 + 2000.00(1.025)^2 &= x + 1500.00(1.025)^5 \\ 1000.00(1.159693) + 2000.00(1.050625) &= x + 1500.00(1.1314082) \\ 1159.69 + 2101.25 &= x + 1697.11 \\ x &= 1563.83 \end{aligned}$$

The final payment is $1563.83.

Programmed Solution

E_1 (Set P/Y = 4) [2nd] (CLR TVM) 1000 [±] [PV] 10 [I/Y] 6 [N] [CPT] [FV] [1159.693418] [STO] 1

E_2 [2nd] (CLR TVM) 2000 [±] [PV] 10 [I/Y] 2 [N] [CPT] [FV] [2101.25] [STO] 2

E_3 [2nd] (CLR TVM) 1500 [±] [PV] 10 [I/Y] 5 [N] [CPT] [FV] [1697.112319] [STO] 3

RCL 1 + RCL 2 − RCL 3 = x

$1159.693418 + 2101.25 - 1697.112319 = x$

$x = 1563.831099$

EXAMPLE 3.5G

Scheduled debt payments of $750.00 due seven months ago, $600.00 due two months ago, and $900.00 due in five months are to be settled by two equal replacement payments due now and three months from now respectively. Determine the size of the equal replacement payments at 9% p.a. compounded monthly.

SOLUTION

Let the size of the equal replacement payments be represented by $$x$ and choose "now" as the focal date.

$I/Y = 9$; $P/Y = 12$; $i = \frac{9\%}{12} = 0.0075$

Figure 3.14 shows the method and data.

First, consider the dated values of the original scheduled debt payments at the chosen focal date.

The due dates of the debt payments of $750.00 and $600.00 are seven months and two months respectively before the focal date. Their dated values at the focal date are $750.00(1.0075)^7$ and $600.00(1.0075)^2$, represented by E_1 and E_2, respectively.

The due date of the scheduled payment of $900.00 is five months after the focal date. Its dated value is $900.00(1.0075)^{-5}$, shown as E_3.

Second, consider the dated values of the replacement payments at the selected focal date.

The first replacement payment due at the focal date is $$x$. The second replacement payment is due three months after the focal date. Its dated value is $x(1.0075)^{-3}$, shown as E_4.

Now equate the dated values of the replacement payments with the dated values of the original scheduled debt payments to set up the equation of values.

$$
\begin{aligned}
x + x(1.0075)^{-3} &= 750.00(1.0075)^7 + 600.00(1.0075)^2 + 900.00(1.0075)^{-5} \\
x + 0.9778333x &= 750.00(1.0536961) + 600.00(1.0150562) + 900.00(0.9633292) \\
1.9778333x &= 790.27 + 609.03 + 867.00 \\
1.9778333x &= 2266.30 \\
x &= \frac{2266.30}{1.9778333} \\
x &= 1145.85
\end{aligned}
$$

The size of the two equal payments is $1145.85.

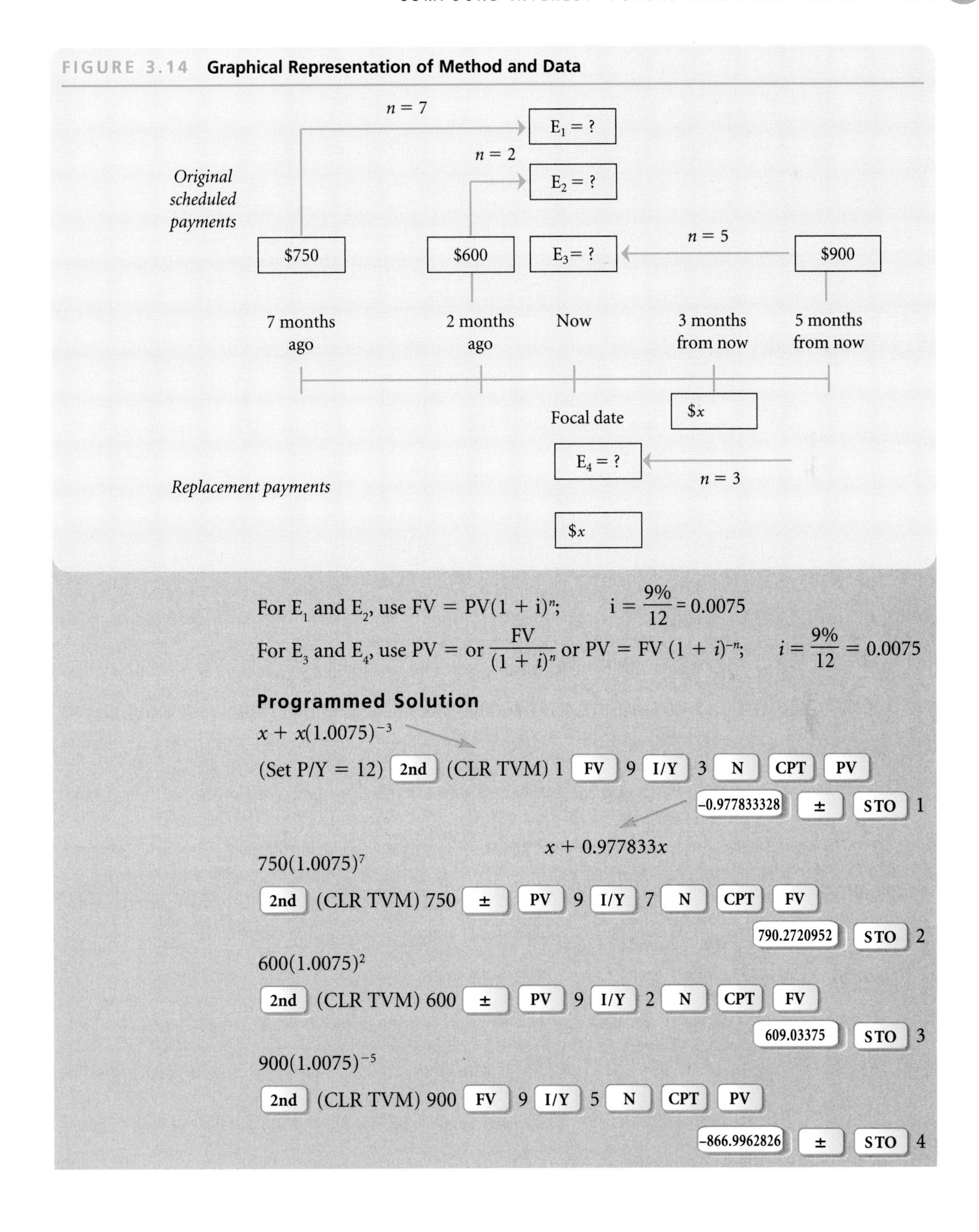

FIGURE 3.14 **Graphical Representation of Method and Data**

1 + RCL 1 = STO 5 = RCL 2 + RCL 3 + RCL 4 = STO 6

or

$$x + 0.9778333x = 790.2720952 + 609.03375 + 866.996283$$

$$1.977833x = 2266.302128$$

$$x = 1145.85$$ (RCL 6 ÷ RCL 5 +)

Note: In $x(1.0075)^{-3}$, FV is not known. To obtain the factor $(1.0075)^{-3}$, use FV = 1.

EXAMPLE 3.5H

Two scheduled payments, one of \$4000 due in three months with interest at 9% compounded quarterly and the other of \$3000 due in eighteen months with interest at 8.5% compounded semi-annually, are to be discharged by making two equal replacement payments. What is the size of the equal replacement payments if the first is due one year from now, the second two years from now, and money is now worth 10% compounded monthly?

SOLUTION

Let the size of the equal replacement payments be represented by \$$x$ and choose "one year from now" as the focal date.

Figure 3.15 illustrates the problem.

Since the two scheduled payments are interest-bearing, first determine their maturity value.

The maturity value of \$4000 due in three months at 9% compounded quarterly = $4000(1.0225)^1$ = \$4090.00, shown as E_1.

The maturity value of \$3000 due in eighteen months at 8.5% compounded semi-annually = $3000(1.0425)^3$

= 3000(1.1329955) = \$3398.99, shown as E_3.

Now determine the dated values of the two maturity values at the selected focal date subject to 10% compounded monthly.

The first scheduled payment matures nine months *before* the selected focal date. Its dated value = $4090.00(1.008\dot{3})^9$ = 4090.00(1.077549) = \$4407.18, shown as E_2.

The second scheduled payment matures six months *after* the selected focal date. Its dated value = $3398.99(1.008\dot{3})^{-6}$ = 3398.99(0.951427) = \$3233.89, shown as E_4.

The dated values of the two replacement payments at the selected focal date are \$$x$ and \$$x(1.008\dot{3})^{-12}$, shown as E_5. Therefore, the equation of values is

$$x + x(1.008\dot{3})^{-12} = 4407.18 + 3233.89$$

$$x + 0.905212x = 7641.07$$

$$x = \frac{7641.07}{1.905212}$$

$$x = 4010.61$$

The size of the two equal replacement payments is \$4010.61.

FIGURE 3.15 **Graphical Representation of Method and Data**

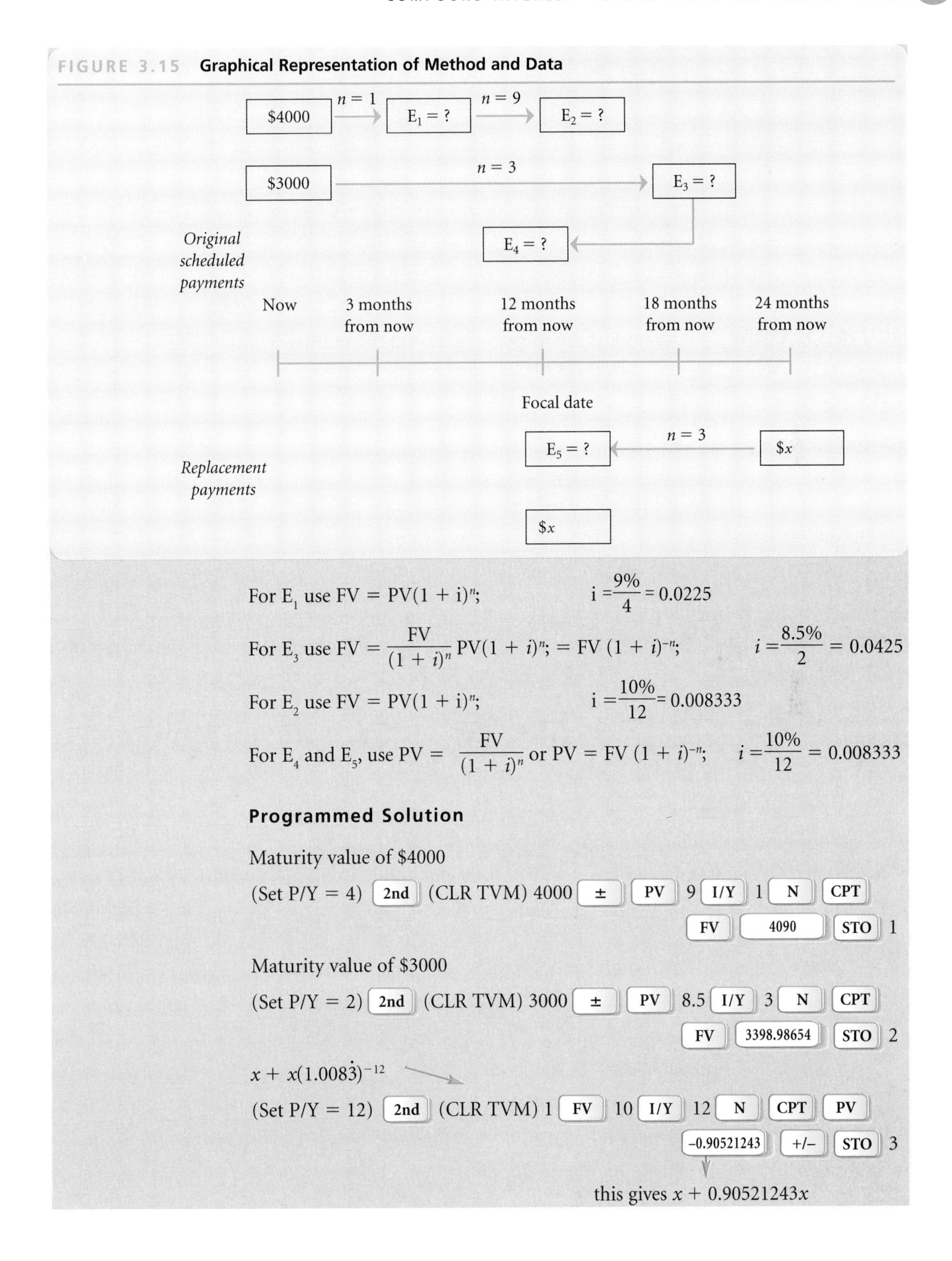

For E_1 use $FV = PV(1 + i)^n$; $\quad i = \frac{9\%}{4} = 0.0225$

For E_3 use $FV = \frac{FV}{(1+i)^n} PV(1 + i)^n; = FV(1 + i)^{-n}$; $\quad i = \frac{8.5\%}{2} = 0.0425$

For E_2 use $FV = PV(1 + i)^n$; $\quad i = \frac{10\%}{12} = 0.008333$

For E_4 and E_5, use $PV = \frac{FV}{(1+i)^n}$ or $PV = FV(1 + i)^{-n}$; $\quad i = \frac{10\%}{12} = 0.008333$

Programmed Solution

Maturity value of $4000

(Set P/Y = 4) 2nd (CLR TVM) 4000 ± PV 9 I/Y 1 N CPT FV 4090 STO 1

Maturity value of $3000

(Set P/Y = 2) 2nd (CLR TVM) 3000 ± PV 8.5 I/Y 3 N CPT FV 3398.98654 STO 2

$x + x(1.008\dot{3})^{-12}$

(Set P/Y = 12) 2nd (CLR TVM) 1 FV 10 I/Y 12 N CPT PV −0.90521243 +/− STO 3

this gives $x + 0.90521243x$

$4090.00(1.008\dot{3})^9$ RCL 1

2nd (CLR TVM) 4090 ± PV 10 I/Y 9 N CPT FV

4407.176326 STO 4

$3398.99(1.008\dot{3})^{-6}$ RCL 2

2nd (CLR TVM) 3398.986547 FV 10 I/Y 6 N CPT PV

-3233.885954 +/– STO 5

1 + RCL 3 = STO 6   RCL 4 + RCL 5 = STO 7

$$x + 0.905212x = 4407.176326 + 3233.885594$$
$$1.905212x = 7641.062280$$
$$x = 4010.61$$

(RCL 7 ÷ RCL 6 +)

EXAMPLE 3.5I

What is the size of the equal payments that must be made at the end of each of the next five years to settle a debt of $5000.00 due in five years if money is worth 9% p.a. compounded annually?

SOLUTION

Select as the focal date "five years from now." Let the equal payments be represented by $$x$ and let the dated values of the first four payments be represented by E1, E2, E3, and E4 respectively as shown in Figure 3.16.

Then the equation of values may be set up.

$$5000.00 = x + E_4 + E_3 + E_2 + E_1$$
$$5000.00 = x + x(1.09) + x(1.09)^2 + x(1.09)^3 + x(1.09)^4$$
$$5000.00 = x[1 + (1.09) + (1.09)^2 + (1.09)^3 + (1.09)^4]$$
$$5000.00 = x(1 + 1.09 + 1.1881 + 1.295029 + 1.411582)$$
$$5000.00 = 5.984711x$$
$$x = 835.46$$

The size of the equal payments is $835.46.

Programmed Solution

(Set P/Y = 1) E_1 2nd (CLR TVM) 1 ± PV 9 I/Y 4 N CPT FV

1.411582 STO 1

E_2 2nd (CLR TVM) 1 ± PV 9 I/Y 3 N CPT FV

1.295029 STO 2

E_3 2nd (CLR TVM) 1 ± PV 9 I/Y 2 N CPT FV

1.1881 STO 3

E_4 2nd (CLR TVM) 1 ± PV 9 I/Y 1 N CPT FV

1.09 STO 4

RCL 1 + RCL 2 + RCL 3 + RCL 4 + 1 = 5.984711 STO

or

$$1.411582x + 1.295029x + 1.1881x + 1.09x + x = \$5000.00$$
$$5.984711x = \$5000.00$$
$$x = \$835.46 \quad (5000 \div \text{RCL } 5 =)$$

FIGURE 3.16 **Graphical Representation of Method and Data**

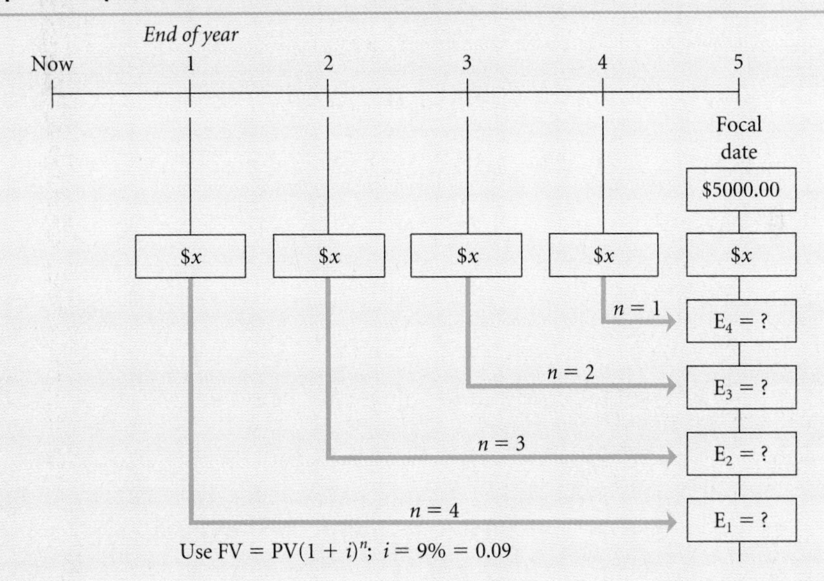

EXAMPLE 3.5J

What is the size of the equal payments that must be made at the end of each of the next five quarters to settle a debt of $3000.00 due now if money is worth 12% p.a. compounded quarterly?

SOLUTION

Select as the focal date "now." Let the size of the equal payments be represented by $x and let the dated values of the five payments be represented by E1, E2, E3, E4, and E5 respectively as shown in Figure 3.17.

FIGURE 3.17 **Graphical Representation of Method and Data**

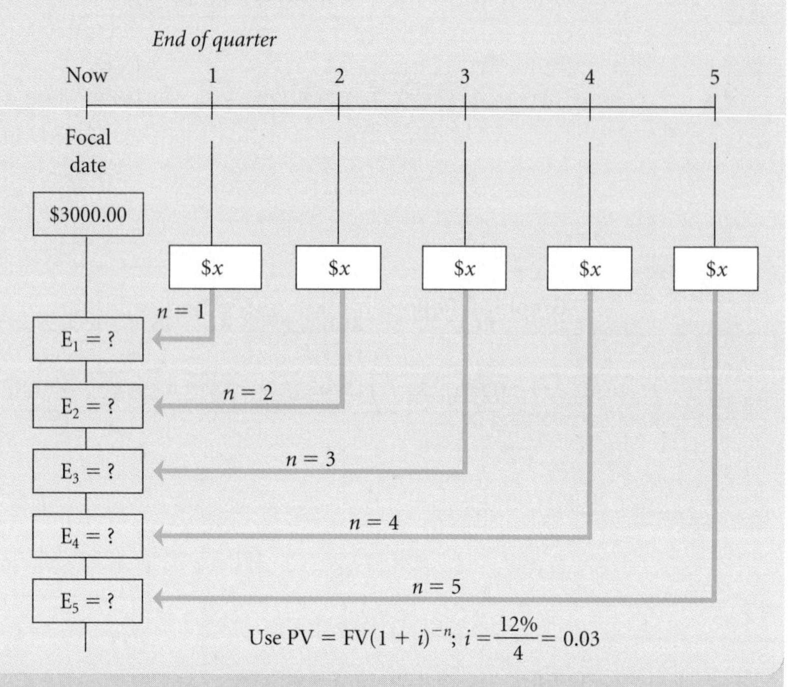

Then the equation of values may be set up.

$$3000.00 = E_1 + E_2 + E_3 + E_4 + E_5$$

$$3000.00 = x(1.03)^{-1} + x(1.03)^{-2} + x(1.03)^{-3} + x(1.03)^{-4} + x(1.03)^{-5}$$
$$3000.00 = x[(1.03)^{-1} + (1.03)^{-2} + (1.03)^{-3} + (1.03)^{-4} + (1.03)^{-5}]$$
$$3000.00 = x(0.970874 + 0.942596 + 0.915142 + 0.888487 + 0.862609)$$
$$3000.00 = 4.5797074x$$
$$x = \frac{3000.00}{4.579707}$$
$$x = 655.06$$

The size of the equal payments is $655.06.

Programmed Solution

(Set P/Y = 4) E_1 [2nd] (CLR TVM) 1 [FV] 12 [I/Y] 1 [N] [CPT] [PV] [-0.970874] [+/−] [STO] 1

E_2 [2nd] (CLR TVM) 1 [FV] 12 [I/Y] 2 [N] [CPT] [PV] [-0.942596] [+/−] [STO] 2

E_3 [2nd] (CLR TVM) 1 [FV] 12 [I/Y] 3 [N] [CPT] [PV] [-0.915142] [+/−] [STO] 3

E_4 [2nd] (CLR TVM) 1 [FV] 12 [I/Y] 4 [N] [CPT] [PV] [-0.888487] [+/–] [STO] 4

E_1 [2nd] (CLR TVM) 1 [FV] 12 [I/Y] 5 [N] [CPT] [PV] [-0.862609] [+/–] [STO] 5

Remember to treat these payments as positive amounts.

[RCL] 1 + [RCL] 2 + [RCL] 3 + [RCL] 4 + [RCL] 5 = 4.594707 [STO] 6

$$0.970874x + 0.942596x + 0.915142x + 0.888487x + 0.862609x = \$3000.00$$

$$4.579707x = \$3000.00$$

$$x = \$655.06 \quad (3000 \div \text{[RCL]}\ 6 =)$$

Exercise 3.5

If you choose, you can use Excel's ***Future Value (FV)*** or ***Present Value (PV)*** functions to answer the questions below. Refer to **FV** and **PV** on the Spreadsheet Template Disk to learn how to use these Excel functions.

Find the equivalent single replacement payment on the given focal date for each of the following eight situations.

	Scheduled Payments	Int. Rate	Frequency of Conversion	Focal Date
1.	$5000.00 due in 2 years	6%	monthly	5 years from now
2.	$1600.00 due in 18 months	8%	quarterly	42 months from now
3.	$3400.00 due in 4 years	10%	semi-annually	1 year from now
4.	$2700.00 due in 60 months	7%	quarterly	6 months from now
5.	$800.00 due in 6 months and $700.00 due in 15 months	9.5%	monthly	2 years from now
6.	$1000.00 due in 9 months and $1200.00 due in 18 months	8.8%	quarterly	3 years from now
7.	$400.00 due in 3 years and $600.00 due in 5 years	11%	semi-annually	now
8.	$2000.00 due in 20 months and $1500.00 due in 40 months	10.5%	monthly	9 months from now

B. Find the equivalent replacement payments for each of the following.

1. Scheduled payments of \$2000.00 due now and \$2000.00 due in four years are to be replaced by a payment of \$2000.00 due in two years and a second payment due in seven years. Determine the size of the second payment if interest is 10.5% compounded annually and the focal date is seven years from now.

2. Scheduled loan payments of \$1500.00 due in six months and \$1900.00 due in 21 months are rescheduled as a payment of \$2000.00 due in three years and a second payment due in 45 months. Determine the size of the second payment if interest is 7% compounded quarterly and the focal date is 45 months from now.

3. Two debts—the first of \$800.00 due six months ago and the second of \$1400.00 borrowed one year ago for a term of three years at 6.5% compounded annually—are to be replaced by a single payment one year from now. Determine the size of the replacement payment if interest is 7.5% compounded quarterly and the focal date is one year from now.

4. A payment of \$500.00 is due in six months with interest at 12% compounded quarterly. A second payment of \$800.00 is due in 18 months with interest at 10% compounded semi-annually. These two payments are to be replaced by a single payment nine months from now. Determine the size of the replacement payment if interest is 9% compounded monthly and the focal date is nine months from now.

5. Scheduled payments of \$800.00 due two years ago and \$1000.00 due in five years are to be replaced by two equal payments. The first replacement payment is due in four years and the second payment is due in eight years. Determine the size of the two replacement payments if interest is 12% compounded semi-annually and the focal date is four years from now.

6. Loan payments of \$3000.00 due one year ago and \$2500.00 due in four years are to be rescheduled by two equal payments. The first replacement payment is due now and the second payment is due in six years. Determine the size of the two replacement payments if interest is 6.9% compounded monthly and the focal date is now.

7. Scheduled payments of \$900.00 due in three months with interest at 11% compounded quarterly and \$800.00 due in thirty months with interest at 11% compounded quarterly are to be replaced by two equal payments. The first replacement payment is due today and the second payment is due in three years. Determine the size of the two replacement payments if interest is 9% compounded monthly and the focal date is today.

8. Scheduled payments of \$1400.00 due today and \$1600.00 due with interest at 11.5% compounded annually in five years are to be replaced by two equal payments. The first replacement payment is due in 18 months and the second payment is due in four years. Determine the size of the two replacement payments if interest is 11% compounded quarterly and the focal date is 18 months from now.

C. Solve each of the following problems.

1. A loan of $4000.00 is due in five years. If money is worth 7% compounded annually, find the equivalent payment that would settle the debt
   (a) now; (b) in 2 years; (c) in 5 years; (d) in 10 years.

2. A debt payment of $5500 is due in 27 months. If money is worth 8.4% p.a. compounded quarterly, what is the equivalent payment
   (a) now? (b) 15 months from now?
   (c) 27 months from now? (d) 36 months from now?

3. A debt can be paid by payments of $2000.00 scheduled today, $2000.00 scheduled in three years, and $2000.00 scheduled in six years. What single payment would settle the debt four years from now if money is worth 10% compounded semi-annually?

4. Scheduled payments of $600.00, $800.00, and $1200.00 are due in one year, three years, and six years respectively. What is the equivalent single replacement payment two-and-a-half years from now if interest is 7.5% compounded monthly?

5. Scheduled payments of $400.00 due today and $700.00 due with interest at 4.5% compounded monthly in eight months are to be settled by a payment of $500.00 six months from now and a final payment in fifteen months. Determine the size of the final payment if money is worth 6% compounded monthly.

6. Scheduled payments of $1200.00 due one year ago and $1000.00 due six months ago are to be replaced by a payment of $800.00 now, a second payment of $1000.00 nine months from now, and a final payment eighteen months from now. What is the size of the final payment if interest is 10.8% compounded quarterly?

7. An obligation of $8000.00 due one year ago is to be settled by four equal payments due at the beginning of each of the next four years respectively. What is the size of the equal payments if interest is 8% compounded semi-annually?

8. A loan of $3000.00 borrowed today is to be repaid in three equal installments due in one year, three years, and five years respectively. What is the size of the equal installments if money is worth 7.2% compounded monthly?

9. Scheduled payments of $500.00 each are due at the end of each of the next five years. If money is worth 11% compounded annually, what is the single equivalent replacement payment
   (a) five years from now? (b) now?

10. What is the size of the equal payments that must be made at the end of each of the next four years to settle a debt of $3000.00 subject to interest at 10% p.a. compounded annually
    (a) due four years from now? (b) due now?

Review Exercise

1. What is the accumulated value of $500.00 in fifteen years at 6% compounded
   (a) annually? (b) quarterly? (c) monthly?

2. What is the amount of $10 000.00 at 10.5% compounded monthly
   (a) in four years?
   (b) in eight-and-one-half years?
   (c) in twenty years?

3. Landmark Trust offers five-year investment certificates at 7.5% compounded semi-annually.
   (a) What is the value of a $2000 certificate at maturity?
   (b) How much of the maturity value is interest?

4. Western Savings offers three-year term deposits at 9.25% compounded annually while your credit union offers such deposits at 9.0% compounded quarterly. If you have $5000 to invest, what is the maturity value of your deposit
   (a) at Western Savings?
   (b) at your credit union?

5. Find the future value and the compound interest of
   (a) $1800.00 invested at 8% compounded quaterly for 15.5 years;
   (b) $1250.00 invested at 6.5% compounded monthly for 15 years.

6. If $6000.00 is invested for six years and seven months at 6% compounded semi-annually, what is the interest that the investment earns?

7. Compute the maturity value of a $5000.00 promissory note dated November 15, 1995 and due on June 15, 2005, if interest is 8% compounded quarterly.

8. An investment of $2000.00 is made for three years, four months at 4.5% compounded semi-annually. What is the amount of interest?

9. Determine the sum of money that will grow to $14 000.00 in four years, eight months at 5% compounded quarterly.

10. Determine the proceeds of a non-interest-bearing note with a maturity value of $9000.00 three years and ten months before the due date if the interest rate is 7% compounded semi-annually.

11. Determine the discounted value now of $5200.00 due in forty months if money is worth 6.5% compounded quarterly.

12. Find the present value and the compound discount of
    (a) $3600.00 due in 9 years if interest is 8% compounded semi-annually;
    (b) $9000.00 due in 5 years if money is worth 6.8% compounded quarterly.

13. The Ram Company borrowed $20 000.00 at 10% compounded semi-annually and made payments toward the loan of $8000.00 after two years and $10 000.00 after three-and-a-half years. How much is required to pay off the loan one year after the second payment?

14. Ted deposited $1750.00 in an RRSP on March 1, 2000 at 3% compounded quarterly. Subsequently the interest rate was changed to 4% compounded monthly on September 1, 2002 and to 4.5% compounded semi-annually on June 1, 2004. What was the value of the RRSP deposit on December 1, 2006, if no further changes in interest were made?

15. An investment of $2500 is accumulated at 5% compounded quarterly for two-and-one-half years. At that time the interest rate is changed to 6% compounded monthly. How much is the investment worth two years after the change in interest rate?

16. To ensure that funds are available to repay the principal at maturity, a borrower deposits $2000 each year for three years. If interest is 6% compounded quarterly, how much will the borrower have on deposit four years after the first deposit was made?

17. Cindy started a registered retirement savings plan on February 1, 1997, with a deposit of $2500. She added $2000 on February 1, 1998, and $1500 on February 1, 2003. What is the accumulated value of her RRSP account on August 1, 2007, if interest is 5% compounded quarterly?

18. A demand loan of $8000 is repaid by payments of $3000 after fifteen months, $4000 after thirty months, and a final payment after four years. If interest was 8% for the first two years and 9% for the remaining time, and compounding is quarterly, what is the size of the final payment?

19. A non-interest-bearing note for $1500.00 is due on June 30, 2008. The note is discounted at 10% compounded quarterly on September 30, 2004. What are the proceeds of the note?

20. Find the present value and the compound dis count of $4000 due in seven years and six months if interest is 8.8% compounded quarterly.

21. Find the principal that will accumulate to $6000 in fifteen years at 5% compounded monthly.

22. Find the proceeds of a non-interest-bearing promissory note for $75 000 discounted 42 months before maturity at 6.5% compounded semi-annually.

23. A ten-year promissory note for $1750.00 dated May 1, 2001 bearing interest at 4% compounded semi-annually is discounted on August 1, 2007 to yield 6% compounded quarterly. Determine the proceeds of the note.

24. A seven-year, $10 000 promissory note bearing interest at 8% compounded quarterly is discounted four years after the date of issue at 7% compounded semi-annually. What are the proceeds of the note?

25. A $40 000, 15-year promissory note dated June 1, 1998, bearing interest at 12% compounded semi-annually is discounted on September 1, 2006 at 11% compounded quarterly. What are the proceeds of the note?

26. A fifteen-year promissory note for $16 500.00 bearing interest at 12% compounded semi-annually is discounted at 9% compounded monthly three years and four months after the date of issue. Compute the proceeds of the note.

27. An eight-year promissory note for $20 000.00 dated May 2, 2005, bearing interest at 10% compounded quarterly, is discounted on September 2, 2007 at 9.5% compounded semi-annually. Determine the proceeds of the note.

28. Three years and five months after its date of issue, a six-year promissory note for $3300.00 bearing interest at 7.5% compounded monthly is discounted at 7% compounded semi-annually. Find the proceeds of the note.

29. A sum of money has a value of $3000 eighteen months from now. If money is worth 6% compounded monthly, what is its equivalent value

(a) now? **(b)** one year from now?

(c) three years from now?

30. Payments of $1000, $1200, and $1500 are due in six months, eighteen months, and thirty months from now, respectively. What is the equivalent single payment two years from now if money is worth 9.6% compounded quarterly?

31. An obligation of $10 000 is due one year from now with interest at 10% compounded semi-annually. The obligation is to be settled by a payment of $6000 in six months and a final payment in fifteen months. What is the size of the second payment if interest is now 9% compounded monthly?

32. Waldon Toys owes $3000 due in two years with interest at 11% compounded semi-annually and $2500 due in fifteen months at 9% compounded quarterly. If the company wants to discharge these debts by making two equal payments, the first one now and the second eighteen months from now, what is the size of the two payments if money is now worth 8.4% compounded monthly?

33. Debt payments of $400.00 due today, $500.00 due in eighteen months, and $900.00 due in three years are to be combined into a single payment due two years from now. What is the size of the single payment if interest is 8% p.a. compounded quarterly?

34. Debt payments of $2600.00 due one year ago and $2400.00 due two years from now are to be replaced by two equal payments due one year from now and four years from now respectively. What is the size of the equal payments if money is worth 9.6% p.a. compounded semi-annually?

35. A loan of $7000.00 taken out two years ago is to be repaid by three equal installments due now, two years from now, and three years from now, respectively. What is the size of the equal installments if interest on the debt is 12% p.a. compounded monthly?

Self-Test

1. What sum of money invested at 4% compounded quarterly will grow to $3300 in 11 years?

2. Find the compound interest earned by $1300 invested at 7.5% compounded monthly for seven years.

3. Determine the compounding factor for a sum of money invested for 14.5 years at 7% compounded semi-annually.

4. Determine the maturity value of $1400.00 due in 71 months compounded annually at 7.75%.

5. Five years after Anne deposited $3600 in a savings account that earned interest at 4.8% compounded monthly, the rate of interest was changed to 6% compounded semi-annually. How much was in the account twelve years after the deposit was made?

6. A debt can be repaid by payments of $4000 today, $4000 in five years, and $3000 in six years. What single payment would settle the debt one year from now if money is worth 7% compounded semi-annually?

7. A $10 200 debt will accumulate for five years at 11.6% compounded semi-annually. For how much will the debt sell three years after it was incurred if the buyer of the debt charges 10% compounded quarterly?

8. What is the present value of $5900 payable in 15 years if the current interest rate is 7.5% compounded semi-annually?

9. Determine the compound discount on $8800 due in 7.5 years if interest is 9.6% compounded monthly.

10. Two debt payments, the first for $800 due today and the second for $600 due in nine months with interest at 10.5% compounded monthly, are to be settled by a payment of $800 six months from now and a final payment in 24 months. Determine the size of the final payment if money is now worth 9.5% compounded quarterly.

11. A note dated July 1, 1998 promises to pay $8000 with interest at 7% compounded quarterly on January 1, 2007. Find the proceeds from the sale of the note on July 1, 2002 if money is then worth 8% compounded semi-annually.

12. Adam borrowed $5000 at 10% compounded semi-annually. He repaid $2000 after two years and $2500 after three years. How much will he owe after five years?

13. A debt of $7000 due today is to be settled by three equal payments due three months from now, 15 months from now, and 27 months from now, respectively. What is the size of the equal payments at 11% compounded quarterly?

14. Seven years and two months after its date of issue, an eleven-year promissory note for $8200.00 bearing interest at 13.5% compounded monthly is discounted at 10.5% compounded semi-annually. Find the proceeds of the note.

15. Compute the proceeds of a non-interest-bearing note for $1100.00 three years and seven months before the due date if money is worth 7.5% compounded annually.

Challenge Problems

1. Jean-Guy Renoir wanted to leave some money to his grandchildren in his will. He decided that they should each receive the same amount of money when they each turn 21. When he died, his grandchildren were 19, 16, and 13, respectively. How much will they each receive when they turn 21 if Jean-Guy left a lump sum of $50 000 to be shared among them equally? Assume the interest rate will remain at 7.75% p.a. compounded semi-annually from the time of Jean-Guy's death until the youngest grandchild turns 21.

2. Miranda has $1000 to invest. She has narrowed her options to two four-year certificates, A and B. Certificate A pays interest at 8% p.a. compounded semi-annually the first year, 8% p.a. compounded quarterly the second year, 8% p.a. compounded monthly the third year, and 8% p.a. compounded daily the fourth year. Certificate B pays 8% p.a. compounded daily the first year, 8% p.a. compounded monthly the second year, 8% p.a. compounded quarterly the third year, and 8% p.a. compounded semi-annually the fourth year.

(a) What is the value of each certificate at the end of the four years?

(b) How do the values of certificates A and B compare with the value of a third certificate that pays interest at 7% compounded daily for the full four-year term?

Case Study 3.1 What's in Your Best Interest?

» Marika just received a bonus from her employer and decides to deposit the bonus into a savings account at her bank. Her personal banking representative showed her that there were four different savings accounts that she should consider. All of them pay compound interest.

The Daily Interest Savings Account has an interest rate of 2.50%; compound interest is calculated on the minimum balance in the account each day and is paid monthly. The Monthly Interest Savings Account has an interest rate of 2.75%; compound interest is calculated on the minimum balance in the account during the month and is paid monthly. The Investment Savings Account has an interest rate of 3.10%; compound interest is calculated on the minimum monthly balance and paid monthly, but only when the balance remains above $5000.00 for the whole month. If the balance drops below $5000.00 at any time during the month, no interest is paid for that month. Compound interest on the Basic Savings Account is calculated on the minimum balance in the account during the six-month periods ending April 30 and October 31, and is paid on those dates. The Basic Savings Account has an interest rate of 3.00%.

The personal banking representative advised Marika that she should also consider the number of transactions made in the account each month or year when choosing an account. She thanked the representative and went home to consider her options.

QUESTIONS

1. Suppose Marika's bonus was $4000.00 and she planned to invest the money immediately in a bank account for one year. Assume that the bank's interest rates will stay the same during the year.
   (a) For each of the bank's four savings accounts, how much interest would Marika earn in one year's time?
   (b) Which savings account would pay the most interest?

2. Suppose Marika's bonus was $7000.00 and she planned to leave the money in a savings account for one year. Assume that the bank's interest rates will stay the same during the year.
   (a) For each of the bank's four savings accounts, how much interest would Marika earn in one year's time if she left the money in the account for one year?
   (b) Which savings account would pay the most interest?
   (c) Suppose Marika knew she would have to withdraw $3000.00 after ten months. Which savings account would pay the most interest?

3. Suppose Marika had $7000.00 to put into a savings account. She planned to leave the money in the account for one year, then withdraw $3000.00. She would then leave the remaining balance in the account for one more year. Which savings account would pay the most interest over the two-year period? Assume that the bank's interest rates will stay the same over the next two years.

4. For the savings accounts at your bank, credit union, or trust company, find the features and interest rates offered for each. If you had $3000.00 to deposit for one year, which account would you choose?

Case Study 3.2 Planning Ahead

» Brunner Company, a successful Canadian manufacturer, has been growing steadily. The past year was unusually profitable, so management has decided to set aside $2 000 000 to expand the company's factory over the next five years.

Management has developed two different plans for expanding over the next five years, Plan A and Plan B. Plan A would require equal amounts of $550 000 one year from now, two years from now, four years from now, and five years from now. Plan B would require $200 000 now, $500 000 one year from now, $800 000 three years from now, and $850 000 five years from now.

The company has decided to fund the expansion with only the $2 000 000 and any interest it can earn on it. Before deciding which plan to use, the company asked its treasurer to predict the rates of interest it can earn on the $2 000 000. The treasurer expects that Brunner Company can invest the $2 000 000 and earn interest at a rate of 5.0% p.a. compounded semi-annually during Year 1, 6.0% p.a. compounded semi-annually during Years 2 and 3, 6.5% p.a. compounded semi-annually during Year 4, and 6.75% p.a. compounded semi-annually during Year 5. The company can withdraw part of the money from this investment at any time without penalty.

QUESTIONS

1. **(a)** Could Brunner Company meet the cash requirement of Plan A by investing the $2 000 000 as described above? (Use "now" as the focal date.)

(b) What is the exact difference between the cash required and the cash available from the investment?

2. **(a)** Could Brunner Company meet the cash requirements of Plan B by investing the $2 000 000 as described above? (Use "now" as the focal date.)

(b) What is the difference between the cash required and the cash available from the investment?

3. **(a)** Suppose Plan A was changed so that it required equal amounts of $550 000 now, one year from now, two years from now, and four years from now. Could Brunner Company meet the cash requirements of the new Plan A by investing the $2 000 000 as described above? (Use "now" as the focal date.)

(b) What is the difference between the cash required and the cash available from the investment?

4. Suppose the treasurer found another way to invest the $2 000 000 that earned interest at a rate of 5.8% compounded quarterly for the next five years.

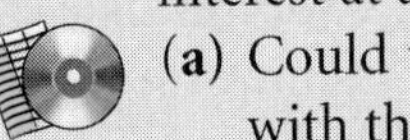

(a) Could the company meet the cash requirements of the original Plan A with this new investment? (Show all your calculations.)

(b) Could the company meet the cash requirements of Plan B with this new investment? (Show all your calculations.)

(c) If the company could meet the cash requirements of both plans, which plan would the treasurer recommend? In other words, which plan would have the lower present value?

SUMMARY OF FORMULAS

Formula 3.1A

$S = P(1 + i)^n$ Finding the future value (or maturity value) when the original principal, the rate of interest, and the time period are known

restated as

$FV = PV(1 + i)^n$

Formula 3.1B

$P = \frac{S}{(1 + i)^n}$ Finding the present value (or principal or proceeds or discounted value) when the future value, the rate of interest, and the time period are known

restated as

$PV = \frac{FV}{(1 + i)^n}$

Formula 3.1C

$P = S(1 + i)^{-n}$ Finding the present value by means of the discount factor (the reciprocal of the compounding factor)

restated as

$PV = FV(1 + i)^{-n}$

Formula 3.2

$i = \frac{j}{m}$ Finding the periodic rate of interest

GLOSSARY

Accumulation factor see **Compounding factor**

Amount see **Future value**

Comparison date see **Focal date**

Compound discount the difference between a given future amount and its present value (or proceeds or discounted value) at a specified time *(p. 103)*

Compound interest a procedure for computing interest whereby interest earned during an interest period is added onto the principal at the end of the interest period *(p. 80)*

Compounding factor the factor $(1 + i)^n$ found in compound interest formulas *(p. 83)*

Compounding frequency the number of times interest is compounded during a given time period (usually one year) *(p. 83)*

Compounding period the time between two successive interest dates *(p. 83)*

Conversion frequency see **Compounding frequency**

Conversion period see **Compounding period**

Discount factor the factor $(1 + i)^{-n}$; the reciprocal of the compounding factor *(p. 105)*

Discounted value see **Present value**

Discounting the process of computing the present value (or proceeds or discounted value) of a future sum of money *(p. 103)*

Equivalent values the dated values of an original sum of money *(p. 116)*

Focal date a specific date chosen to compare the time values of one or more dated sums of money *(p. 116)*

Future value the sum of money to which a principal will grow at compound interest in a specific number of compounding or conversion periods at a specified periodic rate of interest *(p. 82)*

Maturity value see **Future value**

Nominal rate of interest the stated rate at which the compounding is done one or more times per year; usually stated as an annual rate *(p. 83)*

Periodic rate of interest value of interest is obtained by dividing the nominal annual rate by the number of compounding periods per year *(p. 83)*

Present value the principal at any time that will grow at compound interest to a given future value over a given number of compounding periods at a given rate of interest *(p. 103)*

Proceeds see **Present value**

USEFUL INTERNET SITES

www.globefund.com/
Globefund.com This popular site for mutual fund information and analysis is hosted by the *Globe and Mail* Website

www.fin.gc.ca/fin-eng.html
Department of Finance This site has information on the preparation of the federal government's budget that shows what is happening to taxes, as well as information related to interest rates and the economy.

www.cannex.com
CANNEX This site has a large list of comparative interest rates for GICs and term deposits offered by financial institutions in Canada, the United States, and Australia.

4 Compound Interest—Further Topics

OBJECTIVES

Upon completing this chapter, you will be able to do the following with compound interest by using an electronic calculator:

1. Determine the number of conversion periods and find equated dates.
2. Compute periodic and nominal rates of interest.
3. Compute effective and equivalent rates of interest.

Some advertisements that come in the mail look too good to be true. For example, a finance company sent an advertisement offering residential and commercial mortgages, unsecured loans, and lines of credit. It included the following payment examples:

\$ 5 000 : \$ 39.58	\$15 000 : \$118.75
\$ 7 500 : \$ 59.38	\$20 000 : \$158.33
\$10 000 : \$ 79.17	\$25 000 : \$197.92

We should read such ads critically. This ad does not specify the interest rate being charged or whether the interest is simple or compound. It does not specify whether the rates of interest are the same for each type of loan. It does not specify the term of the loans or how often you make the payments. It is important to know this information to assess the ad and compare its offer with loans available from other sources.

INTRODUCTION

In the previous chapter, we considered future value and present value when using compound interest. In this chapter, we will look at other aspects of compound interest, including finding the number of conversion periods, computing equated dates and equivalent and effective rates of interest.

For the calculations in this and the following chapters, we will use our electronic calculators. We can save time on our calculations by using the memory of the calculator when working with these functions. The number of digits retained in memory is almost always greater than the number of digits displayed. Thus, we might get slightly different results if we use the memory rather than rekey the displayed digits. However, we can ignore such differences because they are insignificant. For the worked examples in this text, we have used the memory whenever it was convenient to do so.

4.1 Finding n and Related Problems

A. Finding the number of conversion periods

If the principal PV, the future value FV, and the periodic rate of interest i are known, the number of conversion periods n can be determined by substituting the known values in Formula 3.1A and solving for n.

$$FV = PV(1 + i)^n$$ —— Formula 3.1A

Excel You can use Excel's ***Number of Compounding Periods (NPER)*** function to find the number of conversion periods. Refer to **NPER** on the Spreadsheet Template Disk to learn how to use this Excel function.

EXAMPLE 4.1A

In how many years will $2000.00 grow to $2440.38 at 4% compounded quarterly?

SOLUTION

PV = 2000.00; FV = 2440.38; I/Y = 4; P/Y = 4; $i = \frac{4\%}{4} = 1\% = 0.01$

$2440.38 = 2000.00(1.01)^n$ —— substituting in Formula 3.1A

$(1.01)^n = 1.22019$

$n \ln 1.01 = \ln 1.22019$ —— solve for n using the natural logarithm

$0.009950n = 0.199007$ —— obtain the numerical values using the LN key

$$n = \frac{0.199007}{0.009950}$$

$= 19.999998 = 20$ (quarters)

Number of years $= \frac{20}{4} = 5$

Programmed Solution

You can use preprogrammed financial calculators to find n by the same procedure previously used to find FV, PV.

(Set P/Y = 4) [2nd] (CLR TVM) 2000 [±] [PV]

2440.38 [FV] 4 [I/Y] [CPT] [N] [19.99999671]

At 10% compounded quarterly, \$2000.00 will grow to \$2440.38 in 20 quarters or five years.

Note: Another way to solve Example 4.1A is first to rearrange Formula 3.1A to solve for n:

$$\begin{aligned} FV &= PV(1+i)^n \\ (1+i)^n &= \frac{FV}{PV} \\ n\ln(1+i) &= \ln\left(\frac{FV}{PV}\right) \\ n &= \frac{\ln\left(\frac{FV}{PV}\right)}{\ln(1+i)} \end{aligned}$$

Recall that the total number of conversion periods is n. To convert to years if compounding semi-annually, divide n by 2; if compounding quarterly, divide by 4; if compounding monthly, divide by 12; and if compounding daily, divide by 365. Remember, you do not have to memorize this equation if you understand the principles of formula rearrangement.

EXAMPLE 4.1B

How long does it take for money to double
(i) at 5% p.a.?
(ii) at 10% p.a.?

SOLUTION

While neither PV nor FV is given, any sum of money may be used as principal. For this calculation, a convenient value for the principal is \$1.00.

PV = 1.00; FV = 2.00; I/Y = 5; P/Y = 1

(i) At 5% p.a., $i = 5\% = 0.05$

$$\begin{aligned} 2 &= 1(1.05)^n \\ 1.05^n &= 2 \\ n\ln 1.05 &= \ln 2 \\ 0.048790n &= 0.693147 \\ n &= 14.206699 \text{ (years)} \end{aligned}$$

Programmed Solution

(Set P/Y = 1) [2nd] (CLR TVM) 1 [±] [PV] 2 [FV] 5 [I/Y]

[CPT] [N] [14.20669908] (years)

At 5% p.a., money doubles in approximately 14 years and 3 months.

(ii) At 10% p.a., I/Y = 10; P/Y = 1; $i = 10\% = 0.10$

$$
\begin{aligned}
2 &= 1(1 + 0.10)^n \\
1.10^n &= 2 \\
n \ln 1.10 &= \ln 2 \\
0.095310n &= 0.693147 \\
n &= 7.272541 \text{ (years)}
\end{aligned}
$$

Programmed Solution

[2nd] (CLR TVM) 1 [±] [PV] 2 [FV]

10 [I/Y] [CPT] [N] [7.272540897] (years)

At 10% p.a., money doubles in approximately 7 years and 4 months.

DID YOU KNOW?

The Rule of 70

Did you know that there is a quick way to estimate the number of conversion periods needed to double an amount of money? It is known as the *Rule of 70*. According to this rule, the number of conversion periods required to double money is 70 divided by the periodic rate of interest *i*.

Suppose we want to estimate how long it will take to double money if the interest rate is 10% compounded semi-annually. Applying the Rule of 70, we find it will take about 7 years (that is 70/5 half-year conversion periods). By comparison, if we calculate the time using the standard formulas, we get a result of about 7.103 years (that is, about 14.207 half-year conversion periods).

EXAMPLE 4.1C

How long will it take for money to triple at 6% compounded monthly?

SOLUTION

Let PV = 1; then FV = 3; I/Y = 6; P/Y = 12; $i = \frac{6\%}{12} = 0.5\% = 0.005$

$$
\begin{aligned}
3 &= 1(1.005)^n \\
1.005^n &= 3 \\
n \ln 1.005 &= \ln 3 \\
0.004988n &= 1.098612 \\
n &= 220.271 \text{ (months)}
\end{aligned}
$$

Programmed Solution

(Set P/Y = 12) [2nd] (CLR TVM) 1 [±] [PV] 3 [FV]

6 [I/Y] [CPT] [N] [220.2713073] (years)

At 6% compounded monthly, money triples in approximately 18 years and 4 months.

B. Equated date

Chapter 3, Section 3.5, considered the concept of *equivalence* of values when using the compound interest method. In solving problems of equivalence, the unknown value was always the size of a payment at the selected focal date. While this type is the most frequently arising problem, occasionally the value to be found is the focal date or the interest rate.

The **equated date** is the date on which a single sum of money is equal to the sum of two or more dated sums of money. To find an equated date, an equation of values can be set up by the same technique used in Section 3.5. However, solving the equation for n requires the same technique as used in Section 4.1A. Since this method involves the use of logarithms, you can solve the problem using an electronic calculator as long as the calculator is equipped with the natural logarithm function (LN key).

You may also use a financial calculator that is preprogrammed with the time value of money (TVM) worksheet. After entering all of the relevant data into the calculator, press CPT N. Note that the units of N calculated are in the compounding period. For example, if you are compounding quarterly, your answer for N is in quarters. If you wish to determine a specific date to solve your problem, use the DATE worksheet within the calculator. An explanation of how to use this worksheet is included on the CD attached to this text.

Excel

EXAMPLE 4.1D

A financial obligation requires the payment of \$2000.00 in 6 months, \$3000.00 in 15 months, and \$5000.00 in 24 months. When can the obligation be discharged by the single payment equal to the sum of the required payments if money is worth 9% p.a. compounded monthly?

SOLUTION

The single payment equal to the sum of the required payments is \$2000.00 plus \$3000.00 plus \$5000.00, or \$10 000.00. Select as the focal date "now." Let the number of compounding periods from the focal date to the equated date be represented by n. Since the compounding is done monthly, n will be a number of months, I/Y = 9, P/Y = 12, and $i = \frac{9}{12}\% = 0.75\% = 0.0075$. The method and data are shown graphically in Figure 4.1.

Let E_1, E_2, and E_3 represent the equivalent values of the original payments at the focal date as shown in Figure 4.1.

Let E_4 represent the equivalent value of the single payment of \$10 000.00 at the focal date.

The equation of values can now be set up.

$$E_4 = E_1 + E_2 + E_3$$

$$10\,000.00(1.0075)^{-n} = 2000.00(1.0075)^{-6} + 3000.00(1.0075)^{-15} + 5000.00(1.0075)^{-24}$$

$$10\,000.00(1.0075)^{-n} = 2000.00(0.956158) + 3000.00(0.893973) + 5000.00(0.835831)$$

$$10\,000.00(1.0075)^{-n} = 1912.32 + 2681.92 + 4179.16$$

$$10\,000.00(1.0075)^{-n} = 8773.40$$

$$(1.0075)^{-n} = \frac{8773.40}{10\,000.00}$$

$$(1.0075)^{-n} = 0.87734$$

$$-n(\ln 1.0075) = \ln 0.87734$$

$$-n(0.007472) = -0.130861$$

$$n = \frac{0.130861}{0.007472}$$

$$n = 17.513477$$

The equated date is about 17.5 months from now.

FIGURE 4.1 **Graphical Representation of Method and Data**

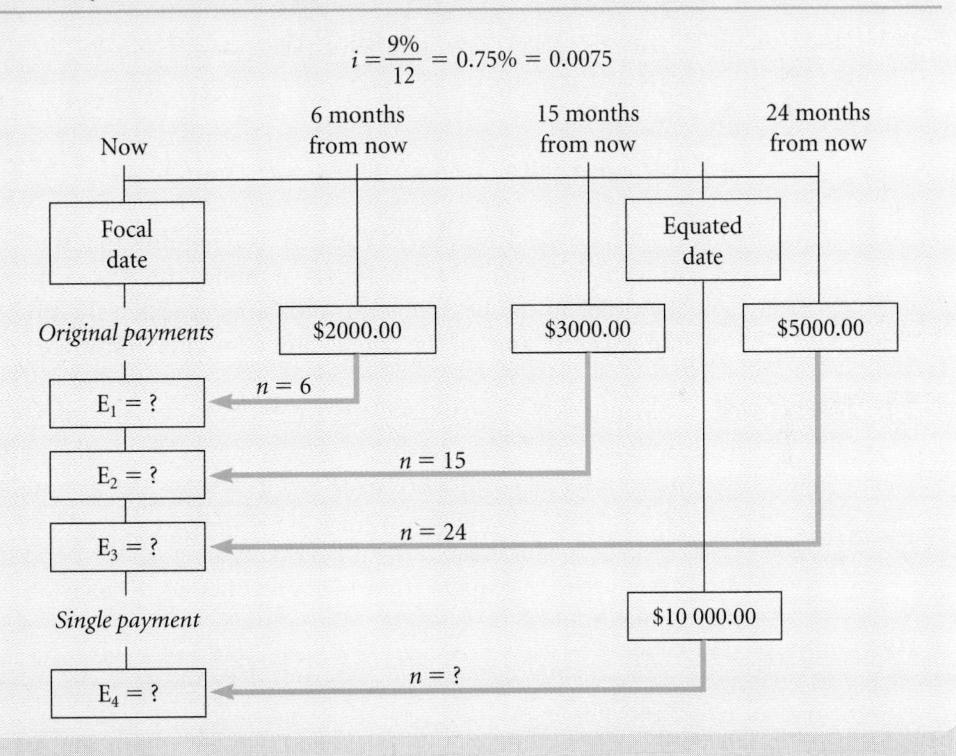

Programmed Solution

First simplify the equation $E_4 = E_1 + E_2 + E_3$.

$10\,000(1.0075)^{-n} = 2000.00(1.0075)^{-6} + 3000(1.0075)^{-15} + 5000(1.0075)^{-24}$

$2000(1.0075)^{-6}$

(Set P/Y = 12) [2nd] (CLR TVM) 2000 [FV]

9 [I/Y] 6 [N] [CPT] [PV] [-1912.316036] [±] [STO] 1

$3000(1.0075)^{-15}$

[2nd] (CLR TVM) 3000 [FV] 9 [I/Y] 15 [N]

[CPT] [PV] [-2681.917614] [±] [STO] 2

$5000(1.0075)^{-24}$

2nd (CLR TVM) 5000 FV 9 I/Y

24 N CPT PV -4179.15702 ± STO 3

RCL 1 + RCL 2 + RCL 3 =

or

$10\,000(1.0075)^{-n} = 1912.316036 + 2681.917614 + 4179.15702$

Remember to remove the negative (cash outflow) signs.

$10\,000(1.0075)^{-n} = 8773.39067$

Now solve the simplified equation

$10\,000(1.0075)^{-n} = 8773.39067$

in which FV = 10 000; PV = 8773.39067; $i = 0.75\%$

2nd (CLR TVM) 10 000 FV 8773.39067

± PV 9 I/Y CPT N 17.51358137

The equated date is about 17.5 months from now.

While the definition of equated date requires that the single sum of money be equal to the sum of the dated sums of money considered, the method used in solving the problem can be applied to problems in which the single sum of money does not equal the sum of the dated sums of money considered.

EXAMPLE 4.1E

A loan is to be repaid by three equal payments of \$1500.00 due now, two years from now, and four years from now, respectively. When can the obligation be paid off by a single payment of \$5010.00 if interest is 10% compounded annually?

SOLUTION

Select as the focal date "now." Let the number of compounding periods from the focal date to the equated date be represented by n. Since the compounding is done annually, n will be a number of years, I/Y = 10, P/Y = 1, and $i = 10\% = 0.10$.

Let E_1, E_2, and E_3 represent the equivalent values of the original payments at the focal date.

$$E_1 = 1500.00 = 1500.00$$
$$E_2 = 1500.00(1.10)^{-2} = 1500.00(0.826446) = 1239.67$$
$$E_3 = 1500.00(1.10)^{-4} = 1500.00(0.6830135) = 1024.52$$
$$E_1 + E_2 + E_3 = 3764.19$$

Let E_4 represent the equivalent value of the single payment of \$5010.00 at the focal date.

$$E_4 = 5010.00(1.10)^{-n}$$
$$E_4 = E_1 + E_2 + E_3$$
$$5010.00(1.10)^{-n} = 3764.19$$
$$(1.10)^{-n} = 0.7513353$$
$$-n(\ln 1.10) = \ln 0.7513353$$
$$-0.095310n = -0.285903$$
$$n = 3$$

Programmed Solution

$E_4 = E_1 + E_2 + E_3$

$5010.00(1.10)^{-n} = 1500.00 + 1500(1.10)^{-2} + 1500(1.10)^{-4}$

$1500(1.10)^{-2}$

(Set P/Y = 1) 2nd (CLR TVM) 1500 FV 10 I/Y 2 N CPT PV -1239.669421 ± STO 1

$1500(1.10)^{-4}$

2nd (CLR TVM) 1500 FV 10 I/Y 4 N CPT PV -1024.520183 ± STO 2

RCL1 + RCL2 =

$5010.00(1.10)^{-n} = 1500.00 + 1239.67 + 1024.52$

$5010.00(1.10)^{-n} = 3764.19$

2nd (CLR TVM) 5010 FV 3764.19 ± PV 10 I/Y CPT N 2.999713326

The single payment should be made three years from now.

EXAMPLE 4.1F

A loan of $2000.00 taken out today is to be repaid by a payment of $1200.00 in six months and a final payment of $1000.00. If interest is 12% compounded monthly, when should the final payment be made?

SOLUTION

Let the focal point be "now"; I/Y = 12; P/Y = 12; $i = \frac{12\%}{12} = 1.0\% = 0.01$.

$$\begin{aligned}
2000.00 &= 1200.00(1.01)^{-6} + 1000.00(1.01)^{-n}\\
2000.00 &= 1200.00(0.942045) + 1000.00(1.01)^{-n}\\
2000.00 &= 1130.45 + 1000.00(1.01)^{-n}\\
869.55 &= 1000.00(1.01)^{-n}\\
(1.01)^{-n} &= 0.86955\\
-n(\ln 1.01) &= \ln 0.86955\\
-0.009950n &= -0.139779\\
n &= 14.0478 \text{ (months)}
\end{aligned}$$

Programmed Solution

$2000.00 = 1200.00(1.01)^{-6} + 1000(1.01)^{-n}$

$1200.00(1.01)^{-6}$

(Set P/Y = 12) 2nd (CLR TVM) 1200 FV 12 I/Y 6 N CPT PV -1130.454282

$$\begin{aligned}
2000.00 &= 1130.454282 + 1000(1.01)^{-n}\\
869.545717 &= 1000(1.01)^{-n}
\end{aligned}$$

869.545717 ± PV 1000 FV 12 I/Y CPT N 14.048213 (months)

$n = 427$ *days* ——— $\left(\frac{14.048213}{12}\right)(365)$

The final payment should be made in 427 days.

EXERCISE 4.1

If you choose, you can use Excel's ***Number of Compounding Periods (NPER)*** function to answer the questions indicated below. Refer to **NPER** on the Spreadsheet Template Disk to learn how to use this Excel function.

A.

1. Determine the number of compounding periods for each of the following six investments.

	Principal	Future Value	Interest Rate	Frequency of Conversion
(a)	$2600.00	$6437.50	7%	annually
(b)	1240.00	1638.40	4%	quarterly
(c)	560.00	1350.00	9%	monthly
(d)	3480.00	4762.60	8%	semi-annually
(e)	950.00	1900.00	7.5%	quarterly
(f)	1300.00	3900.00	6%	semi-annually

2. Find the equated date at which the original payments are equivalent to the single payment for each of the following four sets of payments.

	Original Payments	Interest Rate	Frequency of Conversion	Single Payment
(a)	$400 due in 9 months and $700 due in 21 months	12%	quarterly	$1256.86
(b)	$1200 due today and $2000 due in 5 years	8%	semi-annually	$3808.70
(c)	$1000 due 8 months ago, $1200 due in 6 months, and $1500 due in 16 months	9%	monthly	$3600.00
(d)	$600 due in 2 years, $800 due in 3.5 years, and $900 due in 5 years	10%	quarterly	$1800.00

B. Answer each of the following questions.

1. How long will it take $400.00 to accumulate to $760.00 at 7% p.a. compounded semi-annually?
2. In how many days will $580.00 grow to $600.00 at 4.5% p.a. compounded monthly?
3. In how many years will money quadruple at 8% compounded quarterly?
4. In how many months will money triple at 9% compounded semi-annually?
5. If an investment of $800.00 earned interest of $320.00 at 6% compounded monthly, for how many years was the money invested?
6. A loan of $2000.00 was repaid together with interest of $604.35. If interest was 8% compounded quarterly, for how many months was the loan taken out?

7. If you borrowed \$1000.00 on May 1, 2000, at 10% compounded semi-annually and interest on the loan amounts to \$157.63, on what date is the loan due?

8. A promissory note for \$600.00 dated May 15, 2002 requires an interest payment of \$150.00 at maturity. If interest is at 9% compounded monthly, determine the due date of the note.

9. A non-interest-bearing promissory note for \$1500.00 was discounted at 5% p.a. compounded quarterly. If the proceeds of the note were \$1375.07, how many months before the due date was the note discounted?

10. A five-year, \$1000.00 note bearing interest at 9% compounded annually was discounted at 12% compounded semi-annually yielding proceeds of \$1416.56. How many months before the due date was the discount date?

11. A contract requires payments of \$4000.00 today, \$5000.00 in three years, and \$6000.00 in five years. When can the contract be fulfilled by a single payment equal to the sum of the required payments if money is worth 9% p.a. compounded monthly?

12. A financial obligation requires the payment of \$500.00 in nine months, \$700.00 in fifteen months, and \$600.00 in 27 months. When can the obligation be discharged by a single payment of \$1600.00 if interest is 10% compounded quarterly?

13. When Brenda bought Sheridan Service from Ken, she agreed to make three payments of \$6000.00 each in one year, three years, and five years respectively. Because of initial cash flow difficulties, Brenda offered to pay \$8000.00 in two years and a second payment of \$10 000.00 at a later date. When should she make the second payment if interest is 9.75% compounded semi-annually?

14. Leo sold a property and is to receive \$3000.00 in six months, \$4000.00 in 24 months, and \$5000.00 in 36 months. The deal was renegotiated after nine months at which time Leo received a payment of \$7000.00; he was to receive a further payment of \$6000.00 later. When should Leo receive the second payment if money is worth 11% compounded quarterly?

4.2 Finding *i* and Related Problems

A. Finding the periodic rate *i* and the nominal annual rate of interest *j*

If the original principal PV, the future value FV, and the number of conversion periods n are known, the periodic rate of interest (conversion rate) i can be determined by substituting in Formula 3.1A, $\text{FV} = \text{PV}(1 + i)^n$ and solving for i. The nominal annual rate of interest j can then be found by multiplying i by the number of conversion periods per year m.

EXAMPLE 4.2A What is the annual compounding rate if $200 accumulates to $318.77 in eight years?

SOLUTION PV = 200.00; FV = 318.77; P/Y = 1; $m = 1$; $n = 8$

$$318.77 = 200.00(1 + i)^8 \quad \text{— } i \text{ is an } annual \text{ rate}$$
$$(1 + i)^8 = 1.59385$$
$$[(1 + i)^8]^{\frac{1}{8}} = 1.59385^{0.125} \quad \text{— raise each side to the power } \frac{1}{8}$$
$$1 + i = 1.59385^{0.125}$$
$$1 + i = 1.060000$$
$$i = 0.060000$$
$$i = 6.0\% \quad \text{— the desired annual rate}$$

The annual compounding rate is 6.0%.

Programmed Solution

You can use preprogrammed financial calculators to find i by the same procedure used previously to determine FV or PV. That is, select the compound interest mode, enter the given variables FV, PV, and N, and retrieve the fourth variable I/Y.

(Set P/Y = 1) [2nd] (CLR TVM) 200 [±] [PV] 318.77 [FV]

8 [N] [CPT] [I/Y] [6.000016] (annual)

Note: It is important to know how to rearrange the terms of an equation. For instance, another way to solve Example 4.2A is first to rearrange Formula 3.1A to solve for i:

$$FV = PV(1 + i)^n$$
$$(1 + i)^n = \frac{FV}{PV}$$
$$1 + i = \left(\frac{FV}{PV}\right)^{\frac{1}{n}}$$
$$i = \left(\frac{FV}{PV}\right)^{\frac{1}{n}} - 1$$

Recall that i is the periodic rate of interest. If interest is calculated m times per year, then the nominal annual rate of interest $j = m(i)$.

Remember, you do not have to memorize this equation if you understand the principles of formula rearrangement.

Note: In the financial calculator that you use, you must determine whether the interest rate to be entered represents i, the periodic rate of interest, or j, the nominal rate of interest. In the BAII Plus calculator the [I/Y] to be used equates to j in the formula. Thus, the variables in the formula $i = j/m$ can be translated into i = [I/Y] / [C/Y] for use in the calculator. The value of I/Y obtained from the calculator is to be expressed as the nominal rate. For example, if C/Y = 4, or quarterly, and the calculator determines that I/Y = 8.5, then the rate is expressed as 8.5% p.a. compounded quarterly.

EXAMPLE 4.2B

Find the nominal annual rate of interest compounded quarterly if \$1200.00 accumulates to \$2064.51 in five years.

SOLUTION

PV = 1200.00; FV = 2064.51; P/Y = 4; $n = 20$; $m = 4$

$$2064.51 = 1200.00(1 + i)^{20} \quad \text{—— } i \text{ is a } \textit{quarterly} \text{ rate}$$
$$(1 + i)^{20} = 1.720425$$
$$1 + i = 1.720425^{0.05} \quad \text{—— raise both sides to the power } \frac{1}{20}\text{, that is, 0.05}$$
$$1 + i = 1.027500$$
$$i = 0.027500$$
$$i = 2.75\%$$

The quarterly compounding rate is 2.75%.

Programmed Solution

(Set P/Y = 4) [2nd] (CLR TVM) 1200 [±] [PV]

2064.51 [FV] 20 [N] [CPT] [I/Y] [11.00]

The nominal annual rate of interest is 11.0% per annum compounded quarterly. The periodic rate of interest would be $i = j/m = 11.0\%/4 = 2.75\%$ per quarter.

EXAMPLE 4.2C

At what nominal rate of interest compounded quarterly will money double in four years?

SOLUTION

While neither PV nor FV are given, any sum of money may be used as principal. For this calculation, a convenient value for the principal is \$1.00.

PV = 1; FV = 2; P/Y = 4; $n = 16$; $m = 4$

$$2 = 1(1 + i)^{16} \quad \text{—— } i \text{ is a } \textit{quarterly} \text{ rate}$$
$$(1 + i)^{16} = 2$$
$$1 + i = 2^{\frac{1}{16}}$$
$$1 + i = 2^{0.0625}$$
$$1 + i = 1.044274$$
$$i = 0.44274$$
$$i = 4.4274\%$$

Programmed Solution

(Set P/Y = 4) [2nd] (CLR TVM) 1 [±] [PV] 2 [FV] 16 [N]

[CPT] [I/Y] [17.709513]

The nominal annual rate is 17.709513% per annum compounded quarterly. The quarterly rate is 17.709513%/4 = 4.427%.

EXAMPLE 4.2D Suppose \$1000.00 earns interest of \$93.81 in one year.

(i) What is the nominal annual rate of interest compounded annually?
(ii) What is the nominal annual rate of interest compounded monthly?

SOLUTION (i) PV = 1000.00; I = 93.81; FV = PV + I = 1093.81; P/Y = 1; $n = 1$

$$1093.81 = 1000.00(1 + i)^1$$ — *i* is an *annual* rate ($m = 1$)

$$1 + i = 1.09381$$
$$i = 0.09381$$
$$i = 9.381\%$$

Programmed Solution

(Set P/Y = 1) [2nd] (CLR TVM) 1000 [±] [PV]

1093.81 [FV] 1 [N] [CPT] [I/Y] [9.381]

The annual rate of interest is 9.381%.

Notice that both the nominal rate and the periodic rate of interest are 9.381%, since interest is compounded annually.

(ii) P/Y = 12

$$1093.81 = 1000.00(1 + i)^{12}$$ — *i* is a *monthly* rate ($m = 12$)

$$(1 + i)^{12} = 1.09381$$
$$1 + i = 1.09381^{\frac{1}{12}}$$
$$1 + i = 1.09381^{0.083333}$$
$$1 + i = 1.007500$$
$$i = 0.007500$$
$$i = 0.75\%$$

Programmed Solution

(Set P/Y = 12) [2nd] (CLR TVM) 1000 [±] [PV]

1093.81 [FV] 12 [N] [CPT] [I/Y] [9.000286]

The nominal annual rate of interest compounded monthly is 9% p.a. The periodic rate of interest is 9%/12 = 0.75% per month.

Excel You can use Excel's ***Compound Interest Rate per Period (RATE)*** function to find the periodic rate of interest *i*. Refer to **RATE** on the Spreadsheet Template Disk to learn how to use this Excel function.

EXERCISE 4.2

A. Answer each of the following.

Find the nominal annual rate of interest for each of the following investments.

	Principal	Future Value	Time Due	Frequency of Conversion
1.	$1400.00	$1905.21	7 years	annually
2.	2350.00	3850.00	5 years	quarterly
3.	690.00	1225.00	6 years	monthly
4.	1240.00	2595.12	12 years	semi-annually
5.	3160.00	5000.00	4 years; 9 months	quarterly
6.	900.00	1200.00	3 years; 8 months	monthly

B. Solve each of the following.

1. What is the nominal annual rate of interest compounded quarterly at which $420.00 will accumulate to $1000.00 in nine years and six months?

2. A principal of $2000.00 compounded monthly amounts to $2800.00 in 7.25 years. What is the nominal annual rate of interest?

3. At what nominal annual rate of interest will money double itself in
(a) six years, nine months if compounded quarterly?
(b) nine years, two months if compounded monthly?

4. What is the nominal annual rate of interest at which money will triple itself in 12 years
(a) if compounded annually?
(b) if compounded semi-annually?

5. Yin Li deposited $800.00 into a savings account that compounded interest monthly. What nominal annual rate compounded monthly was earned on the investment if the balance was $952.75 in five years?

6. An investment of $4000.00 earned interest semi-annually. If the balance after $6^1/_2$ years was $6,000.00, what nominal annual rate compounded semi-annually was charged?

7. Surinder borrowed $1200.00 and agreed to pay $1400.00 in settlement of the debt in three years, three months. What quarterly rate of interest was charged on the debt?

8. A debt of $600.00 was to be repaid in 15 months. If $750.14 was repaid, what was the monthly compounded rate of interest charged?

4.3 Effective and Equivalent Interest Rates

A. Effective rate of interest

In Example 4.2D, compounding at an annual rate of interest of 9.381% has the same effect as compounding at 9.0% p.a. compounded monthly since, in both cases, the interest amounts to $93.81.

The annual rate of 9.381% is called the **effective rate of interest**. This rate is defined as the rate of interest compounded annually that yields the same amount of interest as a nominal annual rate of interest compounded a number of times per year other than one.

Converting nominal rates of interest to effective rates is the method used for comparing nominal rates of interest. Since the effective rates of interest are the equivalent rates of interest compounded annually, they may be obtained for any set of nominal rates by computing the accumulated value of $1 after one year for each of the nominal rates under consideration.

Effective Rates Using the BAII Plus

The BAII Plus is programmed in the 2nd function to quickly and efficiently calculate effective interest rates by inputting the nominal rate and the number of compounds. You can also calculate the nominal rate if you know the effective rate.

To go from nominal to effective the process is:

1. 2nd (IConv) (2-key).
2. Enter the nominal rate.
3. Arrow down to C/Y= and enter the number of times interest compounds in a year.
4. Arrow up to Eff= and press CPT.

To go from effective to nominal the process is:

1. 2nd (IConv).
2. Arrow down to Eff = and enter the effective rate.
3. Arrow down to C/Y= and enter the compounding number relating to the nominal rate you are converting to.
4. Arrow down to Nom= and press CPT.

Equivalent Rates Using the BAII Plus

The BAII Plus can be used to calculate equivalent interest rates by using the effective rate as a constant to make equivalent calculations.

The process is

1. 2nd (IConv).
2. Enter any nominal rate.
3. Arrow down to C/Y = and enter the compounds relating to the nominal rate entered in step 2.
4. Arrow up to Eff= and press CPT.
5. Arrow down and change C/Y= to the compounds you want to convert to.

6. Arrow down to Nom= and press CPT.
7. Repeat the process as many times as required to calculate all equivalent rates you are interested in.

EXAMPLE 4.3A

Assume you are given a choice of a term deposit paying 7.2% compounded monthly or an investment certificate paying 7.25% compounded semi-annually. Which rate offers the higher rate of return?

SOLUTION

The investment certificate offers the higher nominal rate of return while the term deposit offers the higher compounding frequency. Because of the different compounding frequencies, the two nominal rates are not directly comparable. To determine which nominal rate offers the higher rate of return, we need to determine the effective rates for the two given rates.

For the term deposit:
I/Y = 7.2; P/Y = 12; $i = 0.6\% = 0.006$; $m = 12$;
the accumulated value of $1 after one year,
$FV = 1(1.006)^{12} = 1.074424$

The decimal fraction 0.074424 is the interest earned in one year and represents the effective rate of interest = 7.44242%.

For the investment certificate:
I/Y = 7.25; P/Y = 2; $i = 3.625\% = 0.03625$; $m = 2$;
the accumulated value of $1 after one year,
$FV = 1(1.03625)^2 = 1.073814$

The decimal fraction 0.073814 represents the effective rate of interest = 7.3814%.

Since the term deposit has the higher effective rate, it offers the higher rate of return.

Programmed Solution

For the term deposit:

$(1 + i)^{12} = (1.0006)^{12} = 1.0744242$

(Set P/Y = 12) 2nd (CLR TVM) 1 ± PV 7.2 I/Y
12 N CPT FV 1.074424

Alternatively

2nd (IConv) Nom = 7.2; C/Y = 12; Eff = CPT 7.4424

For the investment certificate:

$(1 + i)^2 = (1.03625)^2 = 1.073814$

(Set P/Y = 2) 2nd (CLR TVM) 1 ± PV 7.25 I/Y
2 N CPT FV 1.073814

Alternatively

[2nd] (IConv) Nom = 7.25; C/Y = 2; Eff = [CPT] 7.3814

The nominal annual rate is 7.4424% for the term deposit and 7.3814% for the investment certificate.

EXAMPLE 4.3B

How much better is a rate of 8.4% compounded monthly than 8.4% compounded quarterly?

SOLUTION

For 8.4% compounded monthly:
I/Y = 8.4; P/Y = 12; $i = 0.7\% = 0.007$; $m = 12$;
the accumulated value of $1 after one year,
$\text{FV} = 1(1.007)^{12} = 1.087311$

The decimal fraction 0.087311 is the interest earned in one year and represents the effective rate of interest = 8.7311%.

For 8.4% compounded quarterly:
I/Y = 8.4; P/Y = 4; $i = 2.1\% = 0.021$; $m = 4$;
the accumulated value of $1 after one year,
$\text{FV} = 1(1.021)^{4} = 1.086683$

The decimal fraction 0.086683 represents the effective rate of interest = 8.6683%.

Difference = 8.7311% − 8.6683% = 0.0628%.

Programmed Solution

For 8.4% compounded monthly:
$(1 + i)^{12} = (1.007)^{12} = 1.087311$

(Set P/Y = 12) [2nd] (CLR TVM) 1 [±] [PV] 8.4 [I/Y] 12 [N] [CPT] [FV] 1.087311

Alternatively

[2nd] (IConv) Nom = 8.4; C/Y = 12; Eff = [CPT] 8.7311

For 8.4% compounded quarterly:

$(1 + i)^{4} = (1.021)^{4} = 1.086683$

(Set P/Y = 4) [2nd] (CLR TVM) 1 [±] [PV] 8.4 [I/Y] 4 [N] [CPT] [FV] 1.086683

Alternatively

[2nd] (IConv) Nom = 8.4; C/Y = 4; Eff = [CPT] 8.6683

The effective rate is 8.7311% with 8.4% compounded monthly and 8.6683% for 8.4% compounded quarterly.

The effective rates of interest can also be determined by using Formula 4.1 obtained from the method of calculation used in Examples 4.3A and 4.3B.

The formula is obtained as follows.

Let the nominal annual rate of interest be compounded m times per year and let the interest rate per conversion period be i.
Then the accumulated amount after one year is $FV_1 = PV(1 + i)^m$.
Let the corresponding effective annual rate of interest be f.
Then the accumulated amount after one year is $FV_1 = PV(1 + f)^1$.

$PV(1 + f)^1 = PV(1 + i)^m$ ——— the amounts are equal by definition
$1 + f = (1 + i)^m$ ——— divide both sides by PV

$$f = (1 + i)^m - 1$$ ——— Formula 4.1

EXAMPLE 4.3C

Determine the effective rate of interest corresponding to 9% p.a. compounded
(i) monthly;
(ii) quarterly;
(iii) semi-annually;
(iv) annually;
(v) daily.

SOLUTION

(i) $i = \left(\frac{9\%}{12}\right) = 0.0075; \quad m = 12$

$f = (1 + i)^m - 1$ ——— using Formula 4.1
$= (1 + 0.0075)^{12} - 1$
$= 1.093807 - 1$
$= 0.093807$
$= 9.381\%$

(ii) $i = \left(\frac{9\%}{4}\right) = 0.0225; \quad m = 4$

$f = (1.0225)^4 - 1$
$= 1.093083 - 1$
$= 0.093083$
$= 9.308\%$

(iii) $i = \left(\frac{9\%}{2}\right) = 0.045; \quad m = 2$

$f = (1.045)^2 - 1$
$= 1.092025 - 1$
$= 9.2025\%$

(iv) $i = 9\% = 0.09; \quad m = 1$

$f = (1.09)^1 - 1$
$= 9.000\%$

(v) $i = \left(\frac{9\%}{365}\right) = 0.000246; \quad m = 365$

$f = (1.000246)^{365} - 1$
$= 1.094172 - 1$
$= 9.417\%$

Summary of Results

For a nominal annual rate of 9% p.a., effective rates are

when compounding annually ($m = 1$)	$f = 9.000\%$
when compounding semi-annually ($m = 2$)	$f = 9.2025\%$
when compounding quarterly ($m = 4$)	$f = 9.308\%$
when compounding monthly ($m = 12$)	$f = 9.381\%$
when compounding daily ($m = 365$)	$f = 9.417\%$

Programmed Solution

For 9% compounded monthly:

$(1 + i)^{12} = (1.0075)^{12} = 1.093807$

(Set P/Y = 12) **2nd** (CLR TVM) 1 **±** **PV** 9 **I/Y**

12 **N** **CPT** **FV** 1.093807

Alternatively

2nd (IConv) Nom = 9; C/Y = 12; Eff = **CPT** 9.3807

For 9% compounded quarterly:

$(1 + i)^4 = (1.0225)^4 = 1.093083$

(Set P/Y = 4) **2nd** (CLR TVM) 1 **±** **PV** 9 **I/Y**

4 **N** **CPT** **FV** 1.093083

Alternatively

2nd (IConv) Nom = 9; C/Y = 4; Eff = **CPT** 9.3083

For 9% compounded semi-annually:

$(1 + i)^2 = (1.045)^2 = 1.092025$

(Set P/Y = 2) **2nd** (CLR TVM) 1 **±** **PV** 9 **I/Y**

2 **N** **CPT** **FV** 1.092025

Alternatively

2nd (IConv) Nom = 9; C/Y = 2; Eff = **CPT** 9.2025

For 9% compounded annually:

$(1 + i)^1 = (1.09)^1 = 1.09$

(Set P/Y = 1) **2nd** (CLR TVM) 1 **±** **PV** 9 **I/Y**

1 **N** **CPT** **FV** 1.09

Alternatively
2nd (IConv) Nom = 9; C/Y = 1; Eff = **CPT** 9

For 9% compounded daily:
$(1 + i)^{365} = (1.0002466)^{365} = 1.094172$ (i has been rounded)

(Set P/Y = 365) **2nd** (CLR TVM) 1 **±** **PV** 9 **I/Y**
365 **N** **CPT** **FV** **1.094162**

Alternatively
2nd (IConv) Nom = 9; C/Y = 365; Eff = **CPT** 9.4162

POINTERS AND PITFALLS

For nominal annual interest rates and effective rates, the following two points are always true:

1. The nominal annual rate is the effective rate of interest *only* if the number of conversion periods per year is 1, that is, if compounding annually.
2. For a given nominal annual rate, the effective rate of interest increases as the number of conversion periods per year increases.

EXAMPLE 4.3D

You have money to invest in interest-earning deposits. You have determined that suitable deposits are available at your bank paying 6.5% p.a. compounded semi-annually, at a local trust company paying 6.625% p.a., and at your credit union paying 6.45% p.a. compounded monthly. What institution offers the best rate of interest?

SOLUTION

Since the methods of conversion differ, the interest rates are not directly comparable. To make the rates comparable, determine the effective rates of interest corresponding to the nominal annual rates.

For the bank:

$$i = \left(\frac{6.5\%}{2}\right) = 0.0325; \quad m = 2$$

$$f = (1 + 0.0325)^2 - 1 = 1.066056 - 1 = 0.066056 = 6.606\%$$

For the trust company:

$$i = 6.625 = 0.06625; \quad m = 1$$

$$f = i = 6.625\%$$

For the credit union:

$$i = \left(\frac{6.45\%}{12}\right) = 0.005375; \quad m = 12$$

$$f = (1.005375)^{12} - 1 = 1.066441 - 1 = 0.066441 = 6.644\%$$

While the nominal rate offered by the credit union is lowest, the corresponding effective rate of interest is highest due to the higher frequency of conversion. The rate offered by the credit union is best.

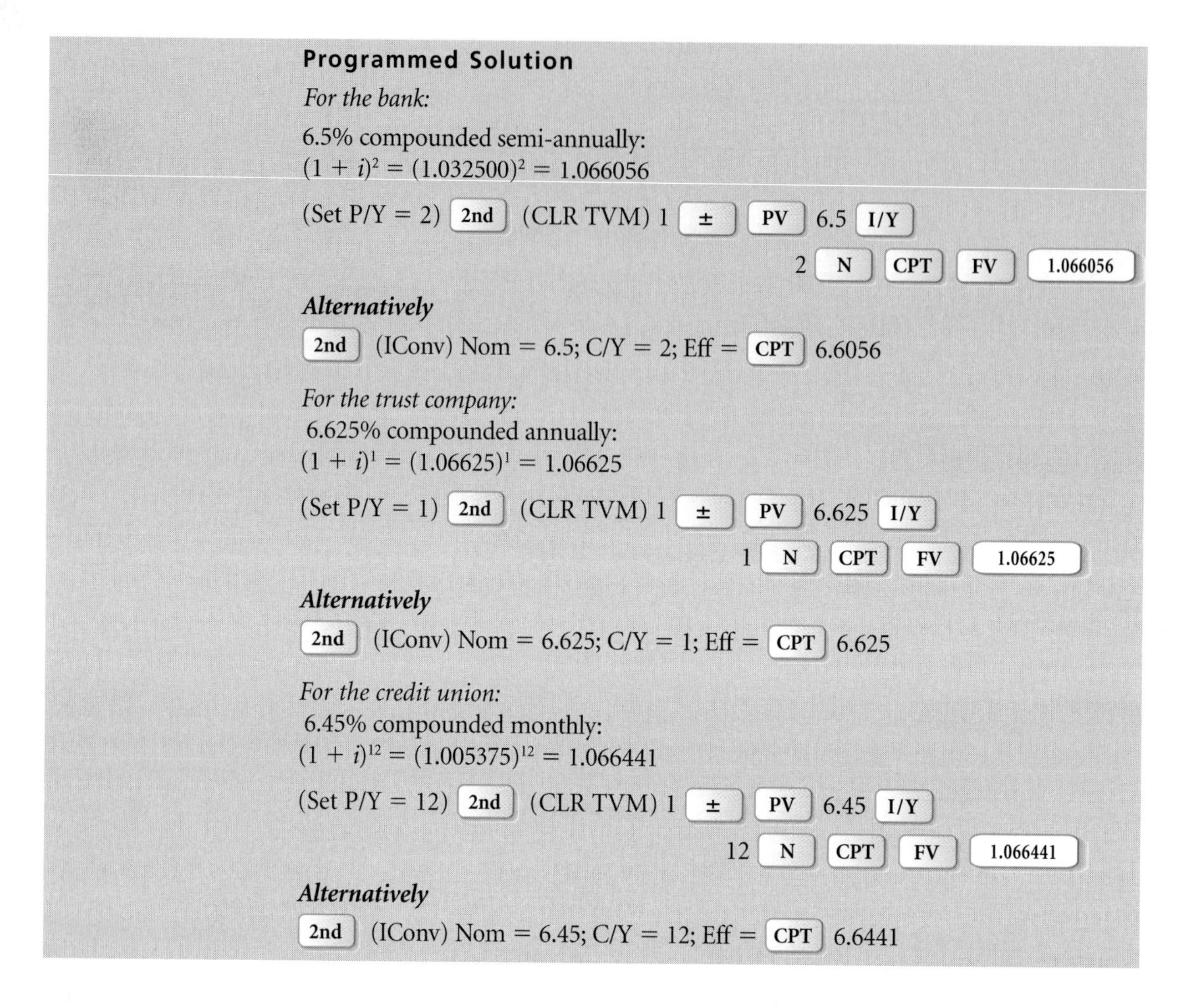

Programmed Solution

For the bank:

6.5% compounded semi-annually:
$(1 + i)^2 = (1.032500)^2 = 1.066056$

(Set P/Y = 2) 2nd (CLR TVM) 1 ± PV 6.5 I/Y
2 N CPT FV 1.066056

Alternatively

2nd (IConv) Nom = 6.5; C/Y = 2; Eff = CPT 6.6056

For the trust company:
6.625% compounded annually:
$(1 + i)^1 = (1.06625)^1 = 1.06625$

(Set P/Y = 1) 2nd (CLR TVM) 1 ± PV 6.625 I/Y
1 N CPT FV 1.06625

Alternatively

2nd (IConv) Nom = 6.625; C/Y = 1; Eff = CPT 6.625

For the credit union:
6.45% compounded monthly:
$(1 + i)^{12} = (1.005375)^{12} = 1.066441$

(Set P/Y = 12) 2nd (CLR TVM) 1 ± PV 6.45 I/Y
12 N CPT FV 1.066441

Alternatively

2nd (IConv) Nom = 6.45; C/Y = 12; Eff = CPT 6.6441

B. Equivalent rates

Notice that in Example 4.2D, two different nominal annual rates of interest (9.381% compounded annually and 9% compounded monthly) produced the same future value ($1093.81) for a given principal ($1000.00) after one year. Interest rates that increase a given principal to the same future value over the same period of time are called **equivalent rates**—9.381% compounded annually and 9% compounded monthly are equivalent rates.

EXAMPLE 4.3E

Find the future value after one year of $100.00 accumulated at
(i) 12.55% compounded annually;
(ii) 12.18% compounded semi-annually;
(iii) 12.00% compounded quarterly;
(iv) 11.88% compounded monthly.

SOLUTION

	(i)	(ii)	(iii)	(iv)
Principal (PV)	100.00	100.00	100.00	100.00
Nominal rate	12.55%	12.18%	12.00%	11.88%
i	0.1255	0.0609	0.03	0.0099
n	1	2	4	12
Future value (FV)	$100.00(1.1255)^1$ $= 100.00(1.1255)$ $= \$112.55$	$100.00(1.0609)^2$ $= 100.00(1.125509)$ $= \$112.55$	$100.00(1.03)^4$ $= 100.00(1.125509)$ $= \$112.55$	$100.00(1.0099)^{12}$ $= 100.00(1.125487)$ $= \$112.55$

Note: The four different nominal annual rates produce the same future value of \$112.55 for the same principal of \$100.00 over the same time period of one year. By definition, the four nominal rates are equivalent rates.

We can find equivalent rates by equating the accumulated values of \$1 for the rates under consideration based on a selected time period, usually one year.

EXAMPLE 4.3F

Find the nominal annual rate compounded semi-annually that is equivalent to an annual rate of 6% compounded annually.

SOLUTION

Let the semi-annual rate of interest be represented by i; P/Y = 2.
For PV = 1, $n = 2$, the accumulated value $FV_1 = (1 + i)^2$.
For the given nominal rate compounded annually the accumulated value $FV_2 = (1 + 0.06)^1$.
By definition, to be equivalent, $FV_1 = FV_2$.

$$(1 + i)^2 = 1.06$$
$$1 + i = 1.06^{0.5}$$
$$1 + i = 1.029563$$
$$i = 0.029563 \text{ — semi-annual rate}$$

Programmed Solution

$$1\ (1 + i)^2 = 1.06$$

(1 → PV; 1.06 → FV)

(Set P/Y = 2) [2nd] (CLR TVM) 1 [±] [PV] 1.06 [FV]
2 [N] [CPT] [I/Y] [5.912603]

Alternatively

[2nd] (IConv) Eff = 6.0; C/Y = 2; Nom = [CPT] 5.912603

The nominal annual rate compounded semi-annually is 5.91%.

The periodic rate is 5.912603%/2 = 2.956301% per semi-annual period.

EXAMPLE 4.3G

What nominal annual rate compounded quarterly is equivalent to 8.4% p.a. compounded monthly?

SOLUTION

Let the quarterly rate be i; PV = 1; $n = 4$.
The accumulated value of \$1 after one year $FV_1 = (1 + i)^4$.

For the given rate I/Y = 8.4; P/Y = 12; $i = \frac{8.4\%}{12} = 0.7\% = 0.007$; $n = 12$.

The accumulated value of \$1 after one year $FV_2 = (1.007)^{12}$.
To be equivalent, $FV_1 = FV_2$.

$$(1 + i)^4 = (1.007)^{12}$$
$$1 + i = (1.007)^3$$
$$1 + i = 1.0211473$$
$$i = 0.0211473 \text{ ——— quarterly rate}$$

Programmed Solution

(Set P/Y = 12) [2nd] (CLR TVM) 1 [±] [PV] 8.4 [I/Y]
12 [N] [CPT] [FV] [1.087310662]

(Set P/Y = 4) [2nd] (CLR TVM) 1 [±] [PV] 1.087310662 [FV]
4 [N] [CPT] [I/Y] 8.458937%

Alternatively

[2nd] (IConv) Nom = 8.4; C/Y = 12; Eff = [CPT] 8.731066

C/Y = 4; Nom = [CPT] 8.458937

The nominal annual rate compounded quarterly is 8.46%. The periodic rate is 2.114734% per quarterly period.

EXAMPLE 4.3H

Peel Credit Union offers premium savings deposits at 8% interest paid semi-annually. The board of directors wants to change to monthly payment of interest. What nominal annual rate compounded monthly should the board set to maintain the same yield?

SOLUTION

Let the monthly rate be i; P/Y = 12; $n = 12$; PV = 1.
The accumulated value of \$1 after one year $FV_1 = (1 + i)^{12}$.
For the existing rate, P/Y = 2; $n = 2$, $i = 0.04$.
The accumulated value of \$1 in one year $FV_2 = (1.04)^2$.
To maintain the same yield the two rates must be equivalent.

$$(1 + i)^{12} = (1.04)^2$$
$$1 + i = (1.04)^{\frac{1}{6}}$$
$$1 + i = 1.006558$$
$$i = 0.006558 \text{ ——— monthly rate}$$

Programmed Solution

$(1 + i)^{12} = (1.04)^2$

$(1.04)^2$

(Set P/Y = 2) [2nd] (CLR TVM) 1 [±] [PV] 8 [I/Y]

2 [N] [CPT] [FV] [1.0816]

$1(1 + i)^{12} = 1.0816$

(Set P/Y = 12) [2nd] (CLR TVM) 1 [±] [PV] 1.0816 [FV]

12 [N] [CPT] [I/Y] [7.869836%]

Alternatively

[2nd] (IConv) Nom = 8.0; C/Y = 2; Eff = [CPT] 8.16

C/Y = 12; Nom = [CPT] 7.869836

The nominal rate compounded monthly is 7.87%. The periodic rate is 7.869836%/12 = 0.655820% per monthly period.

EXERCISE 4.3

If you choose, you can use Excel's ***Effective Annual Interest Rate (EFFECT)*** or ***RATE*** functions to answer the questions indicated below. Refer to **EFFECT** or **RATE** on the Spreadsheet Template Disk to learn how to use these functions.

A. Answer each of the following.

1. Find the nominal rate of interest compounded annually equivalent to each of the following.

(a) 12.5% compounded semi-annually

(b) 6% compounded monthly

(c) 7.2% compounded quarterly

(d) 10.2% compounded monthly

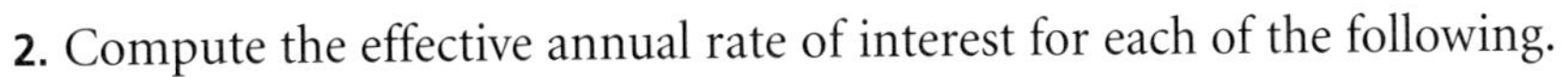

2. Compute the effective annual rate of interest for each of the following.

(a) 9.5% compounded semi-annually

(b) 10.5% compounded quarterly

(c) 5.0% compounded monthly

(d) 7.2% compounded monthly

(e) 3.6% compounded quarterly

(f) 8.2% compounded semi-annually

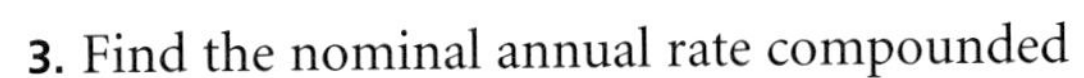

3. Find the nominal annual rate compounded

(a) quarterly that is equivalent to 9% compounded semi-annually

(b) monthly that is equivalent to 6.5% compounded quarterly

(c) monthly that is equivalent to 7.5% compounded semi-annually

(d) semi-annually that is equivalent to 4.25% compounded quarterly

B. Solve each of the following.

1. What is the effective annual rate of interest if $100.00 grows to $150.00 in six years compounded quarterly?
2. What is the effective annual rate of interest if $450.00 grows to $750.00 in three years, five months compounded monthly?
3. If $1100.00 accumulates to $1350.00 in four years, six months compounded semi-annually, what is the effective annual rate of interest?
4. An amount of $2300.00 earns $500.00 interest in three years, two months. What is the effective annual rate if interest compounds monthly?
5. Find the nominal annual rate of interest compounded quarterly that is equal to an effective rate of 9.25%.
6. What nominal annual rate of interest compounded semi-annually is equivalent to an effective rate of 6.37%?
7. If the effective rate of interest on an investment is 6.4%, what is the nominal rate of interest compounded monthly?
8. What is the nominal rate of interest compounded quarterly if the effective rate of interest on an investment is 5.3%?
9. The Central Bank pays 7.5% compounded semi-annually on certain types of deposits. If interest is compounded monthly, what nominal rate of interest will maintain the same effective rate of interest?
10. The treasurer of National Credit Union proposes changing the method of compounding interest on premium savings accounts to daily compounding. If the current rate is 6% compounded quarterly, what nominal rate should the treasurer suggest to the board of directors to maintain the same effective rate of interest?
11. Sofia made a deposit of $600 into a bank account that earns interest at 3.5% compounded monthly. The deposit earns interest at that rate for five years.
    (a) Find the balance of the account at the end of the period.
    (b) How much interest is earned?
    (c) What is the effective rate of interest?
12. Ying invested $5000 into an account earning 2.75% interest compounding daily for two years.
    (a) Find the balance of the account at the end of the period.
    (b) How much interest is earned?
    (c) What is the effective rate of interest?
13. An RRSP earns interest at 4.25% compounded quarterly. An amount of $1200 is invested into the RRSP and earned interest for ten years.
    (a) Find the balance of the account at the end of the period.
    (b) How much interest is earned?
    (c) What is the effective rate of interest?

14. Josef invested \$1750 into an RRSP that earned interest at 5% compounded semi-annually for eight years.

(**a**) Find the balance of the account at the end of the period.

(**b**) How much interest is earned?

(**c**) What is the effective rate of interest?

BUSINESS MATH NEWS BOX

Bank of Montreal's RateOptimizer Investment

The Bank of Montreal's Website presents a number of fixed income products available for RRSP contributions. One of the products is the RateOptimizer Investment. A sound investment strategy is to put money into longer-term investments to get a relatively high interest rate but still have a part of the investment available to you on a regular basis. This way, you can invest that money in products paying higher returns if higher-return investment opportunities arise. The RateOptimizer Investment requires a \$5000 minimum investment. The total amount invested is divided by five and invested equally in five retirement investment certificates (RICs) with terms of one to five years. Each year 20% of the total original amount invested matures and is available for use, or it can be automatically reinvested back into the RateOptimizer Investment for a five-year term at the five-year RIC rate in effect on the reinvestment date.

On May 24, 2003, the following RateOptimizer Investment rates were in effect:

Term	Rate
1 year	2.250
2 years	2.750
3 years	3.000
4 years	3.250
5 years	3.600

QUESTIONS

1. What is the maturity value of each RIC if the interest rates are compounded yearly and paid at the end of each term, and the \$5000 minimum investment is made?
2. What is the maturity value of each RIC if the interest rates are compounded quarterly and paid at the end of each term, and the investment is \$10 000?

Source: Bank of Montreal's RateOptimizer Investment, from "RRSP investment: locked-in long term." Bank of Montreal, www.bmo.com/gic/products/registered/savings/long_term/rateoptimizer_ric.html. May 2003. Reproduced with permission.

Review Exercise

1. At what nominal rate of interest compounded monthly will \$400 earn \$100 interest in four years?

2. At what nominal rate of interest compounded quarterly will \$300 earn \$80 interest in six years?

3. Find the equated date at which payments of \$500 due six months ago and \$600 due today could be settled by a payment of \$1300 if interest is 9% compounded monthly.

4. Find the equated date at which two payments of \$600 due four months ago and \$400 due today could be settled by a payment of \$1100 if interest is 7.25% compounded semi-annually.

5. In what period of time will money triple at 10% compounded semi-annually?

6. In what period of time will money double at 8% compounded monthly?

7. What nominal rate of interest compounded monthly is equivalent to an effective rate of 6.2%?

8. What nominal rate of interest compounded quarterly is equivalent to an effective rate of 5.99%?

9. Find the nominal annual rate of interest

(a) at which \$2500 will grow to \$4000 in eight years compounded quarterly;

(b) at which money will double in five years compounded semi-annually;

(c) if the effective annual rate of interest is 9.2% and compounding is done monthly;

(d) that is equivalent to 8% compounded quarterly.

10. Find the nominal annual rate of interest

(a) at which \$1500 will grow to \$1800 in four years compounded monthly;

(b) at which money will double in seven years compounded quarterly;

(c) if the effective annual rate of interest is 7.75% and compounding is done monthly;

(d) that is equivalent to 6% compounded quarterly.

11. Compute the effective annual rate of interest

(a) for 4.5% compounded monthly;

(b) at which \$2000 will grow to \$3000 in seven years compounded quarterly.

12. Compute the effective annual rate of interest

(a) for 6% compounded monthly;

(b) at which \$1100 will grow to \$2000 in seven years compounded monthly.

13. What is the nominal annual rate of interest compounded monthly that is equivalent to 8.5% compounded quarterly?

14. What is the nominal annual rate of interest compounded quarterly that is equivalent to an effective annual rate of 5%?

15. Patrick had \$2000 to invest. Which of the following options should he choose?

(a) 4% compounded annually

(b) 3.75% compounded semi-annually

(c) 3.5% compounded quarterly

(d) 3.25% compounded monthly

16. (a) How many years will it take for \$7500 to accumulate to \$9517.39 at 3% compounded semi-annually?

(b) Over what period of time will money triple at 9% compounded quarterly?

(c) How long will it take for a loan of \$10 000 to amount to \$13 684 at 10.5% compounded monthly?

17. Mattu had agreed to make two payments—a payment of \$2000 due in nine months and a payment of \$1500 in a year. If Mattu makes a payment of \$1800 now, when should he make a second payment of \$1700 if money is worth 8% compounded quarterly?

18. A four-year, $3200 promissory note with interest at 7% compounded monthly was dis counted at 9% compounded quarterly yielding proceeds of $3870.31. How many months before the due date was the discount date?

19. A financial obligation requires the payment of $2000 now, $2500 in six months, and $4000 in one year. When will a single payment of $9000 discharge the obligation if interest is 6% compounded monthly?

20. Gitu owes two debt payments—a payment of $5000 due in six months and a payment of $6000 due in fifteen months. If Gitu makes a payment of $5000 now, when should he make a second payment of $6000 if money is worth 11% compounded semi-annually?

21. Payment of a debt of $10 000 incurred on December 1, 2000 with interest at 9.5% compounded semi-annually is due on December 1, 2003. If a payment of $7500 is made on December 1, 2002, on what date should a second payment of $7500 be made if money is worth 12% compounded quarterly?

22. A seven-year, $1500 promissory note with interest at 10.5% compounded semi-annually was discounted at 12% compounded quarterly yielding proceeds of $2150. How many months before the due date was the discount date?

Self-Test

1. A ten-year, $9200 promissory note with interest at 6% compounded monthly is discounted at 5% compounded semi-annually yielding proceeds of $12 915.60. How many months before the due date was the date of discount?

2. An amount of $1400 was invested for 71 months, maturing to $2177.36. What annually compounded rate was earned?

3. Determine the effective annual rate of interest equivalent to 5.4% compounded monthly.

4. How many months from now can a payment of $1000 due twelve months ago and a payment of $400 due six months from now be settled by a payment of $1746.56 if interest is 10.2% compounded monthly?

5. At what nominal rate of interest compounded semi-annually will $6900 earn $3000 interest in five years?

6. In how many years will money double at 7.2% compounded quarterly?

7. What is the nominal rate of interest compounded semi-annually that is equivalent to an effective rate of 10.25%?

8. Seven years and two months after its date of issue, an eleven-year promissory note for $8200 bearing interest at 13.5% compounded monthly is discounted to yield $24 253.31. What semi-annually compounded discount rate was used?

9. A non-interest-bearing note for $1100 is discounted three years and seven months before the due date. What annually compounded rate of interest yields proceeds of $848.88?

Challenge Problems

1. Olga deposited $800 in an investment certificate paying 9% compounded semi-annually. On the same day, her sister Ursula deposited $600 in an account paying 7% compounded semi-annually. To the nearest day, when will the future value of Olga's investment be equal to twice the future value of Ursula's investment?

2. A financial institution is advertising a new three-year investment certificate. The interest rate is 7.5% compounded quarterly the first year, 6.5% compounded monthly the second year, and 6% compounded daily the third year. What rate of interest compounded semi-annually for three years would a competing institution have to offer to match the interest produced by this investment certificate?

Case Study 4.1 Choosing a Credit Card

» Joseph Chu is trying to decide which credit card to apply for. He has just received some advertising from several large credit card companies, some of which feature low introductory rates. He has estimated that, based on his good credit rating, he could apply for any of the credit cards being offered.

Credit card A has a $29 annual fee and advertises the relatively low interest rate of 10.5%. A $2.50 fee is charged for each cash advance, plus the applicable interest charges.

Credit card B has no annual fee and advertises a 19.5% interest rate on purchases and cash advances.

Credit card C advertises an introductory promotional interest rate of 5.9% on cash advances and balance transfers for the first five months, if you make your minimum monthly payments on time. The regular annual interest rate is 16.99% on purchases, and the regular annual interest rate is 18.99% on cash advances.

For all three credit cards, interest will not be assessed if the monthly statement amount is paid in full by the payment due date and no cash advances have been taken during the billing period. If the new balance is not paid in full, interest will be charged (1) on the outstanding balance from the statement closing date and (2) on future purchases from the day the purchases are posted to the account. On cash advance transactions, interest is always charged from the date the cash advance is taken.

Currently, Joseph has a credit card with a major Canadian department store. For this card, D, there is an outstanding balance of $1000. Its annual interest rate is 28.8%. If he receives one of the other three cards, he would transfer the department store credit card balance to the new card.

QUESTIONS

1. Joseph maintains an average daily balance of $1000 for the first year, based solely on purchases and balance transfers. For each of the four credit cards,
   (a) calculate the interest charge and fees that he would have to pay over a twelve-month period;

(b) determine the effective annual rate of interest over the first twelve-month period; and

(c) decide which credit card he should choose.

2. For credit card C, if the introductory interest rate fell to 3.9% and the regular annual interest rate rose to 18.99% on purchases, what would the effective annual rate of interest be over the twelve-month period?

3. If you have a credit card, how do its rates and conditions compare with those of the cards described above?

Case Study 4.2 Comparing Car Loans

» After reading consumer car guides and receiving advice from family and friends, Naina has chosen the new car she wants to buy. She now wants to research her financing options to choose the best way to pay for the car.

Naina knows that with taxes, licence, delivery, and dealer preparation fees, her car will cost $23 000. She has saved $5000 toward the purchase price but must borrow the rest. She has narrowed her financing choices to three options: dealer financing, credit union financing, and bank financing.

(i) The car dealer has offered 48-month financing at 9.4% compounded monthly.

(ii) The credit union has offered 36-month financing at 9.5% compounded quarterly. It has also offered 48-month financing at 9.6% compounded quarterly.

(iii) The bank has offered 36-month financing at 9.6% compounded semi-annually. It has also offered 48-month financing at 9.7% compounded semi-annually.

Naina wants to choose the financing option that offers the best interest rate. However, she also wants to explore the financing options that allow her to pay off her car loan more quickly.

QUESTIONS

1. Naina wants to compare the 48-month car loan options offered by the car dealer, the credit union, and the bank.

   (a) What is the effective annual rate of interest for each 48-month option?

   (b) How much interest will Naina save by choosing the best option as against the worst option?

2. Suppose Naina wants to try to pay off her car loan within three years.

   (a) What is the effective annual rate of interest for both of the 36-month options?

   (b) How much interest will Naina save by choosing the better option?

3. If you wanted to get a car loan today, what are the rates of interest for 36-month and 48-month terms? Are car dealers currently offering better interest rates than the banks or credit unions? If so, why?

SUMMARY OF FORMULAS

Formula 3.1A

$FV = PV(1 + i)^n$ Finding the future value of a sum of money when n is a fractional value using the exact method

Formula 3.1B

$PV = \frac{FV}{(1 + i)^n}$ Finding the present value (discounted value or proceeds) when n is a fractional value using the exact method

Formula 3.1C

$PV = FV(1 + i)^{-n}$

Formula 4.1

$f = (1 + i)^m - 1$ Finding the effective rate of interest f for a nominal annual rate compounded m times per year

GLOSSARY

Effective rate of interest the annual rate of interest that yields the same amount of interest per year as a nominal rate compounded a number of times per year *(p. 156)*

Equated date the date on which a single sum of money is equal to the sum of two or more dated sums of money *(p. 146)*

Equivalent rates interest rates that accumulate a given principal to the same future value over the same period of time *(p. 162)*

USEFUL INTERNET SITES

mathforum.org/dr.math

The Rule of 70 Visit the "Ask Dr. Math" site to read about how to use the Rule of 70 or ask other math-related questions.

www.rate.net

Rate.net Rate.net tracks over 11 000 financial institutions in 175 markets nationwide. The site analyzes interest rate performance and financial stability.

www.webfin.com/en/mymoney/rates

Credit Card Rates Go to Rates and then click on Credit Cards. Compare credit card rates and terms for major Canadian financial institutions and department stores on this independent Website.

5 Ordinary Simple Annuities

OBJECTIVES

Upon completing this chapter, you will be able to do the following:

1. Distinguish between types of annuities based on term, payment date, and conversion period.
2. Compute the future value (or accumulated value) FV for ordinary simple annuities.
3. Compute the present value (or discounted value) PV for ordinary simple annuities.
4. Compute the payment PMT for ordinary simple annuities.
5. Compute the number of periods N for ordinary simple annuities.
6. Compute the interest rate I/Y for ordinary simple annuities.

We have seen that simple and compound interest are used in many personal and business situations. There are also many other situations where we pay or receive regular, equal amounts of money. Many make regular, equal payments for rent, insurance, car loans, student loans, and mortgages. Many also receive regular, equal amounts of money, such as wages, salaries, and pensions. These are all examples of annuities. The same principles apply to all of the examples above. Annuity formulas and calculations enable us to answer questions such as "How much money will I have in five years if I deposit $100 per month into a savings account?" and "If I make regular payments toward the purchase of a car, how much am I really paying for it?"

INTRODUCTION

An annuity is a series of payments, usually of equal size, made at periodic time intervals. The word *annuity* implies yearly payments but the term applies to all periodic payment plans, the most frequent of which require annual, semi-annual, quarterly, or monthly payments. As we mentioned above, practical applications of annuities are widely encountered in the finances of businesses and individuals alike. Various types of annuities are identified based on the term of an annuity, the date of payment, and the length of the conversion period. In this chapter, we will deal with ordinary simple annuities.

5.1 Introduction to Annuities

A. Basic concepts

An **annuity** is a series of payments, usually of equal size, made at periodic intervals. The length of time between the successive payments is called the **payment interval** or **payment period**. The length of time from the beginning of the first payment interval to the end of the last payment interval is called the **term of an annuity**. The size of each of the regular payments is the **periodic rent** and the sum of the periodic payments in one year is the **annual rent**.

B. Types of annuities

Several time variables affect annuities, and annuities are classified according to the time variable considered. Depending on whether the term of the annuity is fixed or indefinite, annuities are classified as **annuities certain** or **contingent annuities**.

Typical examples of annuities certain (annuities for which the term is fixed, that is, for which both the beginning date and the ending date are known) include rental payments for real estate, lease payments on equipment, installment payments on loans, mortgage payments, and interest payments on bonds and debentures.

Examples of contingent annuities (annuities for which the beginning date or the ending date or both are uncertain) are life insurance premiums and pension payments or payments from an RRSP converted into a life annuity. The ending date is unknown for these annuities since they terminate with the death of the recipient. Some contingent annuities are the result of clauses in wills, where the beginning date of periodic payments to a beneficiary is unknown, or of payments from a trust fund for the remaining life of a surviving spouse, since neither the beginning date nor the ending date is known.

A special type of annuity is the **perpetuity**, an annuity for which the payments continue forever. Perpetuities result when the size of the period rent is equal to or less than the periodic interest earned by a fund, such as a scholarship fund or an endowment fund to a university.

Variations in the date of payment are another way to classify annuities certain. If payments are made at the end of each payment period, we are dealing with an

ordinary annuity. If, on the other hand, payments are made at the beginning of each payment period, we are dealing with an **annuity due**.

Typical examples of ordinary annuities are installment payments on loans, mortgage payments, and interest payments on bonds and debentures. Rent payments on real estate and lease payments on equipment rentals are examples of annuities due.

Deferring the first payment for a specified period of time gives rise to a **deferred annuity**. This type may be either an ordinary annuity or an annuity due, depending on whether the future payments are at the beginning or at the end of each payment interval.

A third time variable used to classify annuities is the length of the conversion period relative to the payment period. We distinguish between **simple annuities** and **general annuities,** depending on whether or not the conversion period coincides with the payment interval.

A simple annuity is an annuity in which the conversion period coincides with the payment interval. An example is when there are monthly payments on a loan for which the interest is compounded monthly. Since the compound interest period is the same as the payment period, this is a simple annuity, and P/Y and C/Y are equal.

A general annuity is an annuity in which the conversion period and the payment interval do not coincide. Typically, mortgages on homes are compounded semi-annually but repaid by monthly payments. Often, lending institutions offer the borrower the option of making weekly or bi-weekly payments. This is an example of a general annuity, since the conversion period is different from the payment period.

We will introduce ordinary simple annuities in this chapter. In Chapter 6, we will introduce ordinary general annuities. Other annuities, such as annuities due, deferred annuities, and perpetuities, will be introduced in Chapter 7.

EXAMPLE 5.1A

Classify each of the following annuities by

(i) term; (ii) date of payment; (iii) conversion period.

(a) Deposits of $150.00 earning interest at 12% compounded quarterly are made at the beginning of each quarter for four years.

SOLUTION

(i) annuity certain (the term is fixed: four years)
(ii) annuity due (payments are made at the beginning of each quarter)
(iii) simple annuity (the quarterly conversion period equals the quarterly payment period)

(b) Payments of $200.00 are made at the end of each month for five years. Interest is 9% compounded semi-annually.

SOLUTION

(i) annuity certain (the term is fixed: five years)
(ii) ordinary annuity (payments are made at the end of each month)
(iii) general annuity (semi-annual conversion period does not match the monthly payment period)

(c) A fund of $10 000.00 is deposited in a trust account earning interest compounded annually. Starting five years from the date of deposit, the interest earned for the year is to be paid out as a scholarship.

SOLUTION

(i) perpetuity (the payments can go on forever)
(ii) deferred annuity (the first payment is deferred for five years)
(iii) simple annuity (the annual conversion period equals the annual interest period)

(d) In his will, Dr. C. directed that part of his estate be invested in a trust fund earning interest compounded quarterly. His surviving wife was to be paid, for the remainder of her life, $2000.00 at the end of every three months starting three months after his death.

SOLUTION

(i) contingent annuity (both the starting date and the ending date are uncertain)
(ii) ordinary annuity (payments at the end of every three months)
(iii) simple annuity (the quarterly conversion period equals the quarterly payment period)

EXERCISE 5.1

A. Classify each of the following by (a) term; (b) date of payment; (c) conversion period.

1. Payments of $50.00 are made at the beginning of each month for five years at 5% compounded semi-annually.

2. Deposits of $500.00 are made at the end of each quarter for nine years earning interest at 7% compounded quarterly.

3. A fund with an initial deposit of $50 000.00 is set up to provide annual scholarships to eligible business students in an amount not exceeding the annual interest earned by the fund. Scholarship payments are to begin three years from the date of deposit. Interest earned by the fund is compounded semi-annually.

4. The Saskatoon Board of Education introduced a long-term disability plan for its employees. The plan provides for monthly payments equal to 90% of regular salary starting one month after the beginning of the disability. Assume that the plan is subject to monthly compounding.

5. Gary invested $10 000.00 in an account paying interest compounded monthly with the provision that equal monthly payments be made to him from the account for fifteen years at the beginning of each month starting ten years from the date of deposit.

6. Ms. Baka set up a trust fund earning interest compounded semi-annually to provide equal monthly support payments for her surviving husband starting one month after her death.

5.2 Ordinary Simple Annuity—Finding Future Value FV

A. Future value of a series of payments—basic computation

EXAMPLE 5.2A Find the future value of deposits of \$2000.00, \$4000.00, \$5000.00, \$1000.00, and \$3000.00 made at the end of each of five consecutive years respectively at 6% compounded annually, just after the last deposit was made.

SOLUTION The series of deposits can be represented on a time graph as shown:

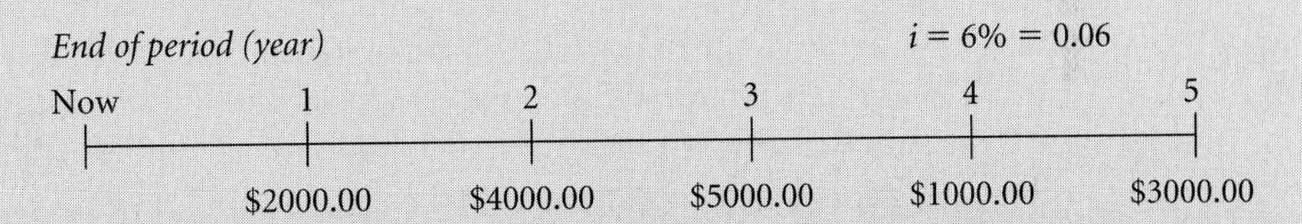

To find the future value of the series of deposits, we need to determine the combined value of the five deposits, including interest, at the focal point five years from now. This can be done using Formula 3.1A, $FV = PV(1 + i)^n$. A graphical representation of the method and data is shown in Figure 5.1 below.

FIGURE 5.1 **Graphical Representation of Method and Data**

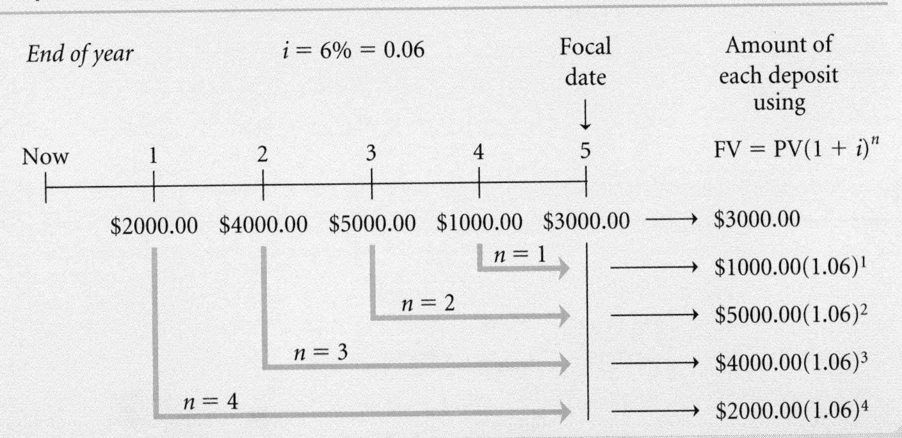

Explanations regarding the amount of each deposit:
The deposit of \$3000.00 has just been made and has a value of \$3000.00 at the focal date. The deposit of \$1000.00 has been in for one year ($n = 1$) and has earned interest for one year at the focal date. Similarly, the deposit of \$5000.00 has been in for two years ($n = 2$), the deposit of \$4000.00 for three years ($n = 3$), and the deposit of \$2000.00 for four years ($n = 4$).

The problem can now be solved by computing the future value of the individual deposits and adding.

Deposit 5	3000.00	=	\$3 000.00
Deposit 4	$1000.00(1.06)^1 = 1000.00(1.06)$	=	1 060.00

Deposit 3	$5000.00(1.06)^2 = 5000.00(1.1236)$	=	5 618.00
Deposit 2	$4000.00(1.06)^3 = 4000.00(1.191016)$	=	4 764.06
Deposit 1	$2000.00(1.06)^4 = 2000.00(1.262477)$	=	2 524.95
	TOTAL	=	$16 967.01

EXAMPLE 5.2B

Find the future value of five deposits of $3000.00 each made at the end of each of five consecutive years respectively at 6% compounded annually, just after the last deposit has been made.

SOLUTION

This example is basically the same as Example 5.2A except that all deposits are equal in size. The problem can be solved in the same way.

While the approach to solving the problem is fundamentally the same as in Example 5.2A, the fact that the deposits are *equal in size* permits a useful mathematical simplification. The equal deposit of $3000.00 can be taken out as a common factor and the individual compounding factors can be added. This method avoids computing the amount of each individual deposit. In other words, it avoids computing separately $3000.00(1)$, $3000.00(1.06)^1$, $3000.00(1.06)^2$, and so on.

FIGURE 5.2 **Graphical Representation of Method and Data**

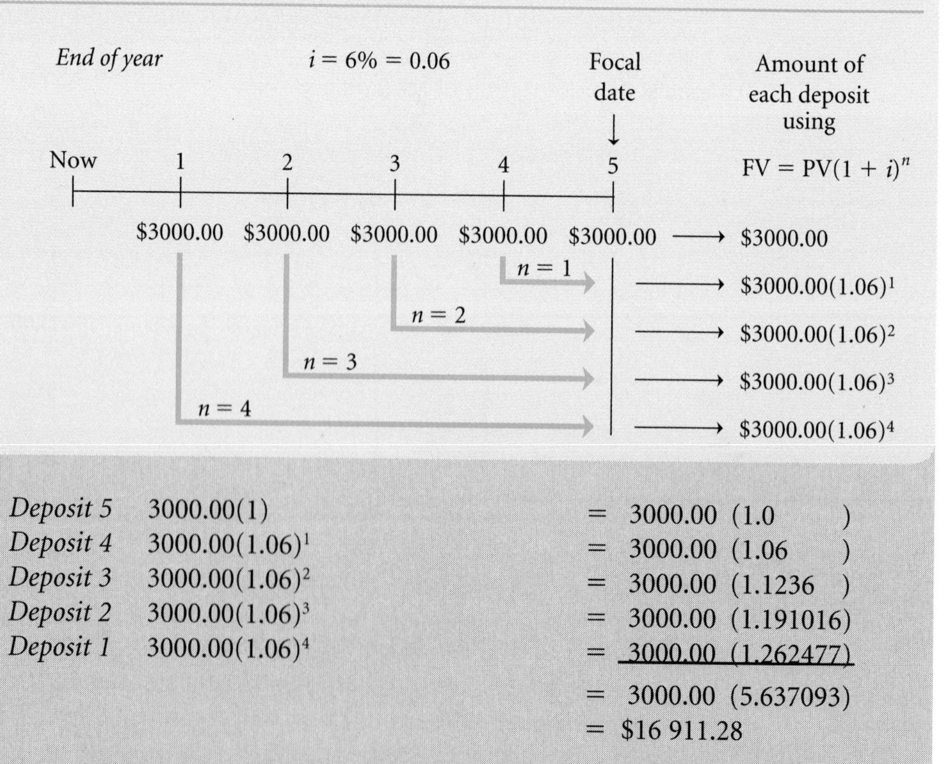

Deposit 5	$3000.00(1)$	=	$3000.00\ (1.0)$
Deposit 4	$3000.00(1.06)^1$	=	$3000.00\ (1.06)$
Deposit 3	$3000.00(1.06)^2$	=	$3000.00\ (1.1236)$
Deposit 2	$3000.00(1.06)^3$	=	$3000.00\ (1.191016)$
Deposit 1	$3000.00(1.06)^4$	=	$3000.00\ (1.262477)$
		=	$3000.00\ (5.637093)$
		=	$16 911.28

Since the deposits in Example 5.2B are equal in size and are made at the end of each period and the payment interval is the same as the compounding period (one year), the problem is an ordinary simple annuity. Finding the sum of the accumulated values of the individual payments at the end of the term of the annuity is defined as finding the **future value of an annuity**.

B. Formula for finding the future value of an ordinary simple annuity

Because annuities are geometric progressions, the following formula has been developed for finding the accumulated value of this type of series of payments.

$$S_n = R\left[\frac{(1+i)^n - 1}{i}\right]$$ —— Formula 5.1 —— future value of an ordinary simple annuity

where S_n = the future value (accumulated value) of an ordinary simple annuity;
R = the size of the periodic payment (rent);
i = the interest rate per conversion period;
n = the number of periodic payments (which for simple annuities is also the number of conversion periods).

The factor $\frac{(1+i)^n - 1}{i}$ is called the **compounding** or **accumulation factor for annuities** or the **accumulated value of one dollar per period.**

EXAMPLE 5.2C

Find the accumulated value of quarterly payments of $50.00 made at the end of each quarter for ten years just after the last payment has been made if interest is 8% compounded quarterly.

SOLUTION

Since the payments are of equal size made at the end of each quarter and compounding is quarterly, the problem is an ordinary simple annuity.

$R = 50.00$; I/Y = 8; P/Y = 4; $i = \frac{8\%}{4} = 2\% = 0.02$; $n = 10(4) = 40$

$$S_n = 50.00\left[\frac{(1+0.02)^{40} - 1}{0.02}\right]$$ —— substituting in Formula 5.1

$$= 50.00\left(\frac{2.208040 - 1}{0.02}\right)$$

$$= 50.00\left(\frac{1.208040}{0.02}\right)$$

$$= 50.00(60.401983)$$

$$= \$3020.10$$

EXAMPLE 5.2D

You deposit $10.00 at the end of each month for five years in an account paying 6% compounded monthly.

(i) What will be the balance in your account at the end of the five-year term?
(ii) How much of the balance will you have contributed?
(iii) How much is interest?

SOLUTION

(i) $R = 10.00$; I/Y = 6; P/Y = 12; $i = \frac{6\%}{12} = 0.5\% = 0.005$; $n = 60$

$$S_n = 10.00\left[\frac{(1+i)^n - 1}{i}\right]$$

$$= 10.00\left(\frac{1.005^{60} - 1}{0.005}\right)$$

$$= 10.00\left[\frac{(1.348850 - 1)}{0.005}\right]$$
$$= 10.00(69.770031)$$
$$= \$697.70$$

(ii) Your contribution is $10.00 per month for 60 months, or 10.00(60) = $600.00.

(iii) Since your contribution is $600.00, the interest earned is 697.70 − 600.00 = $97.70.

C. Restatement of the ordinary simple annuity formula

The algebraic symbols used in the ordinary simple annuity formula correspond to the calculator keys of four preprogrammed financial calculators. To make it easier to relate annuity formulas directly to the symbols used on financial calculator keys and in spreadsheet software such as Excel, we will make the following changes to restate this and upcoming annuity formulas:

1. Replace A_n with PV_n as the symbol for Present Value;
2. Replace S_n with FV_n as the symbol for Future Value;
3. Replace R with PMT as the symbol for Periodic Payment.

From now on

Formula 5.1, $S_n = R\left[\frac{(1+i)^n - 1}{i}\right]$ will be presented as $FV_n = PMT\left[\frac{(1+i)^n - 1}{i}\right]$

D. Using preprogrammed financial calculators

In the same way that you can for compound interest calculations, you can use preprogrammed financial calculators to solve annuity problems efficiently by entering given values and retrieving the answer.

Specific function keys on preprogrammed financial calculators correspond to the five variables used in ordinary simple annuity calculations. Because different models of financial calculators vary in their operation and labelling of the function keys, Appendix II, "Instructions and Tips for Four Preprogrammed Financial Calculator Models," highlights the relevant variations for students using Texas Instruments' BAII Plus and BA-35 Solar, Sharp's EL-733A, and Hewlett-Packard's 10B calculators. The function keys used for the calculator models presented in Appendix II are shown in Table 5.1.

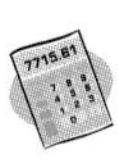

TABLE 5.1 Financial Calculator Function Keys that Correspond to Variables Used in Ordinary Simple Annuity Calculations

		Function Key			
Variable	Algebraic Symbol	TI BAII+	TI BA-35S	Sharp EL-733A	HP 10B
The number of compounding periods	*n*	N	N	*n*	N
The periodic rate of interest[1]	*i*	I/Y + C/Y	%*i*	*i*	I/YR
The periodic annuity payment	PMT or R	PMT	PMT	PMT	PMT
The present value or principal	PV or A_n	PV	PV	PV	PV
The future value or maturity value	FV or S_n	FV	FV	FV	FV

Note: The periodic rate of interest is entered as a percent and not as a decimal equivalent. For example, 8% is entered as "8" not ".08".

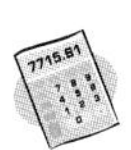

Instructions in this text are given for the Texas Instruments BAII Plus calculator. Refer to Appendix II for instructions for using the Texas Instruments BA-35 Solar, Sharp EL-733A, and Hewlett-Packard 10B calculators.

To begin an ordinary simple annuity calculation, previous data must be cleared. The following key sequence resets all entries to the defaults for each field: 2nd CLR TVM.

The number of periods, or frequency per year, must be determined. Enter this number as the P/Y. If the number of periods per year is specified as "monthly," follow the sequence 2nd P/Y 12 Enter 2nd Quit. Substitute the number for the appropriate number of periods per year in each case. The most common frequencies are monthly, quarterly, semi-annually, or annually. Enter the given values in any order. Pressing CPT followed by the key representing the unknown variable then retrieves the value of the unknown variable.

Note: This assumes that the calculator is in the "END" mode. This is the default. To check, look at the upper right corner of the display screen. If nothing is there, you are in the correct mode. If the letters BGN are there, you must switch modes:

2nd BGN (PMT key)

2nd SET (END will now appear.)

2nd Quit (Back to standard calculator—upper right corner now blank.)

When performing an annuity calculation, usually only one of the present value PV *or* the future value FV is involved. To avoid incorrect answers, the present value PV should be set to zero when determining the future value FV and vice versa. In addition, recall that the Texas Instruments BAII Plus calculator follows the established convention of treating cash inflows (cash received) as positive numbers and cash outflows (cash paid out) as negative numbers. Since periodic annuity payments are considered to be cash outflows, always enter the periodic payment as a negative number. Also, continue to always enter the present value as a negative number (if the present value is other than 0) to ensure your result has a positive value.

To solve Example 5.2D, in which PMT = 10.00, I/Y = 6; P/Y = 12 and $n = 60$, use the following procedure.

	Key in	*Press*	*Display shows*	
(Set P/Y = 12)	0	PV	0	a precaution to avoid incorrect answers
	10	± PMT	10	this enters the periodic payment PMT
	6.0	I/Y	6.0	this enters the interest rate per year
	60	N	60	this enters the number of payments *n*
	CPT	FV	697.700305	this retrieves the wanted amount FV

The future value is $697.70.

E. Applications

EXAMPLE 5.2E Jim West set up a savings plan with City Trust of Victoria whereby he deposits $300.00 at the end of each quarter for eight years. The amount in his account at that time will become a term deposit withdrawable after a further five years. Interest throughout the total time period is 5% compounded quarterly.

(i) How much will be in Jim's account just after he makes his last deposit?
(ii) What will be the balance of his account when he can withdraw the deposit?
(iii) How much of the total at the time of withdrawal did Jim contribute?
(iv) How much is the interest earned?

SOLUTION As Figure 5.3 shows, problems of this type may be solved in stages. The first stage involves finding the future value of an *ordinary annuity*. This amount becomes the principal for the second stage, which involves finding the future value of a *single* sum of money invested for five years.

(i) PMT = 300.00; I/Y = 5; P/Y = 4; $i = \frac{5\%}{4} = 1.25\% = 0.0125$;
$n = 8(4) = 32$

$$FV_1 = 300.00\left[\frac{(1.0125^{32} - 1)}{0.0125}\right]$$ — Formula 5.1

$$= 300.00\left[\frac{(1.488131 - 1)}{0.0125}\right]$$

$$= 300.00(39.050441)$$

$$= \$11\,715.13$$

(ii) $PV = FV_1 = 11\,715.13$; $i = 0.0125$; $n = 5(4) = 20$

$$FV_2 = 11\,715.13(1.0125)^{20}$$ — Formula 3.1A

$$= 11\,715.13(1.282037)$$

$$= \$15\,019.23$$

FIGURE 5.3 **Graphical Representation of Method and Data**

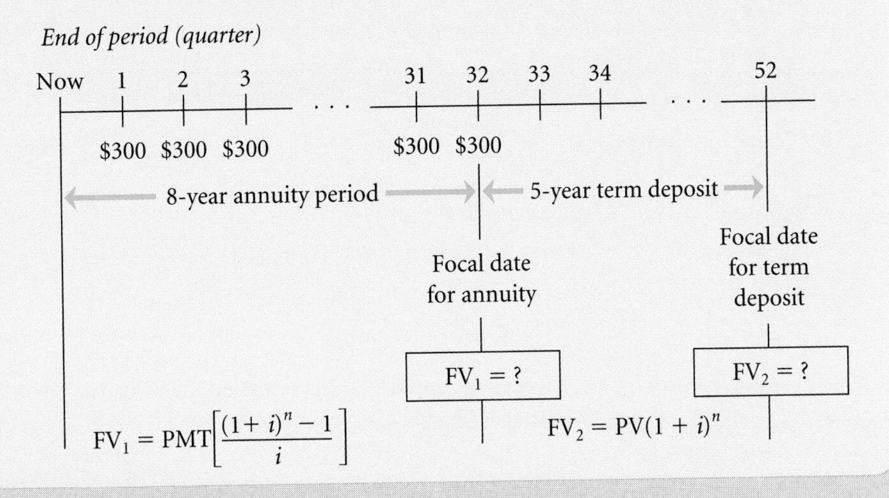

(iii) Jim's contribution = 32(300.00) = \$9600.00.

(iv) The amount of interest earned = 15 019.23 − 9600.00 = \$5419.23.

Programmed Solution for Parts (i) and (ii)

(i) (Set P/Y = 4) 0 [PV] 300 [±] [PMT]
5.0 [I/Y] 32 [N] [CPT] [FV] [11 715.13221]

(ii) 11 715.13221 [±] [PV] 0 [PMT]
5.0 [I/Y] 20 [N] [CPT] [FV] [15 019.23566]

EXAMPLE 5.2F

The Gordons saved for the purchase of their dream home by making deposits of \$1000.00 per year for ten consecutive years in an account with Cooperative Trust in Saskatoon. The account earned interest at 5.75% compounded annually. At the end of the ten-year contribution period, the deposit was left for a further six years earning interest at 5.5% compounded semi-annually.

(i) What down payment were the Gordons able to make on their house?
(ii) How much of the down payment was interest?

SOLUTION

(i) First, find the amount in the account at the end of the term of the ordinary annuity formed by the yearly deposits.

PMT = 1000.00; I/Y = 5.75; P/Y = 1; $i = 5.75\% = 0.0575$; $n = 10$

$$
\begin{aligned}
FV_1 &= 1000.00\left[\frac{(1.0575^{10} - 1)}{0.0575}\right] \\
&= 1000.00\left[\frac{1.749056 - 1}{0.0575}\right] \\
&= 1000.00(13.027064) \\
&= \$13\,027.06
\end{aligned}
$$

Second, compute the accumulated value of FV_1 in six years.

PV = FV_1 = 13 027.06; I/Y = 5.5; P/Y = 2; $i = \frac{5.5\%}{2} = 2.75\% = 0.0275$; $n = 12$

$$
\begin{aligned}
FV_2 &= 13\,027.06(1.0275)^{12} \\
&= 13\,027.06(1.384784) \\
&= \$18\,039.66
\end{aligned}
$$

The Gordons made a down payment of \$18 039.66.

(ii) Since the Gordons contributed (1000.00)(10) = \$10 000.00, the amount of interest in the down payment is \$8039.66.

Programmed Solution for Part (i)

(Set P/Y = 1) 0 [PV] 1000 [±] [PMT]

5.75 [I/Y] 10 [N] [CPT] [FV] [13 027.06408]

(Set P/Y = 2) 13 027.06408 [±] [PV]

0 [PMT] 5.5 [I/Y] 12 [N] [CPT] [FV] [18 039.66698]

EXAMPLE 5.2G

Marise has contributed \$1500.00 per year for the last twelve years into an RRSP deposit account with her bank in Windsor. Interest earned by these deposits was 4.5% compounded annually for the first eight years and 5.5% compounded annually for the last four years. Five years after the last deposit, she converted her RRSP into a registered retirement income fund (RRIF). How much was the beginning balance in the RRIF if interest for those five years remained at 5.5%?

SOLUTION

As Figure 5.4 shows, the problem may be divided into two simple annuities. The first simple annuity covers the deposits for the first eight years; the second simple annuity covers the next four payments.

The focal date for the first annuity is at the end of Year 8 (focal date 1).

The accumulated value (future value) of this simple annuity is computed using Formula 5.1.

$$\begin{aligned} FV_1 &= 1500.00\left[\frac{(1.045^8 - 1)}{0.045}\right] \\ &= 1500.00\left[\frac{1.422101 - 1}{0.045}\right] \\ &= 1500.00(9.380014) \\ &= \$14\,070.02 \end{aligned}$$

FV_1 then accumulates for nine years (to the end of Year 17) at 5.5% to obtain FV_4 at the focal date for the beginning balance in the RRIF (focal date 3).

$$\begin{aligned} FV_4 &= 14\,070.02(1.055)^9 \\ &= 14\,070.02(1.619094) \\ &= \$22\,780.69 \end{aligned}$$

FIGURE 5.4 Graphical Representation of Method and Data

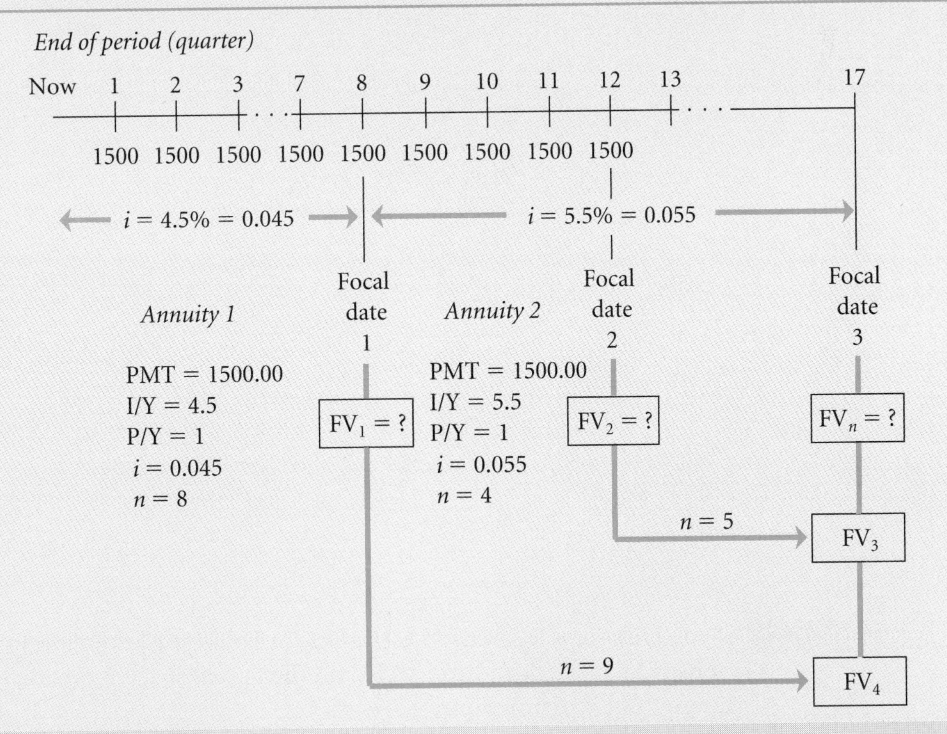

The focal date for the second simple annuity is at the end of Year 12 (focal date 2).

$$\begin{aligned} FV_2 &= 1500.00\left[\frac{(1.055^4 - 1)}{0.055}\right] \\ &= 1500.00\left[\frac{1.238825 - 1}{0.055}\right] \end{aligned}$$

$$= 1500.00(4.342266)$$
$$= \$6513.40$$

FV_2 then accumulates for five years (to the end of year 17) to obtain FV_3 at focal date 3.

$$FV_3 = 6513.40(1.055)^5$$
$$= 6513.40(1.306960)$$
$$= \$8512.75$$

The beginning balance FV_n in the RRIF is then obtained by adding FV_3 and FV_4.

$$FV_n = 8512.75 + 22\,780.69$$
$$= \$31\,293.44$$

Programmed Solution

FV_1

(Set P/Y = 1) 0 PV 1500 ± PMT 4.5 I/Y 8 N CPT FV 14 070.02043

FV_4

14 070.02043 ± PV 0 PMT 5.5 I/Y 9 N CPT FV 22 780.6895 STO1

FV_2

0 PV 1500 ± PMT 5.5 I/Y 4 N CPT FV 6513.399563

FV_3

6513.399563 ± PV 0 PMT 5.5 I/Y 5 N CPT FV 8512.752734 STO2

RCL2 + RCL1

$FV_n = FV_3 + FV_4 = 8512.75 + 22\,780.69 = \$31\,293.44$

EXERCISE 5.2

Excel If you choose, you can use Excel's ***Future Value (FV)*** function to answer the questions indicated below. Refer to **FV** on the Spreadsheet Template Disk to learn how to use this Excel function.

A. Find the future value of the ordinary simple annuity for each of the following six series of payments.

	Periodic Payment	Payment Interval	Term	Interest Rate	Conversion Period
1.	$1500.00	1 quarter	$7\frac{1}{2}$ years	5%	quarterly
2.	$20.00	1 month	6.75 years	6%	monthly
3.	$700.00	6 months	20 years	7%	semi-annually
4.	$10.00	1 month	15 years	9%	monthly
5.	$320.00	3 months	8 years, 9 months	10.4%	quarterly
6.	$2000.00	$\frac{1}{2}$ year	11 years, 6 months	8.8%	semi-annually

B. Answer each of the following questions.

1. Find the accumulated value of payments of $200.00 made at the end of every three months for twelve years if money is worth 5% compounded quarterly.
2. What will deposits of $60.00 made at the end of each month amount to after six years if interest is 4.8% compounded monthly?
3. How much interest is included in the future value of an ordinary simple annuity of $1500.00 paid every six months at 7% compounded semi-annually if the term of the annuity is fifteen years?
4. Jane Alleyre made ordinary annuity payments of $15.00 per month for sixteen years earning 9% compounded monthly. How much interest is included in the future value of the annuity?
5. Saving for his retirement 25 years from now, Jimmy Olsen set up a savings plan whereby he will deposit $25.00 at the end of each month for the next 15 years. Interest is 3.6% compounded monthly.

   (a) How much money will be in Mr. Olsen's account on the date of his retirement?
   (b) How much will Mr. Olsen contribute?
   (c) How much is interest?

6. Mr. and Mrs. Wolf have each contributed $1000.00 per year for the last ten years into RRSP accounts earning 6% compounded annually. Suppose they leave their accumulated contributions for another five years in the RRSP at the same rate of interest.

   (a) How much will Mr. and Mrs. Wolf have in total in their RRSP accounts?
   (b) How much did the Wolfs contribute?
   (c) How much will be interest?

7. Ms. Pitt has made quarterly payments of $1375.00 at the end of each quarter into an RRSP for the last seven years earning interest at 7% compounded quarterly. If she leaves the accumulated money in the RRSP for another three years at 8% compounded semi-annually, how much will she be able to transfer at the end of the three years into a registered retirement income fund?
8. For the last six years Joe Borelli has made deposits of $300.00 at the end of every six months earning interest at 5% compounded semi-annually. If he leaves the accumulated balance for another ten years at 6% compounded quarterly, what will the balance be in Joe's account?

5.3 Ordinary Simple Annuity—Finding Present Value PV

A. Present value of series of payments—basic computation

EXAMPLE 5.3A Find the single sum of money whose value now is equivalent to payments of \$2000.00, \$4000.00, \$5000.00, \$1000.00, and \$3000.00 made at the end of each of five consecutive years respectively at 6% compounded annually.

SOLUTION The series of payments can be represented on a time graph.

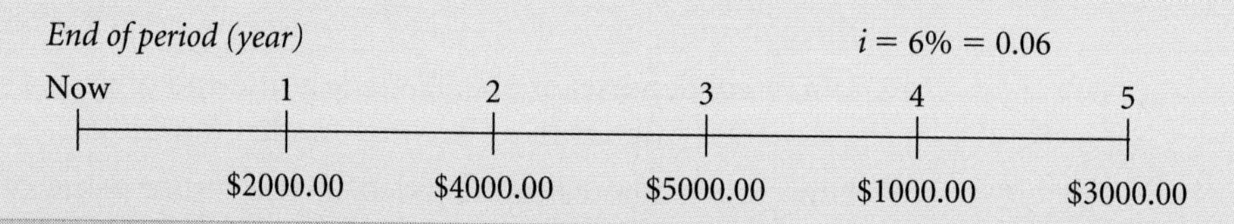

To find the present value of the series of payments, we need to determine the combined present value of the five payments at the focal point "now." This can be done using Formula 3.1C, $PV = FV(1 + i)^{-n}$. A graphical representation of the method and data is shown in Figure 5.5.

The solution to the problem can be completed by computing the present value of the individual payments and adding.

FIGURE 5.5 **Graphical Representation of Method and Data**

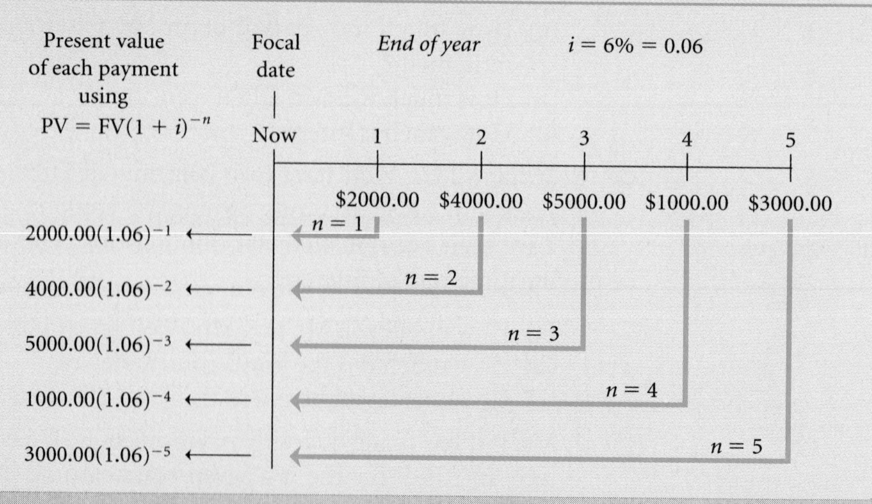

Payment 1	$2000.00(1.06)^{-1}$	=	2000.00(0.943396)	=	\$1 886.79
Payment 2	$4000.00(1.06)^{-2}$	=	4000.00(0.889996)	=	3 559.99
Payment 3	$5000.00(1.06)^{-3}$	=	5000.00(0.839619)	=	4 198.10
Payment 4	$1000.00(1.06)^{-4}$	=	1000.00(0.792094)	=	792.09
Payment 5	$3000.00(1.06)^{-5}$	=	3000.00(0.747258)	=	2 241.77
			TOTAL		\$12 678.74

EXAMPLE 5.3B Find the present value of five payments of $3000.00 made at the end of each of five consecutive years respectively if money is worth 6% compounded annually.

SOLUTION This example is basically the same as Example 5.3A except that all payments are equal in size.

FIGURE 5.6 **Graphical Representation of Method and Data**

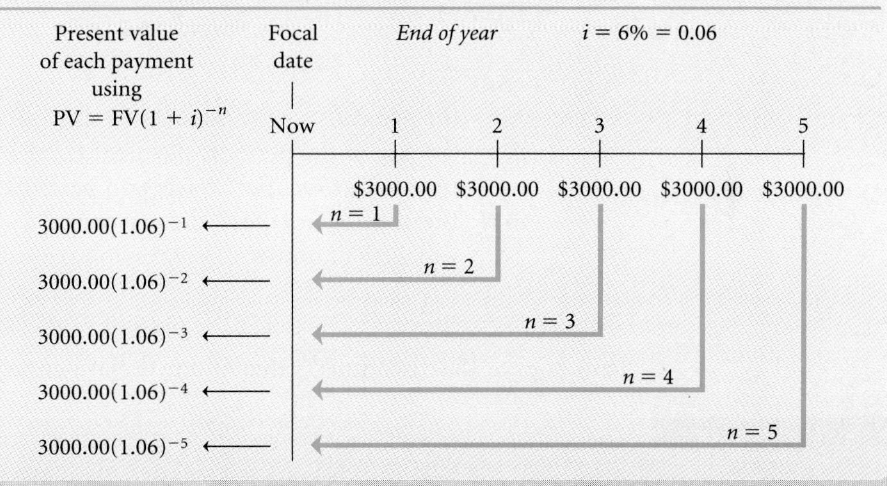

While the approach to solving the problem is fundamentally the same as in Example 5.3A, the fact that the payments are equal in size permits the same mathematical simplification used for Example 5.2B. The equal payment of $3000.00 can be taken out as a common factor and the individual discount factors can be added.

Payment 1 $3000.00(1.06)^{-1} = 3000.00\ (0.943396)$
Payment 2 $3000.00(1.06)^{-2} = 3000.00\ (0.889996)$
Payment 3 $3000.00(1.06)^{-3} = 3000.00\ (0.839619)$
Payment 4 $3000.00(1.06)^{-4} = 3000.00\ (0.792094)$
Payment 5 $3000.00(1.06)^{-5} = 3000.00\ (0.747258)$

$= 3000.00\ (4.2123638)$
$= \$12\ 637.09$

Since the payments in Example 5.3B are equal in size and are made at the end of each period, and the payment period is the same as the compounding period (one year), the problem is an ordinary simple annuity. Finding the sum of the present values of the individual payments at the beginning of the term of an annuity is defined as finding the present value of an annuity. It is useful to carry the mathematical simplification beyond simply taking out the common factor 3000.00.

B. Formula for finding the present value of an ordinary simple annuity

Because annuities are geometric progressions, the following formula for finding the present value of an ordinary simple annuity has been developed.

$$A_n = R\left[\frac{1-(1+i)^{-n}}{i}\right]$$ — **Formula 5.2** — **present value of an ordinary simple annuity**

where A_n = the present value (discounted value) of an ordinary simple annuity;
R = the size of the periodic payment (rent);
i = the interest rate per conversion period;
n = the number of periodic payments (which for simple annuities equals the number of conversion periods).

The factor $\frac{1-(1+i)^{-n}}{i}$ is called the present value factor or discount factor for annuities or the discounted value of one dollar per period.

EXAMPLE 5.3C

Find the present value at the beginning of the first payment period of payments of $50.00 made at the end of each quarter for ten years, if interest is 8% compounded quarterly.

SOLUTION

Because the payments are of equal size made at the end of each quarter and compounding is quarterly, the problem is an ordinary simple annuity. Since the focal date is the beginning of the term of the annuity, the formula for finding the present value of an ordinary simple annuity applies.

$R = 50.00$; I/Y = 8; P/Y = 4; $i = \frac{8\%}{4} = 0.02$; $n = 10(4) = 40$

$$\begin{aligned} A_n &= 50.00\left[\frac{1-(1+0.02)^{-40}}{0.02}\right] \quad \text{——— substituting in Formula 5.2} \\ &= 50.00\left(\frac{1-0.452890}{0.02}\right) \\ &= 50.00\left(\frac{0.547110}{0.02}\right) \\ &= 50.00(27.355480) \\ &= \$1367.77 \end{aligned}$$

EXAMPLE 5.3D

Suppose you want to withdraw $100.00 at the end of each month for five years from an account paying 4.5% compounded monthly.

(i) How much must you have on deposit at the beginning of the month in which the first withdrawal is made at the end of the month?
(ii) How much will you receive in total?
(iii) How much of what you will receive is interest?

SOLUTION

(i) R = 100.00; I/Y = 4.5; P/Y = 12; $i = \frac{4.5\%}{12} = 0.375\% = 0.00375$; $n = 60$

$$\begin{aligned} A_n &= 100.00\left[\frac{1-(1+i)^{-n}}{i}\right] \\ &= 100.00\left[\frac{(1-1.00375^{-60})}{0.00375}\right] \\ &= 100.00\left[\frac{(1-0.798852)}{0.00375}\right] \\ &= 100.00\left(\frac{0.201148}{0.00375}\right) \\ &= 100.00(53.639381) \\ &= \$5363.94 \end{aligned}$$

(ii) Total receipts will be \$100.00 per month for 60 months or \$6000.00.

(iii) Since the initial balance must be \$5363.94, the interest received will be 6000.00 − 5363.94 = \$636.06.

C. Restatement of the ordinary simple annuity formula

To make it easier to relate annuity formulas directly to the symbols used on financial calculator keys and in spreadsheet software such as Excel, we will make the changes noted on page 190 to restate Formula 5.2. From now on

Formula 5.2, $A_n = R\left[\frac{1-(1+i)^{-n}}{i}\right]$, will be presented as

$$PV_n = PMT\left[\frac{1-(1+i)^{-n}}{i}\right]$$

D. Present value using preprogrammed financial calculators

Refer to Appendix II for instructions for using the Texas Instruments BA-35 Solar, Sharp EL-733A, and Hewlett-Packard 10B calculators to perform annuity calculations. The instructions given in this text are for the Texas Instruments BAII Plus calculator.

To find the present value of the ordinary simple annuity in Example 5.3D in which R or PMT = 100.00, I/Y = 4.5, P/Y = 12, and N = 60, proceed as follows.

	Key in	*Press*	*Display shows*	
(Set P/Y= 12)	0	FV	0	a precaution
	100	± PMT	100	
	4.5	I/Y	4.5	
	60	N	60	
	CPT	PV	5363.938035	pressing the PV key retrieves the unknown present value PV_n

You must have \$5363.94 on deposit.

E. Applications

When buying a home or vehicle, most people do not have enough money saved to pay the entire price. However, a small initial payment, called a **down payment**, is often accepted in the meantime. A mortgage loan from a financial institution is needed to supply the balance of the purchase price. The larger the down payment, the lesser the amount that needs to be borrowed. The amount of the loan is the *present value of the future periodic payments.*

The *cash value* is the price of the property at the date of purchase and represents the dated value of all payments at that date.

CASH VALUE = DOWN PAYMENT + PRESENT VALUE OF THE PERIODIC PAYMENTS

EXAMPLE 5.3E

Mr. and Mrs. Hong bought a vacation property for \$3000.00 down and \$1000.00 every half-year for twelve years. If interest is 7% compounded semi-annually, what was the cash value of the property?

SOLUTION

Since the first half-yearly payment is due at the end of the first six-month period and compounding is semi-annual, the present value of the periodic payments is the present value of an ordinary simple annuity.

PMT = 1000.00; I/Y = 7; P/Y = 2; $i = \frac{7\%}{2} = 3.5\% = 0.035$; $n = 12(2) = 24$

$$\begin{aligned} PV_n &= 1000.00\left[\frac{(1 - 1.035^{-24})}{0.035}\right] \\ &= 1000.00\left[\frac{(1 - 0.437957)}{0.035}\right] \\ &= 1000.00\left(\frac{0.562043}{0.035}\right) \\ &= 1000.00(16.058368) \\ &= \$16\,058.37 \end{aligned}$$

Programmed Solution

(Set P/Y = 2) 0 [FV] 1000 [±] [PMT]

7.0 [I/Y] 24 [N] [CPT] [PV] [16 058.3676]

The cash value = 3000.00 + 16 058.37 = \$19 058.37.

DID YOU KNOW?

With the purchase of a \$100 000 home, you could save over \$25 000 in interest costs by making a 25% down payment instead of a 5% down payment.

Total Selling Price: \$100 000

Down Payment Percentage	Down Payment Amount	Mortgage Principal	Total Interest Paid*
5%	\$5 000	\$95 000	\$122 516
10%	\$10 000	\$90 000	\$116 068
25%	\$25 000	\$75 000	\$96 723

*Assumes a constant interest rate of 8% compounded semi-annually and monthly payments repaid over a 25-year amortization period.

EXAMPLE 5.3F

Armand Rice expects to retire in seven years and would like to receive \$500.00 per month for ten years starting at the end of the first month after his retirement. To achieve this goal, he deposited part of the proceeds of \$50 000.00 from the sale of a property into a fund earning 10.5% compounded monthly.

(i) How much must be in the fund at the date of his retirement?
(ii) How much of the proceeds did he deposit in the fund?
(iii) How much does he expect to receive from the fund?
(iv) How much of what he will receive is interest?

SOLUTION

As Figure 5.7 shows, problems of this type should be solved in stages. The first stage involves finding the present value of an annuity. This sum of money becomes the future amount for the second stage, which involves finding the present value of that future amount at the date of deposit.

(i) PMT = 500.00; I/Y = 10.5; P/Y = 12; $i = \frac{10.5\%}{12} = 0.875\% = 0.00875$; $n = 120$

$$\begin{aligned} PV_n &= 500.00\left[\frac{(1 - 1.00875^{-120})}{0.00875}\right] \\ &= 500.00\left[\frac{1 - 0.35154}{0.00875}\right] \\ &= 500.00(74.109758) \\ &= \$37\,054.88 \end{aligned}$$

(ii) $FV = PV_n = 37\,054.88$; $i = 0.00875$; $n = 84$

$$\begin{aligned} PV &= 37\,054.88(1.00875)^{-84} \\ &= 37\,054.88(0.481041) \\ &= \$17\,824.91 \end{aligned}$$

FIGURE 5.7 **Graphical Representation of Method and Data**

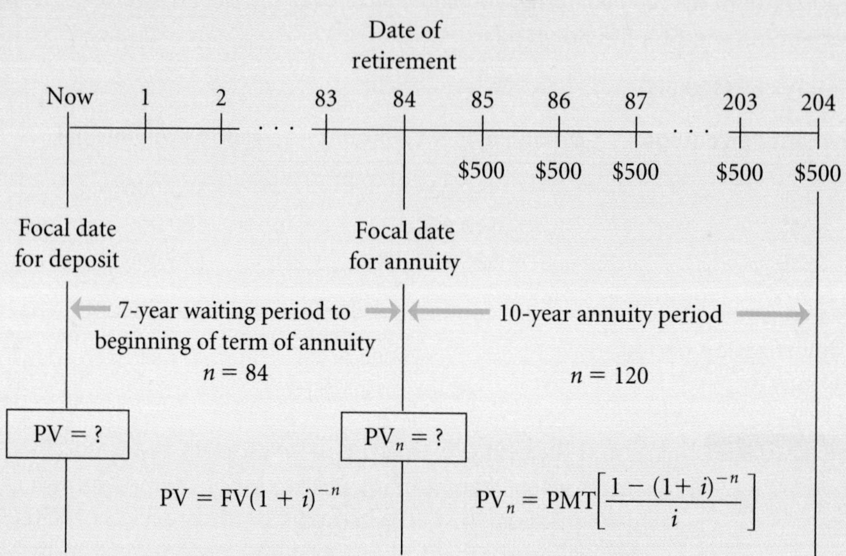

(iii) He expects to receive 500.00(120) = \$60 000.00.

(iv) Interest received will be 60 000.00 − 17 824.91 = \$42 175.09.

Programmed Solution for Parts (i) and (ii)

(i) (Set P/Y = 12) 0 FV 500 ± PMT 10.5 I/Y 120 N CPT PV 37 054.87916

(ii) 37 054.87916 FV 0 PMT 10.5 I/Y 84 N CPT PV −17 824.9119

EXAMPLE 5.3G Sheila Davidson borrowed money from her credit union and agreed to repay the loan in blended monthly payments of \$161.75 over a four-year period. Interest on the loan was 9% compounded monthly.

(i) How much did she borrow?

(ii) If she missed the first eleven payments, how much would she have to pay at the end of the first year to bring her payments up to date?

(iii) If the credit union demanded payment in full after one year, how much money would Sheila Davidson need?

(iv) If the loan is paid off after one year, what would have been its total cost?

(v) How much of the total loan cost is additional interest paid on the missed payments?

SOLUTION

(i) The amount borrowed is the present value (or discounted value) of the 48 payments, as the time diagram shows.

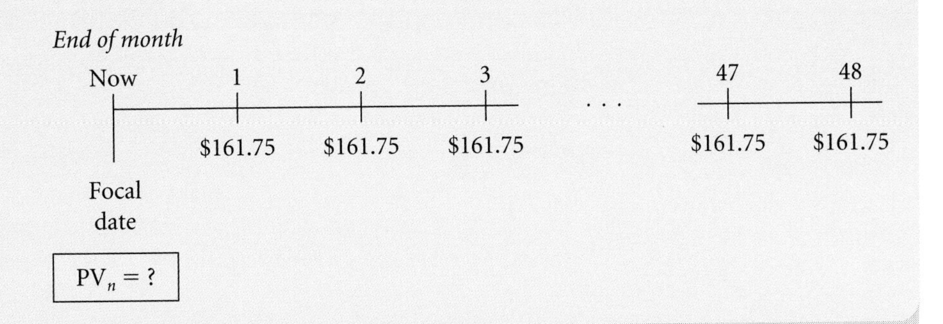

$$\text{PMT} = 161.75;\ \text{I/Y} = 9;\ \text{P/Y} = 12;\ i = \frac{9\%}{12} = 0.75\% = 0.0075;$$

$$n = 4(12) = 48$$

$$\begin{aligned} \text{PV}_n &= 161.75\left[\frac{(1 - 1.0075^{-48})}{0.0075}\right] \quad \text{———— using Formula 5.2} \\ &= 161.75\left[\frac{1 - 0.698614}{0.0075}\right] \\ &= 161.75(40.184782) \\ &= 6499.89 \end{aligned}$$

Programmed Solution

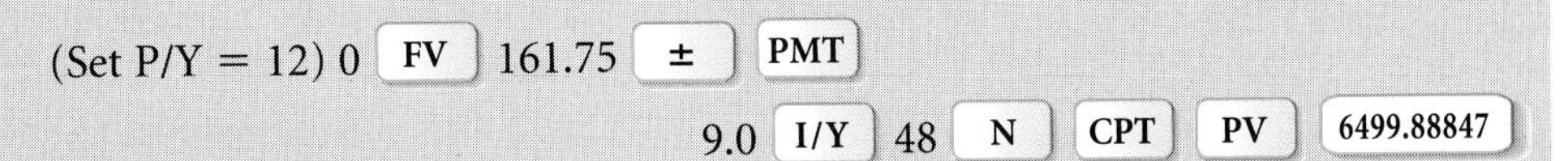

(Set P/Y = 12) 0 FV 161.75 ± PMT

9.0 I/Y 48 N CPT PV 6499.88847

(ii) As the diagram below shows, Sheila Davidson must pay the accumulated value of the first twelve payments to bring her payments up to date after one year.

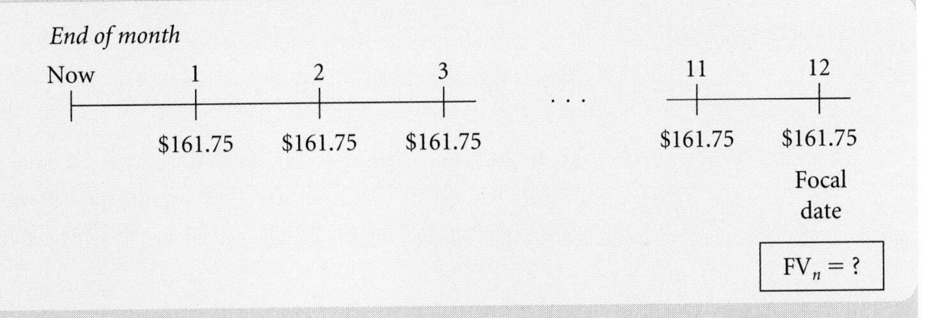

$$\text{PMT} = 161.75; \quad i = 0.0075; \quad n = 12$$

$$\begin{aligned} \text{FV}_n &= 161.75\left[\frac{(1.0075^{12} - 1)}{0.0075}\right] \quad \text{———— using Formula 5.1} \\ &= 161.75\left[\frac{1.093807 - 1}{0.0075}\right] \\ &= 161.75(12.507586) \\ &= \$2023.10 \end{aligned}$$

Programmed Solution

0 (PV) 161.75 (±) (PMT) 9.0 (I/Y) 12 (N) (CPT) (FV) 2023.102093

(iii) The sum of money required to pay off the loan in full is the sum of the accumulated values of the first 12 payments plus the discounted values of the remaining 36 payments.

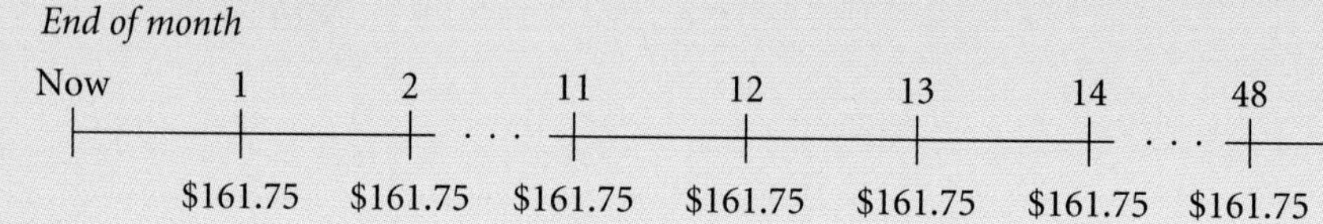

Focal date

Accumulated value of first 12 payments → | ← Discounted value of remaining 36 payments

The accumulated value of the first 12 payments, as computed in part (ii) above, is $2023.10. The discounted value of the remaining payments is found using Formula 5.2.

PMT = 161.75; $i = 0.0075$; $n = 36$

$$
\begin{aligned}
PV_n &= 161.75\left[\frac{(1 - 1.0075^{-36})}{0.0075}\right] \\
&= 161.75\left[\frac{1 - 0.764149}{0.0075}\right] \\
&= 161.75(31.446805) \\
&= \$5086.52
\end{aligned}
$$

Programmed Solution

0 (FV) 161.75 (±) (PMT) 9.0 (I/Y) 36 (N) (CPT) (PV) 5086.520749

The amount of money needed is 5086.52 + 2023.10 = $7109.62.

(iv) The total cost of the loan if paid off after one year is 7109.62 − 6499.89 = $609.73.

(v) To bring the payments up to date, $2023.10 is needed. Since the normal amount paid during the first year would have been 161.75(12) = $1941.00, the additional interest paid is 2023.10 − 1941.00 = $82.10.

EXERCISE 5.3

If you choose, you can use ***Excel's Present Value (PV)*** function to answer the questions indicated below. Refer to **PV** on the Spreadsheet Template Disk to learn how to use this Excel function.

Determine the present value of the ordinary simple annuity for each of the following series of payments.

	Periodic Payment	Payment Interval	Term	Interest Rate	Compounding Period
1.	$1600.00	6 months	$3\frac{1}{2}$ years	8.5%	semi-annually
2.	$700.00	1 quarter	4 years, 9 months	5%	quarterly
3.	$4000.00	1 year	12 years	7.5%	annually
4.	$45.00	1 month	18 years	6.6%	monthly
5.	$250.00	3 months	14 years, 3 months	4.4%	quarterly
6.	$80.00	1 month	9.25 years	7.2%	monthly

Answer each of the following questions.

1. Find the present value of payments of $375.00 made at the end of every six months for fifteen years if money is worth 7% compounded semi-annually.

2. What is the discounted value of payments of $60.00 made at the end of each month for nine years if interest is 4.5% compounded monthly?

3. You want to receive $600.00 at the end of every three months for five years. Interest is 7.6% compounded quarterly.

(a) How much would you have to deposit at the beginning of the five-year period?

(b) How much of what you receive is interest?

4. An installment contract for the purchase of a car requires payments of $252.17 at the end of each month for the next three years. Suppose interest is 8.4% p.a. compounded monthly.

(a) What is the amount financed?

(b) How much is the interest cost?

5. For home entertainment equipment, Ted paid $400.00 down and signed an installment contract that required payments of $69.33 at the end of each month for three years. Suppose interest is 10.8% compounded monthly.

(a) What was the cash price of the equipment?

(b) How much was the cost of financing?

6. Elynor bought a vacation property for $2500.00 down and quarterly mortgage payments of $550.41 at the end of each quarter for five years. Interest is 8% compounded quarterly.

(a) What was the purchase price of the property?

(b) How much interest will Elynor pay?

7. Ed intends to retire in eight years. To supplement his pension he would like to receive $450.00 every three months for fifteen years. If he is to receive the first payment three months after his retirement and interest is 5% p.a. compounded quarterly, how much must he invest today to achieve his goal?

8. Planning for their son's college education, Vivien and Adrian Marsh opened an account paying 6.3% compounded monthly. If ordinary annuity payments of $200.00 per month are to be paid out of the account for three years starting seven years from now, how much did the Marshes deposit?

9. Kimiko signed a mortgage requiring payments of $234.60 at the end of every month for six years at 7.2% compounded monthly.
   (a) How much was the original mortgage balance?
   (b) If Kimiko missed the first five payments, how much would she have to pay after six months to bring the mortgage payments up to date?
   (c) How much would Kimiko have to pay after six months to pay off the mortgage?
   (d) If the mortgage were paid off after six months, what would the total interest cost be?
   (e) How much of the total interest cost is additional interest because of the missed payments?

10. Field Construction agreed to lease payments of $642.79 on construction equipment to be made at the end of each month for three years. Financing is at 9% compounded monthy.
   (a) What is the value of the original lease contract?
   (b) If, due to delays, the first eight payments were deferred, how much money would be needed after nine months to bring the lease payments up to date?
   (c) How much money would be required to pay off the lease after nine months?
   (d) If the lease were paid off after nine months, what would the total interest be?
   (e) How much of the total interest would be due to deferring the first eight payments?

BUSINESS MATH NEWS BOX

Who Wants to Be a Millionaire?

The host of the popular TV show replied, "Everyone!"

Everyone wants to be a millionaire. That is why this program, introduced in 1999, was so popular. By answering a series of questions, contestants could win $100, then choose to stop or to take a chance on answering the next question correctly. They could then win $200, $500, $1000, or more—up to one million dollars. They could also lose much of their current winnings. Everyone has a dream of winning enough money to achieve financial independence.

What would you do with financial independence? How much money would you need to have invested to achieve it? How does one amass a million dollars in investments? In his book, *6 Steps to $1 Million,* Gordon Pape suggests that you can save your way to a million dollars. He suggests starting by establishing a disciplined savings plan, searching out and taking advantage of whatever tax breaks are available to you, and making wise investments.

He recommends that whenever you receive any extra money, you save a portion of it, such as 50%. This extra money might come as a gift, a refund, or an increase in your pay. As you save, use your money effectively by paying off debt such as credit cards, loans, or your mortgage, contributing to a registered retirement savings plan, or starting or increasing your investment portfolio.

If you save a small amount on a regular basis, you will be amazed how your savings will grow. For example, not taking taxes into account, if you save $10 per month, and invest the money where you will realize a 5% average annual return, after 20 years you will have invested $4110.34. And if instead of $10 you save $50, after 20 years you will have invested $20 551.68.

QUESTIONS

1. The information above shows how savings will grow over twenty years at an average annual rate of return of 5%. What compounding period do the numbers given represent most closely: annual, semi-annual, quarterly, or monthly? Use the example of $10 savings per month to calculate your answer.
2. If you saved $100 per month and invested at a 5% average annual return, how much would you have after 20 years? How does your answer compare with the $50-per-month example given above?
3. If you continued to save $100 per month for another ten years, how much would you have at the end of that time period? How does your answer compare with the $50-per-month example given above?
4. To what amount would your savings grow if you invested the following amounts at 6% compounded monthly for twenty years? How does your answer compare with the amounts described above?
   (a) $10 per month
   (b) $50 per month
   (c) $100 per month

Hint: Use a compounding interval of 12 representing a month.

Source: Adapted from Gordon Pape, *6 Steps to $1 Million*, © 2001 Gordon Pape Enterprises Ltd. Reprinted by permission of Gordon Pape Enterprises Ltd.

5.4 Ordinary Simple Annuities—Finding the Periodic Payment PMT

A. Finding the periodic payment PMT when the future value of an annuity is known

If the future value of an annuity FV_n, the number of conversion periods n, and the conversion rate i are known, you can find the periodic payment PMT by substituting the given values in the appropriate future value formula.

For ordinary simple annuities, use Formula 5.1.

$$FV_n = PMT\left[\frac{(1+i)^n - 1}{i}\right]$$ — Formula 5.1

When using a preprogrammed financial calculator, you can find PMT by entering the five known values (FV, N, I/Y, P/Y, and C/Y) and pressing CPT PMT. Recall that the PMT amount will be negative since payments are considered to be cash outflows.

EXAMPLE 5.4A

What deposit made at the end of each quarter will accumulate to \$10 000.00 in four years at 4% compounded quarterly?

SOLUTION

$FV_n = 10\,000.00$; I/Y = 4; P/Y = 4; $i = \frac{4\%}{4} = 1.0\% = 0.01$; $n = 16$

$$10\,000.00 = PMT\left(\frac{1.01^{16} - 1}{0.01}\right) \text{ — substituting in Formula 5.1}$$

$$10\,000.00 = PMT\left(\frac{1.172579 - 1}{0.01}\right)$$

$$10\,000.00 = PMT(17.25786)$$

$$PMT = \frac{10\,000.00}{17.25786}$$

$$PMT = \$579.45$$

As we have emphasized throughout this book, beginning in Chapter 2, understanding how to rearrange terms in a formula is a very important skill. By knowing how to do this, you avoid having to memorize equivalent forms of the same formula.

When using a scientific calculator, you can find PMT by first rearranging the terms of the future value formulas as shown below. Then substitute the three known values (FV, n, and i) into the rearranged formula and use the calculator to solve for PMT.

$$FV_n = PMT\left[\frac{(1 + i)^n - 1}{i}\right] \text{ — Formula 5.1}$$

$$PMT = \frac{FV_n}{\left[\frac{(1 + i)^n - 1}{i}\right]} \text{ — divide both sides by } \left[\frac{(1 + i)^n - 1}{i}\right]$$

$$PMT = \frac{FV_n i}{(1 + i)^n - 1} \text{ — dividing by a fraction is the same as inverting the fraction and multiplying}$$

$FV_n = 10\,000.00$; $i = 0.01$; $n = 16$

$$PMT = \frac{10\,000.00(0.01)}{1.01^{16} - 1} \text{ — substituting in rearranged Formula 5.1}$$

$$PMT = \$579.45$$

EXAMPLE 5.4B

If you want to have \$5000.00 on deposit in your bank account in three years, how much must you deposit at the end of each month if interest is 4.5% compounded monthly?

SOLUTION

$FV_n = 5000.00$; I/Y = 4.5; P/Y = 12; $i = 0.375\%$; $n = 36$

(Set P/Y = 12) [2nd] (CLR TVM) 0 [PV] 5000 [FV]

4.5 [I/Y] 36 [N] [CPT] [PMT] [−129.984622]

The monthly deposit required is \$129.98.

B. Finding the periodic payment PMT when the present value of an annuity is known

If the present value of an annuity PV_n, the number of conversion periods n, and the conversion rate i are known, you can find the periodic payment PMT by substituting the given values in the appropriate present value formula.

For ordinary simple annuities, use Formula 5.2.

$$PV_n = PMT\left[\frac{1-(1+i)^{-n}}{i}\right]$$ — Formula 5.2

When using a scientific calculator, you can find PMT by first rearranging the terms of the present value formulas above. Then substitute the three known values (PV, n, and i) into the rearranged formula and solve for PMT.

When using a preprogrammed financial calculator, you can find PMT by entering the five known values (PV, N, P/Y, C/Y and I/Y) and pressing **CPT** **PMT**

.

EXAMPLE 5.4C

What semi-annual payment is required to pay off a loan of $8000.00 in ten years if interest is 10% compounded semi-annually?

SOLUTION

$PV_n = 8000.00$; I/Y = 10; P/Y = 2; $i = \frac{10\%}{2} = 5\% = 0.05$; $n = 10(2) = 20$

$$8000.00 = PMT\left(\frac{1-1.05^{-20}}{0.05}\right)$$ — substituting in Formula 5.2

$$8000.00 = PMT\left(\frac{1-0.376889}{0.05}\right)$$

$$8000.00 = PMT(12.462210)$$

$$PMT = \frac{8000.00}{12.462210}$$

$$PMT = \$641.94$$

Alternatively:

You can first rearrange the terms of Formula 11.2 to solve for PMT.

$$PV_n = PMT\left[\frac{1-(1+i)^{-n}}{i}\right]$$ — Formula 5.2

$$PMT = \frac{PV_n}{\left[\frac{1-(1+i)^{-n}}{i}\right]}$$ — divide both sides by $\left[\frac{1-(1+i)^{-n}}{i}\right]$

$$PMT = \frac{PV_n i}{1-(1+i)^{-n}}$$ — dividing by a fraction is the same as inverting the fraction and multiplying

$PV_n = 8000.00$; $i = 0.05$; $n = 20$

$$PMT = \frac{8000.00(0.05)}{1-1.05^{-20}}$$

$$PMT = \$641.94$$

EXAMPLE 5.4D

Derek bought a new car valued at \$9500.00. He paid \$2000.00 down and financed the remainder over five years at 9% compounded monthly. How much must Derek pay each month?

SOLUTION

$PV_n = 9500.00 - 2000.00 = 7500.00$; I/Y = 9; P/Y = 12; $i = \frac{9\%}{12} = 0.75\%$; $n = 60$

(Set P/Y = 12) 0 [FV] 7500 [±] [PV]

9 [I/Y] 60 [N] [CPT] [PMT] [155.6876642]

Derek's monthly payment is \$155.69.

C. Applications

EXAMPLE 5.4E

Cecile Tremblay, age 37, expects to retire at age 62. To plan for her retirement, she intends to deposit \$1500.00 at the end of each of the next 25 years in a registered retirement savings plan. After her last contribution, she intends to convert the existing balance into a registered retirement income fund from which she expects to make 20 equal annual withdrawals. If she makes the first withdrawal one year after her last contribution and interest is 6.5% compounded annually, how much is the size of the annual withdrawal?

SOLUTION

As the following time diagram shows, the problem can be broken into two steps.

STEP 1 Compute the *accumulated* value FV_n of the 25 annual deposits of \$1500.00 into the RRSP.

PMT = 1500.00; I/Y = 6.5; P/Y = 1; $i = 6.5\% = 0.065$; $n = 25$

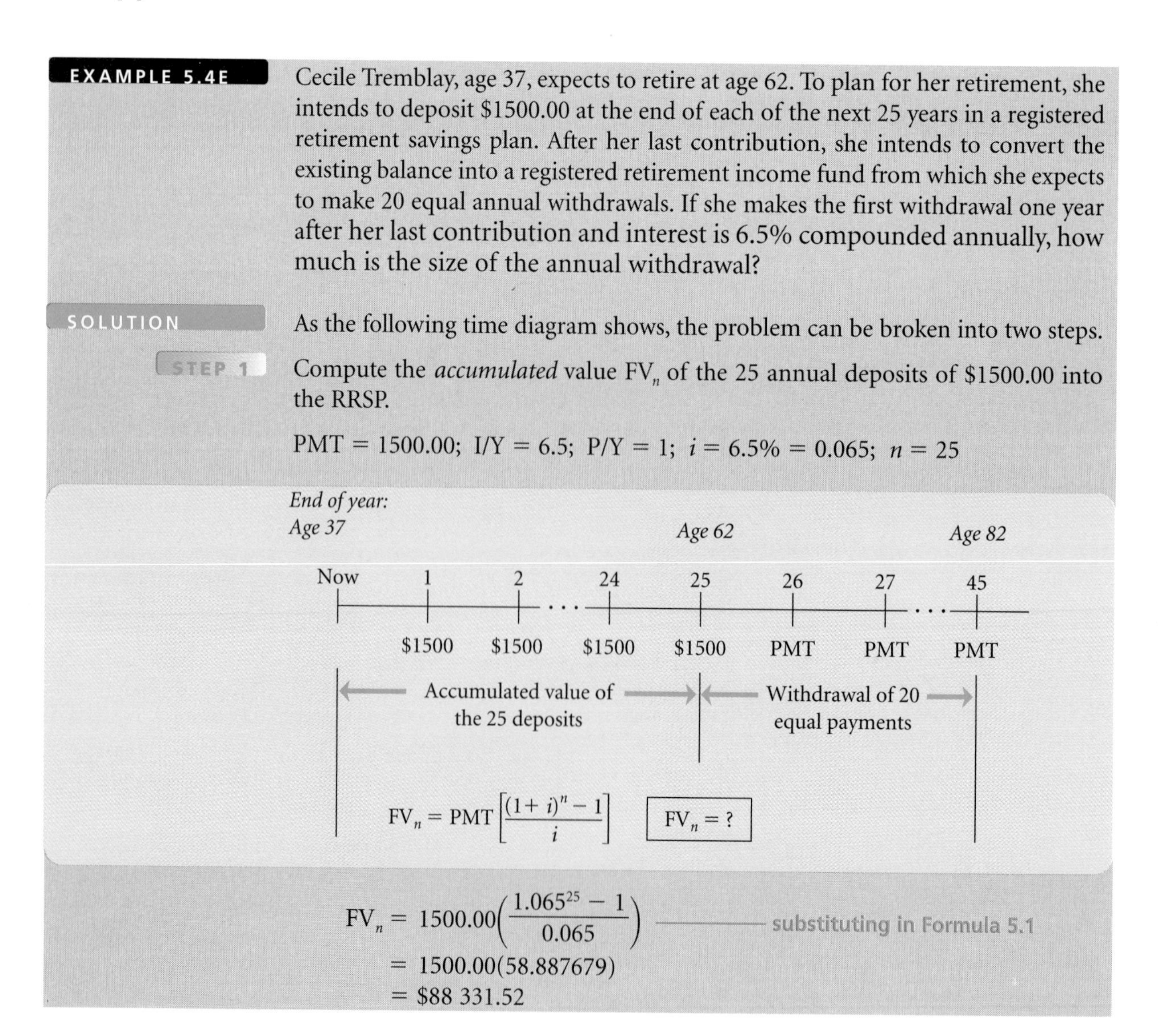

$$FV_n = 1500.00\left(\frac{1.065^{25} - 1}{0.065}\right)$$ ——— substituting in Formula 5.1

$$= 1500.00(58.887679)$$
$$= \$88\ 331.52$$

STEP 2 Compute the annual payment that can be withdrawn from the RRIF that has an initial balance of $88 331.52.

$PV_n = 88\ 331.52;\quad i = 0.065;\quad n = 20$

$$88\ 331.52 = PMT\left(\frac{1 - 1.065^{-20}}{0.065}\right) \text{ ——— using Formula 5.2}$$

$$88\ 331.52 = PMT(11.018507)$$

$$PMT = 8016.65$$

Programmed Solution

STEP 1 (Set P/Y = 1) 0 [PV] 1500 [±] [PMT] 6.5 [I/Y] 25 [N] [CPT] [FV] 88 331.51788

STEP 2 88 331.518 [+/–] [PV] 0 [FV] 6.5 [I/Y] 20 [N] [CPT] [PMT] 8016.650159

The sum of money that Cecile can withdraw each year is $8016.65.

EXERCISE 5.4

Excel If you choose, you can use Excel's *Payment (PMT)* function to answer all of the questions below. Refer to *PMT* on the Spreadsheet Template Disk to learn how to use this Excel function.

A. For each of the following ten ordinary annuities, determine the size of the periodic rent.

	Future Value	Present Value	Payment Period	Term of Annuity	Interest Rate	Conversion Period
1.	$15 000.00		6 months	7 years, 6 mos	5.5%	semi-annually
2.	$6 000.00		1 quarter	9 years, 9 mos	8%	quarterly
3.		$12 000.00	12 months	15 years	4.5%	annually
4.		$7 000.00	6 months	12.5 years	7.5%	semi-annually
5.	$8 000.00		3 months	6 years	6.8%	quarterly
6.		$20 000.00	1 month	20 years	12%	monthly
7.	$45 000.00		1 month	10 years	9%	monthly
8.		$35 000.00	1 quarter	15 years	4%	quarterly
9.		$20 000.00	6 months	8 years	7%	semi-annually
10.	$16 500.00		3 months	15 years	5.75%	quarterly

B. Answer each of the following questions.

1. What deposit made at the end of each quarter for fifteen years will accumulate to $20 000.00 at 6% compounded quarterly?

2. What payment is required at the end of each month for five years to repay a loan of $8000.00 at 8.4% compounded monthly?
3. A contract can be fulfilled by making an immediate payment of $7500.00 or equal payments at the end of every six months for ten years. What is the size of the semi-annual payments at 9.6% compounded semi-annually?
4. What payment made at the end of each month for eighteen years will amount to $16 000.00 at 9.6% compounded monthly?
5. What payment is required at the end of each month for twelve years to amortize a $32 000.00 mortgage if interest is 6.5% compounded monthly?
6. How much must be deposited at the end of each quarter for nine years to accumulate to $11 000.00 at 4% compounded quarterly?
7. What payment made at the end of every six months for fifteen years will accumulate to $18 000.00 at 5% compounded semi-annually?
8. Hunan bought a car priced at $15 300.00 for 15% down and equal monthly payments for four years. If interest is 8% compounded monthly, what is the size of the monthly payment?
9. Ruben bought a boat valued at $16 500.00 on the installment plan requiring a $2000.00 down payment and equal monthly payments for five years. If the first payment is due one month after the date of purchase and interest is 7.5% compounded monthly, what is the size of the monthly payment?
10. How much does a depositor have to save at the end of every three months for seven years to accumulate $3500.00 if interest is 3.75% compounded quarterly?
11. How much would you have to pay into an account at the end of every six months to accumulate $10 000.00 in eight years if interest is 3% compounded semi-annually?
12. Ontario Credit Union entered a lease contract valued at $7200.00. The contract provides for payments at the end of each quarter for three years. If interest is 6.5% compounded quarterly, what is the size of the quarterly payment?
13. Olivia bought a car priced at $10 600.00 for 10% down and the balance in equal monthly payments over four years at 7.2% compounded monthly. How much does Olivia have to pay each month?
14. The Watsons bought a rental property valued at $50 000.00 by paying 20% down and mortgaging the balance over twenty-five years through equal quarterly payments at 10% compounded quarterly. What was the size of the quarterly payments?
15. George plans to deposit $1200.00 at the end of every six months for fifteen years into an RRSP account. After the last deposit, he intends to convert the existing balance into an RRIF and withdraw equal amounts at the end of every six months for twenty years. If interest is expected to be 7.5% compounded semi-annually, how much will George be able to collect every six months?

16. Starting three months after her grandson Robin's birth, Mrs. Devine made deposits of $60.00 into a trust fund every three months until Robin was twenty-one years old. The trust fund provides for equal withdrawals at the end of each quarter for four years, beginning three months after the last deposit. If interest is 4.75% compounded quarterly, how much will Robin receive every three months?

17. On the day of his daughter's birth, Mr. Dodd deposited $2000.00 in a trust fund with his credit union at 5% compounded quarterly. Following her eighteenth birthday, the daughter is to receive equal payments at the end of each month for four years while she is at college. If interest is to be 6.0% compounded monthly after the daughter's eighteenth birthday, how much will she receive every month?

18. Equal payments are to be made at the end of each month for fifteen years with interest at 9% compounded monthly. After the last payment, the fund is to be invested for seven years at 10% compounded quarterly and have a maturity value of $20 000.00. What is the size of the monthly payment?

19. Mr. Talbot received a retirement bonus of $18 000.00, which he deposited in an RRSP. He intends to leave the money for fourteen years, then transfer the balance into an RRIF and make equal withdrawals at the end of every six months for twenty years. If interest is 6.5% compounded semi-annually, what will be the size of each withdrawal?

5.5 Finding the Term *n* of an Annuity

A. Finding the term *n* when the future value of an annuity is known

If the future value of an annuity FV_n, the periodic payment PMT, and the conversion rate i are known, you can find the term of the annuity n by substituting the given values in the appropriate future value formula.

For ordinary simple annuities, use Formula 5.1.

$$FV_n = PMT\left[\frac{(1+i)^n - 1}{i}\right]$$ —— Formula 5.1

DID YOU KNOW?

By law, you must collapse your RRSP by the end of the year you turn 69. One way of avoiding a large income tax expense in that year is to purchase a life annuity. The monthly payments you receive end at your death, and the monthly amounts are fixed. Another way to avoid a large income tax expense when you collapse your RRSP is to create an RRIF. Unlike a life annuity, with an RRIF you decide the amount of the monthly payment you receive (although Canada Customs and Revenue Agency sets a minimum percentage you must withdraw each year). You can also change the amount over time. It is possible to withdraw all of your funds while you still have years to live. Did you know you can outlive an RRIF, but you can't outlive a life annuity?

When using a scientific calculator, you can find n by first rearranging the terms of the future value formulas above. Then substitute the three known values (FV, PMT, and i) into the rearranged formula and solve for n.

When using a preprogrammed financial calculator, you can find n by entering the five known values (FV, PMT, P/Y, C/Y, and I/Y) and pressing CPT N.

EXAMPLE 5.5A

How long will it take for $200.00 deposited at the end of each quarter to amount to $5726.70 at 6% compounded quarterly?

SOLUTION

$FV_n = 5726.70$; $I/Y = 6$; $P/Y = 4$; $i = \frac{6\%}{4} = 1.5\% = 0.015$; $PMT = 200.00$

$$5726.70 = 200.00\left(\frac{1.015^n - 1}{0.015}\right) \text{ — substituting in Formula 5.1}$$

$$28.6335 = \frac{1.015^n - 1}{0.015} \text{ — divide both sides by 200.00}$$

$$0.429503 = 1.015^n - 1 \text{ — multiply both sides by 0.015}$$

$$1.015^n = 1.429503 \text{ — add 1 to both sides}$$

$$n \ln 1.015 = \ln 1.429503 \text{ — solve for } n \text{ using natural logarithms}$$

$$0.014889n = 0.357327$$

$$n = \frac{0.357327}{0.014889}$$

$$n = 24 \text{ (quarters)}$$

It will take six years for $200.00 per quarter to grow to $5726.70.

Alternatively:

You can first rearrange the terms of Formula 5.1 to solve for n.

$$FV_n = PMT\left[\frac{(1+i)^n - 1}{i}\right] \text{ — Formula 5.1}$$

$$\frac{FV_n}{PMT} = \frac{(1+i)^n - 1}{i} \text{ — divide both sides by PMT}$$

$$(1+i)^n = \left(\frac{FV_n i}{PMT}\right) + 1 \text{ — multiply both sides by } i \text{ and add 1 to both sides}$$

$$n \ln(1+i) = \ln[(FV_n i/PMT) + 1] \text{ — solve for } n \text{ using natural logarithms}$$

$$n = \frac{\ln[FV_n i/PMT + 1]}{1_n(1+i)} \text{ — divide both sides by } \ln(1+i)$$

$FV_n = 5726.70$; $i = 0.015$; $PMT = 200.00$

$$n = \frac{\ln[(5726.70 \times 0.015/200.00) + 1]}{\ln 1.015}$$

$$n = \frac{0.357327}{0.014889}$$

$$n = 24 \text{ (quarters)}$$

POINTERS AND PITFALLS

For annuity problems in which the period of investment or loan repayment must be determined, once n has been calculated, simply divide n by m to calculate the period of investment or loan repayment in years. To illustrate, the solution in Example 5.5A is $n = 24$. Since $m = 4$ (the question states "at the end of each quarter"), the period of investment is $n \div m$, or $24 \div 4 = 6$ years.

EXAMPLE 5.5B

In how many months will your bank account grow to $3000.00 if you deposit $150.00 at the end of each month and the account earns 9% compounded monthly?

SOLUTION

$FV_n = 3000.00$; PMT = 150.00; I/Y = 9; P/Y = 12; $i = \frac{9\%}{12} = 0.75\%$

(Set P/Y = 12) 0 [PV] 3000 [FV]

150 [±] [PMT] 9 [I/Y] [CPT] [N] [18.7047]

months

It will take about 19 months to accumulate $3000.00.

Interpretation of Result

When FV_n, PMT, and i are known, it is unlikely that n will be a whole number. The fractional time period of 0.7 month indicates that the accumulated value of 18 deposits of $150.00 will be less than $3000.00, while the accumulated value of 19 deposits will be more than $3000.00. This point can be verified by computing FV_{18} and FV_{19}.

$$FV_{18} = 150.00\left(\frac{1.0075^{18} - 1}{0.0075}\right) = 150.00(19.194718) = \$2879.21$$

$$FV_{19} = 150.00\left(\frac{1.0075^{19} - 1}{0.0075}\right) = 150.00(20.338679) = \$3050.80$$

The definition of an annuity does not provide for making payments at unequal time intervals. The appropriate answer to problems in which n is a fractional value is a whole number. The usual approach is to round upwards so that in this case $n = 19$.

Rounding upwards implies that the deposit made at the end of the nineteenth month is smaller than the usual deposit of $150.00. The method of computing the size of the final deposit or payment when the term of the annuity is a fractional value rounded upwards is considered in Chapter 8.

B. Finding the term n when the present value of an annuity is known

If the present value PV_n, the periodic payment PMT, and the conversion rate i are known, you can find the term of the annuity n by substituting the given values in the present value formula.

$$PV_n = PMT\left[\frac{1-(1+i)^{-n}}{i}\right] \quad \text{——— Formula 5.2}$$

When using a scientific calculator, you can find n by first rearranging the terms of the present value formulas above. Then substitute the three known values (PV, PMT, and i) into the rearranged formula and solve for n.

When using a preprogrammed financial calculator, you can find n by entering the five known values (PV, PMT, P/Y, C/Y, and I/Y) and pressing CPT N.

Note: When using the Texas Instruments BAII Plus financial calculator, you *must* enter *either* the PV or PMT as a negative amount, *but not both.* Due to the sign conventions used by this calculator, entering *both* PV and PMT as negative amounts or as positive amounts will lead to an *incorrect* final answer. The calculator does *not* indicate that the answer is incorrect or that an entry error was made. (This has not been an issue until now because either PV or PMT was 0 in all the examples we discussed.) To avoid incorrect answers, *always* enter the PV amount as a negative number and the PMT amount as a positive number when FV = 0 and you are calculating n. However, if PV = 0, enter PMT as a negative number and FV as a positive number. The examples in this text follow these rules. Refer to Appendix II to check whether this step is necessary if you use the Texas Instruments BA Solar, the Sharp EL-733A, or the Hewlett-Packard 10B calculator.

EXAMPLE 5.5C How many quarterly payments of \$600.00 are required to repay a loan of \$5400.00 at 6% compounded quarterly?

SOLUTION $PV_n = 5400.00$; PMT = 600.00; I/Y = 6; P/Y = 4; $i = \frac{6\%}{4} = 1.5\% = 0.015$

$$5400.00 = 600.00\left(\frac{1-1.015^{-n}}{0.015}\right) \quad \text{——— substituting in Formula 5.2}$$

$$9.00 = \frac{1-1.015^{-n}}{0.015} \quad \text{——— divide both sides by 600.00}$$

$$0.135 = 1 - 1.015^{-n} \quad \text{——— multiply both sides by 0.015}$$

$$1.015^{-n} = 0.865$$

$$-n \ln 1.015 = \ln 0.865 \quad \text{——— solve for } n \text{ using natural logarithms}$$

$$-0.014889n = -0.145026$$

$$n = \frac{0.145026}{0.014889}$$

$$n = 9.740726$$

$$n = 10 \text{ quarters}$$

10 quarterly payments are required to repay the loan.

Alternatively:

You can first rearrange the terms of Formula 5.2 to solve for n.

$$PV_n = PMT\left[\frac{1-(1+i)^{-n}}{i}\right]$$ —— Formula 5.2

$$\frac{PV_n}{PMT} = \frac{1-(1+i)^{-n}}{i}$$ —— divide both sides by PMT

$$\left(\frac{PV_n i}{PMT}\right) - 1 = -(1+i)^{-n}$$ —— multiply both sides by i and subtract 1 from both sides

$$(1+i)^{-n} = 1 - \left(\frac{PV_n i}{PMT}\right)$$ —— multiply both sides by −1

$$-n\ln(1+i) = \ln\left[1 - \left(\frac{PV_n i}{PMT}\right)\right]$$ —— solve for n using natural logarithms

$$n = \frac{\ln\left[1 - \left(\frac{PV_n i}{PMT}\right)\right]}{-\ln(1+i)}$$ —— divide both sides by $-\ln(1+i)$

$PV_n = 5400.00$; $PMT = 600.00$; $i = 0.015$

$$n = \frac{\ln\left[1 - \frac{(5400.00)(0.015)}{600.00}\right]}{-\ln(1.015)}$$ —— substituting in rearranged Formula 5.2

$$n = \frac{-0.145026}{-0.014889}$$

$$n = 9.740718$$

n = 10 quarters

EXAMPLE 5.5D

On his retirement, Art received a gratuity of \$8000.00 from his employer. Taking advantage of the existing tax legislation, he invested the money in an annuity that provides for semi-annual payments of \$1200.00 at the end of every six months. If interest is 6.25% compounded semi-annually, how long will the annuity exist?

SOLUTION

$PV_n = 8000.00$; PMT = 1200.00; I/Y = 6.25; P/Y = 2; $i = \frac{6.25\%}{2} = 3.125\%$

(Set P/Y = 2) 0 [FV] 8000 [±] [PV]

1200 [PMT] 6.25 [I/Y] [CPT] [N] [7.59188]

half-year periods

The annuity will be in existence for four years. Art will receive seven payments of \$1200.00 and a final payment that will be less than \$1200.00.

EXERCISE 5.5

Excel If you choose, you can use Excel's ***Number of Compounding Periods (NPER)*** function to answer all of the questions below. Refer to **NPER** on the Spreadsheet Template Disk to learn how to use this Excel function.

Find the term of each of the following ten ordinary annuities. (State your answer in years and months.)

	Future Value	Present Value	Periodic Rent	Payment Interval	Interest Rate	Conversion Period
1.	$20 000.00		$800.00	1 year	7.5%	annually
2.	$17 000.00		$35.00	1 month	9%	monthly
3.		$14 500.00	$190.00	1 month	5.25%	monthly
4.		$5 000.00	$300.00	3 months	4%	quarterly
5.	$3 600.00		$175.00	6 months	7.4%	semi-annually
6.		$9 500.00	$740.00	1 quarter	5.2%	quarterly
7.		$21 400.00	$1660.00	6 months	4.5%	semi-annually
8.	$13 600.00		$140.00	3 months	8%	quarterly
9.	$7 200.00		$90.00	1 month	3.75%	monthly
10.		$9 700.00	$315.00	3 months	11%	quarterly

Answer each of the following questions.

1. How long would it take you to save $4500.00 by making deposits of $50.00 at the end of every month into a savings account earning 6% compounded monthly?
2. In what period of time could you pay back a loan of $3600.00 by making monthly payments of $96.00 if interest is 10.5% compounded monthly?
3. How long will it take to save $5000.00 by making deposits of $60.00 at the end of every month into an account earning interest at 6% monthly?
4. For how long will Amir have to make payments of $300.00 at the end of every three months to repay a loan of $5000.00 if interest is 7% compounded quarterly?
5. A deposit of $4000.00 is made today. For how long can $500.00 be withdrawn from the account at the end of every three months starting three months from now if interest is 4% compounded quarterly?
6. Suppose $646.56 is deposited at the end of every six months into an account earning 6.5% compounded semi-annually. If the balance in the account four years after the last deposit is to be $20 000.00, how many deposits are needed?
7. For how long can $1000.00 be withdrawn at the end of each month from an account containing $36 000.00 if interest is 6.4% compounded monthly?
8. Josie borrowed $8000.00 compounded monthly to help finance her education. She contracted to repay the loan in monthly payments of $300.00 each. If the payments are due at the end of each month and interest is 4% compounded monthly, how long will Josie have to make monthly payments?
9. A mortgage of $26 500.00 is to be repaid by making payments of $1560.00 at the end of every six months. If interest is 7% compounded semi-annually, what is the term of the mortgage?
10. A car loan of $12 000.00 is to be repaid with end-of-month payments of $292.96. If interest is 8% compounded monthly, how long is the term of the loan?

5.6 Finding the Periodic Rate of Interest *i* Using Programmed Financial Calculators

A. Finding the periodic rate of interest *i* for simple annuities

Preprogrammed financial calculators are especially helpful when solving for the conversion rate i. Determining i without a financial calculator is extremely time consuming. However, it *can* be done by hand, as illustrated in Appendix A on the CD-ROM.

When the future value or present value, the periodic payment PMT, and the term N of an annuity are known, the periodic rate of interest i can be found by entering the three known values into a preprogammed financial calculator. (Remember, if both PV and PMT are non-zero, enter PV only as a negative amount.) For ordinary simple annuities, retrieve the answer by pressing CPT I/Y. This represents the nominal annual rate of interest j. By dividing j by the number of compounding periods per year m, you can obtain the periodic interest rate i.

EXAMPLE 5.6A

Compute the nominal annual rate of interest at which $100.00 deposited at the end of each month for ten years will amount to $15 000.00.

SOLUTION

$FV_n = 15\,000.00$; PMT = 100.00; P/Y = 12; $n = 120$; $m = 12$

(Set P/Y = 12) 0 PV 15 000 FV 100 ± PMT

120 N CPT I/Y 4.350057

allow several seconds for the computation

The nominal annual rate of interest is 4.35% approximately. The monthly conversion rate is 4.350057/12 = 0.362505 per month approximately.

EXAMPLE 5.6B

A loan of $6000.00 is paid off over five years by monthly payments of $120.23. What is the nominal annual rate of interest on the loan?

SOLUTION

$PV_n = 6000.00$; PMT = 120.23; $n = 60$; $m = 12$

(Set P/Y = 12) 0 FV 6000 ± PV 120.23 PMT

60 N CPT I/Y 7.500810

allow several seconds for the computation

The nominal annual rate of interest is 7.5% approximately. The monthly compounding rate is 7.500810/12 = 0.625067%.

EXERCISE 5.6

A. Compute the nominal annual rate of interest for each of the following eight ordinary simple annuities.

	Future Value	Present Value	Periodic Rent	Payment Interval	Term	Conversion Period
1.	$9 000.00		$230.47	3 months	8 years	quarterly
2.	$4 800.00		$68.36	1 month	5 years	monthly
3.		$7 400.00	$119.06	1 month	7 years	monthly
4.		$6 980.00	$800.00	6 months	5 years	semi-annually
5.	$70 000.00		$1014.73	1 year	25 years	annually
6.		$42 000.00	$528.00	1 month	10 years	monthly
7.		$28 700.00	$2015.00	6 months	15 years	semi-annually
8.	$36 000.00		$584.10	3 months	12 years	quarterly

B. Answer each of the following questions.

1. Compute the nominal annual rate of interest at which $350.00 paid at the end of every three months for six years accumulates to $12 239.76.
2. What is the nominal annual rate of interest if a four-year loan of $6000.00 is repaid by monthly payments of $144.23?
3. Rita converted an RRSP balance of $199 875.67 into an RRIF that will pay her $1800.00 at the end of every month for nine years. What is the nominal annual rate of interest?
4. Katrina contributed $2500.00 every year into an RRSP for ten years. What nominal annual rate of interest will the RRSP earn if the balance in Katrina's account just after she made her last contribution was $33 600.00?
5. A car valued at $11 400.00 can be purchased for 10% down and monthly payments of $286.21 for three-and-a-half years. What is the effective annual cost of financing?
6. Property worth $50 000.00 can be purchased for 20% down and quarterly mortgage payments of $1000.00 for 25 years. What effective annual rate of interest is charged?
7. What nominal annual rate of interest compounded monthly was paid if contributions of $250.00 made into an RRSP at the end of every month amounted to $35 000.00 after ten years?
8. Compute the nominal annual rate of interest compounded monthly at which $400.00 paid at the end of the month for eight years accumulates to $45 000.00.
9. What is the nominal annual rate of interest compounded quarterly if a loan of $21 500.00 is repaid in seven years by payments of $1000.00 made at the end of every three months?
10. A property worth $35 000.00 is purchased for 10% down and semi-annual payments of $2100.00 for twelve years. What is the effective annual rate of interest if interest is compounded semi-annually?

Review Exercise

1. Payments of $360.00 are made into a fund at the end of every three months for twelve years. The fund earns interest at 7% compounded quarterly.
   (a) What will be the balance in the fund after twelve years?
   (b) How much of the balance is deposits?
   (c) How much of the balance is interest?

2. A trust fund is set up to make payments of $950.00 at the end of each month for seven-and-a-half years. Interest on the fund is 7.8% compounded monthly.
   (a) How much money must be deposited into the fund?
   (b) How much will be paid out of the fund?
   (c) How much interest is earned by the fund?

3. How much interest is included in the accumulated value of $75.90 paid at the end of each month for four years if interest is 9% compounded monthly?

4. If a loan was repaid by quarterly payments of $320.00 in five years at 8% compounded quarterly, how much money had been borrowed?

5. How long will it take to build up a fund of $10 000.00 by saving $300.00 every six months at 4.5% compounded semi-annually?

6. What is the term of a mortgage of $35 000.00 repaid by monthly payments of $475.00 if interest is 7.5% compounded monthly?

7. Suppose you would like to have $10 000.00 in your savings account and interest is 8% compounded quarterly. How much must you deposit every three months for five years if the deposits are made at the end of each quarter?

8. Equal sums of money are withdrawn monthly from a fund of $20 000.00 for fifteen years. If interest is 9% compounded monthly, what is the size of each withdrawal if the withdrawal is made at the end of each month?

9. If you contribute $1500.00 into an RRSP every six months for twelve years and interest on the deposits is 8% compounded semi-annually, how much would the balance in the RRSP be seven years after the last contribution?

10. Doris purchased a piano with $300.00 down and monthly payments of $124.00 for two-and-a-half years at 9% compounded monthly. What was the purchase price of the piano?

11. A contract valued at $11 500.00 requires payment of $1450.00 at the end of every six months. If interest is 10.5% compounded semi-annually, what is the term of the contract?

12. What nominal annual rate of interest is paid on quarterly RRSP contributions of $1100.00 made for fifteen years if the balance just after the last contribution is $106 000.00?

13. What nominal annual rate of interest was charged on a loan of $5600.00 repaid in monthly installments of $121.85 in four-and-a-half years?

14. Glenn has made contributions of $250.00 every three months into an RRSP for ten years. Interest for the first four years was 4% compounded quarterly. Since then the interest rate has been 5% compounded quarterly. How much will Glenn have in his RRSP three years after the last contribution?

15. Avi expects to retire in twelve years. Beginning one month after his retirement he would like to receive $500.00 per month for twenty years. How much must he deposit into a fund today to be able to do so if the rate of interest on the deposit is 6% compounded monthly?

16. A contract is signed requiring payments of $750.00 at the end of every three months for eight years. How much is the cash value of the contract if money is worth 9% compounded quarterly?

17. The amount of $10 000.00 is put into a five-year term deposit paying 7.5% compounded semi-annually. After five years the deposit is converted into an ordinary annuity of equal semi-annual payments of $2000.00 each. If interest remains the same, what is the term of the annuity?

18. Mirielle has deposited $125.00 at the end of each month for 15 years at 7.5% compounded monthly. After her last deposit she converted the balance into an ordinary annuity paying $1200.00 every three months for twelve years. If interest on the annuity is compounded quarterly, what is the effective annual rate of interest paid by the annuity?

19. A contract is signed requiring payments of $750.00 at the end of every three months for eight years. How much is the cash value of the contract if money is worth 10.5% compounded quarterly?

20. A savings plan requiring quarterly deposits of $400.00 for twenty years provides for a lump-sum payment of $92 000.00 just after the last deposit has been made.

(a) What is the effective annual rate of interest on the savings plan?

(b) If, instead of the lump sum, monthly ordinary annuity payments of $1350.00 may be accepted at the same nominal rate of interest (correct to two decimals) but compounded monthly, what is the term of the annuity?

Self-Test

1. You won $100 000 in a lottery and you want to set some of that sum aside for ten years. After ten years, you would like to receive $2400 at the end of every three months for eight years. How much of your winnings must you set aside if interest is 5.5% compounded quarterly?
2. A sum of money is deposited at the end of every month for ten years at 7.5% compounded monthly. After the last deposit, interest for the account is to be 6% compounded quarterly and the account is to be paid out by quarterly payments of $4800.00 over six years. What is the size of the monthly deposit?
3. Compute the nominal annual rate of interest on a loan of $48 000.00 repaid in semi-annual installments of $4000.00 in ten years.
4. A loan of $14 400.00 is to be repaid in quarterly payments of $600.00. How many payments are required to repay the loan at 10.5% compounded quarterly?
5. The amount of $57 426.00 is invested at 6% compounded monthly for six years. After the initial six-year period, the balance in the fund is converted into an annuity due paying $3600.00 every three months. If interest on the annuity is 5.9% compounded quarterly, what is the term of the annuity in months?
6. A loan was repaid in seven years by monthly payments of $450. If interest was 12% compounded monthly, how much interest was paid?
7. Ms. Simms made quarterly deposits of $540 into a savings account. For the first five years interest was 5% compounded quarterly. Since then the rate of interest has been 5.5% compounded quarterly. How much is the account balance after thirteen years?

8. How much interest is included in the accumulated value of $3200 paid at the end of every six months for four years if the interest rate is 6.5% compounded semi-annually?

9. What is the size of semi-annual deposits that will accumulate to $67 200.00 after eight years at 6.5% compounded semi-annually?

Challenge Problems

1. After winning some money at a casino, Tony is considering purchasing an annuity that promises to pay him $300 at the end of each month for 12 months, then $350 at the end of each month for 24 months, and then $375 at the end of each month for 36 months. If the first payment is due at the end of the first month and interest is 7.5% compounded monthly over the life of the annuity, find Tony's purchase price.

2. On March 1, 2003, Yves decided to save for a new truck. He deposited $500.00 at the end of every three months in a bank account earning interest at 5% compounded quarterly. He made his first deposit on June 1, 2003. On June 1, 2005, Yves decided that he needed the money to go to college, so on September 1, 2005, he stopped making deposits and started withdrawing $300.00 at the end of each quarter until December 1, 2006. How much is left in his account after the last withdrawal if his bank account interest rate changed to 6.5% compounded quarterly on March 1, 2006?

Case Study 5.1 Saving for Your Dream

» Molly Sawatsky dreams of spending an extended holiday in Australia. If she can save enough money, she wants to have six months to travel with friends. She has done some research and determined that she would need approximately $1200 in Canadian currency per month to achieve her dream.

At her current salary, she plans to save $100 at the end of each month, investing it at 4% interest compounded monthly. She hopes to be able to take this trip three years from now.

QUESTIONS

1. If Molly saves $100 per month for three years, will she have enough to pay for her trip?

2. If Molly saves the $100 per month, and earns an interest rate of 6% compounded monthly, will she have enough to pay for her trip?

3. If Molly saves $200 per month for three years, with interest at 4% compounded monthly, how much will she have in her vacation investment account?

Case Study 5.2 Getting the Picture

» Suzanne had a summer job working in the business office of Blast-It TV and Stereo, a local chain of home electronics stores. When Petr Jacobssen, the owner of the chain, heard she had completed one year of business courses, he asked Suzanne to calculate the profitability of two new large-screen TVs. He plans to offer a special payment plan for the two new models to attract customers to his stores. He wants to promote heavily the more profitable TV.

When Petr gave Suzanne the information about the two TVs, he told her to ignore all taxes when making her calculations. The cost of TV A to the company is $1720.00 and the cost of TV B to the company is $1680.00, after all trade discounts have been taken. The company plans to sell TV A for a $300.00 down payment and $210.00 per month for twelve months, beginning one month from the date of the purchase. The company plans to sell TV B for a $100.00 down payment and $160.00 per month for eighteen months, beginning one month from the date of purchase. The monthly payments for both TVs reflect an interest rate of 13.5% compounded monthly.

Petr wants Suzanne to calculate the profit of TV A and TV B as a percent of the TV's cost to the company. To calculate profit, Petr deducts overhead (which he calculates as 15% of cost) and the cost of the item from the selling price of the item. When he sells items that are paid for at a later time, he calculates the selling price as the *cash value* of the item. (Remember that cash value equals the down payment plus the present value of the periodic payments.)

Suzanne realized that she could calculate the profitability of each TV by using her knowledge of ordinary annuities. She went to work on her assignment to provide Petr with the information he requested.

QUESTIONS

1. **(a)** What is the cash value of TV A? Round your answer to the nearest dollar.
(b) What is the cash value of TV B? Round your answer to the nearest dollar.

2. **(a)** Given Petr's system of calculations, how much overhead should be assigned to TV A?
(b) How much overhead should be assigned to TV B?

3. **(a)** According to Petr's system of calculations, what is the profit of TV A as a percent of its cost?
(b) What is the profit of TV B as a percent of its cost?
(c) Which TV should Suzanne recommend be more heavily promoted?

4. Three months later, due to Blast-It's successful sales of TV A and TV B, the suppliers of each model gave the company new volume discounts. For TV A, Blast-It received a discount of 8% off its current cost. For TV B, the company received a discount of 5% off its current cost. The special payment plans for TV A and TV B will stay the same. Under these new conditions, which TV should Suzanne recommend be more heavily promoted?

SUMMARY OF FORMULAS

Formula 3.1A

$FV = PV(1 + i)^n$ — Finding the future value of a compound amount (maturity value) when the original principal, the rate of interest, and the time period are known

Formula 3.1C

$PV = FV(1 + i)^{-n}$ — Finding the present value by means of the discount factor (the reciprocal of the compounding factor)

Formula 4.1

$f = (1 + i)^m - 1$ — Finding the effective rate of interest f for a nominal annual rate compounded m times per year

Formula 5.1

$$S_n = R\left[\frac{(1 + i)^n - 1}{i}\right]$$

Finding the future value (accumulated value) of an ordinary simple annuity

restated as

$$FV_n = PMT\left[\frac{(1 + i)^n - 1}{i}\right]$$

Formula 5.2

$$A_n = R\left[\frac{1 - (1 + i)^{-n}}{i}\right]$$

Finding the present value (discounted value) of an ordinary simple annuity

restated as

$$PV_n = PMT\left[\frac{1 - (1 + i)^{-n}}{i}\right]$$

GLOSSARY

Accumulated value of one dollar per period *see* **Accumulation factor for annuities**

Accumulation factor for annuities the factor $\frac{(1 + i)^n - 1}{i}$ *(p. 179)*

Annual rent the sum of the periodic payments in one year *(p. 174)*

Annuity a series of payments, usually equal in size, made at equal periodic time intervals *(p. 173)*

Annuity certain an annuity for which the term is fixed *(p. 174)*

Annuity due an annuity in which the periodic payments are made at the beginning of each payment interval *(p. 175)*

Compounding factor for annuities *see* **Accumulation factor for annuities**

Contingent annuity an annuity in which the term is uncertain; that is, either the beginning date of the term or the ending date of the term or both are unknown *(p. 174)*

Deferred annuity an annuity in which the first payment is delayed for a number of payment periods *(p. 175)*

Down payment the portion of the purchase price that is supplied by the purchaser as an initial payment *(p. 192)*

Future value of an annuity the sum of the accumulated values of the periodic payments at the end of the term of the annuity *(p. 178)*

General annuity an annuity in which the conversion (or compounding) period is different from the payment interval *(p. 175)*

Ordinary annuity an annuity in which the payments are made at the end of each payment interval *(p. 175)*

Payment interval the length of time between successive payments *(p. 174)*

Payment period *see* **Payment interval**

Periodic rent the size of the regular periodic payment *(p. 174)*

Perpetuity an annuity for which the payments continue forever *(p. 174)*

Simple annuity an annuity in which the conversion period is the same as the payment interval *(p. 175)*

Term of an annuity the length of time from the beginning of the first payment interval to the end of the last payment interval *(p. 174)*

USEFUL INTERNET SITES

www.gordonpape.com

Building Wealth on the Net Visit this site to read Gordon Pape's latest financial advice or subscribe to the *Internet Wealth Builder*. Pape is one of Canada's best-known and highly regarded investment advisors.

www.nmfn.com

The Longevity Game Click on Learning Centre/The Longevity Game. Predict your life expectancy with this game found on Northwestern Mutual Life Insurance Company's site. Life expectancy is used to calculate life insurance rates.

www.money.msn.ca

RRSP This Canadian site provides information on advantages and disadvantages of RRSPs, including a calculator to determine tax savings.

6 Ordinary General Annuities

OBJECTIVES

Upon completing this chapter, you will be able to do the following:

1. Compute the future value (or accumulated value) for ordinary general annuities.
2. Compute the present value (or discounted value) for ordinary general annuities.
3. Compute the payment for ordinary general annuities.
4. Compute the number of periods for ordinary general annuities.
5. Compute the interest rate for ordinary general annuities.

We often encounter situations where the frequency of the payment in an annuity is not the same as the frequency of the compounding when interest is applied. In these cases, we are dealing with a general annuity. When the payment is made at the end of the period, we have an ordinary general annuity. It is necessary to understand these annuities so that we can make informed decisions about different payment options. When considering borrowing money to buy a car, or negotiating a mortgage, we must determine when it is best for us to make the payments and how much we can afford. The timing of the payments is crucial in planning our cash flow.

INTRODUCTION

In Chapter 5, we calculated the present value, future value, payment, term, and interest rate for ordinary simple annuities. In this chapter, we will calculate the present value, future value, payment, term, and interest rate for ordinary general annuities. We will analyze the relationship between the payment interval and the interest conversion period. We will introduce new formulas to calculate the future value and the present value of an ordinary general annuity. We will rearrange these formulas to determine the payment, term, and interest rate for these annuities. We will also use preprogrammed calculators as an alternative to determing the results.

6.1 Ordinary General Annuities—Finding the Future Value

A. Basic concepts and computation

Chapter 5 considered ordinary simple annuities in detail. Simple annuities are a special case in which the payment interval and the interest conversion period are the same length. However, interest is often compounded more or less frequently than payments are made. In Canada, for example, residential mortgages are usually compounded semi-annually while payments are made monthly.

Annuities in which the length of the interest conversion period is different from the length of the payment interval are called *general annuities*.

The basic method of solving problems involving interest uses equivalent sets of financial obligations at a selected focal date. Thus, when dealing with any kind of annuity, including general annuities, the essential tool is an equation of value. This basic approach is used to make the basic computations and develop useful formulas.

EXAMPLE 6.1A

What is the accumulated value of \$100.00 deposited at the end of every six months for three years if interest is 4% compounded annually?

SOLUTION

Since the payments are made semi-annually while the compounding is done annually, this annuity is classified as a general annuity. Furthermore, since the payments are at the end of each payment interval, the annuity is an ordinary general annuity. While the difference in the length of the payment period compared to the length of the compounding period introduces a mathematical complication, the basic approach to finding the amount of the ordinary general annuity is the same as that used in finding the amount of an ordinary simple annuity.

The basic solution and data for the problem are shown graphically in Figure 6.1. Since deposits are made at the end of every six months for three years, there are six deposits of \$100.00 at the times indicated. Because interest is compounded annually, I/Y = 4%, P/Y = 2, C/Y = 1, $i = 4\% = 0.04$, and there are three conversion periods.

FIGURE 6.1 **Graphical Representation of Method and Data**

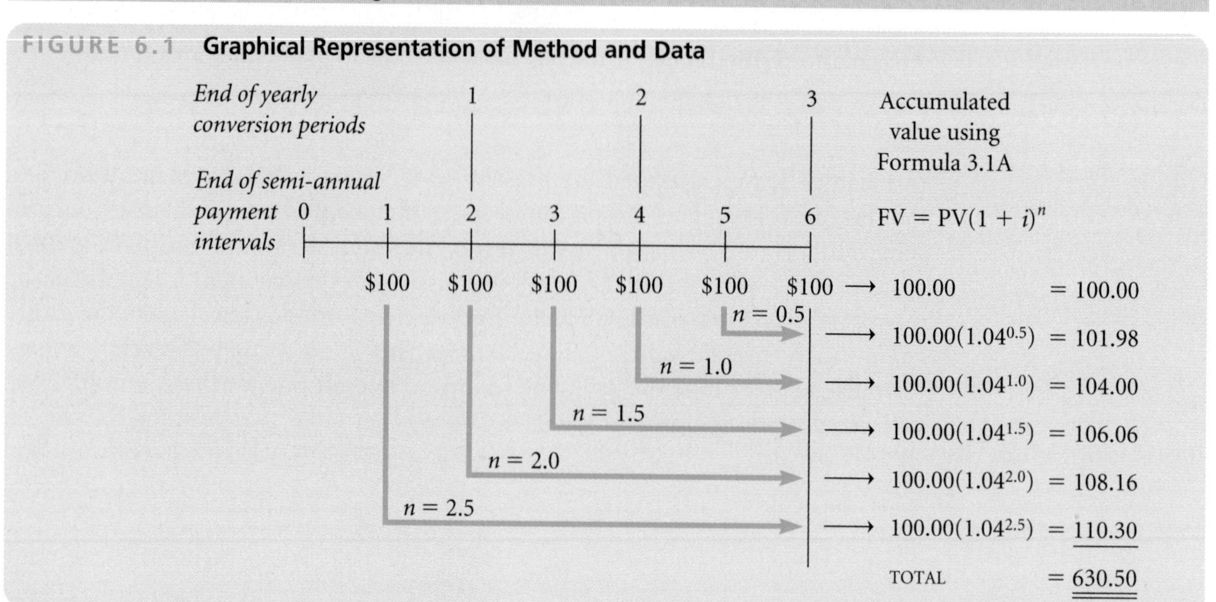

The focal point is at the end of Year 3. The last deposit is made at the focal date and has a value of \$100.00 on that date. The fifth deposit is made after 2.5 years and, in terms of conversion periods, has accumulated for a half conversion period ($n = 0.5$); its accumulated value is $100.00(1.04^{0.5}) = \$101.98$ at the focal date. The fourth deposit is made after two years and has accumulated for one conversion period ($n = 1.0$); its accumulated value at the focal date is $100.00(1.04^{1.0}) = \$104.00$. Similarly, the accumulated value of the third deposit is $100.00(1.04^{1.5}) = \$106.06$, while the accumulated value of the second deposit is $100.00(1.04^{2.0}) = \$108.16$. Finally, the accumulated value of the first deposit is $100.00(1.04^{2.5}) = \$110.30$. The total accumulated value after three years is \$630.50.

EXAMPLE 6.1B

What is the accumulated value of deposits of \$100.00 made at the end of each year for four years if interest is 4% compounded quarterly?

SOLUTION

Since the deposits are made at the end of every year for four years, there are four payments of \$100.00 at the times shown in Figure 6.2. Since interest is compounded quarterly, I/Y = 4, P/Y = 4, $i = \frac{4}{4}\% = 1\%$, and there are 4(4) = 16 conversion periods.

FIGURE 6.2 **Graphical Representation of Method and Data**

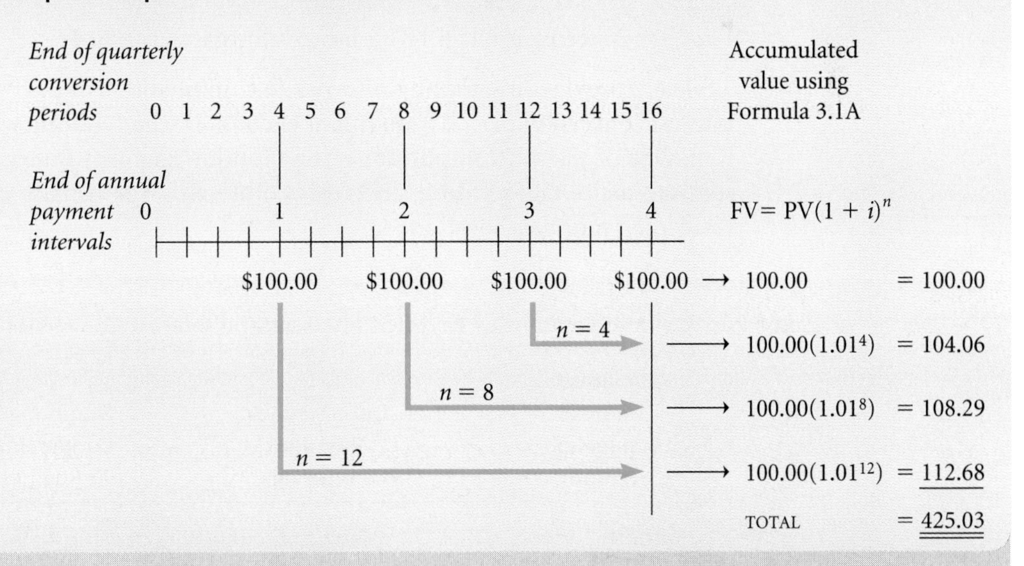

The focal point is the end of Year 4. The last payment is made at the focal point and has a value of \$100.00 at that date. The third payment is made after three years and, in terms of conversion periods, has accumulated for four conversion periods(n = 4); its accumulated value is $100.00(1.01^4) = \$104.06$. The second deposit has accumulated for 8 conversion periods (n = 8); its accumulated value is 100.00(1.018) = \$108.29. The first deposit has accumulated for 12 conversion periods (n = 12); its accumulated value is 100.00(1.0112) = \$112.68. The total accumulated value of the deposits after four years is \$425.03.

B. Relationship between payment interval and interest conversion period

Examples 6.1A and 6.1B illustrate the two possible cases of ordinary general annuities.

CASE 1 The interest conversion period is *longer* than the payment period; each payment interval contains only a fraction of one conversion period.

CASE 2 The interest conversion period is *shorter* than the payment period; each payment period contains more than one conversion period.

The number of interest conversion periods per payment interval, designated by the letter c, can be determined from the following ratio.

$$c = \frac{\text{THE NUMBER OF INTEREST CONVERSION PERIODS PER YEAR}}{\text{THE NUMBER OF PAYMENT PERIODS PER YEAR}}$$

In dealing with general annuities it is important to understand clearly the relationship between the *payment interval* and the number of *interest conversion periods* per payment interval.

For Case 1 (see Example 6.1A), c has a fractional value less than 1.

For Case 2 (see Example 6.1B), c has a value greater than 1.

Table 6.1 provides a sampling of possible combinations of payment intervals and interest conversion periods you might encounter when dealing with general annuities. One of the most important is the monthly payment interval combined with semi-annual compounding since this combination is usually encountered with residential mortgages in Canada.

TABLE 6.1 Some Possible Combinations of Payment Intervals and Interest Conversion Periods

Numerator: Interest Conversion Period	Denominator: Payment Interval	Number of Interest Conversion Periods per Payment Interval
monthly	semi-annually	$c = \frac{12}{2} = 6$
monthly	quarterly	$c = \frac{12}{4} = 3$
quarterly	annually	$c = \frac{4}{1} = 4$
annually	semi-annually	$c = \frac{1}{2} = 0.5$
annually	quarterly	$c = \frac{1}{4} = 0.25$
semi-annually	monthly	$c = \frac{2}{12} = \frac{1}{6}$
quarterly	monthly	$c = \frac{4}{12} = \frac{1}{3}$
annually	monthly	$c = \frac{1}{12}$

C. Computing the effective rate of interest per payment period

In Chapter 4 you were introduced to the concept of the effective rate of interest. It was defined as the nominal rate of interest compounded annually and represented by the symbol f. It can readily be determined from the periodic rate of interest i by means of Formula 4.1, $f = (1 + i)^m - 1$, where m is the number of compounding periods per year.

When dealing with general annuities, it is useful to utilize the **effective rate of interest per payment period**. Depending on the length of the payment interval, the effective interest rate per payment period may be a monthly, quarterly, semi-annual, or annual rate. It can be obtained from the periodic rate of interest i by a formula identical in nature to Formula 4.1.

To avoid confusion, we will distinguish the effective rate of interest per payment period from the effective annual rate of interest f by using the symbol p for the effective rate of interest per payment period. It can be determined by means of Formula 6.1,

$$p = (1 + i)^c - 1 \qquad \text{Formula 6.1}$$

$$\text{where } c = \frac{\text{THE NUMBER OF INTEREST CONVERSION PERIODS PER PAYMENT PERIOD}}{\text{THE NUMBER OF PAYMENT PERIODS PER YEAR}}$$

Note: Formula 6.1 is the same as Formula 4.1 except that f is replaced by p, and m is replaced by c. Also note that when the payment interval is one year, $p = f$.

POINTERS AND PITFALLS

Formula 6.1, which is $p = (1 + i)^c - 1$, is useful in calculating the effective interest rate per rent payment period in general annuity problems, while Formula 4.1, which is the effective rate formula $f = (1+ i)^m - 1$, is useful in making conversions between compound rates of interest and corresponding effective rates of interest.

When using an electronic calculator equipped with a universal power key, the given periodic rate of interest i can be easily converted into the equivalent effective rate of interest per payment period.

EXAMPLE 6.1C

Jean receives annuity payments at the end of every six months. If she deposits these payments in an account earning interest at 9% compounded monthly, what is the effective semi-annual rate of interest?

SOLUTION

Since the payments are made at the end of every six months while interest is compounded monthly,

$$c = \frac{\text{THE NUMBER OF INTEREST CONVERSION PERIODS PER YEAR}}{\text{THE NUMBER OF PAYMENT PERIODS PER YEAR}} = \frac{12}{2} = 6$$

$i = \frac{9\%}{12} = 0.75\% = 0.0075$

$p = (1 + i)^c - 1$
$p = 1.0075^6 - 1$
$p = 1.045852 - 1$
$p = 0.045852 = 4.5852\%$

The effective semi-annual rate of interest is 4.5852%.

EXAMPLE 6.1D

Peel Credit Union pays 6% compounded quarterly on its Premium Savings Accounts. If Roland Catchpole deposits $25.00 in his account at the end of every month, what is the effective monthly rate of interest?

SOLUTION

Since the payments are made at the end of every month while interest is compounded quarterly,

$c = \frac{4}{12} = \frac{1}{3}; \quad i = \frac{6\%}{4} = 1.5\% = 0.015$

$p = 1.015^{\frac{1}{3}} - 1$
$p = 1.0049752 - 1$
$p = 0.0049752 = 0.4975\%$

The effective monthly rate of interest is 0.4975%.

D. Future value of an ordinary general annuity using the effective rate of interest per payment period

Determining the effective rate per payment period p allows us to convert the ordinary general annuity problem into an ordinary simple annuity problem. Consistently with the symbols previously used for developing formulas for ordinary simple annuities, we will use the following notation:

FV_{nc} = the future value (or accumulated value) of an ordinary general annuity;
PMT = the size of the periodic payment;
n = the number of periodic payments;
c = the number of interest conversion periods per payment interval;
i = the interest rate per interest conversion period;
p = the effective rate of interest per payment period.

Substituting p for i in Formula 5.1, we obtain

$$FV_{nc} = PMT\left[\frac{(1 + p)^n - 1}{p}\right] \text{ where } p = (1 + i)^c - 1 \qquad \text{Formula 6.2}$$

EXAMPLE 6.1E

Determine the accumulated value after ten years of payments of $2000.00 made at the end of each year if interest is 6% compounded monthly.

SOLUTION

This problem is an ordinary general annuity.

$PMT = 2000.00; \quad n = 10; \quad c = 12; \quad i = \frac{6\%}{12} = 0.5\% = 0.005$

The effective annual rate

$p = 1.005^{12} - 1 = 1.0616778 - 1 = 0.0616778 = 6.16778\%$

The given ordinary general annuity can be converted into an ordinary simple annuity.

$PMT = 2000.00; \quad n = 10; \quad p = 0.0616778$

$$
\begin{aligned}
FV_{nc} &= 2000.00\left(\frac{1.0616778^{10} - 1}{0.0616778}\right) \text{ — substituting in Formula 6.2} \\
&= 2000.00(13.285114) \\
&= \$26\,570.23
\end{aligned}
$$

The accumulated value after ten years is about $26 570.23.

EXAMPLE 6.1F

Creditview Farms set aside $1250.00 at the end of each month for the purchase of a combine. How much money will be available after five years if interest is 6.0% compounded semi-annually?

SOLUTION

This problem involves an ordinary general annuity.

$PMT = 1250.00; \quad n = 5(12) = 60; \quad c = \frac{2}{12} = \frac{1}{6}; \quad i = \frac{6.0\%}{2} = 3.0\% = 0.03$

The effective monthly rate of interest

$p = 1.03^{\frac{1}{6}} - 1 = 1.004939 - 1 = 0.004939 = 0.4939\%$

$$
\begin{aligned}
FV_{nc} &= 1250.00\left(\frac{1.004939^{60} - 1}{0.004939}\right) \text{ — substituting in Formula 6.2} \\
&= 1250.00(69.638419) \\
&= \$87\,048.02
\end{aligned}
$$

After five years, the amount available is $87 048.02.

E. Using preprogrammed calculators to find the future value of an ordinary general annuity

The use of p rather than i as the rate of interest is the only difference in the programmed solution for the general annuity compared to the programmed solution for a simple annuity.

When using preprogrammed calculators, such as the BAII Plus, the calculator must be set for the number of payment periods per year P/Y and the number of interest compounding periods per year C/Y. To set the calculator for Example 6.1G, follow these steps:

2nd P/Y 2 Enter ↓ C/Y 4 Enter 2nd Quit

To show the nominal interest rate per year, I/Y needs to be entered. In this example, the number is 6, indicating 6%.

Remember also that n denotes the number of payments made. In this example, there are 20 payments (10 years × P/Y of 2). ***Note:*** This assumes the calculator is in the default setting "END."

Note that each time the P/Y is changed, the C/Y is automatically changed to match the P/Y. To make the C/Y different from the P/Y, it must be re-entered separately.

EXAMPLE 6.1G

Find the future value of $2500.00 deposited at the end of every six months for ten years if interest is 6% compounded quarterly.

SOLUTION

PMT = 2500.00; $n = 2(10) = 20$; P/Y = 2; C/Y = 4; $c = \frac{4}{2} = 2$; I/Y = 6; $i = \frac{6\%}{4} = 1.5\% = 0.015$

To use the formula:

STEP 1 Convert i into the effective semi-annual rate of interest.
$p = 1.015^2 - 1 = 1.030225 - 1 = 0.030225 = 3.0225\%$

STEP 2 Using p as the interest rate per payment period, determine the amount of the ordinary annuity.

To use the calculator:

(Set P/Y = 2; C/Y = 4)

Key in	*Press*	*Display shows*	
6	I/Y	6	
0	PV	0	
2500	± PMT	−2500	
20	N	20	the number of payments
CPT	FV	67 329.89319	

The amount on deposit after ten years will be $67 329.89.

EXAMPLE 6.1H

Determine the accumulated value of payments of $1250.00 made at the end of each quarter for eight years if interest is 5.5% compounded annually.

SOLUTION

PMT = 1250.00; $n = 8(4) = 32$; P/Y = 4; C/Y = 1; $c = \frac{1}{4} = 0.25$;
I/Y = 5.5; $i = 5.5\% = 0.055$

To use the formula:

$p = 1.055^{0.25} - 1 = 1.013475 - 1 = 0.013475 = 1.34752\%$

To use the calculator:
(Set P/Y = 4; C/Y = 1) 5.5 [I/Y] 0 [PV]
1250 [±] [PMT] 32 [N] [CPT] [FV] [49 599.22016]

The accumulated value of the payments is $49 599.22.

EXERCISE 6.1

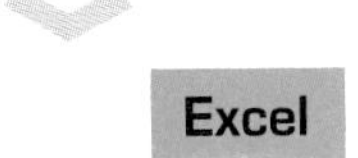

If you choose, you can use Excel's ***Future Value (FV)*** function to answer the questions indicated below. Refer to **FV** on the Spreadsheet Template Disk to learn how to use this Excel function.

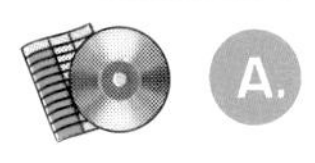

A. Find the future value of each of the following eight ordinary annuities.

	Periodic Payment	Payment Interval	Term	Interest Rate	Conversion Period
1.	$2500.00	6 months	7 years	8%	quarterly
2.	$900.00	3 months	5 years	6%	monthly
3.	$72.00	1 month	15 years	3%	semi-annually
4.	$225.00	3 months	10 years	5%	annually
5.	$1750.00	6 months	12 years	7%	annually
6.	$680.00	1 month	3 years	9%	annually
7.	$7500.00	1 year	4 years	6%	quarterly
8.	$143.00	1 month	9 years	4%	quarterly

B. Answer each of the following questions.

1. Find the future value of payments of $425.00 made at the end of every three months for nine years if interest is 9% compounded monthly.

2. What is the accumulated value of deposits of $1500.00 made at the end of every six months for six years if interest is 6% compounded quarterly?

3. How much will deposits of $15.00 made at the end of each month amount to after ten years if interest is 5% compounded quarterly?

4. What is the future value of payments of $250.00 made at the end of every three months in fifteen years if interest is 7.5% compounded annually?

5. Mr. Tomas has contributed $1000.00 at the end of each year into an RRSP paying 6% compounded quarterly.

(a) How much will Mr. Tomas have in the RRSP after ten years?
(b) After ten years, how much of the amount is interest?

6. Alexa Sanchez saves $5.00 at the end of each month and deposits the money in an account paying 4% compounded quarterly.

(a) How much will she accumulate in 25 years?
(b) How much of the accumulated amount is interest?

7. Edwin Ng has made deposits of $500.00 into his savings account at the end of every three months for ten years. If interest is 4.5% compounded semi-annually and if he leaves the accumulated balance for another five years, what will be the balance in his account then?

8. Mrs. Cook has made deposits of $950.00 at the end of every six months for fifteen years. If interest is 3% compounded monthly, how much will Mrs. Cook have accumulated ten years after the last deposit?

6.2 Ordinary General Annuities—Finding the Present Value PV

A. Present value of an ordinary general annuity using the effective rate of interest per payment period

As with the future value of an ordinary general annuity, we can convert the given periodic rate of interest i into the effective rate of interest per payment period.

$p = (1 + i)^c - 1$

The use of p converts the ordinary general annuity problem into an ordinary simple annuity problem.

Substituting p for i in Formula 5.2, we obtain

$$PV_{nc} = PMT\left[\frac{1 - (1 + p)^{-n}}{p}\right] \quad \text{Formula 6.3}$$

EXAMPLE 6.2A

A loan is repaid by making payments of $2000.00 at the end of every six months for twelve years. If interest on the loan is 8% compounded quarterly, what was the principal of the loan?

SOLUTION

$PMT = 2000.00$; $n = 12(2) = 24$; $P/Y = 2$; $C/Y = 4$; $c = \frac{4}{2} = 2$; $I/Y = 8$;

$i = \frac{8\%}{4} = 2\% = 0.02$

The effective semi-annual rate of interest

$p = 1.02^2 - 1 = 1.0404 - 1 = 0.0404 = 4.04\%$

$$PV_{nc} = 2000.00\left(\frac{1 - 1.0404^{-24}}{0.0404}\right) \quad \text{substituting in Formula 6.3}$$

$$= 2000.00(15.184713)$$

$$= \$30\,369.43$$

The loan principal was $30 369.43.

EXAMPLE 6.2B

A second mortgage requires payments of $370.00 at the end of each month for fifteen years. If interest is 11% compounded semi-annually, what was the amount borrowed?

SOLUTION

PMT = 370.00; $n = 15(12) = 180$; P/Y = 12; C/Y = 2; $c = \frac{2}{12} = \frac{1}{6}$;
I/Y = 11; $i = \frac{11\%}{2} = 5.5\% = 0.055$

The effective monthly rate of interest

$$p = 1.055^{\frac{1}{6}} - 1 = 1.008963 - 1 = 0.008963 = 0.8963\%$$

$$PV_{nc} = 370.00\left(\frac{1 - 1.008963^{-180}}{0.008963}\right)$$ ——— substituting in Formula 6.3

$$= 370.00(89.180057)$$
$$= \$32\,996.62$$

The amount borrowed was $32 996.62.

B. Using preprogrammed calculators to find the present value of an ordinary general annuity

STEP 1 Set P/Y as the number of payment periods per year.

STEP 2 Set C/Y as the number of compounding periods per year and I/Y as the nominal interest rate.

STEP 3 Determine the present value of the ordinary general annuity.

EXAMPLE 6.2C

A contract is fulfilled by making payments of $8500.00 at the end of every year for fifteen years. If interest is 7% compounded quarterly, what is the cash price of the contract?

SOLUTION

PMT = 8500.00; N = 15; P/Y = 1; C/Y = 4; $c = 4$; I/Y = 7;
$i = \frac{7\%}{4} = 1.75\% = 0.0175$

(Set P/Y = 1; C/Y = 4) 7 [I/Y] 0 [FV] 8500 [±] [PMT]
15 [N] [CPT] [PV] [76 516.37862]

The cash price of the contract is $76 516.38.

EXAMPLE 6.2D

A 25-year mortgage on a house requires payments of $619.94 at the end of each month. If interest is 9.5% compounded semi-annually, what was the mortgage principal?

SOLUTION

PMT = 619.94; N = 25(12) = 300; P/Y = 12; C/Y = 2; $c = \frac{2}{12} = \frac{1}{6}$; I/Y = 9.5;
$i = \frac{9.5\%}{2} = 4.75\% = 0.0475$

(Set P/Y = 12; C/Y = 2) 9.5 [I/Y] 0 [FV] 619.94 [±] [PMT]
300 [N] [CPT] [PV] [72 000.01223]

The mortgage principal was $72 000.01.

EXERCISE 6.2

If you choose, you can use Excel's ***Present Value (PV)*** function to answer the questions indicated below. Refer to **PV** on the Spreadsheet Template Disk to learn how to use this Excel function.

A. Find the present value of the following eight ordinary annuities.

	Periodic Payment	Payment Interval	Term	Interest Rate	Conversion Period
1.	$1400.00	3 months	12 years	6%	monthly
2.	$6000.00	1 year	9 years	10%	quarterly
3.	$3000.00	3 months	4 years	6%	annually
4.	$200.00	1 month	2 years	5%	semi-annually
5.	$95.00	1 month	5 years	4.5%	annually
6.	$975.00	6 months	8 years	8%	annually
7.	$1890.00	6 months	15 years	7%	quarterly
8.	$155.00	1 month	10 years	8%	quarterly

B. Answer each of the following questions.

1. Find the present value of payments of $250.00 made at the end of every three months for twelve years if money is worth 3% compounded monthly.

2. What is the discounted value of $1560.00 paid at the end of each year for nine years if interest is 6% compounded quarterly?

3. What cash payment is equivalent to making payments of $825.00 at the end of every three months for 16 years if interest is 7% compounded semi-annually?

4. What is the principal from which $175.00 can be withdrawn at the end of each month for twenty years if interest is 5% compounded quarterly?

5. A property was purchased for $5000.00 down and payments of $2500.00 at the end of every six months for six years. Interest is 6% compounded monthly.

(a) What was the purchase price of the property?
(b) How much is the cost of financing?

6. A car was purchased for $1500.00 down and payments of $265.00 at the end of each month for four years. Interest is 9% compounded quarterly.

(a) What was the purchase price of the car?
(b) How much interest will be paid?

7. Payments of $715.59 are made at the end of each month to repay a 25-year mortgage. If interest is 10% compounded semi-annually, what is the original mortgage principal?

8. A 15-year mortgage is amortized by making payments of $1031.61 at the end of every three months. If interest is 8.25% compounded annually, what was the original mortgage balance?

9. Cedomir Dale expects to retire in seven years. He bought a retirement annuity paying $1200.00 every three months for twenty years. If the first payment is due three months after his retirement and interest is 6.6% compounded monthly, how much did Mr. Dale invest?

10. For her daughter's university education, Georgina Harcourt has invested an inheritance in a fund paying 5.2% compounded quarterly. If ordinary annuity payments of $450.00 per month are to be made out of the fund for four years and the annuity begins twelve years from now, how much was the inheritance?

6.3 Ordinary General Annuities—Finding the Periodic Payment PMT

A. Finding the periodic payment PMT when the future value of a general annuity is known

If the future value of an annuity FV_{nc}, the number of conversion periods n, and the conversion rate i are known, you can find the periodic payment PMT by substituting the given values in the future value Formula 6.2.

$$FV_{nc} = PMT\left[\frac{(1+p)^n - 1}{p}\right] \text{ where } p = (1+i)^c - 1$$ — Formula 6.2

As we have emphasized throughout this book, beginning in Chapter 2, understanding how to rearrange terms in a formula is a very important skill. By knowing how to do this, you avoid having to memorize equivalent forms of the same formula.

When using a scientific calculator, you can find PMT by first rearranging the terms of the future value formulas as shown below. Then substitute the three known values (FV, n, and i) into the rearranged formula and use the calculator to solve for PMT.

$$FV_{nc} = PMT\left[\frac{(1+p)^n - 1}{p}\right]$$ — Formula 6.2

$$PMT = \frac{FV_{nc}}{\left[\frac{(1+p)^n - 1}{p}\right]}$$ — divide both sides by $\left[\frac{(1+p)^n - 1}{p}\right]$

$$PMT = \frac{FV_{nc}\, p}{(1+p)^n - 1}$$ — Formula 6.2a: dividing by a fraction is the same as inverting the fraction and multiplying

When using a preprogrammed financial calculator, you can find PMT by entering the five known values (FV, N, I/Y, P/Y and C/Y) and pressing CPT PMT. Recall that the PMT amount will be negative since payments are considered to be cash outflows.

EXAMPLE 6.3A

What sum of money must be deposited at the end of every three months into an account paying 6% compounded monthly to accumulate to \$25 000.00 in ten years?

SOLUTION

$FV_{nc} = 25\,000.00;\ n = 10(4) = 40;\ P/Y = 4;\ C/Y = 12;\ c = \frac{12}{4} = 3;\ I/Y = 6;$
$i = 0.5\% = 0.005$

The effective quarterly rate of interest

$p = 1.005^3 - 1 = 1.015075 - 1 = 0.015075 = 1.5075\%$

$$25\,000.00 = PMT\left(\frac{1.015075^{40} - 1}{0.015075}\right)$$ ——— substituting in Formula 6.2

$$25\,000.00 = PMT(54.354225)$$

$$PMT = \frac{25\,000.00}{54.354225}$$

$$PMT = \$459.95$$

Using Formula 12.2a, which is Formula 6.2 rearranged,

$$PMT = \frac{FV_{nc}\,p}{(1+p)^n - 1}$$

$$PMT = \frac{25\,000.000(0.015075)}{1.015075^{40} - 1}$$ ——— substituting in Formula 6.2A

$$PMT = \$459.95$$

Programmed Solution

(Set P/Y = 4; C/Y = 12) [2nd] (CLR TVM) 6 [I/Y] 0 [PV]

25 000 [FV] 40 [N] [CPT] [PMT] [-459.945847]

The required quarterly deposit is \$459.95.

B. Finding the periodic payment PMT when the present value of a general annuity is known

If the present value of a general annuity PV_{nc}, the number of conversion periods n, and the conversion rate i are known, you can find the periodic payment PMT by substituting the given values in the present value Formula 6.3.

$$PV_{nc} = PMT\left[\frac{1-(1+p)^{-n}}{p}\right]$$ ——— Formula 6.3

When using a scientific calculator, you can find PMT by first rearranging the terms of the present value formulas. Then substitute the three known values (PV, n, and i) into the rearranged formula and solve for PMT.

$$PV_{nc} = PMT\left[\frac{1-(1+p)^{-n}}{p}\right]$$ ——— Formula 6.3

$$\text{PMT} = \frac{\text{PV}_{nc}}{\left[\frac{1-(1+p)^{-n}}{p}\right]}$$ divide both sides by $\left[\frac{1-(1+p)^{-n}}{p}\right]$

$$\text{PMT} = \frac{\text{PV}_{nc}\,p}{1-(1+p)^{-n}}$$ Formula 6.3A: dividing by a fraction is the same as inverting the fraction and multiplying

When using a preprogrammed financial calculator, you can find PMT by entering the five known values (PV N, I/Y, P/Y, and C/Y) and pressing CPT PMT.

EXAMPLE 6.3B

Mr. and Mrs. White applied to their credit union for a first mortgage of $60 000.00 to buy a house. The mortgage is to be amortized over 25 years and interest on the mortgage is 8.5% compounded semi-annually. What is the size of the monthly payment if payments are made at the end of each month?

SOLUTION

$\text{PV}_{nc} = 60\ 000.00$; $n = 25(12) = 300$; P/Y = 12; C/Y = 2; $c = \frac{2}{12} = \frac{1}{6}$;

I/Y = 8.5; $i = \frac{8.5\%}{2} = 4.25\% = 0.0425$

The effective monthly rate of interest

$p = 1.0425^{\frac{1}{6}} - 1 = 1.006961 - 1 = 0.006961 = 0.6961\%$

$60\ 000.00 = \text{PMT}\left(\frac{1-1.006961^{-300}}{0.006961}\right)$ substituting in Formula 6.3

$60\ 000.00 = \text{PMT}(125.72819)$

$\text{PMT} = \frac{60\ 000.00}{125.72819}$

$\text{PMT} = \$477.22$

Programmed Solution

(Set P/Y = 12; C/Y = 2) 8.5 I/Y 0 FV 60 000 ± PV 300 N CPT PMT 477.218111

The monthly payment due at the end of each month is $477.22.

EXERCISE 6.3

Excel

If you choose, you can use Excel's ***Payment (PMT)*** function to answer all of the questions below. Refer to **PMT** on the Spreadsheet Template Disk to learn how to use this Excel function.

For each of the following ten ordinary general annuities, determine the size of the periodic rent.

	Future Value	Present Value	Payment Period	Term of Annuity	Interest Rate	Conversion Period
1.	$15 000.00		6 months	7 years, 6 mos	5.5%	annually
2.	$6000.00		1 quarter	9 years, 9 mos	8%	monthly
3.		$12 000.00	12 months	15 years	4.5%	semi-annually
4.		$7 000.00	6 months	12.5 years	7.5%	quarterly
5.	$8000.00		3 months	6 years	6.8%	semi-annually
6.		$20 000.00	3 months	20 years	12%	monthly
7.	$45 000.00		6 months	10 years	9%	quarterly
8.		$35 000.00	1 year	15 years	4%	quarterly
9.		$20 000.00	1 month	8 years	7%	semi-annually
10.	$16 500.00		3 months	15 years	5.75%	annually

Answer each of the following questions.

1. What payment made at the end of each quarter for fifteen years will accumulate to $12 000.00 at 6% compounded monthly?

2. What payment is required at the end of each month for five years to repay a loan of $6000.00 at 7% compounded semi-annually?

3. A contract can be fulfilled by making an immediate payment of $9500.00 or equal payments at the end of every six months for eight years. What is the size of the semi-annual payments at 7.4% compounded quarterly?

4. What payment made at the end of each year for eighteen years will amount to $16 000.00 at 4.2% compounded monthly?

5. What payment is required at the end of each month for fifteen years to amortize a $32 000.00 mortgage if interest is 9.5% compounded semi-annually?

6. How much must be deposited at the end of each quarter for ten years to accumulate to $12 000.00 at 6% compounded monthly?

7. What payment made at the end of every three months for twenty years will accumulate to $20 000.00 at 7% compounded semi-annually?

8. Deragh bought a car priced at $9300.00 for 15% down and equal monthly payments for four years. If interest is 8% compounded semi-annually, what is the size of the monthly payment?

9. Equal payments are to be made at the end of each month for fifteen years with interest at 9% compounded quarterly. After the last payment, the fund is to be invested for seven years at 10% compounded quarterly and have a maturity value of $20 000.00. What is the size of the monthly payment?

10. To finance the development of a new product, a company borrowed $30 000.00 at 7% compounded monthly. If the loan is to be repaid in equal quarterly payments over seven years and the first payment is due three months after the date of the loan, what is the size of the quarterly payment?

6.4 Ordinary General Annuities—Finding the Term *n*

A. Finding the term *n* when the future value of a general annuity is known

If the future value of an annuity FV_{nc}, the periodic payment PMT, and the conversion rate i are known, you can find the term of the annuity n by substituting the given values in the future value Formula 6.2.

$$FV_{nc} = PMT\left[\frac{(1+p)^n - 1}{p}\right] \text{ where } p = (1+i)^c - 1 \qquad \text{Formula 6.2}$$

When using a scientific calculator, you can find n by first rearranging the terms of the future value Formula 6.2. Then substitute the three known values (FV, PMT, and p) into the rearranged formula and solve for n.

$$FV_{nc} = PMT\left[\frac{(1+p)^n - 1}{p}\right] \qquad \text{Formula 6.2}$$

$$\frac{FV_{nc}}{PMT} = \frac{(1+p)^n - 1}{p} \qquad \text{divide both sides by PMT}$$

$$(1+p)^n = \left(\frac{FV_{nc}\,p}{PMT}\right) + 1 \qquad \text{multiply both sides by } p \text{ and add 1 to both sides}$$

$$n\ln(1+p) = \ln[(FV_{nc}\,p/PMT) + 1] \qquad \text{solve for } n \text{ using natural logarithms}$$

$$n = \frac{\ln[FV_{nc}\,p/PMT + 1]}{\ln(1+p)} \qquad \text{Formula 6.2B: divide both sides by } \ln(1+p)$$

When using a preprogrammed financial calculator, you can find n by entering the five known values (FV, PMT, I/Y, P/Y, and C/Y) and pressing CPT N.

EXAMPLE 6.4A

What period of time is required for \$125.00 deposited at the end of each month at 11% compounded quarterly to grow to \$15 000.00?

SOLUTION

$FV_{nc} = 15\,000.00$; $PMT = 125.00$; $P/Y = 12$; $C/Y = 4$; $c = \frac{4}{12} = \frac{1}{3}$;

$I/Y = 11$; $i = \frac{11\%}{4} = 2.75\% = 0.0275$

The effective monthly rate of interest

$p = 1.0275^{\frac{1}{3}} - 1 = 1.009084 - 1 = 0.009084 = 0.9084\%$

$$15\,000.00 = 125.00\left(\frac{1.009084^n - 1}{0.009084}\right) \qquad \text{using Formula 6.2}$$

$$120.00 = \frac{(1.009084^n - 1)}{0.009084}$$

$$\begin{aligned}
1.090068 &= 1.009084^n - 1 \\
1.009084^n &= 2.090068 \\
n \ln 1.009084 &= \ln 2.090068 \\
n(0.009043) &= 0.737197 \\
n &= \frac{0.737197}{0.009043} \\
n &= 81.522227 \\
n &= 82 \text{ months approximately}
\end{aligned}$$

With a preprogrammed calculator, the procedure is

(Set P/Y = 12; C/Y = 4) 11 [I/Y] 0 [PV] 15 000 [FV]

125 [±] [PMT] [CPT] [N] [81.522240]

It will take about six years and ten months to accumulate $15 000.00.

B. Finding the term n when the present value of a general annuity is known

If the present value PV_{nc}, the periodic payment PMT, and the conversion rate i are known, you can find the term of the annuity n by substituting the given values in the present value Formula 6.3.

$$PV_{nc} = PMT\left[\frac{1 - (1 + p)^{-n}}{p}\right] \quad \text{——— Formula 6.3}$$

Alternatively, you can first rearrange the terms of Formula 6.3 to solve for n.

$$PV_{nc} = PMT\left[\frac{1 - (1 + p)^{-n}}{p}\right] \quad \text{——— Formula 6.3}$$

$$\frac{PV_{nc}}{PMT} = \frac{1 - (1 + p)^{-n}}{p} \quad \text{——— divide both sides by PMT}$$

$$\left(\frac{PV_{nc}p}{PMT}\right) - 1 = -(1 + p)^{-n} \quad \text{——— multiply both sides by } i \text{ and subtract 1 from both sides}$$

$$(1 + p)^{-n} = 1 - \left(\frac{PV_{nc}p}{PMT}\right) \quad \text{——— multiply both sides by } -1$$

$$-n\ln(1 + p) = \ln\left[1 - \left(\frac{PV_{nc}p}{PMT}\right)\right] \quad \text{——— solve for } n \text{ using natural logarithms}$$

$$n = \frac{\ln\left[1 - \left(\frac{PV_{nc}p}{PMT}\right)\right]}{-\ln(1 + p)} \quad \text{——— Formula 6.3B: divide both sides by } -\ln(1 + p)$$

When using a scientific calculator, you can find n by first rearranging the terms of the present value formulas above. Then substitute the three known values (PV, PMT, and i) into the rearranged formula and solve for n.

When using a preprogrammed financial calculator, you can find n by entering the five known values (PV, PMT, I/Y, P/Y, and C/Y) and pressing CPT N.

Note: When using the Texas Instruments BAII PLUS financial calculator, you *must* enter *either* the PV or PMT as a negative amount, *but not both.* Due to the sign conventions used by this calculator, entering *both* PV and PMT as negative amounts or as positive amounts will lead to an *incorrect* final answer. The calculator does *not* indicate that the answer is incorrect or that an entry error was made. (This has not been an issue until now because either PV or PMT was 0 in all the examples we discussed.) To avoid incorrect answers, *always* enter the PV amount as a negative number and the PMT amount as a positive number when FV = 0 and you are calculating N. However, if PV = 0, enter PMT as a negative number and FV as a positive number. The examples in this text follow these rules. Refer to Appendix II to check whether this step is necessary if you use the Texas Instruments BA Solar, the SHARP EL-733A, or the Hewlett-Packard 10B calculator.

EXAMPLE 6.4B

A business valued at \$96 000.00 is bought for a down payment of 25% and payments of \$4000.00 at the end of every three months. If interest is 9% compounded monthly, for how long will payments have to be made?

SOLUTION

$PV_{nc} = 96\ 000.00(0.75) = 72\ 000.00; \quad PMT = 4000.00;$

$P/Y = 4; \; C/Y = 12; \; c = \frac{12}{4} = 3; \; I/Y = 9; \; i = \frac{9\%}{12} = 0.75\% = 0.0075$

The effective quarterly rate of interest

$p = 1.0075^3 - 1 = 1.022669 - 1 = 0.022669 = 2.2669\%$

$$72\ 000.00 = 4000.00\left(\frac{1 - 1.022669^{-n}}{0.022669}\right) \quad \text{——— using Formula 6.3}$$

$$0.408046 = 1 - 1.022669^{-n}$$

$$1.022669^{-n} = 0.591954$$

$$-n \ln 1.022669 = \ln 0.591954$$

$$-n(0.022416) = -0.524326$$

$$n = 23.390585 \text{ (quarters)}$$

Programmed Solution

(Set P/Y = 4; C/Y = 12) 9 I/Y 0 FV 72 000 ± PV 4000 PMT CPT N 23.390604

Payments will have to be made for six years.

EXERCISE 6.4

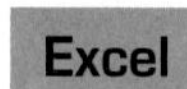

If you choose, you can use Excel's ***Number of Compounding Periods (NPER)*** function to answer all of the questions below. Refer to **NPER** on the Spreadsheet Template Disk to learn how to use this Excel function.

Find the term of each of the following ten ordinary general annuities. (State your answer in years and months.)

	Future Value	Present Value	Periodic Rent	Payment Interval	Interest Rate	Compounding Period
1.	$20 000.00		$800.00	1 year	7.5%	semi-annually
2.	$17 000.00		$35.00	3 months	9%	monthly
3.		$14 500.00	$190.00	1 month	5.25%	quarterly
4.		$5 000.00	$300.00	3 months	4%	semi-annually
5.	$36 00.00		$175.00	6 months	7.4%	annually
6.		$9 500.00	$740.00	1 quarter	5.2%	monthly
7.		$21 400.00	$1660.00	6 months	4.5%	monthly
8.	$13 600.00		$140.00	6 months	8%	quarterly
9.	$72 00.00		$90.00	1 month	3.75%	semi-annually
10.		$11 700.00	$315.00	3 months	11%	annually

B. Answer each of the following questions.

1. How long would it take you to save $5000.00 by making deposits of $100.00 at the end of every month into a savings account earning 6% compounded quarterly?
2. In what period of time could you pay back a loan of $3000.00 by making monthly payments of $90.00 if interest is 10.5% compounded semi-annually?
3. How long will it take to save $15 000.00 by making deposits of $90.00 at the end of every month into an account earning interest at 4% compounded quarterly?
4. For how long will Jack have to make payments of $350.00 at the end of every three months to repay a loan of $6000.00 if interest is 9% compounded monthly?
5. Mirsad is saving $500.00 at the end of each month. How soon can he retire if he wants to have a retirement fund of $120 000.00 and interest is 5.4% compounded quarterly?
6. For how long must contributions of $2000.00 be made at the end of each year to accumulate to $100 000.00 at 6% compounded quarterly?
7. Suppose $370.37 is deposited at the end of every three months into an account earning 6.5% compounded semi-annually. If the balance in the account is to be $20 000.00, how many deposits are needed?
8. For how long can $800.00 be withdrawn at the end of each month from an account originally containing $16 000.00, if interest is 6.8% compounded semi-annually?

9. A mortgage of $120 000.00 is to be repaid by making payments of $750.00 at the end of each month. If interest is 5.75% compounded semi-annually, what is the term of the mortgage?

10. Mr. Deneau accumulated $100 000.00 in an RRSP. He converted the RRSP into an RRIF and started to withdraw $4500.00 at the end of every three months from the fund. If interest is 6.75% compounded monthly, for how long can Mr. Deneau make withdrawals?

6.5 Ordinary General Annuities—Finding the Periodic Interest Rate *i*

A. Finding the periodic rate of interest *i* using preprogrammed financial calculators

Preprogrammed financial calculators are especially helpful when solving for the conversion rate *i*. Determining *i* without a financial calculator is extremely time-consuming. However, it *can* be done by hand, as illustrated in Appendix A on the CD-ROM.

Computer Deals!

Desktop Computer
2.4 GHz Processor
512 MB SDRAM
80 GB Hard Drive
Sound Card and Speakers
56 Kbps Modem
17" Flat Screen Monitor
Software Package
$2145
Or $80.81 for 36 months

Notebook Computer
2.4 GHz Processor
512 MB SDRAM
40 GB Hard Drive
Built-in Sound Card and Speakers
56 Kbps Modem
15" TFT Display
Software Package
$1999
Or $89.26 for 30 months

Peripherals Package
High Resolution Scanner
Digital Camera
Photo Printer
$1495
Or $74.64 for 24 months

NO PST! NO GST!
(We will pay the PST and GST for you.)

Monthly payments begin one month from the date of purchase when you choose the monthly payment option.

QUESTIONS

1. What is the nominal rate of interest compounded monthly you pay if you purchase the Desktop Computer using the monthly payment option?
2. What is the nominal rate of interest compounded monthly you pay if you purchase the Notebook Computer using the monthly payment option?
3. What is the nominal rate of interest compounded monthly you pay if you purchase the Peripherals Package using the monthly payment option?

When the future value or present value, the periodic payment PMT, and the term n of a general annuity are known, retrieve the value of I/Y by pressing CPT I/Y.

EXAMPLE 6.5A

Irina deposited $150.00 in a savings account at the end of each month for 60 months. If the accumulated value of the deposits was $10 000.00 and interest was compounded semi-annually, what was the nominal annual rate of interest?

SOLUTION

FV = 10 000.00; PMT = 150.00; $n = 60$; P/Y = 12; C/Y = 2; $c = \frac{2}{12} = \frac{1}{6}$

(Set P/Y = 12; C/Y = 2) 0 PV 10 000 FV

150 ± PMT 60 N CPT I/Y 4.255410

The nominal annual rate of interest is 4.26% p.a. compounded semi-annually.

EXAMPLE 6.5B

Compute the nominal annual rate of interest compounded monthly at which $500.00 deposited at the end of every three months for ten years will amount to $30 000.00.

SOLUTION

FV = 30 000.00; PMT = 500.00; $n = 40$; P/Y = 4; C/Y = 12; $c = \frac{12}{4} = 3$

(Set P/Y = 4; C/Y = 12) 0 PV 30 000 FV

500 ± PMT 40 N CPT I/Y 7.484516

The nominal annual rate is 7.83% p.a. compounded monthly.

EXERCISE 6.5

A. For each of the following eight ordinary general annuities, determine the nominal annual rate of interest.

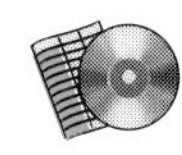

	Future Value	Present Value	Periodic Payment	Payment Interval	Term	Conversion Period
1.	$39 200.00		$23 00.00	1 year	12 years	monthly
2.		$9 600.00	$12 20.00	6 months	5 years	monthly
3.		$62 400.00	$26 00.00	6 months	25 years	annually
4.	$55 500.00		$75.00	1 month	20 years	semi-annually
5.	$6 400.00		$200.00	6 months	9 years	monthly
6.	$25 000.00		$790.00	1 year	15 years	quarterly
7.		$7 500.00	$420.00	3 months	5 years	monthly
8.		$60 000.00	$450.00	1 month	25 years	semi-annually

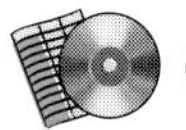

B. Answer each of the following questions.

1. What is the nominal annual rate of interest compounded semi-annually if a four-year loan of $6000.00 is repaid by monthly payments of $144.23?
2. A car valued at $11 400.00 can be purchased for 10% down and monthly payments of $286.21 for three-and-a-half years. What is the nominal rate of interest compounded annually?
3. A property worth $50 000.00 can be purchased for 20% down and quarterly mortgage payments of $1000.00 for 25 years. What nominal rate of interest compounded monthly is charged?
4. A vacation property valued at $25 000.00 was bought for fifteen payments of $2200.00 due at the end of every six months. What nominal annual rate of interest compounded annually was charged?
5. Compute the nominal annual rate of interest compounded monthly at which $400.00 paid at the end of every three months for eight years accumulates to $20 000.00.
6. What is the nominal annual rate of interest compounded quarterly if a loan of $21 500.00 is repaid in seven years by payments of $2000.00 made at the end of every six months?
7. A mortgage of $27 500.00 is repaid by making payments of $280.00 at the end of each month for fifteen years. What is the nominal annual rate of interest compounded semi-annually?
8. A property worth $35 000.00 is purchased for 10% down and semi-annual payments of $2100.00 for twelve years. What is the effective annual rate of interest if interest is compounded quarterly?

Review Exercise

1. Payments of $375.00 made every three months are accumulated at 3% compounded monthly What is their future value after eight years if the payments are made at the end of every three months?
2. What is the accumulated value after twelve years of monthly deposits of $145.00 earning interest at 5% compounded semi-annually if the deposits are made at the end of each month?
3. What single cash payment is equivalent to payments of $3500.00 every six months at 7% compounded quarterly if the payments are made at the end of every six months for fifteen years?
4. What is the principal invested at 6.5% compounded semi-annually from which monthly withdrawals of $240.00 can be made at the end of each month for twenty-five years?
5. Contributions of $500.00 are made at the end of every three months into an RRSP. What is the accumulated balance after twenty years if interest is 6% compounded semi-annually?
6. A 25-year mortgage is amortized by payments of $761.50 made at the end of each month. If interest is 9.5% compounded semi-annually, what is the mortgage principal?
7. Kristan wants to accumulate $18 000.00 into an account earning 7.5% compounded semi-annually. How much must she deposit
   (a) at the end of each month for ten years?
   (b) at the end of each year for eight years?
8. What sum of money can be withdrawn from a fund of $15 750.00 invested at 4.25% compounded semi-annually
   (a) at the end of every month for twelve years?
   (b) at the end of each year for fifteen years?
9. How long will it take for payments of $350.00 to accumulate to $12 000.00 at 3% compounded monthly if made
   (a) at the end of every three months?
   (b) at the end of every six months?
10. A $92 000.00 mortgage with a 25-year term is repaid by making monthly payments. If interest is 5.8% compounded semi-annually, how much are the payments?
11. A debt of $14 000.00 is repaid by making payments of $1500.00. If interest is 9% compounded monthly, for how long will payments have to be made
    (a) at the end of every six months?
    (b) at the end of each year?
12. What is the nominal rate of interest compounded monthly at which payments of $200.00 made at the end of every three months accumulate to $9200.00 in eight years?
13. A debt of $2290.00 is repaid by making pay ments of $198.00. If interest is 16.95% compounded monthly, for how long will quarterly payments have to be made?
14. A $60 000.00 mortgage with a 25-year term is repaid by making monthly payments of $480.00. What is the nominal annual rate of interest compounded semi-annually on the mortgage?
15. Marc invested a bonus of $6000.00 in an RRSP earning 4.9% compounded semi-annually for twenty years. At the end of the twenty years, he rolled the RRSP balance over into an RRIF paying $2000.00 at the end of each quarter starting three months after the date of rollover. If interest on the RRIF is 4.3% compounded monthly, for how long will Marc receive quarterly payments?
16. Satwinder deposited $145.00 at the end of each month for fifteen years at 7.5% compounded monthly. After her last deposit she converted the balance into an ordinary annuity paying $1200.00 every three months for twelve years. If interest on the annuity is compounded semi-annually, what is the nominal rate of interest paid by the annuity?

17. Mrs. Jolly contributes $222.00 at the end of every three months to an RRSP. Interest on the account is 6% compounded monthly.

(a) What will the balance in the account be after eleven years?

(b) How much of the balance will be interest?

(c) If Mrs. Jolly converts the balance after eleven years into an RRIF paying 5% compounded monthly and makes equal quarterly withdrawals for twelve years starting three months after the conversion into her RRIF, what is the size of the quarterly withdrawal?

(d) What is the combined interest earned by the RRSP and the RRIF?

18. How much must be contributed into an RRSP at the end of each year for twenty-five years to accumulate to $100 000.00 if interest is 8% compounded quarterly?

19. For how long must $75.00 be deposited at the end of each month to accumulate to $9500.00 at 6.5% compounded quarterly?

20. A $70 000.00 mortgage is amortized by making monthly payments of $534.95. If interest is 7.2% compounded semi-annually, what is the term of the mortgage?

Self-Test

1. Monthly deposits of $480.00 were made at the end of each month for eight years. If interest is 4.5% compounded semi-annually, what amount can be withdrawn immediately after the last deposit?

2. A loan was repaid in five years by quarterly payments of $1200.00 at 9.5% compounded semi-annually. How much interest was paid?

3. A loan of $6000.00 was repaid by quarterly payments of $450. If interest was 12% compounded monthly, how long did it take to pay back the loan?

4. A mortgage of $95 000.00 is to be amortized by monthly payments over twenty-five years. If the payments are made at the end of each month and interest is 8.5% compounded semi-annually, what is the size of the monthly payments?

5. The amount of $46 200.00 is invested at 9.5% compounded quarterly for four years. After four years the balance in the fund is converted into an annuity. If interest on the annuity is 6.5% compounded semi-annually and payments are made at the end of every three months for seven years, what is the size of the payments?

6. A $45 000.00 mortgage is repaid in twenty years by making monthly payments of $387.72. What is the nominal annual rate of interest compounded semi-annually?

7. For how long would you have to deposit $491.00 at the end of every three months to accumulate $20 000.00 at 6.0% compounded monthly?

8. What is the size of monthly deposits that will accumulate to $67 200.00 after eight years at 6.5% compounded semi-annually?

9. Eva contributed $200.00 every month for five years into an RRSP earning 4.3% compounded quarterly. Six years after the last contribution, she converted the RRSP into an annuity that is to pay her monthly for thirty years. If the first payment is due one month after the conversion into the annuity and interest on the annuity is 5.4% compounded semi-annually, how much will Eva receive every month?

10. Joy would like to receive $6000.00 at the end of every three months for ten years after her retirement. If she retires now and interest is 6.5% compounded semi-annually, how much must she deposit into an account?

Challenge Problems

1. After winning some money at a casino, Tony is considering purchasing an annuity that promises to pay him $300 at the end of each month for 12 months, then $350 at the end of each month for 24 months, and then $375 at the end of each month for 36 months. If the first payment is due at the end of the first month and interest is 7.5% compounded annually over the life of the annuity, find Tony's purchase price.

2. A loan of $5600.00 is to be repaid at 9% compounded annually by making ten payments at the end of each quarter. Each of the last six payments is two times the amount of each of the first four payments. What is the size of each payment?

Case Study 6.1 Cash-Back Options

» Rico Berardini is shopping for a new car and has noticed that many car manufacturers are offering special deals to sell off the current year's cars before the new models arrive. Rico's local car dealer is advertising 5.9% financing for a full 48 months (i.e., 5.9% compounded monthly) or up to $2000.00 cash back on selected vehicles.

The car that Rico wants to buy costs $21 500.00 including taxes, delivery, licence, and dealer preparation. This car qualifies for $1500.00 cash back if Rico pays cash for the car. Rico has a good credit rating and knows that he could arrange a car loan at his bank for the full price of any car he chooses. His other option is to take the dealer financing offered at 5.9% for 48 months.

To compare these two financing options, Rico wants to see which option requires the lower monthly payment. He knows he can use annuity formulas to calculate the monthly payments.

QUESTIONS

1. Suppose Rico buys the car on July 1. What monthly payment must Rico make if he chooses the dealer's 5.9% financing option and pays off the loan over 48 months? (Assume he makes each monthly payment at the end of the month and his first payment is due on July 31.)

2. Suppose the bank offers Rico a 48-month loan with the interest compounded monthly and the payments due at the end of each month. If Rico accepts the bank loan, he can get $1500.00 cash back on this car.

   Rico works out a method to calculate the bank rate of interest required to make bank financing the same cost as dealer financing. First, calculate the monthly rate of interest that would make the monthly bank payments equal to the monthly dealer payments. Then, calculate the effective rate of interest represented by the monthly compounded rate. If the financing from the bank is at a lower rate of interest compounded monthly, choose the bank financing. The reason is that the monthly payments for the bank's financing would be lower than the monthly payments for the dealer's 5.9% financing.

   (a) How much money would Rico have to borrow from the bank to pay cash for this car?

   (b) Using the method above, calculate the effective annual rate of interest and the nominal annual rate of interest required to make the monthly payments for bank financing exactly the same as for dealer financing.

3. Suppose Rico decides to explore the costs of financing a more expensive car. The more expensive car costs $29 900.00 in total and qualifies for the 5.9% dealer financing for 48 months or $2000.00 cash back. What is the highest effective annual rate of interest at which Rico should borrow from the bank instead of using the dealer's 5.9% financing?

Case Study 6.2 Fitness Finances

» Jocelyn is planning to open Fitness Fundamentals, a new health and fitness club. She must decide what to charge for each type of membership and what payment options to offer.

Jocelyn plans to offer two types of memberships. The General membership allows members to use all facilities, and it provides a simple locker room. The Health Club membership allows members to use all facilities, and it provides towels, shower supplies, and a sauna. When members pay for their annual membership when they join, the fee is $450.00 for the General membership and $750.00 for the Health Club membership.

QUESTIONS

1. Jocelyn wants to offer members the option of paying their annual membership fees in twelve equal monthly installments. The first installment would be due on the day of joining.
   (a) To the nearest dollar, how much should Jocelyn charge General members monthly if she wants to make interest of 15% compounded monthly?
   (b) To the nearest dollar, how much should she charge Health Club members monthly if she wants to make interest of 20% compounded monthly?
2. Jocelyn knows that many of the Health Club members will be executives whose companies will pay for their memberships. Many companies prefer to make quarterly payments for annual memberships. For this reason, Jocelyn decides to offer a quarterly payment option. The first installment would be due on the day of joining. To the nearest dollar, how much should Jocelyn charge Health Club members quarterly if she wants to make interest of 20% compounded monthly?
3. As an opening special, Jocelyn wants to offer all members who join during the opening week three free months of membership. To calculate the nominal rate of interest she would earn on these opening special memberships, Jocelyn will add the payments made for the year for a membership, then spread the total payments equally over fifteen months as if the payment period were fifteen months. (For example, if the total monthly payments were $12 \times \$X$, under this scheme Jocelyn would calculate the new monthly payments as $12 \times \$X \div 15$.) She would then calculate the nominal rate of interest earned on these new monthly payments over the fifteen-month membership period.
   (a) What is the nominal rate of interest earned on the opening special for the General membership with the monthly payment option?
   (b) What is the nominal rate of interest earned on the opening special for the Health Club membership with the monthly payment option?
   (c) What is the nominal rate of interest earned on the opening special for the Health Club membership with the quarterly payment option?

SUMMARY OF FORMULAS

Formula 6.1

$p = (1 + i)^c - 1$

Finding the effective rate of interest per payment period p for a nominal annual rate of interest compounded c times per payment interval

Formula 6.2

$$FV_{nc} = PMT\left[\frac{(1 + p)^n - 1}{p}\right] \text{ where } p = (1 + i)^c - 1$$

Finding the future value of an ordinary general annuity using the effective rate of interest per payment period

Formula 6.2A

$$PMT = \frac{FV_{nc}p}{(1 + p)^n - 1}$$

Finding the payment of an ordinary general annuity using the effective rate of interest per payment period when the future value is known

Formula 6.2B

$$n = \frac{\ln[FV_n p/PMT + 1]}{\ln(1 + p)}$$

Finding the number of payments of an ordinary general annuity using the effective rate of interest per payment period when the future value is known

Formula 6.3

$$PV_{nc} = PMT\left[\frac{1 - (1 + p)^{-n}}{p}\right]$$

Finding the present value of an ordinary general annuity using the effective rate of interest per payment period

Formula 6.3A

$$PMT = \frac{PV_{nc}p}{1 - (1 + p)^{-n}}$$

Finding the payment of an ordinary general annuity using the effective rate of interest per payment period when the present value is known

Formula 6.3B

$$n = \frac{\ln[1 + (PV_n p/PMT)]}{-\ln(1 + p)}$$

Finding the number of payments of an ordinary general annuity using the effective rate of interest per payment period when the present value is known

GLOSSARY

Effective rate of interest per payment period the rate of interest earned during the payment period that yields the same amount of interest as a nominal annual rate compounded c times per year, where c is the number of payment periods per year *(p. 223)*

USEFUL INTERNET SITES

www.money.canoe.ca

RRIF: Savings Find current information and rates of return on RRIFs from several financial institutions and financial services providers on the Canoe Money site.

www.fidelity.ca

Growth Calculator Click on Investor Centre, Education/Planning, Calculators, and then Growth Calculator. This interactive chart on the Fidelity Investments site allows you to calculate both simple and compound earnings on investment capital by entering values for a series of variables.

www.butterworths.ca

The Bottom Line Click on The Bottom Line from the list of choices. *The Bottom Line* is a monthly Canadian magazine for finance professionals.

7 Annuities Due, Deferred Annuities, and Perpetuities

OBJECTIVES

Upon completing this chapter, you will be able to do the following:

1. Compute the future value, present value, periodic payment, term, and interest rate for simple annuities due.
2. Compute the future value, present value, periodic payment, term, and interest rate for general annuities due.
3. Compute the future value, present value, periodic payment, term, and interest rate for ordinary deferred annuities and deferred annuities due.
4. Compute the present value, periodic payment, and interest rate for ordinary perpetuities, perpetuities due, and deferred perpetuities.

When you pay rent on an apartment or a house, your rent is due at the beginning of the month. This series of regular, equal payments represents a type of annuity called an annuity due. Its payment schedule differs from that of an ordinary annuity (which we discussed in Chapters 5 and 6), where payments were due at the end of each month.

In some situations, a series of regular, equal payments may not begin until some future time. These payment situations are known as deferred annuities. For example, if you won a large amount of money today and you wanted a series of regular, equal payments to begin five years from now, you would set up a deferred annuity. Both ordinary annuities and annuities due can be deferred.

INTRODUCTION

In the previous chapters, we considered ordinary annuities, both simple and general, in which the payments are made at the end of each payment period.

In this chapter we will consider other annuities resulting from variations in the payment dates and the length of time that the payments continue. These include annuities due (in which payments are made at the beginning of the period), deferred annuities (in which the first payment is made after the first payment interval has been completed), and perpetuities (in which payments continue indefinitely).

7.1 Simple Annuities Due

A. Future value of a simple annuity due

By definition, an **annuity due** is an annuity in which the periodic payments are made at the beginning of each payment interval.

Finding the future value of such an annuity is similar to finding the future value of an ordinary annuity. In fact, the future value of an annuity due is closely related to the future value of an ordinary annuity. This same close relationship also holds for the present value.

EXAMPLE 7.1A

Find the accumulated value (future value) at the end date of the last payment period of deposits of \$3000.00 each made at the beginning of five consecutive years respectively at 6% compounded annually.

SOLUTION

As for any problem involving a series of payments, the method of solution and the data can be shown on a time diagram.

FIGURE 7.1 **Graphical Representation of Method and Data**

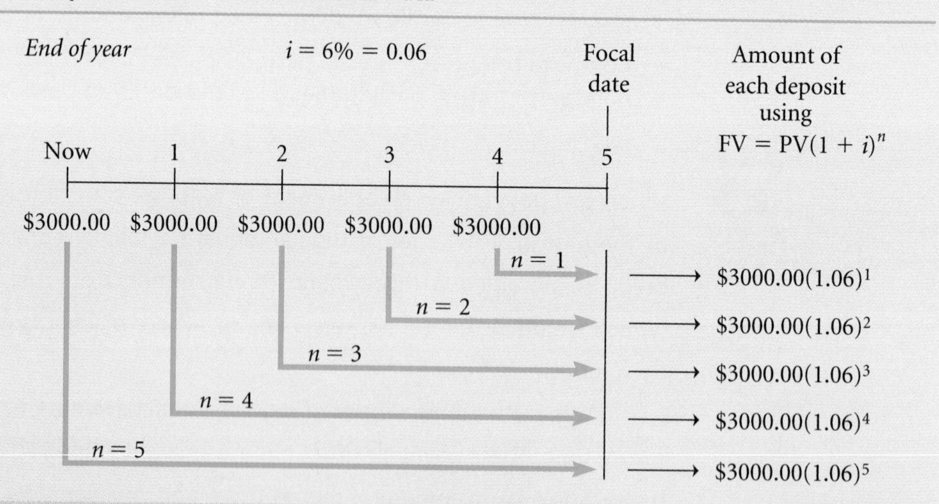

As shown in Figure 7.1, the first deposit is located at the beginning of Year 1, which is the same as "now"; the second deposit is located at the beginning of Year 2, which is the same as the end of Year 1; the third at the beginning of Year 3; the fourth at the beginning of Year 4; and the fifth and last deposit at the beginning of Year 5, which is also the beginning of the last payment period. The focal date, however, is located at the *end* of the last payment period.

The accumulated values of the individual deposits are obtained by using Formula 3.1A, $FV = PV(1 + i)^n$. Finding the combined total of the five accumulated values is made easier by taking out the common factors 3000.00 and 1.06 as follows:

$$\begin{aligned}
&\textit{Deposit 5}\ \ 3000.00(1.06)^1 = \\
&\textit{Deposit 4}\ \ 3000.00(1.06)^2 = \\
&\textit{Deposit 3}\ \ 3000.00(1.06)^3 = 3000.00(1.06)\begin{bmatrix}(1.0)\\(1.06)\\(1.06)^2\\(1.06)^3\\(1.06)^4\end{bmatrix} = 3000.00(1.06)\begin{bmatrix}(1.0)\\(1.06)\\(1.1236)\\(1.191016)\\(1.262477)\end{bmatrix}\\
&\textit{Deposit 2}\ \ 3000.00(1.06)^4 = \\
&\textit{Deposit 1}\ \ 3000.00(1.06)^5 = \\
&\qquad = 3000.00(1.06)(5.637093)\\
&\qquad = 16\,911.28(1.06)\\
&\qquad = \$17\,925.96
\end{aligned}$$

Note: This example is the same as Example 5.2B except that the deposits are made at the beginning of each payment period rather than at the end. The answer to Example 5.2B was \$16 911.28. We could have obtained the answer to Example 7.1A simply by multiplying \$16 911.28 by 1.06. It appears that the future value of the annuity due can be obtained by multiplying the future value of the ordinary annuity by the factor $(1 + i)$.

The general notation for simple annuities due is the same as for ordinary simple annuities except that the accumulated value (future value) of the annuity due is represented by the symbol FV_n(due).

The formula for the future value of a simple annuity due is:

$$S_n(\text{due}) = R(1+i)\left[\frac{(1+i)^n - 1}{i}\right]$$

restated as

$$FV_n(\text{due}) = PMT(1+i)\left[\frac{(1+i)^n - 1}{i}\right]$$ — **Formula 7.1**

Note: Formula 7.1, the future value of a simple annuity due, differs from Formula 5.1, the future value of an ordinary simple annuity, only by the factor $(1 + i)$.

FUTURE VALUE OF A SIMPLE ANNUITY DUE $= (1 + i) \times$ FUTURE VALUE OF THE ORDINARY SIMPLE ANNUITY

The relationship between an annuity due and the corresponding ordinary annuity is graphically illustrated in the comparison of the line diagrams.

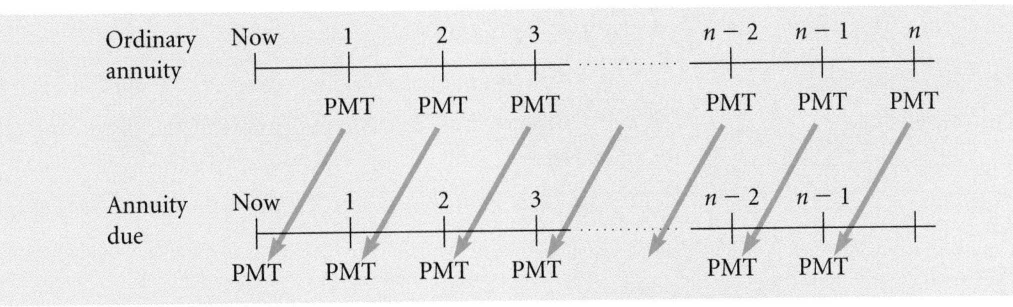

The two line graphs show the shift of the payments by one period. In an annuity due, every payment earns interest for one more period than in an ordinary annuity and this explains the factor $(1 + i)$.

EXAMPLE 7.1B

Find the accumulated value at the end of the last payment period of quarterly payments of $50.00 made at the beginning of each quarter for ten years if interest is 6% compounded quarterly.

SOLUTION

Since the payments are of equal size made at the beginning of each period, the payment series is an annuity due and since the focal date is the end of the last payment period, the future value of the annuity due is to be found.

$\text{PMT} = 50.00;\ \text{P/Y} = 4;\ \text{C/Y} = 4;\ \text{I/Y} = 6;\ i = \dfrac{6\%}{4} = 0.015;\ n = 10(4) = 40$

$$\begin{aligned}\text{FV}_n(\text{due}) &= 50.00(1.015)\left(\frac{1.015^{40} - 1}{0.015}\right) \quad \text{——— substituting in Formula 7.1}\\ &= 50.00(1.015)(54.267894)\\ &= 2713.39(1.015)\\ &= \$2754.10\end{aligned}$$

EXAMPLE 7.1C

You deposit $100.00 at the beginning of each month for five years in an account paying 4.2% compounded monthly.

(i) What will the balance in your account be at the end of five years?
(ii) How much of the balance will you have contributed?
(iii) How much of the balance will be interest?

SOLUTION

(i) $\text{PMT} = 100.00;\ \text{P/Y} = 12;\ \text{C/Y} = 12;\ \text{I/Y} = 4.2;$

$i = \dfrac{4.2\%}{12} = 0.35\% = 0.0035;\quad n = 5(12) = 60$

$$\begin{aligned}\text{FV}_n(\text{due}) &= 100.00(1.0035)\left(\frac{1.0035^{60} - 1}{0.0035}\right)\\ &= 100.00(1.0035)(66.635949)\\ &= 6663.5949(1.0035)\\ &= 6686.92\end{aligned}$$

(ii) Your contribution is $(100.00)(60) = \$6000.00$.

(iii) The interest earned $= 6686.92 - 6000.00 = \$686.92$.

POINTERS AND PITFALLS

To distinguish between problems dealing with ordinary annuities and annuities due, look for key words or phrases that signal one type of annuity or the other.

Ordinary annuities:	"payments (or deposits) made at the *end* of each (or every)…"
Annuities due:	"payments (or deposits) made at the *beginning* of each (or every)…"
	"first payment is due on the date of sale (or signing)"
	"payable in advance"

B. Present value of a simple annuity due

EXAMPLE 7.1D Find the present value of five payments of $3000.00 each made at the beginning of each of five consecutive years respectively if money is worth 6% compounded annually.

SOLUTION As Figure 7.2 shows, the present value of the individual payments is obtained using Formula 3.1C, $PV = FV(1 + i)^{-n}$. The sum of the individual present values is easier to find when the common factor 3000.00 is taken out.

FIGURE 7.2 **Graphical Representation of Method and Data**

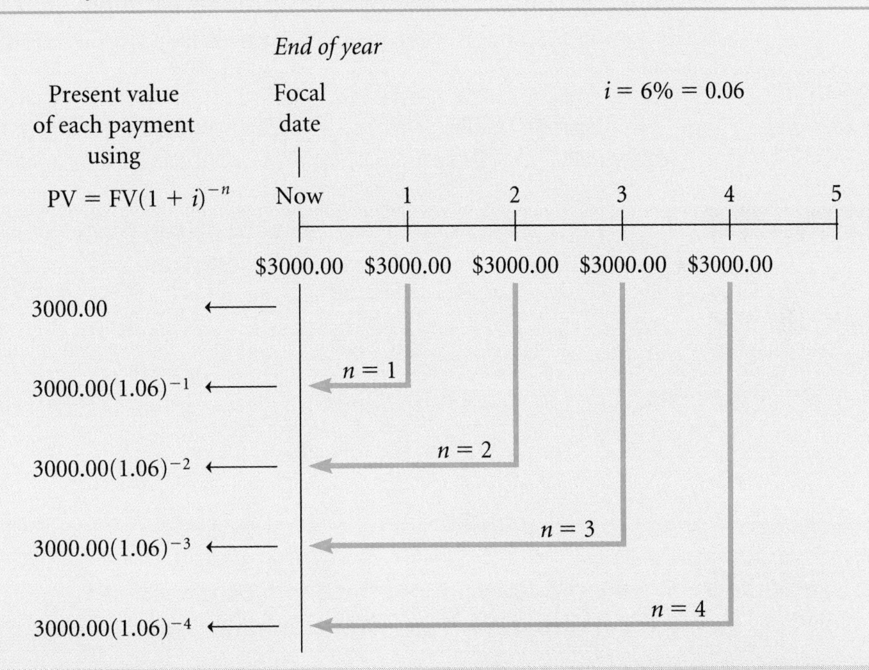

$$\begin{array}{ll}
\textit{Payment 1} & 3000.00(1.06)^{0} \\
\textit{Payment 2} & 3000.00(1.06)^{-1} \\
\textit{Payment 3} & 3000.00(1.06)^{-2} \\
\textit{Payment 4} & 3000.00(1.06)^{-3} \\
\textit{Payment 5} & 3000.00(1.06)^{-4}
\end{array}
= 3000.00\begin{bmatrix}(1.0)\\(1.06)^{-1}\\(1.06)^{-2}\\(1.06)^{-3}\\(1.06)^{-4}\end{bmatrix}
= 3000.00\begin{bmatrix}(1.000000)\\(0.943396)\\(0.889996)\\(0.839619)\\(0.792094)\end{bmatrix}$$

$$= 3000.00\ (4.465105)$$

$$= \$13\ 395.32$$

Example 7.1D is the same as Example 5.3B, except that the payments are made at the beginning of each payment period. The answer to Example 5.3B was $12 637.09. If this amount is multiplied by 1.06, the result is $13 395.32, the answer to Example 7.1D.

This result implies that we could have obtained the present value of the annuity due in Example 7.1D by multiplying the present value of the ordinary annuity in Example 7.3B by the factor 1.06, which is the factor $(1 + i)$.

The present value of an annuity due is represented by the symbol PV_n (due). The formula for the present value of a simple annuity due is:

$$A_n(\text{due}) = R(1 + i)\left[\frac{1 - (1 + i)^{-n}}{i}\right]$$

restated as

$$PV_n(\text{due}) = PMT(1 + i)\left[\frac{1 - (1 + i)^{-n}}{i}\right]$$ —— **Formula 7.2**

Note: Formula 7.2, the present value of a simple annuity due, differs from Formula 5.2, the present value of an ordinary simple annuity, only by the factor $(1 + i)$.

PRESENT VALUE OF A SIMPLE ANNUITY DUE $= (1 + i) \times$ PRESENT VALUE OF THE ORDINARY SIMPLE ANNUITY

EXAMPLE 7.1E

Find the present value of payments of \$50.00 made at the beginning of each quarter for ten years if interest is 6% compounded quarterly.

SOLUTION

$PMT = 50.00;\quad i = \frac{6\%}{4} = 1.5\% = 0.015;\quad n = 10(4) = 40$

$$\begin{aligned} PV_n(\text{due}) &= 50.00(1.015)\left[\frac{1 - (1.015)^{-40}}{(0.015)}\right] \quad \text{—— substituting in Formula 7.2} \\ &= 50.00(1.015)(29.915845) \\ &= 1495.79(1.015) \\ &= \$1518.23 \end{aligned}$$

EXAMPLE 7.1F

What is the cash value of a three-year lease of office facilities renting for \$536.50 payable at the beginning of each month if money is worth 9% compounded monthly?

SOLUTION

Since the payments are at the beginning of each payment period, the problem involves an annuity due, and since we want the cash value, the present value of the annuity due is required.

$PMT = 536.50;\quad i = \frac{9\%}{12} = 0.75\% = 0.0075;\quad n = 3(12) = 36$

$$\begin{aligned} PV_n(\text{due}) &= 536.50(1.0075)\left[\frac{1 - (1.0075)^{-36}}{0.0075}\right] \\ &= 536.50(1.0075)(31.446805) \\ &= 16\,871.21(1.0075) \\ &= \$16\,997.74 \end{aligned}$$

The cash value of the lease is \$16 997.74.

C. Using preprogrammed financial calculators

You can easily determine the future value or the present value of an annuity due using a preprogrammed financial calculator. One method is to find the corresponding value for an ordinary simple annuity and multiply by $(1 + i)$. If you are using the Texas Instruments BAII Plus, set the calculator in "BGN" mode (since annuity due payments are made at the beginning of each payment period), then solve for the unknown variable the same way as you do for ordinary simple annuities. Using the Texas Instruments BAII Plus, follow this key sequence to set the calculator in "BGN" mode:

Key in	*Press*	*Display shows*	
2nd	(BGN)	END *or* BGN	checks the mode
2nd	(SET)	BGN	if previously in "END" mode. BGN will appear in the upper right corner of the display.
		or	
		END	if previously in "BGN" mode. The display will be blank. Press 2nd (SET) again so that BGN appears in the display.
2nd	(QUIT)	0	returns to the standard calculation mode

"BGN" will appear in the upper right corner of the display.

Refer to Appendix II for instruction for annuity due calculations if you are using the Texas Instruments BA Solar, the Sharp EL-733A, or the Hewlett-Packard 10B calculator.

EXAMPLE 7.1G

Payments of $425 are to be made at the beginning of each quarter for 10 years. If money is worth 6% compounded quarterly, determine

(i) the accumulated value of the payments;
(ii) the present value of the payments.

SOLUTION

(i) PMT = 425; I/Y = 6; P/Y = 4; $i = \frac{6\%}{4} = 1.5\%$; $n = 40$

Using the Texas Instruments BAII Plus, ensure the calculator is in "BGN" mode, then follow this key sequence to retrieve the *future value* of an annuity due.

(Set P/Y = 4) 0 PV 425 ± PMT 6.0 I/Y
40 N CPT FV 23 409.81274

The accumulated value of the annuity due is $23 409.81.

(ii) Using the Texas Instruments BA II PLUS, ensure the calculator is in "BGN" mode, then follow this key sequence to retrieve the *present value* of an annuity due.

0 [FV] 425 [±] [PMT] 6.0 [I/Y] 40 [N] [CPT] [PV] [12 904.94772]

The present value of the annuity due is $12 904.95.

EXAMPLE 7.1H

Frank deposited monthly rent receipts of $250.00 due at the beginning of each month in a savings account paying 4.5% compounded monthly for four years. Frank made no further deposits after four years but left the money in the account.

(i) What will the balance be twelve full years after he made the first deposit?
(ii) How much of the total will be due to rent?
(iii) How much will be interest?

SOLUTION

(i) First determine the balance at the end of four years. This problem involves finding the future value of a simple annuity due.

$$\text{PMT} = 250.00;\ \text{I/Y} = 4.5;\ \text{P/Y} = 12;\ i = \frac{4.5\%}{12} = 0.375\%;\ n = 48$$

$$\begin{aligned}\text{FV}_n(\text{due}) &= 250.00(1.00375)\left(\frac{1.00375^{48} - 1}{0.00375}\right)\\ &= 250.00(1.00375)(52.483834)\\ &= 13\ 120.96(1.00375)\\ &= \$13\ 170.16\end{aligned}$$

Now accumulate $13 170.16 for another eight years.

$$\text{PV} = 13\ 170.16;\quad i = 0.00375;\quad n = 8(12) = 96$$

$$\begin{aligned}\text{FV} &= 13\ 170.16(1.00375)^{96} \quad \text{——— substituting in Formula 3.1A}\\ &= 13\ 170.16(1.4323647)\\ &= \$18\ 864.47\end{aligned}$$

Programmed Solution

("BGN" mode) (Set P/Y = 12) 0 [PV] 250 [±] [PMT]

4.5 [I/Y] 48 [N] [CPT] [FV] [13 170.16209]

13 170.16209 [±] [PV] 0 [PMT] 4.5 [I/Y]

96 [N] [CPT] [FV] [18 864.47466]

The balance in the account after twelve years is $18 864.47.

(ii) The rent receipts in the total are 250.00(48) = $12 000.00.

(iii) Interest in the balance is 18 864.47 − 12 000.00 = $6864.47.

D. Finding the periodic payment PMT of a simple annuity due

If the future value of an annuity FV_n(due), the number of conversion periods n, and the conversion rate i are known, you can find the periodic payment PMT by substituting the given values in the future value Formula 7.1.

$$FV_n(\text{due}) = PMT(1 + i)\left[\frac{(1 + i)^n - 1}{i}\right] \quad \text{——— Formula 7.1}$$

If the present value of an annuity PV_n(due), the number of conversion periods n, and the conversion rate i are known, you can find the periodic payment PMT by substituting the given values in the present value Formula 7.2.

$$PV_n(\text{due}) = PMT(1 + i)\left[\frac{1 - (1 + i)^{-n}}{i}\right] \quad \text{— Formula 7.2}$$

When using a scientific calculator, you can find PMT by first rearranging the terms of the above formulas. Then substitute the three known values (FV_{DUE} or PV_{DUE}, n, and i) into the appropriate rearranged formula and solve for PMT.

When using a preprogrammed financial calculator, you can find PMT by entering the five known values (FV_{DUE} or PV_{DUE}, V, N, I/Y, P/Y, and C/Y) and pressing CPT PMT.

EXAMPLE 7.1I

What semi-annual payment must be made into a fund at the beginning of every six months to accumulate to $9600.00 in ten years at 7% compounded semi-annually?

SOLUTION

$FV_n(\text{due}) = 9600.00;\ i = \frac{7\%}{2} = 3.5\% = 0.035;\ n = 10(2) = 20$

$$9600.00 = PMT(1.035)\left(\frac{1.035^{20} - 1}{0.035}\right) \quad \text{——— substituting in Formula 6.1}$$

$$9600.00 = PMT(1.035)(28.279682)$$

$$9600.00 = PMT(29.269471)$$

$$PMT = \frac{9600.00}{29.269471}$$

$$PMT = \$327.99$$

Programmed Solution

("BGN" mode) 0 PV 9600 FV 3.5 I/Y 20 N CPT PMT -327.986799

The semi-annual payment is $327.99.

EXAMPLE 7.1J

What monthly rent payment at the beginning of each month for four years is required to fulfill a lease contract worth $7000.00 if money is worth 7.5% compounded monthly?

SOLUTION

$PV_n(\text{due}) = 7000.00$; P/Y = 12; C/Y = 12; I/Y = 7.5;

$i = \frac{7.5\%}{12} = 0.625\% = 0.00625$; $n = 4(12) = 48$

$$7000.00 = PMT(1.00625)\left(\frac{1 - 1.00625^{-48}}{0.00625}\right)$$ — substituting in Formula 7.2

$$7000.00 = PMT(1.00625)(41.358371)$$
$$7000.00 = PMT(41.616861)$$
$$PMT = \frac{7000.00}{41.616861}$$
$$PMT = \$168.20$$

Programmed Solution

("BGN" mode) (Set P/Y = 12; C/Y = 12) 0 [FV] 7000 [±] [PV]

7.5 [I/Y] 48 [N] [CPT] [PMT] [168.201057]

The monthly rent payment due at the beginning of each month is $168.20.

EXAMPLE 7.1K

How much will you have to deposit into an account at the beginning of every three months for twelve years if you want to have a balance of $100 000.00 twenty years from now and interest is 8% compounded quarterly?

SOLUTION

First, find the balance that you must have in the account at the end of the term of the annuity (after twelve years).

FV = 100 000.00; P/Y = 4; C/Y = 4; I/Y = 8; $i = \frac{8\%}{4} = 2\% = 0.02$;
$n = 8(4) = 32$

$$PV = 100\,000.00(1.02)^{-32}$$
$$= 100\,000.00(0.530633)$$
$$= \$53\,063.33$$

Next, find the quarterly payment needed at the beginning of each quarter to accumulate to $53 063.33.

$FV_n(\text{due}) = 53\,063.33$; P/Y = 4; C/Y = 4; I/Y = 8; $i = \frac{8\%}{4} = 0.02$;
$n = 12(4) = 48$

$$53\,063.33 = PMT(1.02)\left(\frac{1.02^{48} - 1}{0.02}\right)$$
$$53\,063.33 = PMT(1.02)(79.353519)$$
$$53\,063.33 = PMT(80.940590)$$
$$PMT = \frac{53.063.33}{80.940590}$$
$$PMT = 53\,063.33(0.012355)$$
$$PMT = \$655.58$$

Programmed Solution

(Set P/Y = 4; C/Y = 4) 0 [PMT] 100 000 [FV]

8 [I/Y] 32 [N] [CPT] [PV] [-53 063.33035]

("BGN" mode) ± 53 063.33035 FV 0 PV

8 I/Y 48 N CPT PMT −655.583689

The quarterly deposit at the beginning of each payment period is $655.58.

E. Finding the term n of a simple annuity due

If the future value of an annuity FV_n (due), the periodic payment PMT, and the conversion rate i are known, you can find the term of the annuity n by substituting the given values in the future value Formula 7.1.

$$FV_n(\text{due}) = PMT(1+i)\left[\frac{(1+i)^n - 1}{i}\right]$$ —— Formula 7.1

If the present value PV_n(due), the periodic payment PMT, and the conversion rate i are known, you can find the term of the annuity n by substituting the given values in the present value Formula 7.2.

$$PV_n(\text{due}) = PMT(1+i)\left[\frac{1-(1+i)^{-n}}{i}\right]$$ —— Formula 7.2

When using a scientific calculator, you can find n by first rearranging the terms of the appropriate formula above. Then substitute the three known values (FV_{DUE} or PV_{DUE}, PMT, and i) into the rearranged formula and solve for n.

When using a preprogrammed financial calculator, you can find n by entering the five known values (PV, PMT, I/Y, P/Y, and C/Y) and pressing CPT N.

EXAMPLE 7.1L Over what length of time will $75.00 deposited at the beginning of each month grow to $5000.00 at 4.5% compounded monthly?

SOLUTION $FV_n(\text{due}) = 5000.00$; PMT = 75.00; P/Y = 12; C/Y = 12; I/Y = 4.5;

$$i = \frac{4.5\%}{12} = 0.375\% = 0.00375$$

$$5000.00 = 75.00(1.00375)\left(\frac{1.00375^n - 1}{0.00375}\right)$$ —— substituting in Formula 6.1

$$5000.00 = 20\,075.00(1.00375^n - 1)$$

$$\frac{5000.00}{20\,075.00} = 1.00375^n - 1$$

$$0.249066 + 1 = 1.00375^n$$

$$n \ln 1.00375 = \ln 1.249066$$

$$n(0.003743) = 0.222396$$

$$n = \frac{0.222396}{0.0037430}$$

$$n = 59.416748$$

As discussed in Example 5.5, n should be rounded upwards.

$$n = 60 \text{ (months)}$$

Programmed Solution

("BGN" mode)(Set P/Y = 12; C/Y = 12) 0 [PV] 5000 [FV]

75 [±] [PMT] 4.5 [I/Y] [CPT] [N] [59.416748]

It will take five years to accumulate \$5000.00.

EXAMPLE 7.1M

Surrey Credit Union intends to accumulate a building fund of \$150 000.00 by depositing \$4125.00 at the beginning of every three months at 7% compounded quarterly. How long will it take for the fund to reach the desired amount?

SOLUTION

FV_n(due) = 150 000.00; PMT = 4125.00; P/Y = 4; C/Y = 4; I/Y = 7;

$i = \frac{7\%}{4} = 1.75\%$

Programmed Solution

("BGN" mode) [PV] 150 000 (Set P/Y = 4; C/Y = 4) 0 [FV]

4125 [±] [PMT] 7 [I/Y] [CPT] [N] [28.00020963]

$n = 28$ quarters (approximately)

It will take seven years to build up the fund.

EXAMPLE 7.1N

For how long can you withdraw \$480.00 at the beginning of every three months from a fund of \$9000.00 if interest is 10% compounded quarterly?

SOLUTION

PV_n(due) = 9000.00; PMT = 480.00; P/Y = 4; C/Y = 4; I/Y = 10;

$i = \frac{10\%}{4} = 2.5\% = 0.025$

$$9000.00 = 480.00(1.025)\left(\frac{1 - 1.025^{-n}}{0.025}\right)$$ — substituting in Formula 7.2

$$9000.00 = 19\,680.00(1 - 1.025^{-n})$$

$$\frac{9000.00}{19\,680.00} = 1 - 1.025^{-n}$$

$$0.457317 = 1 - 1.025^{-n}$$

$$1.025^{-n} = 1 - 0.457317$$

$$1.025^{-n} = 0.542683$$

$$-n \ln 1.025 = \ln 0.542683$$

$$-n(0.024693) = -0.611230$$

$$n = \frac{0.611230}{0.024693}$$

$$n = 24.753561$$

$$n = 25 \text{ (quarters)}$$

Programmed Solution

(Set P/Y = 4; C/Y = 4) ("BGN" mode) 9000 [±] [PV]

0 [FV] 480 [PMT] 10 [I/Y] [CPT] [N] [24.75355963]

$n = 28$ quarters (approximately)

Withdrawals of \$480.00 can be made for six years and three months. (The last withdrawal will be less than \$480.00.)

EXAMPLE 7.1O

A lease contract valued at $7800.00 is to be fulfilled by rental payments of $180.00 due at the beginning of each month. If money is worth 9% compounded monthly, what should the term of the lease be?

SOLUTION

$PV_n(\text{due}) = 7800.00$; PMT = 180.00; P/Y = 12; I/Y = 9; C/Y = 12;

$i = \frac{9\%}{12} = 0.75\%$

Programmed Solution

("BGN" mode) (Set P/Y = 12; C/Y = 12) 0 [FV] 7800 [±] [PV]

180 [PMT] 9 [I/Y] [CPT] [N] [52.12312544]

$n = 53$ months (approximately)

The term of the lease should be four years and five months.

F. Finding the periodic rate of interest of a simple annuity due

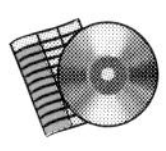

Preprogrammed financial calculators are especially helpful when solving for the conversion rate i. Determining i without a financial calculator is extremely time-consuming. However, it *can* be done by hand, as illustrated in Appendix A on the CD-ROM.

When the future value or present value, the periodic payment PMT, and the term n of an annuity due are known, the periodic rate of interest i can be found by entering the five known values into a preprogrammed financial calculator. (Remember, if both PV and PMT are non-zero, enter PV *only* as a negative amount.) For simple annuities due, retrieve the answer by being in "BGN" mode and pressing [CPT] [I/Y]. This is the nominal annual rate of interest.

EXAMPLE 7.1P

Compute the nominal annual rate of interest at which $100.00 deposited at the beginning of each month for ten years will amount to $15 000.00.

SOLUTION

$FV_n(\text{due}) = 15\,000$; PMT = 100.00; P/Y = 12; C/Y = 12; $n = 120$; $m = 12$

("BGN" mode) (Set P/Y = 12; C/Y = 12) 0 [PV] 15 000 [FV]

100 [±] [PMT] 120 [N] [CPT] [I/Y] [4.282801]

allow several seconds for the computation

The nominal annual rate of interest is 4.28% compounded monthly.
The monthly conversion rate is 4.282801%/12 = 0.356900%.

EXAMPLE 7.1Q

A lease agreement valued at $7500.00 requires payment of $450.00 at the beginning of every quarter for five years. What is the nominal annual rate of interest charged?

SOLUTION

$PV_n(\text{due}) = 7500.00$; $PMT = 450.00$; $P/Y = 4$; $C/Y = 4$; $n = 20$; $m = 4$

("BGN" mode) (Set P/Y 5 4; C/Y 5 4) 0 FV 7500 ± PV

450.00 PMT 20 N CPT I/Y 8.032647

allow several seconds for the computation

The nominal annual rate of interest is 8.03% compounded quarterly.
The quarterly compounding rate is 8.032647/4 = 2.008162%.

EXERCISE 7.1

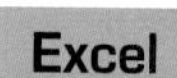

If you choose, you can use Excel's ***Present Value (PV)*** function or ***Future Value (FV)*** function to answer the questions indicated below. Refer to **PV** and **FV** on the Spreadsheet Template Disk to learn how to use these Excel functions.

Find the future value and the present value of each of the following six simple annuities due.

	Periodic Payment	Payment Interval	Term	Interest Rate	Conversion Period
1.	$3000.00	3 months	8 years	8%	quarterly
2.	$750.00	1 month	5 years	7.2%	monthly
3.	$2000.00	6 months	12 years	5.6%	semi-annually
4.	$450.00	3 months	15 years	4.4%	quarterly
5.	$65.00	1 month	20 years	9%	monthly
6.	$160.00	1 month	15 years	6%	monthly

Find the periodic payment for each of the following four simple annuities due.

	Future Value	Present Value	Payment Period	Term	Interest Rate	Conversion Period
1.	$20 000.00		3 months	15 years	6%	quarterly
2.		$12 000.00	1 year	8 years	7%	annually
3.		$18 500.00	6 months	12 years	3%	semi-annually
4.	$9 400.00		1 month	5 years	12%	monthly

Find the length of the term for each of the following four simple annuities due.

	Future Value	Present Value	Periodic Payment	Payment Period	Interest Rate	Conversion Period
1.	$5 300.00		$35.00	1 month	6%	monthly
2.		$8400.00	$440.00	3 months	7%	quarterly
3.		$6450.00	$1120.00	1 year	10%	annually
4.	$15 400.00		$396.00	6 months	5%	semi-annually

D. Compute the nominal annual rate of interest for each of the following four simple annuities due.

	Future Value	Present Value	Periodic Rent	Payment Interval	Term	Compounding Period
1.	$70 000.00		$1014.73	1 year	25 years	annually
2.		$42 000.00	$528.00	1 month	10 years	monthly
3.		$28 700.00	$2015.00	6 months	15 years	semi-annually
4.	$36 000.00		$584.10	3 months	12 years	quarterly

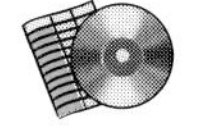

E. Answer each of the following questions.

1. Find the amount of an annuity due of $300.00 payable at the beginning of every month for seven years at 6% compounded monthly.
2. Determine the accumulated value after twelve years of deposits of $360.00 made at the beginning of every three months and earning interest at 7% compounded quarterly.
3. Until he retires sixteen years from now, Mr. Lait plans to deposit $300.00 at the beginning of every three months in an account paying interest at 5% compounded quarterly.
   (a) What will be the balance in his account when he retires?
   (b) How much of the balance will be interest?
4. Joanna contributes $750.00 at the beginning of every six months into an RRSP paying interest at 8% compounded semi-annually.
   (a) How much will her RRSP deposits amount to in twenty years?
   (b) How much of the amount will be interest?
5. Find the present value of payments of $2500.00 made at the beginning of every six months for ten years if money is worth 9.5% compounded semi-annually.
6. What is the discounted value of deposits of $240.00 made at the beginning of every three months for seven years if money is worth 8.8% compounded quarterly?
7. A washer-dryer combination can be purchased from a department store by making monthly credit card payments of $52.50 for two-and-a-half years. The first payment is due on the date of sale and interest is 21% compounded monthly.
   (a) What is the purchase price?
   (b) How much will be paid in installments?
   (c) How much is the cost of financing?
8. Diane Wallace bought a living-room suite on credit, signing an installment contract with a finance company that requires monthly payments of $62.25 for three years. The first payment is made on the date of signing and interest is 24% compounded monthly.

(a) What was the cash price?

(b) How much will Diane pay in total?

(c) How much of what she pays will be interest?

9. The monthly premium on a three-year insurance policy is $64.00 payable in advance. What is the cash value of the policy if money is worth 4.8% compounded monthly?

10. The monthly rent payment on office space is $535.00 payable in advance. What yearly payment made in advance would satisfy the lease if interest is 6.6% compounded monthly?

11. Elspeth McNab bought a boat valued at $12 500.00 on the installment plan requiring equal monthly payments for four years. If the first payment is due on the date of purchase and interest is 7.5% compounded monthly, what is the size of the monthly payment?

12. How much does a depositor have to save at the beginning of every three months for nine years to accumulate $35 000.00 if interest is 8% compounded quarterly?

13. Payments on a seven-year lease valued at $12 200.00 are to be made at the beginning of each month during the last five years of the lease. If interest is 9% compounded monthly, what is the size of the monthly payments?

14. Julia deposited $1500.00 in an RRSP at the beginning of every six months for twenty years. The money earned interest at 6.25% compounded semi-annually. After twenty years, she converted the RRSP into an RRIF from which she wants to withdraw equal amounts at the beginning of each month for fifteen years. If interest on the RRIF is 6.6% compounded monthly, how much does she receive each month?

15. Mr. Clark wants to receive payments of $900.00 at the beginning of every three months for twenty years starting on the date of his retirement. If he retires in twenty-five years, how much must he deposit in an account at the beginning of every three months if interest on the account is 5.25% compounded quarterly?

16. Quarterly payments of $1445.00 are to be made at the beginning of every three months on a lease valued at $25 000.00. What should the term of the lease be if money is worth 8% compounded quarterly?

17. Tom is saving $600.00 at the beginning of each month. How soon can he retire if he wants to have a retirement fund of $120 000.00 and interest is 5.4% compounded monthly?

18. If you save $75.00 at the beginning of every month for ten years, for how long can you withdraw $260.00 at the beginning of each month starting ten years from now, assuming that interest is 6% compounded monthly?

19. Ali deposits \$450.00 at the beginning of every three months. He wants to build up his account so that he can withdraw \$1000.00 every three months starting three months after the last deposit. If he wants to make the withdrawals for fifteen years and interest is 10% compounded quarterly, for how long must Ali make the quarterly deposits?

20. What nominal annual rate of interest was paid if contributions of \$250.00 made into an RRSP at the beginning of every three months amounted to \$14 559.00 after ten years?

21. A vacation property valued at \$25 000.00 was bought for fifteen payments of \$2200.00 due at the beginning of every six months. What nominal annual rate of interest was charged?

22. An insurance policy provides a benefit of \$250 000.00 twenty years from now. Alternatively, the policy pays \$4220.00 at the beginning of each year for twenty years. What is the effective annual rate of interest paid?

7.2 General Annuities Due

A. Future value of a general annuity due

As with a simple annuity due, the future value of a **general annuity due** is greater than the future value of the corresponding ordinary general annuity by the interest on it for one payment period.

Since the interest on a general annuity for one payment period is $(1 + i)^c$, or $(1 + p)$,

PRESENT VALUE OF A GENERAL ANNUITY DUE $= (1 + p) \times$ PRESENT VALUE OF THE CORRESPONDING ORDINARY GENERAL ANNUITY

Thus, for the future value of a general annuity due use Formula 7.3:

$$S_{nc}(\text{due}) = R(1 + p)\left[\frac{(1 + p)^n - 1}{p}\right]$$

restated as

$$FV_{nc}(\text{due}) = PMT(1 + p)\left[\frac{(1 + p)^n - 1}{p}\right]$$

where $p = (1 + i)^c - 1$

— Formula 7.3

EXAMPLE 7.2A What is the accumulated value after five years of payments of \$20 000 made at the beginning of each year if interest is 7% compounded quarterly?

SOLUTION

PMT = 20 000.00; $n = 5$; $c = 4$; P/Y = 1; C/Y = 4; I/Y = 7;

$i = \frac{7\%}{4} = 1.75\% = 0.0175$

The effective annual rate of interest

$p = 1.0175^4 - 1 = 1.071859 - 1 = 0.071859 = 7.1859\%$

$$\begin{aligned} FV_{nc}(\text{due}) &= 20\,000.00(1.071859)\left(\frac{1.071859^5 - 1}{0.071859}\right) \\ &= 20\,000.00(1.071859)(5.772109) \\ &= 20\,000.00(6.186888) \\ &= \$123\,737.75 \end{aligned}$$

substituting in Formula 7.3

Programmed Solution

("BGN" mode) (Set P/Y 5 1; C/Y = 4) 7 [I/Y] 0 [PV]

20 000 [±] [PMT] 5 [N] [CPT] [FV] [123 737.7535]

The accumulated value after five years is $123 737.75.

B. Present value of a general annuity due

For a general annuity due, the present value is greater than the present value of the corresponding ordinary general annuity by the interest on it for one payment period.

THE PRESENT VALUE OF A GENERAL ANNUITY DUE $= (1 + p) \times$ THE PRESENT VALUE OF THE CORRESPONDING ORDINARY GENERAL ANNUITY

Thus, for the present value of a general annuity due use Formula 7.4. Using the effective rate of interest per payment period,

$$A_{nc}(\text{due}) = R(1 + p)\left[\frac{1 - (1 + p)^{-n}}{p}\right]$$

restated as

$$PV_{nc}(\text{due}) = PMT(1 + p)\left[\frac{1 - (1 + p)^{-n}}{p}\right]$$

where $p = (1 + i)^c - 1$

Formula 7.4

EXAMPLE 7.2B

A three-year lease requires payments of $1600.00 at the beginning of every three months. If money is worth 9.0% compounded monthly, what is the cash value of the lease?

SOLUTION

PMT = 1600.00; $n = 3(4) = 12$; P/Y = 4; C/Y = 12; $c = \frac{12}{4} = 3$; I/Y = 9;

$i = \frac{9.0\%}{12} = 0.75\% = 0.0075$

The effective quarterly rate of interest

$p = 1.0075^3 - 1 = 1.022669 - 1 = 0.022669 = 2.26692\%$

$$PV_{nc}(\text{due}) = 1600.00(1.022669)\left(\frac{1 - 1.022669^{-12}}{0.022669}\right) \text{ — substituting in Formula 7.4}$$

$$= 1600.00(1.022669)(10.404043)$$
$$= 1600.00(10.639894)$$
$$= \$17\,023.83$$

Programmed Solution

("BGN" mode) (Set P/Y = 4; C/Y = 12) 9 [I/Y] 0 [FV]

1600.00 [±] [PMT] 12 [N] [CPT] [PV] [17 023.83049]

The cash value of the lease is $17 023.83.

C. Finding the periodic payment PMT of a general annuity due

If the future value of an annuity FV_{nc}(due), the number of conversion periods n, and the conversion rate i are known, you can find the periodic payment PMT by substituting the given values in the future value formula.

$$FV_{nc}(\text{due}) = PMT(1 + p)\left[\frac{(1 + p)^n - 1}{p}\right]$$

where $p = (1 + i)^c - 1$ — Formula 7.3

If the present value of an annuity PV_{nc}(due), the number of conversion periods n, and the conversion rate i are known, you can find the periodic payment PMT by substituting the given values in the present value formula.

$$PV_{nc}(\text{due}) = PMT(1 + p)\left[\frac{1 - (1 + p)^{-n}}{p}\right]$$

where $p = (1 + i)^c - 1$ — Formula 7.4

When using a scientific calculator, you can find PMT by first rearranging the terms of the appropriate formulas above. Then substitute the three known values (FV_{nc}(due), PV_{nc}(due), n, and i) into the rearranged formula and solve for PMT.

When using a preprogrammed financial calculator, you can find PMT by entering the five known values (FV_{nc}(due) or PV_{nc}(due), N, I/Y, P/Y, and C/Y) and pressing [CPT] [PMT].

EXAMPLE 7.2C

What semi-annual payment must be made into a fund at the beginning of every six months to accumulate to $9600.00 in ten years at 7% compounded annually?

SOLUTION

$FV_n(\text{due}) = 9600.00$; $P/Y = 2$; $C/Y = 1$; $c = 1/2$; $I/Y = 7$; $i = 7\%/1 = 0.07$;
$p = 1.07^{1/2} - 1 = 0.034408$

$$9600.00 = PMT(1.034408)\frac{[1.034408^{20} - 1]}{0.034408}$$ — substituting in Formula 7.3

$$9600.00 = PMT(1.034408)(28.108280)$$
$$9600.00 = PMT(29.075430)$$
$$PMT = \frac{9600.00}{29.075430}$$
$$PMT = \$330.18$$

Programmed Solution

("BGN" mode) (Set P/Y = 2; C/Y = 1) 0 [PV] 9600 [FV]
7 [I/Y] 20 [N] [CPT] [PMT] [-330.1755229]

The semi-annual payment is $330.18.

EXAMPLE 7.2D

What deposit made at the beginning of each month will accumulate to $18 000.00 at 5% compounded quarterly at the end of eight years?

SOLUTION

$FV_{nc}(\text{due}) = 18\,000.00$; $n = 8(12) = 96$; $P/Y = 12$; $C/Y = 4$; $c = \frac{4}{12} = \frac{1}{3}$
$I/Y = 5$; $i = \frac{5\%}{4} = 1.25\% = 0.0125$

The effective monthly rate of interest

$$p = 1.0125^{\frac{1}{3}} - 1 = 1.004149 - 1 = 0.004149 = 0.4149\%$$

$$18\,000.00 = PMT(1.004149)\left(\frac{1.004149^{96} - 1}{0.004149}\right)$$ — substituting in Formula 7.3

$$18\,000.00 = PMT(1.004149)(117.638106)$$
$$18\,000.00 = PMT(118.126236)$$
$$PMT = \frac{18\,00.00}{118.126236}$$
$$PMT = \$152.38$$

Programmed Solution

("BGN" mode) (Set P/Y = 12; C/Y = 4) 5 [I/Y] 0 [PV]
18 000 [FV] 96 [N] [CPT] [PMT] [-152.3793585]

The monthly deposit is $152.38.

EXAMPLE 7.2E

What monthly payment must be made at the beginning of each month on a five-year lease valued at \$100 000.00 if interest is 10% compounded semi-annually?

SOLUTION

$PV_{nc}(\text{due}) = 100\,000.00;\; n = 5(12) = 60;\; P/Y = 12;\; C/Y = 2;\; c = \frac{2}{12} = \frac{1}{6};$

$I/Y = 10;\; i = \frac{10\%}{2} = 5\% = 0.05$

The effective monthly rate of interest

$p = 1.05^{\frac{1}{6}} - 1 = 1.008165 - 1 = 0.008165 = 0.8165\%$

$$100\,000.00 = PMT(1.008165)\left(\frac{1 - 1.008165^{-60}}{0.008165}\right)$$ — substituting in Formula 7.4

$$100\,000.00 = PMT(1.008165)(47.286470)$$
$$100\,000.00 = PMT(47.672557)$$
$$PMT = \frac{100\,000.00}{47.672557}$$
$$PMT = \$2097.64$$

Programmed Solution

("BGN" mode) (Set P/Y = 12; C/Y = 2) 10 [I/Y] 0 [FV]

100 000 [±] [PV] 60 [N] [CPT] [PMT] [2097.642904]

The monthly payment due at the beginning of each month is \$2097.64.

D. Finding the term *n* of a general annuity due

If the future value of an annuity FV_{nc} (due), the periodic payment PMT, and the conversion rate i are known, you can find the term of the annuity n by substituting the given values in the future value Formula 7.3.

$$FV_{nc}(\text{due}) = PMT(1 + p)\left[\frac{(1 + p)^n - 1}{p}\right]$$

where $p = (1 + i)^c - 1$ — Formula 7.3

If the present value PV_{nc} (due), the periodic payment PMT, and the conversion rate i are known, you can find the term of the annuity n by substituting the given values in the present value Formula 7.4.

$$PV_{nc}(\text{due}) = PMT(1 + p)\left[\frac{1 - (1 + p)^{-n}}{p}\right]$$

where $p = (1 + i)^c - 1$ — Formula 7.4

When using a scientific calculator, you can find n by first rearranging the terms of the appropriate formula. Then substitute the three known values (FV_{nc}(due) or PV_{nc}(due), PMT, and i) into the rearranged formula and solve for n.

When using a preprogrammed financial calculator, you can find n by entering the five known values (FV_{nc} (due) or PV_{nc} (due), PMT, I/Y, P/Y, and C/Y) and pressing CPT N.

EXAMPLE 7.2F

Ted Davis wants to accumulate \$140 000.00 in an RRSP by making annual contributions of \$5500.00 at the beginning of each year. If interest on the RRSP is 11% compounded quarterly, for how long will Ted have to make contributions?

SOLUTION

$FV_{nc}(\text{due}) = 140\,000.00$; PMT = 5500.00; P/Y = 1; C/Y = 4; $c = 4$;

I/Y = 11; $i = \frac{11\%}{4} = 2.75\% = 0.0275$

The effective annual rate of interest

$p = 1.0275^4 - 1 = 1.114621 - 1 = 0.114621 = 11.4621\%$

$$140\,000.00 = 5500.00(1.114621)\left(\frac{1.114621^n - 1}{0.114621}\right) \quad \text{using Formula 7.3}$$

$$140\,000.00 = 53\,484.101(1.114621^n - 1)$$

$$2.617600 = 1.114621^n - 1$$

$$1.114621^n = 3.617600$$

$$n \ln 1.114621 = \ln 3.617600$$

$$n(0.108515) = 1.285811$$

$$n = \frac{1.285811}{0.108515}$$

$$n = 11.849186$$

$$n = 12 \text{ years (approximately)}$$

Programmed Solution

("BGN" mode) (Set P/Y = 1; C/Y = 4) 11 I/Y 0 PV

5500 ± PMT 140 000 FV CPT N 11.84918768

Ted will have to contribute for about twelve years.

EXAMPLE 7.2G

Ted Davis, having reached his goal of a \$140 000.00 balance in his RRSP, immediately converts it into an RRIF and withdraws from it \$1650.00 at the beginning of each month. If interest continues at 5.75% compounded quarterly, for how long can he make withdrawals?

SOLUTION

$PV_{nc}(\text{due}) = 140\,000.00$; PMT = 1650.00; P/Y = 12; C/Y = 4; $c = \frac{4}{12} = \frac{1}{3}$;

I/Y = 5.75; $i = \frac{5.75\%}{4} = 1.4375\% = 0.014375$

The effective monthly rate of interest

$p = 1.014375^{\frac{1}{3}} - 1 = 1.004769 - 1 = 0.004769 = 0.4769\%$

$$140\,000.00 = 1650.00(1.004769)\left(\frac{1 - 1.004769^{-n}}{0.004769}\right) \text{ —— using Formula 7.4}$$

$$140\,000.00 = 347\,642.59(1 - 1.004769^{-n})$$

$$0.4027124 = 1 - 1.004769^{-n}$$

$$1.004769^{-n} = 0.597288$$

$$-n \ln 1.004769 = \ln 0.597288$$

$$-n(0.004758) = -0.515357$$

$$n = 108.32388$$

$$n = 109 \text{ months (approximately)}$$

Programmed Solution

("BGN" mode) (Set P/Y = 12; C/Y = 4) 5.75 [I/Y] 0 [FV]

140 000 [±] [PV] 1650 [PMT] [CPT] [N] [108.3238824]

Ted will be able to make withdrawals for nine years and one month.

E. Finding the rate of interest of a general annuity due

When the future value or present value, the periodic payment PMT, and the term n of a general annuity due are known, you can find the nominal interest rate by entering the given values into a preprogrammed calculator. Retrieve the value of I/Y by being in "BGN" mode and pressing [CPT] [I/Y].

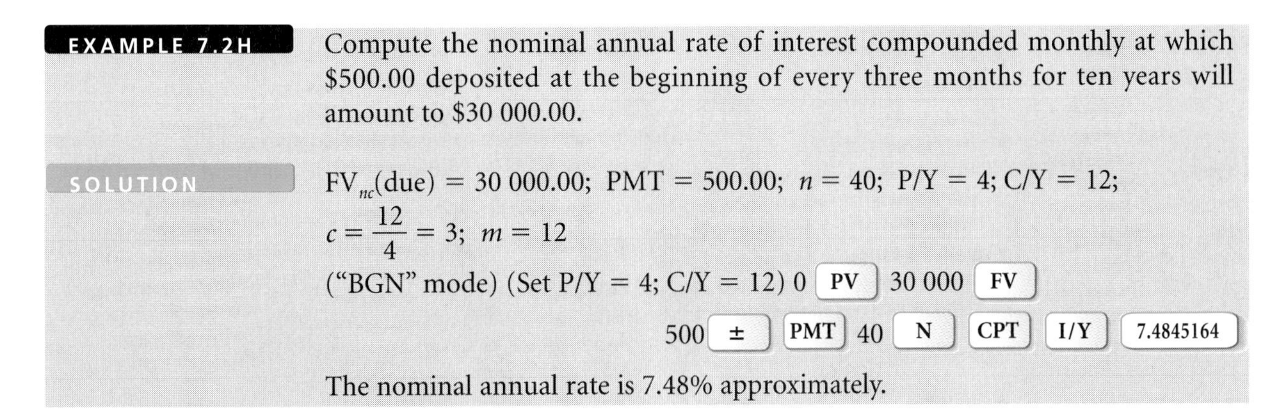

EXAMPLE 7.2H

Compute the nominal annual rate of interest compounded monthly at which \$500.00 deposited at the beginning of every three months for ten years will amount to \$30 000.00.

SOLUTION

$FV_{nc}(\text{due}) = 30\,000.00$; $PMT = 500.00$; $n = 40$; $P/Y = 4$; $C/Y = 12$;

$c = \frac{12}{4} = 3$; $m = 12$

("BGN" mode) (Set P/Y = 4; C/Y = 12) 0 [PV] 30 000 [FV]

500 [±] [PMT] 40 [N] [CPT] [I/Y] [7.4845164]

The nominal annual rate is 7.48% approximately.

EXERCISE 7.2

If you choose, you can use Excel's ***Present Value (PV)*** function or ***Future Value (FV)*** function to answer the questions indicated below. Refer to **PV** and **FV** on the Spreadsheet Template Disk to learn how to use these Excel functions.

For each of the following four annuities due, determine the unknown value represented by the question mark.

	Future Value FV_{nc}(due)	Present Value PV_{nc}(due)	Periodic Payment PMT	Payment Interval	Term	Nominal Rate of Interest	Conversion Period
1.	?		$1500.00	6 months	10 years	5%	quarterly
2.	?		$175.00	1 month	7 years	7%	semi-annually
3.		?	$650.00	3 months	6 years	12%	monthly
4.		?	$93.00	1 month	4 years	4%	quarterly

Find the periodic payment for each of the following four annuities due.

	Future Value	Present Value	Payment Period	Term	Interest Rate	Conversion Period
1.	$16 500.00		1 year	10 years	4%	quarterly
2.	$9 200.00		3 months	5 years	5%	semi-annually
3.		$10 000.00	3 months	3 years	6%	monthly
4.		$24 300.00	1 month	20 years	9%	semi-annually

Find the length of the term for each of the following four annuities due.

	Future Value	Present Value	Periodic Payment	Payment Period	Interest Rate	Conversion Period
1.	$32 000.00		$450.00	6 months	7.5%	monthly
2.	$7 500.00		$150.00	3 months	11%	annually
3.		$12 500.00	$860.00	3 months	9%	monthly
4.		$45 000.00	$540.00	1 month	4%	semi-annually

For each of the following four annuities due, determine the nominal annual rate of interest.

	Future Value	Present Value	Periodic Payment	Payment Interval	Term	Conversion Period
1.	$6 400.00		$200.00	6 months	9 years	monthly
2.	$25 000.00		$790.00	1 year	15 years	quarterly
3.		$7 500.00	$420.00	3 months	5 years	monthly
4.		$60 000.00	$450.00	1 month	25 years	semi-annually

Answer each of the following questions.

1. Bomac Steel sets aside $5000.00 at the beginning of every six months in a fund to replace erecting equipment. If interest is 6% compounded quarterly, how much will be in the fund after five years?
2. Jamie Dean contributes $125.00 at the beginning of each month into an RRSP paying interest at 6.5% compounded semi-annually. What will be the accumulated balance in the RRSP at the end of 25 years?
3. What is the cash value of a lease requiring payments of $750.00 at the beginning of each month for three years if interest is 8% compounded quarterly?

4. Gerald Carter and Marysia Wokawski bought a property by making semi-annual payments of $2500.00 for seven years. If the first payment is due on the date of purchase and interest is 9% compounded quarterly, what is the purchase price of the property?
5. How much would you have to pay into an account at the beginning of every six months to accumulate $10 000.00 in eight years if interest is 7% compounded quarterly?
6. Teachers' Credit Union entered a lease contract valued at $5400.00. The contract provides for payments at the beginning of each month for three years. If interest is 5.5% compounded quarterly, what is the size of the monthly payment?
7. Sarah Ling has saved $85 000.00. If she decides to withdraw $3000.00 at the beginning of every three months and interest is 6.125% compounded annually, for how long can she make withdrawals?
8. For how long must contributions of $1600.00 be made at the beginning of each year to accumulate to $96 000.00 at 10% compounded quarterly?
9. What is the nominal annual rate of interest compounded annually on a lease valued at $21 600.00 if payments of $680.00 are made at the beginning of each month for three years?
10. An insurance policy provides for a lump-sum benefit of $50 000.00 fifteen years from now. Alternatively, payments of $1700.00 may be received at the beginning of each of the next fifteen years. What is the effective annual rate of interest if interest is compounded quarterly?

BUSINESS MATH NEWS BOX

Tax-Smart Investing

In the investment world, things aren't always what they seem. For example, because you pay income tax, earning interest at 5% doesn't mean that you'll keep all of the $5 you make on every $100 you invest. In fact, if your taxable income is about $40 000, only $3 goes into your pocket. The other $2 goes to the government.

So how do you improve your returns on investments held outside your RRSP? By keeping more, after-tax, of what your investments earn.

Not all investment income is taxed in the same manner. Interest income is fully taxable. Capital gains and Canadian dividends, however, are taxed more lightly.

Capital gains generally arise when you sell a capital asset for more than your purchase price. Only half of the gain is subject to tax. So, if your marginal tax rate is 40%, the effective rate on capital gains is half of that, or 20%.

Dividends from Canadian corporations also get special treatment. If your marginal tax rate is 40%, Canadian dividends are effectively taxed at about 25% as a result of the dividend tax credit.

The table on the next page illustrates the results of these different tax treatments.

What $500 of investment income is really worth

For the average Canadian taxpayer,* capital gains income from Canadian corporations is the most tax-effective way to invest. Here's where $500 of investment income goes:

Investment Income	Interest	Capital Gains	Dividends
To Canada Customs and Revenue Agency	$198.90	$99.45	$121.10
In your pocket	$301.10	$400.55	$378.90

*Assumes federal tax bracket of 26% and provincial tax rate of 50%.

QUESTIONS

1. (a) Using the numbers given in the table, calculate (to two decimal places) the percent of the $500.00 investment income that the investor receives after taxes if the investment income is
   (i) interest;
   (ii) capital gains;
   (iii) dividends.

   (b) How much do the percents you calculated above differ from the rates stated in the text of the article?

2. Suppose you had to make a choice. You could receive $1500.00 in dividends every six months for five years, starting today. Or you could receive $1500.00 in interest every six months for five years, starting today. Assume that income tax is deducted before you receive your investment income. Assume the tax rates you have calculated from the table will be the same for five years and money will be worth 5% compounded semi-annually for the next five years. Using five years from now as the focal date, how much more after-tax investment income would you have if you chose the dividends?

3. Suppose you had a different choice. You could receive $1500.00 in capital gains every six months for five years, starting today. Or, you could receive the interest income described in Question 2. How much more after-tax investment income would you have if you chose the capital gains rather than the interest income? (Use the same assumptions as in Question 2.)

Sources: Investor's Group, used with permission. New rates from the February 27, 2003, Federal Budget and Canada Customs and Revenue Agency.

7.3 Deferred Annuities

A. Basic concepts and computation of an ordinary simple deferred annuity

A **deferred annuity** is one in which the first payment is made at a time *later* than the end of the first payment interval. The time period from the time referred to as "now" to the starting point of the term of the annuity is called the **period of deferment**. The number of compounding periods in the period of deferment is designated by the letter symbol *d*. The future value of a deferred ordinary simple annuity (designated by the symbol FV_n(defer)) is the accumulated value of the periodic payments at the end of the term of the annuity.

EXAMPLE 7.3A

Payments of $500.00 are due at the end of each year for ten years. If the annuity is deferred for four years and interest is 6% compounded annually, determine the future value of the deferred annuity.

SOLUTION

FIGURE 7.3 **Graphical Representation of Method and Data**

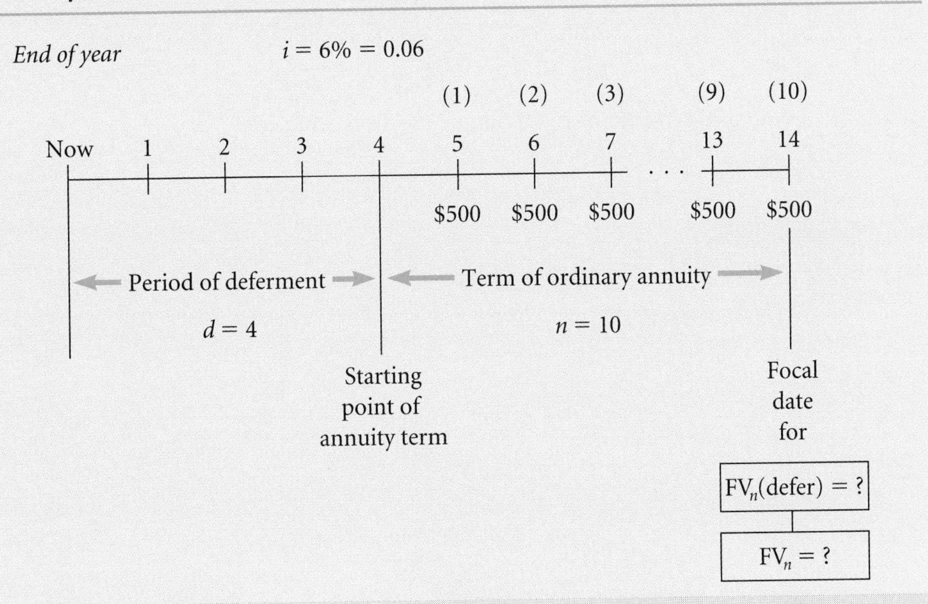

PMT = 500.00; P/Y = 1; C/Y = 1; I/Y = 6; $i = 6\% = 0.06$; $n = 10$; $d = 4$

$$FV_n(\text{defer}) = FV_n = 500.00\left(\frac{1.06^{10} - 1}{0.06}\right) \quad \text{——— using Formula 5.1}$$

$$= 500.00(13.180795)$$

$$= \$6590.40$$

Programmed Solution

("END" mode) (Set P/Y = 1; C/Y = 1) 0 [PV] 500 [±] [PMT]

10 [N] 6 [I/Y] [CPT] [FV] [6590.397471]

Note: The period of deferment does *not* affect the solution to the problem of finding the future value of a deferred annuity: $FV_n(\text{defer}) = FV_n$. Therefore, the problem of finding the future value of a deferred annuity is identical to the problem of finding the future value of an annuity. No further consideration is given in this text to the problem of finding the future value of a deferred annuity.

The present value of a deferred annuity is the discounted value of the periodic payment at the beginning of the period of deferment.

The present value of a deferred simple annuity is designated by the symbol PV_n (defer).

The present value of a deferred general annuity is designated by the symbol PV_{nc}(defer).

EXAMPLE 7.3B

Payments of $500.00 are due at the end of each year for ten years. If the annuity is deferred for four years and interest is 6% compounded annually, determine the present value of the deferred annuity.

SOLUTION

FIGURE 7.4 **Graphical Representation of Method and Data**

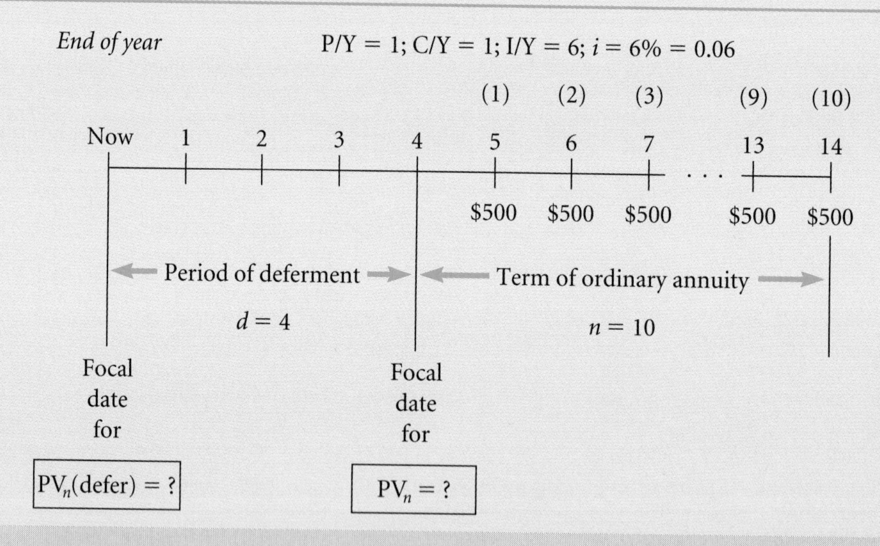

The problem of finding the present value of a deferred annuity can be divided into two smaller problems. We have used this approach in solving Example 5.3F in Chapter 5. We use it again in this problem to find the present value of the deferred annuity.

First, find the present value of the ordinary annuity (focal date at the beginning of the term of the annuity).

$$PV_n = 500.00\left(\frac{1 - 1.06^{-10}}{0.06}\right) \quad \text{using Formula 5.2}$$
$$= 500.00(7.360087)$$
$$= \$3680.04$$

Second, find the present value of PV_n at the focal date "now."

$$PV_n(\text{defer}) = PV = 3680.04(1.06^{-4}) \quad \text{using Formula 3.1C}$$
$$= 3680.04(0.792094)$$
$$= \$2914.94$$

Programmed Solution

(Set P/Y = 1; C/Y = 1) 0 [FV] 500 [±] [PMT]

6 [I/Y] 10 [N] [CPT] [PV] [-3680.043526]

0 [PMT] 3680.043526 [FV] 6 [I/Y] 4 [N] [CPT] [PV] [-2914.939157]

EXAMPLE 7.3C

Mr. Peric wants to receive payments of $800.00 at the end of each month for ten years after his retirement. If he retires in seven years, how much must Mr. Peric invest now if interest on the investments is 6% compounded monthly?

SOLUTION

The monthly payments after retirement form an ordinary annuity.

$\text{PMT} = 800.00;\ \text{P/Y} = 12;\ \text{C/Y} = 12;\ \text{I/Y} = 6;\ i = \frac{6\%}{12} = 0.5\% = 0.005;$

$n = 10(12) = 120$

The period of deferment is 7 years, so $d = 7(12) = 84$.
The amount to be invested now is the present value of the deferred annuity.

$$\begin{aligned} PV_n &= 800.00\left(\frac{1 - 1.005^{-120}}{0.005}\right) \\ &= 800.00(90.073453) \\ &= 72\,058.76 \end{aligned}$$

$$\begin{aligned} PV_n(\text{defer}) &= 72\,058.76(1.005)^{-84} \\ &= 72\,058.76(0.657735) \\ &= 47\,395.55 \end{aligned}$$

Programmed Solution

(Set P/Y = 12; C/Y = 12) 0 [FV] 800 [±] [PMT]

6 [I/Y] 120 [N] [CPT] [PV] [72 058.76266]

72 058.76266 [FV] 0 [PMT] 6 [I/Y] 84 [N] [CPT] [PV] [-47 395.55497]

Mr. Peric must invest $47 395.55.

You can find the periodic payment PMT for deferred annuities by first determining the future value of the known present value at the end of the period of deferment. Then substitute in the appropriate annuity formula.

EXAMPLE 7.3D

Find the size of the payment required at the end of every three months to repay a five-year loan of $25 000.00 if the payments are deferred for two years and interest is 6% compounded quarterly.

SOLUTION

The payments form a deferred ordinary annuity.

$PV_n(\text{defer}) = 25\,000.00;\ \text{P/Y} = 4;\ \text{C/Y} = 4;\ \text{I/Y} = 6;\ i = \frac{6\%}{4} = 1.5\% = 0.015;$
$n = 5(4) = 20;\ d = 2(4) = 8$

Value of \$25 000.00 at the end of the period of deferment,

$$FV = 25\,000.00(1.015)^8 \quad \text{——— using Formula 3.1A}$$
$$= 25\,000.00(1.126493)$$
$$= 28\,162.32$$

$$28\,162.32 = PMT\left(\frac{1 - 1.015^{-20}}{0.015}\right) \quad \text{——— substituting in Formula 5.2}$$
$$28\,162.32 = 17.168639PMT$$
$$PMT = \frac{28\,162.32}{17.168639}$$
$$PMT = 1640.34$$

Programmed Solution

(Set P/Y = 4; C/Y = 4) 2nd (CLR TVM) 25 000 ± PV
6 I/Y 8 N CPT FV 28 162.31466

28 162.31466 ± PV 0 FV 6 I/Y 20 N CPT PMT 1640.334742

The size of the required payment is \$1640.33.

You can find the term for deferred ordinary annuities by the same approach used to determine the periodic payment PMT.

EXAMPLE 7.3E

For how long can you pay \$500.00 at the end of each month out of a fund of \$10 000.00, deposited today at 10.5% compounded monthly, if the payments are deferred for nine years?

SOLUTION

The payments form a deferred ordinary annuity.

PV_n(defer) = 10 000.00; PMT = 500.00; P/Y = 12; C/Y = 12; I/Y = 10.5;

$i = \frac{10.5\%}{12} = 0.875\% = 0.00875$; $d = 9(12) = 108$

The value of the \$10 000 at the end of the period of deferment,

$$FV = 10\,000.00(1.00875)^{108}$$
$$= 10\,000.00(2.562260)$$
$$= 25\,622.60$$

$$25\,622.60 = 500.00\left(\frac{1 - 1.00875^{-n}}{0.00875}\right)$$
$$25\,622.60 = 57\,142.857(1 - 1.00875^{-n})$$
$$\frac{25\,622.60}{57\,142.857} = 1 - 1.00875^{-n}$$
$$0.448400 = 1 - 1.00875^{-n}$$
$$1.00875^{-n} = 0.551605$$
$$-n \ln 1.00875 = \ln 0.551605$$
$$-n(0.008712) = -0.594924$$

$$n = \frac{0.5949239}{0.0087119}$$
$$n = 68.288334$$
$$n = 69 \text{ (months)}$$

Programmed Solution

(Set P/Y 5 12; C/Y 5 12) 0 PMT 10 000 ± PV 10.5 I/Y 108 N CPT FV 25 622.59753

± 25 622.59753 PV 0 FV 10.5 I/Y 500 PMT CPT N 68.2883325

Payments can be made for five years and nine months.

B. Computation of an ordinary general deferred annuity

The same principles apply to an ordinary general deferred annuity as apply to an ordinary simple deferred annuity.

EXAMPLE 7.3F

Payments of $1000.00 are due at the end of each year for five years. If the payments are deferred for three years and interest is 10% compounded quarterly, what is the present value of the deferred payments?

SOLUTION

STEP 1

Find the present value of the ordinary general annuity.

$\text{PMT} = 1000.00;\ n = 5;\ \text{P/Y} = 1;\ \text{C/Y} = 4;\ c = 4;\ \text{I/Y} = 10;\ i = \frac{10\%}{4} = 2.5\% = 0.025$

The effective annual rate of interest

$p = 1.025^4 - 1 = 1.103813 - 1 = 0.103813 = 10.3813\%$

$$\text{PV}_{nc} = 1000.00\left(\frac{1 - 1.103813^{-5}}{0.103813}\right) \quad \text{——— substituting in Formula 6.3}$$
$$= 1000.00(3.754149)$$
$$= \$3754.15$$

Programmed Solution

(Set P/Y = 1; C/Y = 4) 10 I/Y 0 FV 1000 ± PMT 5 N CPT PV 3754.148977

STEP 2

Find the present value of PV_{nc} at the beginning of the period of deferment.

$\text{FV} = 3754.15$ is the present value of the general annuity PV_{nc};
$d = 3$ is the number of deferred payment intervals;
$p = 10.38129\%$ is the effective rate of interest per payment interval.

$$PV_{nc}(\text{defer}) = PV = 3754.15(1.103813^{-3})$$ — substituting in Formula 3.1C

$$= 3754.15(0.743556)$$

$$= \$2791.42$$

Programmed Solution

3754.15 [FV] 10 [I/Y] 0 [PMT] 3 [N] [CPT] [PV] [-2791.420326]

The present value of the deferred payments is $2791.42.

EXAMPLE 7.3G

Mr. Kovacs deposited a retirement bonus of $31 500.00 in an income averaging annuity paying $375.00 at the end of each month. If payments are deferred for nine months and interest is 6% compounded quarterly, for what period of time will Mr. Kovacs receive annuity payments?

SOLUTION

$PV_{nc}(\text{defer}) = 31\,500.00$; $PMT = 375.00$; $d = 9$; $P/Y = 12$; $C/Y = 4$;

$c = \frac{4}{12} = \frac{1}{3}$; $I/Y = 6$; $i = \frac{6\%}{4} = 1.5\% = 0.015$

The effective monthly rate of interest

$$p = 1.015^{\frac{1}{3}} - 1 = 1.0049752 - 1 = 0.0049752 = 0.49752\%.$$

First find the accumulated value at the end of the period of deferment.

$$FV = 31\,500.00(1.004975)^9$$

$$= 31\,500.00(1.045678)$$

$$= 32\,938.87$$

Programmed Solution

(Set P/Y = 12; C/Y 5 4) 6 [I/Y] 0 [PMT]

31 500 [±] [PV] 9 [N] [CPT] [FV] [32 938.86881]

Then find the number of payments for an ordinary annuity with a present value of $32 938.87.

$$32\,938.87 = 375.00\left(\frac{1 - 1.004975^{-n}}{0.004975}\right)$$

$$32\,938.87 = 75\,373.854(1 - 1.004975^{-n})$$

$$0.4370066 = 1 - 1.004975^{-n}$$

$$1.00497^{-n} = 0.562993$$

$$-n \ln 1.00497 = \ln 0.562993$$

$$-n(0.004963) = -0.574487$$

$$n = \frac{0.574487}{0.004963}$$

$$n = 115.75639$$

$$n = 116 \text{ months}$$

Programmed Solution

[±] 32 938.869 [PV] 0 [FV] 6 [I/Y] 375 [PMT] [CPT] [N] [115.7572503]

Mr. Kovacs will receive payments for nine years and eight months.

EXAMPLE 7.3H

A contract that has a cash value of $36 000.00 requires payments at the end of every three months for six years. If the payments are deferred for three years and interest is 9% compounded semi-annually, what is the size of the quarterly payments?

SOLUTION

$PV_{nc}(\text{defer}) = 36\ 000.00$; $n = 6(4) = 24$; $d = 3(4) = 12$; P/Y = 4; C/Y = 2;

$c = \frac{2}{4} = 0.5$; I/Y = 9; $i = \frac{9\%}{2} = 4.5\% = 0.045$

The effective quarterly rate of interest

$p = 1.045^{0.5} - 1 = 1.022252 - 1 = 0.022252 = 2.2252\%$

First determine the accumulated value of the deposit at the end of the period of deferment.

$$
\begin{aligned}
FV &= 36\ 000.00(1.022252)^{12} \quad \text{——— substituting in Formula 3.1A} \\
&= 36\ 000.00(1.302260) \\
&= 46\ 881.36
\end{aligned}
$$

Programmed Solution

(Set P/Y = 4; C/Y = 2) 9 [I/Y] 0 [PMT] 36 000 [±] [PV]

12 [N] [CPT] [FV] [46 881.36449]

Now determine the periodic payment for the ordinary general annuity whose present value is $46 881.36.

$$
46\ 881.36 = PMT\left(\frac{1 - 1.022252^{-24}}{0.022252}\right) \quad \text{——— substituting in Formula 6.3}
$$

$$
46\ 881.36 = PMT(18.440075)
$$

$$
PMT = \frac{46\ 881.36}{18.440075}
$$

$$
PMT = \$2542.36
$$

Programmed Solution

46 881.36 [±] [PV] 0 [FV] 9 [I/Y] 24 [N] [CPT] [PMT] [2542.363174]

The required quarterly payment is $2542.36.

C. Computation of a simple deferred annuity due

The same principles apply to a simple deferred annuity due as apply to a simple annuity due.

EXAMPLE 7.3I

Mei Willis would like to receive annuity payments of \$2000.00 at the beginning of each quarter for seven years. The annuity is to start five years from now and interest is 5% compounded quarterly.

(i) How much must Mei invest today?
(ii) How much will Mei receive from the annuity?
(iii) How much of what she receives will be interest?

SOLUTION

(i) First, find the present value of the annuity due (the focal point is five years from now).

PMT = 2000.00; P/Y = 4; C/Y = 4; I/Y = 5; $i = \frac{5\%}{4} = 1.25\% = 0.0125$;
$n = 7(4) = 28$

$$\begin{aligned} PV_n(\text{due}) &= 2000.00(1.0125)\left[\frac{1 - 1.0125^{-28}}{0.0125}\right] \\ &= 2000.00(1.0125)(23.502518) \\ &= 47\,005.035(1.0125) \\ &= \$47\,592.60 \end{aligned}$$

Second, determine the present value of \$47 592.60 (the focal point is "now").

FV = 47 592.60; $i = 1.25\%$; $n = 5(4) = 20$

$$\begin{aligned} PV &= 47\,592.60(1.0125)^{-20} \\ &= 47\,592.60(0.780009) \\ &= \$37\,122.63 \end{aligned}$$

Programmed Solution

("BGN" mode) (Set P/Y = 4; C/Y = 4) 0 [FV] 2000 [±] [PMT]
5 [I/Y] 28 [N] [CPT] [PV] [47 592.5985]

47 592.5985 [FV] 0 [PMT] 5 [I/Y] 20 [N] [CPT] [PV] [-37 122.63367]

Mei will have to invest \$37 122.63.

(ii) Mei will receive 28(2000.00) = \$56 000.00.

(iii) Interest will be 56 000.00 − 37 122.63 = \$18 877.37.

EXAMPLE 7.3J

$2000.00 is to be withdrawn from a fund at the beginning of every three months for twelve years starting ten years from now. If interest is 10% compounded quarterly, what must be the balance in the fund today to permit the withdrawals?

SOLUTION

The withdrawals form an annuity due.

PMT = 2000.00; P/Y = 4; C/Y = 4; I/Y = 10; $i = \frac{10\%}{4} = 2.5\% = 0.025$;
$n = 12(4) = 48$

The period of deferment is 10 years, so $d = (10)(4) = 40$.

$$\begin{aligned} PV_n(\text{due}) &= 2000.00(1.025)\left[\frac{1-(1.025)^{-48}}{0.025}\right] \\ &= 2000(1.025)(27.773154) \\ &= 56\,934.97 \end{aligned}$$

$$\begin{aligned} PV_n(\text{defer}) &= 56\,934.97(1.025)^{-40} \\ &= 56\,934.97(0.372431) \\ &= 21\,204.33 \end{aligned}$$

Programmed Solution

("BGN" mode) (Set P/Y 5 4; C/Y 5 4) 0 [FV] 2000 [±] [PMT]
10 [I/Y] 48 [N] [CPT] [PV] [56 934.9651]

56 934.9651 [FV] 0 [PMT] 10 [I/Y] 40 [N] [CPT] [PV] [−21 204.32456]

The balance in the fund today must be $21 204.33.

EXAMPLE 7.3K

What payment can be made at the beginning of each month for six years if $5000.00 is invested today at 12% compounded monthly and the payments are deferred for ten years?

SOLUTION

The payments form a deferred annuity due.

PV_n(defer) = 5000.00; P/Y = 12; C/Y = 12; I/Y = 12; $i = \frac{12\%}{12} = 1\% = 0.01$;
$n = 6(12) = 72$; $d = 10(12) = 120$

Value of the $5000 at the end of the period of deferment,

$$\begin{aligned} FV &= 5000.00(1.01)^{120} \\ &= 5000.00(3.300387) \\ &= 16\,501.94 \end{aligned}$$

$$16\,501.94 = PMT(1.01)\left(\frac{1-1.01^{-72}}{0.01}\right)$$ ——— substituting in Formula 7.2

$$\begin{aligned} 16\,501.94 &= PMT(1.01)(51.150392) \\ 16\,501.94 &= 51.661896\ PMT \\ PMT &= \frac{16\,501.94}{51.661896} \\ PMT &= 319.42 \end{aligned}$$

Programmed Solution

(Set P/Y = 12; C/Y = 12) 0 [PMT] 5000 [±] [PV]

12 [I/Y] 120 [N] [CPT] [FV] 16 501.93447

("BGN" mode) [±] 16 501.93447 [PV] 0 [FV]

12 [I/Y] 72 [N] [CPT] [PMT] 319.4217778

The monthly payment is $319.42.

EXAMPLE 7.3L

What payment can be received at the beginning of each month for fifteen years if $10 000.00 is deposited in a fund ten years before the first payment is made and interest is 6% compounded monthly?

SOLUTION

First, find the accumulated value of the initial deposit at the beginning of the term of the annuity due ten years after the deposit.

$PV = 10\,000.00$; $P/Y = 12$; $C/Y = 12$; $I/Y = 6$; $i = \frac{6\%}{12} = 0.5\% = 0.005$;
$n = 10(12) = 120$

$$\begin{aligned} FV &= 10\,000(1.005)^{120} \\ &= 10\,000.00(1.819397) \\ &= \$18\,193.97 \end{aligned}$$

Now determine the monthly withdrawal that can be made at the beginning of each month from the initial balance of $18 193.97.

$PV_n(\text{due}) = 18\,193.87$; $i = 0.5\%$; $n = 15(12) = 180$

$$18\,193.97 = PMT(1.005)\left[\frac{1 - (1.005)^{-180}}{0.005}\right]$$

$$18\,193.97 = PMT(1.005)(118.503515)$$

$$PMT = \frac{18\,193.97}{119.096032}$$

$$PMT = \$152.77$$

Programmed Solution

(Set P/Y = 12; C/Y =12) 0 [PMT] 10 000 [±] [PV]

6 [I/Y] 120 [N] [CPT] [FV] 18 193.96734

("BGN" mode) 18 193.96734 [±] [PV] 0 [FV]

6 [I/Y] 180 [N] [CPT] [PMT] 152.7671997

A payment of $152.77 can be made at the beginning of each month.

EXAMPLE 7.3M

A scholarship of $2000.00 per year is to be paid at the beginning of each year from a scholarship fund of $15 000.00 invested at 7% compounded annually. How long will the scholarship be paid if payments are deferred for five years?

SOLUTION

The annual payments form a deferred annuity due.

$PV_n(\text{defer}) = 15\,000.00$; PMT = 2000.00; P/Y = 1; C/Y = 1; I/Y = 7;

$i = 7\% = 0.07$; $d = 5$

The value of the $15 000.00 at the end of the period of deferment,

$$\begin{aligned} FV &= 15\,000.00(1.07)^5 \\ &= 15\,000.00(1.402552) \\ &= 21\,038.28 \end{aligned}$$

$$\begin{aligned} 21\,038.28 &= 2000.00(1.07)\left(\frac{1 - 1.07^{-n}}{0.07}\right) \\ 21\,038.28 &= 30\,571.43(1 - 1.07^{-n}) \\ 0.688168 &= 1 - 1.07^{-n} \\ 1.07^{-n} &= 0.311832 \\ -n \ln 1.07 &= \ln 0.311832 \\ -0.067659n &= -1.165291 \\ n &= \frac{1.165291}{0.067659} \\ n &= 17.223 \\ n &= 18 \text{ years (approximately)} \end{aligned}$$

Programmed Solution

(Set P/Y = 1; C/Y = 1) 0 [PMT] 15 000 [±] [PV]
7 [I/Y] 5 [N] [CPT] [FV] [21 038.27596]

("BGN" mode) [±] 21 038.27596 [PV] 0 [FV]
7 [I/Y] 2000 [PMT] [CPT] [N] [17.223081]

The scholarship fund will provide 17 payments of $2000 and a final payment of less than $2000.00.

D. Computation of a general deferred annuity due

The same principles apply to a general deferred annuity due as apply to a simple annuity due.

EXAMPLE 7.3N

Tom Casey wants to withdraw $925.00 at the beginning of each quarter for twelve years. If the withdrawals are to begin ten years from now and interest is 4.5% compounded monthly, how much must Tom deposit today to be able to make the withdrawals?

SOLUTION

$PMT = 925.00;\ n = 12(4) = 48;\ d = 10(4) = 40;\ P/Y = 4;\ C/Y = 12;$

$I/Y = 4.5;\ c = \frac{12}{4} = 3;\ i = \frac{4.5\%}{12} = 0.375\% = 0.00375$

The effective quarterly rate of interest

$p = 1.00375^3 - 1 = 1.011292 - 1 = 0.011292 = 1.1292\%$

STEP 1 Find the present value of the general annuity due.

$$PV_{nc}(due) = 925.00(1.011292)\left(\frac{1 - 1.011292^{-48}}{0.011292}\right) \quad \text{substituting in Formula 7.4}$$

$$= 925.00(1.011292)(36.898193)$$

$$= \$34\ 516.24$$

Programmed Solution

(Set P/Y = 4; C/Y = 12) ("BGN" mode) 4.5 [I/Y] 0 [FV]

925 [±] [PMT] 48 [N] [CPT] [PV] [34 516.21131]

STEP 2 Find the present value of PV_{nc}(due) at the beginning of the period of deferment.

FV = 34 516.24 is the present value of the general annuity PV_{nc}(due)

d = 40 is the number of deferred payment intervals;

p = 1.12922% is the effective rate of interest per payment interval

$$PV_{nc}(defer) = PV = 34\ 516.24(1.0112922)^{-40} \quad \text{substituting in Formula 3.1C}$$

$$= 34\ 516.24(0.6381661)$$

$$= 22\ 027.09$$

Programmed Solution

34 516.21131 [FV] 4.5 [I/Y] 0 [PMT] 40 [N] [CPT] [PV] [−22 027.03920]

Tom must deposit \$22 027.04 to make the withdrawals. (A small difference in the results is due to rounding.)

EXAMPLE 7.30

A lease contract that has a cash value of \$64 000.00 requires payments at the beginning of each month for seven years. If the payments are deferred for two years and interest is 8% compounded quarterly, what is the size of the monthly payment?

SOLUTION

$PV_{nc}(defer) = 64\ 000.00;\ n = 7(12) = 84;\ P/Y = 12;\ C/Y = 4;$

$d = 2(12) = 24;\ c = \frac{4}{12} = \frac{1}{3};\ I/Y = 8;\ i = \frac{8\%}{4} = 2\% = 0.02$

The effective monthly rate of interest

$p = 1.02^{\frac{1}{3}} - 1 = 1.006623 - 1 = 0.6623\%$

First determine the accumulated value of the cash value at the end of the period of deferment.

$$FV = 64\ 000.00(1.006623)^{24}$$

$$= 64\ 000.00(1.171659)$$

$$= 74\ 986.20$$

Programmed Solution

(Set P/Y = 12; C/Y = 4) 8 [I/Y] 0 [PMT]

64 000 [±] [PV] 24 [N] [CPT] [FV] 74 986.20038

Now determine the periodic payment for the general annuity due whose present value is $74 986.20.

$$74\ 986.20 = \text{PMT}(1.006623)\left(\frac{1 - 1.006623^{-84}}{0.006623}\right)$$
$$74\ 986.20 = \text{PMT}(1.006623)(64.267570)$$
$$74\ 986.20 = \text{PMT}(64.693195)$$
$$\text{PMT} = 1159.10$$

Programmed Solution

("BGN" mode) [±] 74 986.20 [PV] 0 [FV]

8 [I/Y] 84 [N] [CPT] [PMT] 1159.104917

The monthly payment is $1159.10.

EXAMPLE 7.3P

By age 65, Janice Berstein had accumulated $120 000.00 in an RRSP by making yearly contributions over a period of years. At age 69, she converted the existing balance into an RRIF from which she started to withdraw $2000.00 per month. If the first withdrawal was on the date of conversion and interest on the account is 6.5% compounded quarterly, for how long will Janice Berstein receive annuity payments?

SOLUTION

$\text{PV}_{nc}(\text{defer}) = 120\ 000.00$; PMT = 2000.00; $d = 4(12) = 48$;

P/Y = 12; C/Y = 4; $c = \frac{4}{12} = \frac{1}{3}$; I/Y = 6.5; $i = \frac{6.5\%}{4} = 1.625\% = 0.01625$

The effective monthly rate of interest

$$p = 1.01625^{\frac{1}{3}} - 1 = 1.005388 - 1 = 0.005388 = 0.5388\%$$

Since payments are at the beginning of each month, the problem involves a deferred general annuity due.

First find the accumulated value at the end of the period of deferment.

$$\text{FV} = 120\ 000.00(1.005388)^{48}$$
$$= 120\ 000.00(1.294223)$$
$$= 155\ 306.78$$

Programmed Solution

(Set P/Y = 12; C/Y = 4) 6.5 [I/Y] 0 [PMT]

120 000 [±] [PV] 48 [N] [CPT] [FV] 155 306.6971

Then find the number of payments for an annuity due with a present value of $155 306.78.

$$\begin{aligned}
155\,306.78 &= 2000.00(1.005388)\left(\frac{1-1.005388^{-n}}{0.0053888}\right)\\
155\,306.78 &= 373\,222.81(1-1.005388^{-n})\\
0.416124 &= 1-1.005388^{-n}\\
1.005388^{-n} &= 0.583876\\
-n\ln 1.005388 &= \ln 0.583876\\
-n(0.005373) &= -0.538066\\
n &= 100.141\\
n &= 101 \text{ months}
\end{aligned}$$

Programmed Solution

("BGN" mode) [±] 155 306.6971 [PV] 0 [FV]

6.5 [I/Y] 2000 [PMT] [CPT] [N] [100.139798]

Janice Berstein will receive payments for eight years and five months.

EXERCISE 7.3

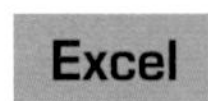

If you choose, you can use Excel's ***Present Value (PV)*** function to answer the questions indicated below. Refer to **PV** on the Spreadsheet Template Disk to learn how to use this Excel function.

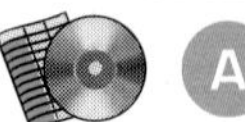

Find the present value of each of the following ten deferred annuities.

	Periodic Payment	Made At	Payment Period	Period of Deferment	Term	Interest Rate	Conversion Period
1.	$45.00	end	1 month	5 years	7 years	12%	monthly
2.	$125.00	end	6 months	8 years	15 years	7%	semi-annually
3.	$225.00	end	1 month	12 years	20 years	10.5%	monthly
4.	$850.00	beginning	1 year	3 years	10 years	7.5%	annually
5.	$720.00	beginning	3 months	6 years	12 years	4%	quarterly
6.	$85.00	beginning	1 month	20 years	15 years	6%	monthly
7.	$720.00	end	3 months	4 years	10 years	12%	monthly
8.	$1500.00	end	1 month	2 years	3 years	5%	semi-annually
9.	$145.00	beginning	6 months	3 years	5 years	8%	quarterly
10.	$225.00	beginning	3 months	6 years	8 years	9%	annually

Answer each of the following questions.

1. Calvin Jones bought his neighbour's farm for $10 000 down and payments of $5000.00 at the end of every three months for ten years. If the payments are deferred for two years and interest is 8% compounded quarterly, what was the purchase price of the farm?

2. Mrs. Bell expects to retire in seven years and would like to receive $800.00 at the end of each month for ten years following the date of her retirement. How much must Mrs. Bell deposit today in an account paying 7.5% compounded semi-annually to receive the monthly payments?

3. The Omega Venture Group needs to borrow to finance a project. Repayment of the loan involves payments of $8500.00 at the end of every three months for eight years. No payments are to be made during the development period of three years. Interest is 9% compounded quarterly.
   (a) How much should the Group borrow?
   (b) What amount will be repaid?
   (c) How much of that amount will be interest?
4. What sum of money invested now will provide payments of $1200.00 at the end of every three months for six years if the payments are deferred for nine years and interest is 10% compounded quarterly?
5. From his savings account, Samuel planned to withdraw amounts at the beginning of every three months for five years starting three years from now. If he started with $7200.00, and the account earned 9% compounded quarterly, what is the amount of each withdrawal?
6. An annuity with a cash value of $14 500.00 earns 7% compounded semi-annually. End-of-period semi-annual payments are deferred for seven years, then continue for ten years. How much is the amount of each payment?
7. A deposit of $20 000.00 is made for a twenty-year term. After the term expires, equal withdrawals are to be made for twelve years at the end of every six months. What is the size of the semi-annual withdrawal if interest is 4.75% compounded semi-annually?
8. On the day of his daughter's birth, Mr. Dodd deposited $2000.00 in a trust fund with his credit union at 5% compounded quarterly. Following her eighteenth birthday, the daughter is to receive equal payments at the end of each month for four years while she is at college. If interest is to be 6.0% compounded monthly after the daughter's eighteenth birthday, how much will she receive every month?
9. To finance the development of a new product, a company borrowed $50 000.00 at 7% compounded quarterly. If the loan is to be repaid in equal quarterly payments over seven years and the first payment is due three years after the date of the loan, what is the size of the quarterly payment?
10. Mr. Talbot received a retirement bonus of $18 000.00, which he deposited in an RRSP. He intends to leave the money for fourteen years, then transfer the balance into an RRIF and make equal withdrawals at the end of every six months for twenty years. If interest is 6.5% compounded semi-annually, what will be the size of each withdrawal?
11. An annuity with a cash value of $8800.00 pays $325.00 at the end of every month. If the annuity earns 6% compounded monthly, and payments begin three years from now, how long will the payments last?
12. A deposit of $4000.00 is made today for a five-year period. For how long can $500.00 be withdrawn from the account at the end of every three months starting three months after the end of the five-year term if interest is 4% compounded quarterly?

13. For how long can $1000.00 be withdrawn at the end of each month from an account containing $16 000.00, if the withdrawals are deferred for six years and interest is 7.2% compounded monthly?

14. Greg borrowed $6500.00 at 8.4% compounded monthly to help finance his education. He contracted to repay the loan in monthly payments of $300.00 each. If the payments are due at the end of each month and the payments are deferred for four years, for how long will Greg have to make monthly payments?

15. Denise Kadawalski intends to retire in twelve years and would like to receive $2400.00 every six months for fifteen years starting on the date of her retirement. How much must Denise deposit in an account today if interest is 6.5% compounded semi-annually?

16. Arlene and Mario Dumont want to set up a fund to finance their daughter's university education. They want to be able to withdraw $400.00 from the fund at the beginning of each month for four years. Their daughter enters university in seven-and-a-half years and interest is 6% compounded monthly.

(**a.**) How much must the Dumonts deposit in the fund today?

(**b.**) What will be the amount of the total withdrawals?

(**c.**) How much of the amount withdrawn will be interest?

17. An investment in a lease offers returns of $2500 per month due at the beginning of each month for five years. What investment is justified if the returns are deferred for two years and the interest required is 12% compounded monthly?

18. An RRIF with a beginning balance of $21 000.00 earns interest at 10% compounded quarterly. If withdrawals of $3485.00 are made at the beginning of every three months, starting eight years from now, how long will the RRIF last?

19. Mrs. Woo paid $24 000.00 into a retirement fund paying interest at 11% compounded semi-annually. If she retires in seventeen years, for how long can Mrs. Woo withdraw $10 000.00 from the fund every six months? Assume that the first withdrawal is on the date of retirement.

20. A lease valued at $32 000.00 requires payments of $4000.00 every three months. If the first payment is due three years after the lease was signed and interest is 12% compounded quarterly, what is the term of the lease?

21. An annuity purchased for $9000.00 makes month-end payments for seven years and earns interest at 5% quarterly. If payments start three years and one month from now, how much is each payment?

22. Sarah has just inherited a $12 650.00 annuity from her grandmother. The annuity earns interest at 6% quarterly and makes payments at the beginning of every six months for four years. If the payments begin in two years, what is the amount of each payment?

23. Ed Ainsley borrowed $10 000.00 from his uncle to finance his post-graduate studies. The loan agreement calls for equal payments at the end of each month for ten years. The payments are deferred for four years and interest is 8% compounded semi-annually. What is the size of the monthly payments?
24. Mrs. McCarthy has paid a single premium of $22 750.00 for an annuity, with the understanding that she will receive $385.00 at the end of each month. How long will the annuity last if it earns 5% compounded semi-annually, and the first payment period starts one year from now?
25. The sale of a property provides for payments of $2000.00 due at the beginning of every three months for five years. If the payments are deferred for two years and interest is 9% compounded monthly, what is the cash value of the property?
26. Dr. Young bought $18 000.00 worth of equipment from Medical Supply Company. The purchase agreement requires equal payments every six months for eight years. If the first payment is due two years after the date of purchase and interest is 7% compounded quarterly, what is the size of the payments?
27. Bhupinder, who has just had his fifty-fifth birthday, invested $3740.00 on that day for his retirement. The investment earns 8% compounded monthly. For how long will he be able to withdraw $1100.00 at the beginning of each year, starting on his sixty-fifth birthday?
28. A property development agreement valued at $45 000.00 requires annual lease payments of $15 000.00. The first payment is due five years after the date of the agreement and interest is 11% compounded semi-annually. For how long will payments be made?
29. A retirement bonus of $23 600.00 is invested in an annuity deferred for twelve years. The annuity provides payments of $4000.00 due at the beginning of every six months. If interest is 10% compounded annually, for how long will annuity payments be made?

7.4 Perpetuities

A. Basic concepts

A **perpetuity** is an annuity in which the periodic payments begin on a fixed date and continue indefinitely. Interest payments on permanently invested sums of money are prime examples of perpetuities. Dividends on preferred shares fall into this category assuming that the issuing corporation has an indefinite life. Scholarships paid perpetually from an endowment fit the definition of perpetuity.

As there is no end to the term, it is *not* possible to determine the future value of a perpetuity. However, the present value of a perpetuity *is* a definite value and this section deals with the present value of simple perpetuities.

DID YOU KNOW?

Did you know that the term *consols* is short for "consolidated annuities"? Consols are government bonds that pay interest to the bondholder but have no maturity date. They are an example of a perpetuity. Since consols have no maturity date, the government will pay interest on these bonds indefinitely. Did you know that, in 1751, the government of Great Britain consolidated various government bond issues to create the first consols?

B. Present value of ordinary perpetuities

We will use the following symbols when dealing with perpetuities:

A = the present value of the perpetuity;
R = the periodic rent (or perpetuity payment);
i = the rate of interest per conversion period;
p = the effective rate of interest per payment period.

FIGURE 7.5 Graphical Representation of an Ordinary Perpetuity

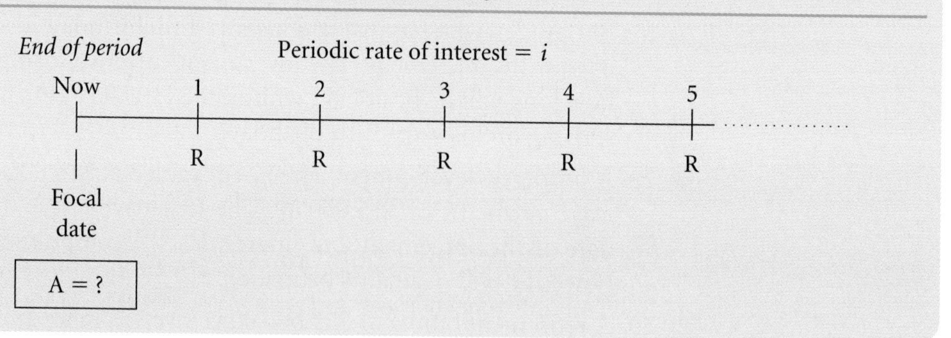

The **perpetuity payment** R is the interest earned by the present value of the perpetuity in one interest period.
When N is determined, the present value of an ordinary simple annuity was calculated using the formula

$$PV_n = PMT\left[\frac{1-(1+i)^{-n}}{i}\right]$$

As N in the formula increases and approaches infinity, the factor $(1 + i)^{-n}$ approaches 0.

Thus, the formula for finding the present value of an ordinary simple perpetuity is:

$$A = \frac{R}{i}$$
restated as
$$PV = \frac{PMT}{i}$$
— Formula 7.5

For an ordinary general perpetuity, the payment R (or PMT) is the interest earned by the present value of the perpetuity in one payment interval. That is,

$$R = pA \quad \text{or} \quad PMT = pPV$$

Thus, the formula for finding the present value of an ordinary general perpetuity is:

$$A = \frac{R}{p}$$
restated as
$$PV = \frac{PMT}{p}$$
where $p = (1 + i)^c - 1$
— Formula 7.6

where

$p =$ the effective rate of interest per payment interval;
$c =$ the number of conversion periods per payment period.

EXAMPLE 7.4A What sum of money invested today at 10% compounded annually will provide a scholarship of \$1500.00 at the end of every year?

SOLUTION $PMT = 1500.00; \quad i = 10\% = 0.10$

$$PV = \frac{1500.00}{0.10} = \$15\,000.00$$ — substituting in Formula 7.5

EXAMPLE 7.4B The maintenance cost for Northern Railroad of a crossing with a provincial highway is \$2000.00 at the end of each month. Proposed construction of an overpass would eliminate the monthly maintenance cost. If money is worth 12% compounded monthly, how much should Northern be willing to contribute toward the cost of construction?

SOLUTION The monthly maintenance expense payments form an *ordinary simple perpetuity.*

$$PMT = 2000.00; \quad i = \frac{12\%}{12} = 0.01$$

$$PV = \frac{2000.00}{0.01} = \$200\,000.00$$

Northern should be willing to contribute \$200 000.00 toward construction.

EXAMPLE 7.4C

What sum of money invested today at 8% compounded quarterly will provide a scholarship of \$2500.00 at the end of each year?

SOLUTION

Since the payments are to continue indefinitely, the payments form a perpetuity. Furthermore, since the payments are made at the end of each payment interval and the interest conversion period is not the same length as the payment interval, the problem involves an *ordinary general perpetuity.*

$\text{PMT} = 2500.00; \quad c = 4; \quad i = \frac{8\%}{4} = 2\% = 0.02$

The effective annual rate of interest

$p = 1.02^4 - 1 = 1.082432 - 1 = 0.082432 = 8.2432\%$

$\text{PV} = \frac{2500.00}{0.082432} = \$30\,327.95$ ——— substituting in Formula 7.6

The required sum of money is \$30 327.95.

EXAMPLE 7.4D

A will gives an endowment of \$50 000.00 to a university with the provision that a scholarship be paid at the end of each year. If the money is invested at 11% compounded annually, how much is the annual scholarship?

SOLUTION

$\text{PV} = 50\,000.00; \quad i = 11\% = 0.11$

By rearranging the terms of Formula 7.5, we get the equation $\text{PMT} = \text{PV}i$.

$\text{PMT} = \text{PV}i = 50\,000.00(0.11) = \$5500.00$

The annual scholarship is \$5500.00.

EXAMPLE 7.4E

The alumni of Peel College collected \$32 000.00 to provide a fund for bursaries. If the money is invested at 7% compounded annually, what is the size of the bursary that can be paid every six months?

SOLUTION

$\text{PV} = 32\,000.00; \quad c = \frac{1}{2}; \quad i = 7\% = 0.07$

The effective semi-annual rate of interest

$p = 1.07^{0.5} - 1 = 1.034408 - 1 = 0.034408 = 3.4408\%$

By rearranging the terms of Formula 7.6, we get the equation $\text{PMT} = p\text{PV}$.

$\text{PMT} = p\text{PV} = 0.034408(32\,000.00) = \$1101.06$

The size of the bursary is \$1101.06.

C. Present value of perpetuities due

A perpetuity due differs from an ordinary perpetuity only in that the first payment is made at the focal date. Therefore, a simple perpetuity due may be treated as consisting of an immediate payment R followed by an ordinary perpetuity. The formula for finding the present value of a simple perpetuity due is:

$$A(\text{due}) = R + \frac{R}{i}$$

restated as

$$PV(\text{due}) = PMT + \frac{PMT}{i}$$

Formula 7.7

As in the case of the simple perpetuity due, the general perpetuity due can be treated as consisting of an immediate payment R followed by an ordinary general perpetuity. Using the symbol A(due) for the present value, the formula for finding the present value of a general perpetuity due is:

$$A(\text{due}) = R + \frac{R}{p}$$

restated as

$$PV(\text{due}) = PMT + \frac{PMT}{p}$$

where $p = (1 + i)^c - 1$

Formula 7.8

EXAMPLE 7.4F

A tract of land is leased in perpetuity at $1250.00 due at the beginning of each month. If money is worth 7.5% compounded monthly, what is the present value of the lease?

SOLUTION

$PMT = 1250.00; \quad i = \frac{7.5\%}{12} = 0.625\% = 0.00625$

$$PV = 1250.00 + \frac{1250.00}{0.00625}$$ ——— substituting in Formula 7.7

$$= 1250.00 + 200\,000.00$$

$$= \$201\,250.00$$

EXAMPLE 7.4G

What is the present value of perpetuity payments of $750.00 made at the beginning of each month if interest is 8.5% compounded semi-annually?

SOLUTION

$PMT = 750.00; \quad c = \frac{2}{12} = \frac{1}{6}; \quad i = \frac{8.5\%}{2} = 4.25\% = 0.0425$

The effective monthly rate of interest

$p = 1.0425^{\frac{1}{6}} - 1 = 1.006961 - 1 = 0.006961 = 0.6961\%$

$$PV(\text{due}) = 750.00 + \frac{750.00}{0.006961} = \$108\,491.59$$ ——— substituting in Formula 7.8

The present value of the perpetuity is $108 491.59.

EXAMPLE 7.4H

How much money must be invested today in a fund earning 5.5% compounded annually to pay annual scholarships of $2000.00 starting

(i) one year from now?

(ii) immediately?

(iii) four years from now?

SOLUTION

PMT = 2000.00; $i = 5.5\% = 0.055$

(i) The annual scholarship payments form an ordinary perpetuity.

$$PV = \frac{2000.00}{0.055} = \$36\,363.64$$

The required sum of money is $36 363.64.

(ii) The annual scholarship payments form a perpetuity due.

$$\begin{aligned} PV &= 2000.00 + \frac{2000.00}{0.055} \\ &= 2000.00 + 36\,363.64 \\ &= \$38\,363.64 \end{aligned}$$

The required sum of money is $38 363.64.

(iii) The annual scholarship payments form an ordinary perpetuity deferred for three years.

$$\begin{aligned} PV(\text{defer}) &= PV(1.055^{-3}) \\ &= 36\,363.64(0.851614) \\ &= \$30\,967.77 \end{aligned}$$

The required sum of money is $30 967.77.

EXAMPLE 7.4I

What sum of money invested today in a fund earning 6.6% compounded monthly will provide perpetuity payments of $395.00 every three months starting

(i) immediately?
(ii) three months from now?
(iii) one year from now?

SOLUTION

PMT = 395.00; $c = \frac{12}{4} = 3$; $i = \frac{6.6\%}{12} = 0.55\% = 0.0055$

The effective quarterly rate of interest

$p = 1.0055^3 - 1 = 1.016591 - 1 = 0.016591 = 1.6591\%$

(i) Because the perpetuity payments are at the beginning of each payment interval, they form a *perpetuity due.*

$$PV(\text{due}) = 395.00 + \frac{395.00}{0.016591} = 395.00 + 23\,808.23 = \$24\,203.23$$

The required sum of money is $24 203.23.

(ii) Since the first payment is three months from now, the perpetuity payments form an *ordinary perpetuity.*

$$PV = \frac{395.00}{0.016591} = \$23\,808.23$$

The required sum of money is $23 808.23.

(iii) If you consider that the payments are deferred for one year, they form a *deferred perpetuity due.*

$$\begin{aligned} PV(\text{defer}) &= PV(\text{due}) \times (1.0055)^{-12} \\ &= 24\,203.23(0.936300) \\ &= \$22\,661.49 \end{aligned}$$

The required sum of money is $22 661.49.

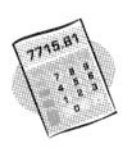

Perpetuities Using a BAII Plus

Perpetuities can be treated like any other annuity on the BAII Plus. While technically there are two missing variables, present value and payments can be calculated using arbitrary values for time and future value. The suggested values to use are a time of 300 years when calculating the value of N and 0 for the future value. Using these values will allow the BAII Plus to mimic perpetuity and allow you to calculate perpetuity values the same way as you would any other annuity.

The process would be:

1. Set the calculator to beginning or end.
2. Set P/Y and C/Y.
3. Input the variables you know.
4. Compute the variable you want to know (present value or payments).

Note: When you are using this method, the answer may be marginally different from the formula due to rounding.

EXERCISE 7.4

Find the present value of each of the following eight perpetuities.

	Perpetuity Payment	Made At:	Payment Interval	Interest Rate	Conversion Period
1.	$1250.00	end	3 months	6.8%	quarterly
2.	$3420.00	end	1 year	8.3%	annually
3.	$5600.00	end	6 months	12%	monthly
4.	$380.00	end	3 months	8%	semi-annually
5.	$985.00	beginning	6 months	4.5%	semi-annually
6.	$125.00	beginning	1 month	5.25%	monthly
7.	$2150.00	beginning	3 months	9%	monthly
8.	$725.00	beginning	1 month	10%	quarterly

Answer each of the following questions.

1. The Xorex Company pays a dividend of $4.25 every three months per preferred share. What is the expected market price per share if money is worth 8% compounded semi-annually?

2. Transcontinental Pipelines is considering a technical process that is expected to reduce annual maintenance costs by $85 000.00. What is the maximum amount of money that could be invested in the process to be economically feasible if interest is 7% compounded quarterly?

3. What is the size of the scholarship that can be paid at the end of every six months from a fund of $25 000.00 if interest is 6.75% compounded quarterly?

4. Alain Rich wants to set up a scholarship fund for his alma mater. The annual scholarship payment is to be $2500.00 with the first such payment due four years after his deposit into the fund. If the fund pays 7.25% compounded annually, how much must Mr. Rich deposit?

5. A rental property provides a monthly income of $1150.00 due at the beginning of every month. What is the cash value of the property if money is worth 6.6% compounded monthly?

6. A rental property provides a net income of $4200.00 at the beginning of every three months. What is the cash value of the property if money is worth 9% compounded monthly?

7. Municipal Hydro offers to acquire a right-of-way from a property owner who receives annual lease payments of $2225.00 due in advance. What is a fair offer if money is worth 5.5% compounded quarterly?

8. The faculty of Eastern College collected $1400.00 for the purpose of setting up a memorial fund from which an annual award is to be made to a qualifying student. If the money is invested at 7% compounded annually and the first annual award payment is to be made five years after the money was deposited, what is the size of the annual award payment?

9. Barbara Katzman bought an income property for $28 000.00 three years ago. She has held the property for the three years without renting it. If she rents the property out now, what should be the size of the monthly rent payment due in advance if money is worth 6% compounded monthly?

10. What monthly lease payment due in advance should be charged for a tract of land valued at $35 000.00 if the agreed interest is 8.5% compounded semi-annually?

Review Exercise

1. Find the future value and the present value of semi-annual payments of $540.00 for seven-and-a-half years if interest is 9.0% compounded semi-annually and the payments are made
   (a) at the end of every six months;
   (b) at the beginning of every six months.

2. Determine the future value and the present value of monthly payments of $50.00 each for eight years at 6% compounded monthly if
   (a) the payments form an annuity due;
   (b) the payments form an ordinary annuity.

3. Robert Deed deposited $100.00 in a trust account on the day of his son's birth and every three months thereafter. If interest paid is 7% compounded quarterly, what will the balance in the trust account be before the deposit is made on the son's twenty-first birthday?

4. Jim Wong makes deposits of $225.00 at the beginning of every three months. Interest earned by the deposits is 3% compounded quarterly.
   (a) What will the balance in Jim's account be after eight years?
   (b) How much of the balance will Jim have contributed?
   (c) How much of the balance is interest?

5. Home entertainment equipment can be purchased by making monthly payments of $82.00 for three and a half years. The first payment is due at the time of purchase and the financing cost is 16.5% compounded monthly.
   (a) What is the purchase price?
   (b) How much will be paid in installments?
   (c) How much is the cost of financing?

6. How long will it take to build up a fund of $10 000.00 by saving $300.00 every six months at 4.5% compounded semi-annually?

7. How long will it take to accumulate $18 000.00 at 6% compounded monthly if $125.00 is deposited in an account at the beginning of every month?

8. For how long must $1000.00 be deposited at the beginning of every year to accumulate to $180 000.00 twelve years after the end of the year in which the last deposit was made if interest is 7.5% compounded annually?

9. Kelly Farms bought a tractor priced at $10 500.00 on February 1. Kelly agreed to make monthly payments of $475.00 beginning December 1 of the same year. For how long will Kelly Farms have to make these payments if interest is 10.5% compounded monthly?

10. Okanagan Vineyards borrowed $75 000.00 on a five-year promissory note. It agreed to make payments of $6000.00 every three months starting on the date of maturity. If interest is 8.5% compounded quarterly, for how long will the company have to make the payments?

11. At what nominal annual rate of interest compounded semi-annually will $1700.00 deposited at the beginning of every six months accumulate to $40 000.00 in nine years?

12. What is the effective annual rate of interest charged on a four-year lease valued at $9600.00 if payments of $235.00 are made at the beginning of each month for the four years?

13. Suppose you would like to have $10 000.00 in your savings account and interest is 8% compounded quarterly. How much must you deposit every three months for five years if the deposits are made
    (a) at the end of each quarter?
    (b) at the beginning of each quarter?

14. Equal sums of money are withdrawn monthly from a fund of $20 000.00 for fifteen years. If interest is 9% compounded monthly, what is the size of each withdrawal
    (a) if the withdrawal is made at the beginning of each month?
    (b) if the withdrawal is made at the end of each month?

15. Mrs. Bean contributes $450.00 at the beginning of every three months to an RRSP. Interest on the account is 6% compounded quarterly.

(a) What will the balance in the account be after seven years?

(b) How much of the balance will be interest?

(c) If Mrs. Bean converts the balance after seven years into an RRIF paying 5% compounded quarterly and makes equal quarterly withdrawals for twelve years starting three months after the conversion into the RRIF, what is the size of the quarterly withdrawal?

(d) What is the combined interest earned by the RRSP and the RRIF?

16. Art will receive monthly payments of $850.00 from a trust account starting on the date of his retirement and continuing for twenty years. Interest is 10.5% compounded monthly.

(a) What is the balance in the trust account on the date of Art's retirement?

(b) How much interest will be included in the payments Art receives?

(c) If Art made equal monthly deposits at the beginning of each month for fifteen years before his retirement, how much did he deposit each month?

17. Alicia Sidlo invested a retirement bonus of $12 500.00 in an RRSP paying 5.95% compounded semi-annually for ten years. At the end of ten years, she rolled the RRSP balance over into an RRIF paying $500.00 at the beginning of each month starting with the date of rollover. If interest on the RRIF is 5.94% compounded monthly, for how long will Alicia receive monthly payments?

18. Mrs. Ball deposits $550.00 at the beginning of every three months. Starting three months after the last deposit, she intends to withdraw $3500.00 every three months for fourteen years. If interest is 8% compounded quarterly, for how long must Mrs. Ball make deposits?

19. Terry saves $50.00 at the beginning of each month for sixteen years. Beginning one month after his last deposit, he intends to withdraw $375.00 per month. If interest is 6% compounded monthly, for how long can Terry make withdrawals?

20. Payments of $375.00 made every three months are accumulated at 3.75% compounded monthly. What is their amount after eight years if the payments are made

(a) at the end of every three months?

(b) at the beginning of every three months?

21. What is the accumulated value after twelve years of monthly deposits of $145.00 earning interest at 5% compounded semi-annually if the deposits are made

(a) at the end of each month?

(b) at the beginning of each month?

22. If you save $25.00 at the beginning of each month and interest is 4% compounded quarterly, how much will you accumulate in thirty years?

23. A property was purchased for quarterly payments of $1350.00 for ten years. If the first payment was made on the date of purchase and interest is 5.5% compounded annually, what was the purchase price of the property?

24. How much must be deposited into an account to accumulate to $32 000.00 at 7% compounded semi-annually

(a) at the beginning of each month for twenty years?

(b) at the end of each year for fifteen years?

25. Kelly Associates are the makers of a ten-year $75 000.00 promissory note bearing interest at 8% compounded semi-annually. To pay off the note on its due date, the company is making payments at the beginning of every three months into a fund paying 7.5% compounded monthly. What is the size of the quarterly payments?

26. A church congregation has raised $37 625.00 for future outreach work. If the money is invested in a fund paying 7% compounded quarterly, what annual payment can be made for ten years from the fund to its mission if the first payment is to be made four years from the date of investment in the fund?

27. What sum of money can be withdrawn from a fund of $16 750.00 invested at 6.5% compounded semi-annually

(a) at the end of every three months for twelve years?

(b) at the beginning of each year for twenty years?

(c) at the end of each month for fifteen years but deferred for ten years?

(d) at the beginning of every three months for twelve years but deferred for twenty years?

(e) at the end of each month in perpetuity?

(f) at the beginning of each year in perpetuity?

28. In what period of time will payments of $450.00 accumulate to $20 000.00 at 6% compounded monthly if made

(a) at the end of every three months?

(b) at the beginning of every six months?

29. Over what period of time will RRSP contributions of $1350.00 made at the beginning of each year amount to $125 000.00 if interest is 7% compounded quarterly?

30. A lease contract valued at $50 000.00 requires semi-annual payments of $5200.00. If the first payment is due at the date of signing the contract and interest is 9% compounded monthly, what is the term of the lease?

31. A debt of $20 000.00 is repaid by making payments of $3500.00. If interest is 9% compounded monthly, for how long will payments have to be made

(a) at the end of every six months?

(b) at the beginning of each year?

(c) at the end of every three months with payments deferred for five years?

(d) at the beginning of every six months with payments deferred for three years?

32. What is the nominal rate of interest compounded quarterly at which payments of $400.00 made at the beginning of every six months accumulate to $8400.00 in eight years?

33. A contract is signed requiring payments of $750.00 at the end of every three months for eight years.

(a) How much is the cash value of the contract if money is worth 10.5% compounded quarterly?

(b) If the first three payments are missed, how much would have to be paid after one year to bring the contract up to date?

(c) If, because of the missed payments, the contract has to be paid out at the end of one year, how much money is needed?

(d) How much of the total interest paid is due to the missed payments?

34. Anne received $45 000.00 from her mother's estate. She wants to set aside part of her inheritance for her retirement nine years from now. At that time she would like to receive a pension supplement of $600.00 at the end of each month for twenty-five years. If the first payment is due one month after her retirement and interest is 6.5% compounded monthly, how much must Anne set aside?

35. Frank invested a retirement bonus of $15 000.00 in an income averaging annuity paying 6% compounded monthly. He withdraws the money in equal monthly amounts over five years. If the first withdrawal is made nine months after the deposit, what is the size of each withdrawal?

36. Aaron deposited $900.00 every six months for twenty years into a fund paying 5.5% compounded semi-annually. Five years after the last deposit he converted the existing balance in the fund into an ordinary annuity paying him equal monthly payments for fifteen years. If interest on the annuity is 6% compounded monthly, what is the size of the monthly payment he will receive?

37. Sally contributed $500.00 every six months for fourteen years into an RRSP earning interest at 6.5% compounded semi-annually. Seven years after the last contribution, Sally converted the RRSP into an RRIF that is to pay her equal quarterly amounts for sixteen years. If the first payment is due three months after the conversion into the RRIF and interest on the RRIF is 7% compounded quarterly, how much will Sally receive every three months?

38. Wendy deposited $500.00 into an RRSP every three months for twenty-five years. Upon her retirement she converted the RRSP balance into an RRIF that is to pay her equal quarterly amounts for twenty years. If the first payment is due three months after her retirement and interest is 9% compounded quarterly, how much will Wendy receive every three months?

39. Ty received a separation payment of $25 000.00 at age thirty-five. He invested that sum of money at 5.5% compounded semi-annually until he was sixty-five. At that time he converted the existing balance into an ordinary annuity paying $6000.00 every three months with interest at 6% compounded quarterly. For how long will the annuity run?

40. Mr. Maxwell intends to retire in ten years and wishes to receive $4800.00 every three months for twenty years starting on the date of his retirement. How much must he deposit now to receive the quarterly payments from an account paying 6% compounded quarterly?

41. Tomac Swim Club bought electronic timing equipment on a contract requiring monthly payments of $725.00 for three years beginning eighteen months after the date of purchase. What was the cash value of the equipment if interest is 7.5% compounded monthly?

42. What sum of money invested today in a retirement fund will permit withdrawals of $800.00 at the end of each month for twenty years if interest is 5.75% compounded semi-annually and the payments are deferred for fifteen years?

43. A lease requires semi-annual payments of $6000.00 for five years. If the first payment is due in four years and interest is 9% compounded monthly, what is the cash value of the lease?

44. A debt of $40 000.00 is to be repaid in installments due at the end of each month for seven years. If the payments are deferred for three years and interest is 7% compounded quarterly, what is the size of the monthly payments?

45. Redden Ogilvie bought his parents' farm for $200 000.00. The transfer agreement requires Redden to make quarterly payments for twenty years. If the first payment is due in five years and the rate of interest is 10% compounded annually, what is the size of the quarterly payments?

46. An annuity provides payments of $4500.00 at the end of every three months. The annuity is bought for $33 500.00 and payments are deferred for twelve years. If interest is 12% compounded monthly, for how long will payments be received?

47. A retirement bonus of $25 000.00 is invested in an income averaging annuity paying $1400.00 every three months. If the interest is 11.5% compounded semi-annually and the first payment is due in one year, for how long will payments be received?

48. $8000.00 was invested at a fixed rate of 5.95% compounded semi-annually for seven years. After seven years, the fund was converted into an ordinary annuity paying $450.00 per month. If interest on the annuity was 6.6% compounded monthly, what was the term of the annuity?

49. What single cash payment made now is equivalent to payments of $3500.00 every six months at 8% compounded quarterly if the payments are made
   (a) at the end of every six months for fifteen years?
   (b) at the beginning of every six months for ten years?
   (c) at the end of every six months for eight years but deferred for four years?

(**d**) at the beginning of every six months for nine years but deferred for three years?

(**e**) at the end of every six months in perpetuity?

(**f**) at the beginning of every six months in perpetuity?

50. What is the principal invested at 4.75% compounded semi-annually from which monthly withdrawals of $240.00 can be made

(**a**) at the end of each month for twenty-five years?

(**b**) at the beginning of each month for fifteen years?

(**c**) at the end of each month for twenty years but deferred for ten years?

(**d**) at the beginning of each month for fifteen years but deferred for twelve years?

(**e**) at the end of each month in perpetuity?

(**f**) at the beginning of each month in perpetuity?

51. An income property is estimated to net $1750.00 per month continually. If money is worth 6.6% compounded monthly, what is the cash price of the property?

52. Preferred shares of Western Oil paying a quarterly dividend are to be offered at $55.65 per share. If money is worth 9% compounded semi-annually, what is the minimum quarterly dividend to make investment in such shares economically feasible?

53. The semi-annual dividend per preferred share issued by InterCity Trust is $7.50. If comparable investments yield 14% compounded quarterly, what should be the selling price of these shares?

54. Western Pipelines pays $8000.00 at the beginning of each year for using a tract of land. What should the company offer the property owner as a purchase price if interest is 9.5% compounded annually?

55. A fund to provide an annual scholarship of $4000.00 is to be set up. If the first payment is due in three years and interest is 11% compounded quarterly, what sum of money must be deposited in the scholarship fund today?

Self-Test

1. Payments of $1080.00 are made into a fund at the beginning of every three months for eleven years. If the fund earns interest at 9.5% compounded quarterly, how much will the balance in the fund be after eleven years?

2. Find the present value of payments of $960.00 made at the beginning of every month for seven years if money is worth 6% compounded monthly.

3. Tim bought a boat valued at $10 104.00 on the installment plan. He made equal semi-annual payments for five years. If the first payment is due on the date of purchase and interest is 10.5% compounded semi-annually, what is the size of the semi-annual payments?

4. Sara Eng wants to withdraw $3000.00 at the beginning of every three months for thirty years starting at the date of her retirement. If she retires in twenty years and interest is 10% compounded quarterly, how much must Ms. Eng deposit into an account every month for the next twenty years starting now?

5. J.J. deposited $1680.00 at the beginning of every six months for eight years into a fund paying 5.5% compounded semi-annually. Fifteen years after the first deposit, he converted the existing balance into an annuity paying him equal monthly payments for twenty years. If the payments are made at the end of each month and interest is 6% compounded monthly, what is the size of the monthly payments?

6. The amount of $57 426.00 is invested at 6% compounded monthly for six years. After the initial six-year period, the balance in the fund is converted into an annuity due paying $3600.00 every three months. If interest on the annuity is 5.9% compounded quarterly, what is the term of the annuity in months?

7. What is the nominal annual rate of interest charged on a lease valued at $3840.00 if payments of $200.00 are made at the beginning of every three months for six years?

8. Deposits of $1400.00 made at the beginning of every three months amount to $40 000.00 after six years. What is the effective annual rate of interest earned by the deposits if interest is compounded quarterly?

9. A lease requires monthly payments of $950.00 due in advance. If interest is 12% compounded quarterly and the term of the lease is five years, what is the cash value of the lease?

10. Sally Smedley and Roberto Jones bought their neighbour's farm for $30 000.00 down and payments of $6000.00 at the end of every six months for six years. What is the purchase price of the farm if the semi-annual payments are deferred for four years and interest is 8.5% compounded semi-annually?

11. Eden would like to receive $3000.00 at the end of every six months for seven years after her retirement. If she retires ten years from now and interest is 6.5% compounded semi-annually, how much must she deposit into an account every six months starting now?

12. Ken acquired his sister's share of their business by agreeing to make payments of $4000.00 at the end of each year for twelve years. If the payments are deferred for three years and money is worth 5% compounded quarterly, what is the cash value of the sister's share of the business?

13. The amount of $46 200.00 is invested at 9.5% compounded quarterly for four years. After four years the balance in the fund is converted into an annuity. If interest on the annuity is 6.5% compounded semi-annually and payments are made at the end of every six months for seven years, what is the size of the payments?

14. What sum of money must be deposited in a trust fund to provide a scholarship of $960.00 payable at the end of each month if interest is 7.5% compounded monthly?

15. A bank pays a quarterly dividend of $0.75 per share. If comparable investments yield 13.5% compounded monthly, what is the sales value of the shares?

16. Western Pipelines pays $480.00 at the beginning of every half-year for using a tract of land. What should the company offer the property owner as a purchase price if interest is 11% compounded semi-annually?

17. Carla plans to invest in a property that after three years will yield $1200.00 at the end of each month indefinitely. How much should Carla be willing to pay if an alternative investment yields 9% compounded monthly?

18. Mr. Smart wants to set up an annual scholarship of $3000.00. If the first payment is to be made in five years and interest is 7.0% compounded annually, how much must Mr. Smart pay into the scholarship fund?

Challenge Problems

1. A regular deposit of $100.00 is made at the beginning of each year for twenty years. Simple interest is calculated at i% per year for the twenty years. At the end of the twenty-year period, the total interest in the account is $840.00. Suppose that interest of i% compounded annually had been paid instead. How much interest would have been in the account at the end of the twenty years?

2. Herman has agreed to repay a debt by using the following repayment schedule. Starting today, he will make $100.00 payments at the beginning of each month for the next two-and-a-half years. He will then pay nothing for the next two years. Finally, after four-and-a-half years, he will make $200.00 payments at the beginning of each month for one year, which will pay off his debt completely. For the first four-and-a-half years, the interest on the debt is 9% compounded monthly. For the final year, the interest is lowered to 8.5% compounded monthly. Find the size of Herman's debt. Round your answer to the nearest dollar.

Case Study 7.1 From Casino to College

» Hermia and Chan Chen have three young daughters. Their daughters are Su, who turned ten years old in April, Joy, who turned seven in January, and Wei, who turned four in March. In May, the Chens visited a charity casino, and Lady Luck was with them. They won a jackpot of $50 000.00. They decided to invest a large part of their winnings that year in order to create future education funds for their three daughters.

Since it will be a number of years before the girls are ready for college or university, Hermia and Chan had to make some assumptions regarding the girls' futures. They have assumed that each daughter will take a three-year course and that the costs of education will continue to increase. Based on these assumptions, Hermia and Chan decided to provide each daughter with a monthly allowance that would cover tuition and some living expenses. Because they were uncertain about the girls' finding summer jobs in the future, Hermia and Chan decided their daughters would receive the allowance for twelve months of the year. Su will receive an allowance of $1000.00 at the beginning of each month, starting September 1 of the year she turns eighteen. Because of the increasing costs of education, Joy will receive an allowance that is 6% more than Su's allowance. She will receive it at the beginning of each month, starting September 1 of the year she turns eighteen. Wei will receive an allowance that is 6% more than Joy's at the beginning of each month, starting September 1 of the year she turns eighteen.

The week after winning the money at the casino, Hermia and Chan visited their local bank manager to set up the investment that would pay the girls the allowances when they were ready for college or university. The bank manager suggested an investment paying interest of 7.5% compounded monthly from now until the three girls had each completed their three years of education. Hermia and Chan thought this sounded reasonable. So on June 1, a week after talking with the bank manager, they deposited the sum of money necessary to finance their daughters' post-secondary educations.

QUESTIONS

1. How much money must Hermia and Chan invest on June 1 to provide the desired allowance to Su?
2. **(a)** How much allowance will Joy receive each month for tuition and living expenses?
   (b) How much money must Hermia and Chan invest on June 1 to provide the desired allowance to Joy?
3. **(a)** How much allowance will Wei receive each month for tuition and living expenses?
   (b) How much money must Hermia and Chan invest on June 1 to provide the desired allowance to Wei?
4. After making their investment on June 1, how much of their winnings do Hermia and Chan have left to spend on themselves?

Case Study 7.2 Setting Up Scholarships

» Recently, Mid North Community College launched a drive to raise money for three new student bursaries. The Student Awards Committee of the college has been working with three community organizations to create and fund these bursaries. The organizations are Friends of Education, Mid North Community Foundation, and Metro Service Club. These organizations are all convinced that the best way to fund the bursaries is to make one large donation to the college. The donation would be invested so that it would grow over time, and regular, annual bursaries would be paid out indefinitely.

Friends of Education agreed to donate a sum of money on September 1 that would allow its annual bursary of $1600.00 to be awarded immediately on September 1. It has agreed to allow the college to choose the best local student to receive its bursary each year.

Mid North Community Foundation would like to earn some interest on its September 1 donation before awarding its first bursary on December 1, three months later. The Foundation has agreed to award one bursary of $1300 per year. It will choose the recipient from all applications received.

Metro Service Club will make its donation on September 1 but it wants to award its bursary of $800 per year starting September 1 next year. This will give the club time to develop the criteria used to choose the recipient of the bursary.

QUESTIONS

1. What sum of money must Friends of Education invest on September 1 if its donation is expected to earn 6% compounded semi-annually?
2. What sum of money must Mid North Community Foundation invest on September 1 if its donation is expected to earn 6.2% compounded quarterly?
3. What sum of money must Metro Service Club invest on September 1 this year if its donation is expected to earn 5.8% compounded monthly?

SUMMARY OF FORMULAS

There are no new formulas in this chapter. Please refer to the formulas in Chapters 5 and 6.

GLOSSARY

Annuity due an annuity in which the periodic payments are made at the beginning of each payment interval *(p. 250)*

Deferred annuity an annuity in which the first payment is made at a time later than the end of the first payment interval *(p. 274)*

General annuity due a general annuity in which the payments are made at the beginning of each payment interval *(p. 265)*

Period of deferment the period from the time referred to as "now" to the starting point of the term of the annuity *(p. 274)*

Perpetuity an annuity in which the periodic payments begin at a fixed date and continue indefinitely *(p. 291)*

Perpetuity payment the interest earned by the present value of the perpetuity in one interest period *(p. 292)*

USEFUL INTERNET SITES

www.tdcanadatrust.com/mutualfunds/resp_cesp.jsp
RESPs Visit the Toronto-Dominion Bank's financial planning centre to read general information and FAQs about RESPs and the products that TD Bank offers.

www.smartmoney.com
SmartMoney.com SmartMoney has daily stock and mutual fund recommendations, hourly market updates, personal finance investing research tools and advice, and up-to-the-minute stock and mutual fund quotes and charts.

www.forbes.com
***Forbes* Magazine** This site provides access to articles on current financial business issues, as well as tools for mutual funds, stocks, and personal finances.

8 Amortization of Loans, Including Residential Mortgages

OBJECTIVES

Upon completing this chapter, you will be able to do the following:

1. Perform computations associated with amortization of debts involving simple annuities, including the size of the periodic payments, outstanding balance, interest due, and principal repaid, and construct complete or partial amortization schedules.
2. Perform computations associated with the amortization of debts involving general annuities, and construct complete or partial amortization schedules.
3. Find the size of the final payment when all payments except the final payment are equal in size.
4. Compute the effective interest rate for fixed-rate residential mortgages.
5. Compute the periodic payments for fixed-rate mortgages and for demand mortgages.
6. Distinguish between regular mortgage payments and rounded mortgage payments.
7. Create statements for various types of residential mortgages.

When budgeting for the future, businesses often consider arranging various loans. If you know the interest rate, you might want to calculate what the loan will cost over various time periods. You can do this by using the formulas for annuities that you are already very familiar with. By knowing what the size of your equal loan payments would be, you can make an informed decision about what loans to arrange.

One of the largest loans most of us will ever have is a mortgage on a house or a condominium. Given the monthly payments, we can calculate the amount of each payment that goes toward principal and the amount that goes toward interest. This information can help us decide what we can afford to buy and how quickly we can pay off the mortgage.

INTRODUCTION

Amortization refers to the repayment of interest-bearing debts by a series of payments, usually equal in size, made at equal intervals of time. The periodic payments, when equal in size, form an annuity whose present value is equivalent to the original loan principal. Mortgages and many consumer loans are repaid by this method. An amortization schedule shows the allocation of each payment to first cover the interest due and then reduce the principal.

8.1 Amortization Involving Simple Annuities

A. Finding the periodic payment

What is **amortization**? An interest-bearing debt is *amortized* if both principal and interest are repaid by a series of equal payments made at equal intervals of time.

The basic problem in amortizing a debt is finding the size of the periodic payment. If the payment interval and the interest conversion period are equal in length, the problem involves finding the periodic payment for a simple annuity. Since debts are generally repaid by making payments at the end of the payment interval, the method and formula for ordinary simple annuities apply.

$$PV_n = PMT\left[\frac{1-(1+i)^{-n}}{i}\right] \quad \text{——— Formula 5.2}$$

EXAMPLE 8.1A A debt of $5000.00 with interest at 9% compounded annually is to be repaid by equal payments at the end of each year for six years. What is the size of the annual payments?

SOLUTION

FIGURE 8.1 **Graphical Representation of Method and Data**

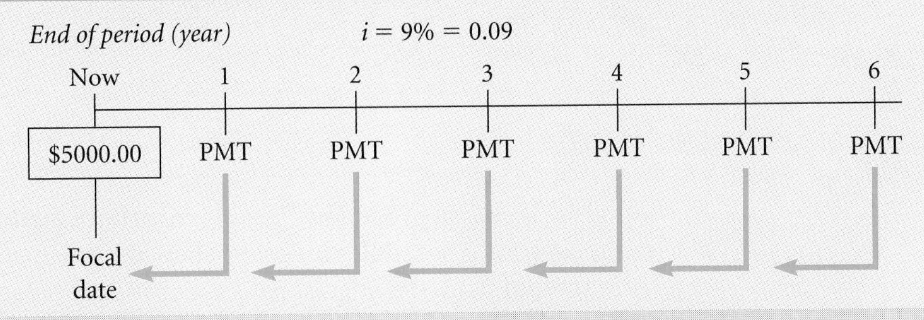

As Figure 8.1 shows, the equal annual payments (designated by PMT) form an ordinary simple annuity in which

$PV_n = 5000.00$; $n = 6$; P/Y = 1; C/Y = 1; I/Y = 9; $i = 9\% = 0.09$

Using "now" as the focal date, you can find the value of the annual payment PMT using Formula 5.2.

$$5000.00 = \text{PMT}\left(\frac{1 - 1.09^{-6}}{0.09}\right)$$ ——— using Formula 5.2

$$5000.00 = \text{PMT}\,(4.485919)$$

$$\text{PMT} = \$1114.60$$

Programmed Solution

("END" mode) (Set P/Y = 1; C/Y = 1) 0 [FV] 5000 [±] [PV] 9 [I/Y] 6 [N] [CPT] [PMT] [1114.598916]

The annual payment is $1114.60.

EXAMPLE 8.1B

A loan of $8000.00 made at 6% compounded monthly is amortized over five years by making equal monthly payments.

(i) What is the size of the monthly payment?
(ii) What is the total amount paid to amortize the loan?
(iii) What is the cost of financing?

SOLUTION

(i) $PV_n = 8000.00$; $n = 5(12) = 60$; P/Y = 12; C/Y = 12; I/Y = 6;

$$i = \frac{6\%}{12} = 0.5\%$$

$$8000.00 = \text{PMT}\left(\frac{1 - 1.005^{-60}}{0.005}\right)$$ ——— using Formula 5.2

$$8000.00 = \text{PMT}(51.725561)$$

$$\text{PMT} = \$154.66$$

Programmed Solution

("END" mode) (Set P/Y = 12; C/Y = 12) 0 [FV] 8000 [±] [PV] 6 [I/Y] 60 [N] [CPT] [PMT] [154.6624122]

(i) The monthly payment is $154.66.
(ii) The total amount paid is 60(154.66) = $9279.60.
(iii) The cost of financing is 9279.60 − 8000.00 = $1279.60.

B. Amortization schedules

As previously discussed in Section 2.6, **amortization schedules** show in detail how a debt is repaid. Such schedules normally show the payment number (or payment date), the amount paid, the interest paid, the principal repaid, and the outstanding debt balance.

1. Amortization Schedule When All Payments Are Equal (Blended Payments)

When all payments are equal, you must first determine the size of the periodic payment, as shown in Examples 8.1A and 8.1B.

EXAMPLE 8.1C

A debt of $5000.00 is amortized by making equal payments at the end of every three months for two years. If interest is 8% compounded quarterly, construct an amortization schedule.

SOLUTION

STEP 1 Determine the size of the quarterly payments.

$PV_n = 5000.00;\; n = 4(2) = 8;\; P/Y = 4;\; C/Y = 4;\; I/Y = 8;\; i = \frac{8\%}{4} = 2.0\%$

$$5000.00 = PMT\left(\frac{1 - 1.02^{-8}}{0.02}\right)$$

$$5000.00 = PMT(7.325481)$$

$$PMT = \$682.55$$

Programmed Solution

("END" mode) (Set P/Y = 4; C/Y = 4) 0 [FV] 5000 [±] [PV] 8 [I/Y] 8 [N] [CPT] [PMT] [682.548996]

STEP 2 Construct the amortization schedule as shown below.

Payment Number	Amount Paid	Interest Paid $i = 0.02$	Principal Repaid	Outstanding Principal Balance
0				5000.00
1	682.55	100.00	582.55	4417.45
2	682.55	88.35	594.20	3823.25
3	682.55	76.47	606.08	3217.17
4	682.55	64.34	618.21	2598.96
5	682.55	51.98	630.57	1968.39
6	682.55	39.37	643.18	1325.21
7	682.55	26.50	656.05	669.16
8	682.54	13.38	669.16	0.00
TOTAL	5460.39	460.39	5000.00	

Explanations regarding the construction of the amortization schedule

1. Payment number 0 is used to introduce the initial balance of the loan.
2. The interest included in the first payment is the periodic interest rate i multiplied by the period's beginning balance, $0.02 \times 5000.00 = \$100.00$. Since the amount paid is $682.55, the amount available for repayment of principal is $682.55 - 100.00 = \$582.55$. The outstanding principal balance after the first payment is $5000.00 - 582.55 = \$4417.45$.

3. The interest included in the second payment is $0.02 \times 4417.45 = \$88.35$. Since the amount paid is \$682.55, the amount available for repayment of principal is $682.55 - 88.35 = \$594.20$. The outstanding principal is $4417.45 - 594.20 = \$3823.25$.

4. Computation of interest, principal repaid, and outstanding balance for payments 3 to 7 are made in a similar manner.

5. The last payment of \$682.54 is slightly different from the other payments as a result of rounding in the amount paid or the interest paid. To allow for such rounding errors, the last payment is computed by adding the interest due in the last payment ($0.02 \times 669.16 = \$13.38$) to the outstanding balance: $669.16 + 13.38 = \$682.54$.

6. The three totals provide useful information and can be used as a check on the accuracy of the schedule.

   (a) The total principal repaid must equal the original outstanding balance;

   (b) The total amount paid is the periodic payment times the number of such payments plus/minus any adjustment in the last payment, $682.55 \times 8 - 0.01 = 5460.40 - 0.01 = \$5460.39$.

   (c) The total interest paid is the difference between the amount paid and the original principal, $5460.39 - 5000.00 = \$460.39$.

Programmed Solution

The amortization schedule can also be built by using the Amortization function within the preprogrammed financial calculator. Using the Texas Instruments BAII Plus, follow these steps:

1. Enter all of the information within the TVM function and compute the payment, PMT, as above.
2. Press **2nd** **Amort** to start the function.
3. On the calculator display, P1 refers to the number of the first period of the range to be specified. For example, to obtain data for periods 2 through 5, P1 would be entered as "2." Remember to press the **Enter** key to enter new data.
4. Press the down arrow **↓** to move to the next cell. P2 refers to the number of the last period of the range to be specified. For example, to obtain data for periods 2 through 5, P2 would be entered as "5." Note that, to obtain information for just one period, both P1 and P2 must be entered, as the same number. For example, to obtain data for period 1, both P1 and P2 must be entered as "1," specifying "from" period 1 and "to" period 1.
5. Press the down arrow **↓** to move to the next cell. BAL is automatically calculated as the loan balance at the end of the period specified in P2. In the amortization schedule above, when period 1 is specified, the BAL calculated is –4417.45. The outstanding balance appears as a negative number because the PV was entered as a negative number.

6. Press the down arrow ↓ to move to the next cell. PRN is automatically calculated as the portion of the principal that was repaid. In the amortization schedule above, the PRN is calculated as 582.55.
7. Press the down arrow ↓ to move to the next cell. INT is automatically calculated as the portion of the period's payment that was interest on the loan. In the amortization schedule above, the INT is calculated as 100.00.

2. Amortization Schedule When All Payments Except the Final Payment Are Equal

When the size of the periodic payment is determined by agreement, usually because it is a convenient round figure rather than a computed blended payment, the size of the final payment will probably be different from the preceding agreed-upon payments. This final payment is obtained in the amortization schedule by adding the interest due on the outstanding balance to the outstanding balance.

This type of loan repayment schedule has been illustrated and explained in Section 2.6. The following example is included for review.

EXAMPLE 8.1D

Bronco Repairs borrowed $15 000.00 from National Credit Union at 10% compounded quarterly. The loan agreement requires payment of $2500.00 at the end of every three months. Construct an amortization schedule.

SOLUTION

$i = \frac{10\%}{4} = 2.5\% = 0.025$

Payment Number	Amount Paid	Interest Paid $i = 0.025$	Principal Repaid	Outstanding Principal Balance
0				15 000.00
1	2 500.00	375.00	2 125.00	12 875.00
2	2 500.00	321.88	2 178.12	10 696.88
3	2 500.00	267.42	2 232.58	8 464.30
4	2 500.00	211.61	2 288.39	6 175.91
5	2 500.00	154.40	2 345.60	3 830.31
6	2 500.00	95.76	2 404.24	1 426.07
7	1 461.72	35.65	1 426.07	0.00
TOTAL	16 461.72	1461.72	15 000.00	

Note: After Payment 6, the outstanding principal is less than the agreed-upon payment. When this happens, the final payment will be the outstanding balance plus the interest due on the outstanding balance.

$1426.07 + (1426.07 \times 0.025) = 1426.07 + 35.65 = \$1461.72$

C. Finding the outstanding principal balance

For various reasons, such as early partial repayment or early full repayment or refinancing, either the borrower or the lender needs to know the outstanding balance at a certain time. This can be done by checking the amortization schedule, if available, or by direct mathematical computation. This computation is also useful for checking the accuracy of the schedule as it is developed.

1. Finding the Outstanding Principal When All Payments Are Equal

EXAMPLE 8.1E For Example 8.1C, compute the outstanding balance just after the third payment has been made.

SOLUTION The loan history showing the quarterly payments of \$682.55 can be represented on a time diagram as shown in Figure 8.2.

FIGURE 8.2 **Graphical Representation of Loan Payments**

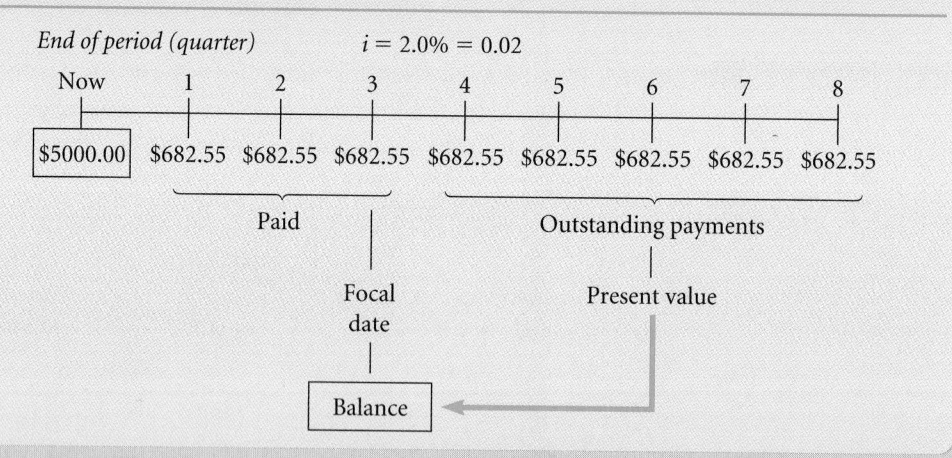

In the same way in which the original loan balance of \$5000.00 equals the present value of the number of payments that are necessary to amortize the loan, the outstanding balance at the end of any payment interval just after a payment has been made is the present value of the outstanding payments.

OUTSTANDING BALANCE = PRESENT VALUE OF THE OUTSTANDING PAYMENTS

To answer this problem, use Formula 5.2. Since three of the eight payments have been made, five payments remain outstanding.

$\text{PMT} = 682.55;\ n = 5;\ \text{P/Y} = 4;\ \text{C/Y} = 4;\ \text{I/Y} = 8;\ i = 2.0\% = 0.02$

$$\begin{aligned} PV_3 &= 682.55\left(\frac{1 - 1.02^{-5}}{0.02}\right) \quad \text{——— using Formula 5.2} \\ &= 682.55(4.713460) \\ &= \$3217.17 \end{aligned}$$

Programmed Solution

("END" mode) (Set P/Y = 4; C/Y = 4) 0 [FV] 682.55 [±] [PMT] 8 [I/Y] 5 [N] [CPT] [PV] [3217.171788]

The outstanding balance after the third payment is $3217.17.

Note: Any difference in the amortization schedule is due to rounding.

EXAMPLE 8.1F

You borrow $7500.00 from a finance company at 13.5% compounded monthly with the agreement to make monthly payments. If the loan is amortized over five years, what is the loan balance after two years?

SOLUTION

First determine the size of the monthly payment.

$PV_n = 7500.00$; $n = 5(12) = 60$; P/Y = 12; C/Y = 12; I/Y = 13.5;

$$i = \frac{13.5\%}{12} = 1.125\% = 0.01125$$

$$\begin{aligned} 7500.00 &= \text{PMT}\left(\frac{1 - 1.01125^{-60}}{0.01125}\right) \\ 7500.00 &= \text{PMT}(43.459657) \\ \text{PMT} &= \$172.57 \end{aligned}$$

Programmed Solution

("END" mode) (Set P/Y = 12; C/Y = 12) 0 [FV] 7500 [±] [PV] 13.5 [I/Y] 60 [N] [CPT] [PMT] [172.573845]

Now determine the balance after two years.

After two years, 24 of the 60 payments have been made; 36 payments remain outstanding.

PMT = 172.57; $n = 36$; $i = 1.125\%$

$$\begin{aligned} PV_{24} &= 172.57\left(\frac{1 - 1.01125^{-36}}{0.01125}\right) \\ &= 172.57(29.467851) \\ &= \$5085.27 \end{aligned}$$

Programmed Solution

("END" mode) (Set P/Y = 12; C/Y = 12) 172.57 [±] [PMT] 0 [FV] 13.5 [I/Y] 36 [N] [CPT] [PV] [−5085.267094]

The outstanding balance after two years is $5085.27.

The method used in Examples 8.1E and 8.1F is called the **prospective method** of finding the outstanding balance because it considers the future prospects of the debt—the payments that remain outstanding.

Alternatively, the outstanding balance can be found by the **retrospective method**, which considers the payments that have been made. This method finds the outstanding balance by deducting the accumulated value of the payments that have been made from the accumulated value of the original debt.

$$\text{OUTSTANDING BALANCE} = \text{ACCUMULATED VALUE OF THE ORIGINAL DEBT} - \text{ACCUMULATED VALUE OF THE PAYMENTS MADE}$$

For Example 8.1E

The accumulated value of the original debt at the end of three payment intervals

$$\begin{aligned} FV &= 5000.00(1.02^3) \quad \text{using Formula 3.1A} \\ &= 5000.00(1.061208) \\ &= 5306.04 \end{aligned}$$

The accumulated value of the payments made

$$\begin{aligned} FV_3 &= 682.55\left(\frac{1.02^3 - 1}{0.02}\right) \\ &= 682.55(3.060400) \\ &= \$2088.88 \end{aligned}$$

Programmed Solution

("END" mode) (Set P/Y = 4; C/Y = 4) [2nd] (CLR TVM) 5000 [±] [PV] 8 [I/Y] 3 [N] [CPT] [FV] [5306.04]

0 [PV] 682.55 [±] [PMT] 8 [I/Y] 3 [N] [CPT] [FV] [2088.87602]

The outstanding balance is 5306.04 − 2088.88 = \$3217.16.

For Example 14.1F

The accumulated value of the debt after two years (24 payments)

$$\begin{aligned} FV &= 7500.00(1.01125^{24}) \\ &= 7500.00(1.307991) \\ &= \$9809.93 \end{aligned}$$

The accumulated value of the 24 payments made

$$\begin{aligned} FV_{24} &= 172.57\left(\frac{1.01125^{24} - 1}{0.01125}\right) \\ &= 172.57(27.376998) \\ &= \$4724.45 \end{aligned}$$

Programmed Solution

("END" mode)(Set P/Y = 12; C/Y = 12) 2nd (CLR TVM) 7500 ± PV 13.5 I/Y 24 N CPT FV 9809.934198

0 PV 172.57 ± PMT 13.5 I/Y 24 N CPT FV 4724.448528

The outstanding balance is 9809.93 − 4724.45 = $5085.48.

Note: The difference in the two methods—$5085.48 versus $5085.27—is due to rounding in the payment.

Because the prospective method is more direct, it is preferred when finding the outstanding balance of loans repaid by installments that are all equal.

2. Finding the Outstanding Balance When All Payments Except the Final Payment Are Equal

When the last payment is different from the other payments, the retrospective method for finding the outstanding balance is preferable.

EXAMPLE 8.1G

For Example 8.1D, compute the outstanding balance after four payments.

SOLUTION

The accumulated value of the original debt after four payments

$$
\begin{aligned}
FV &= 15\,000.00(1.025^4) \\
&= 15\,000.00(1.103813) \\
&= \$16\,557.19
\end{aligned}
$$

The accumulated value of the four payments

$$
\begin{aligned}
S_4 &= 2500.00\left(\frac{1.025^4 - 1}{0.025}\right) \\
&= 2500.00(4.152516) \\
&= \$10\,381.29
\end{aligned}
$$

Programmed Solution

("END" mode) (Set P/Y = 4; C/Y = 4) 2nd (CLR TVM) 15 000 ± PV 10 I/Y 4 N CPT FV 16 557.19336

0 PV 2500 ± PMT 10 I/Y 4 N CPT FV 10 381.28906

The outstanding balance is 16 557.19 − 10 381.29 = $6175.90.

EXAMPLE 8.1H

A debt of \$25 000.00 with interest at 11% compounded semi-annually is amortized by making payments of \$2000.00 at the end of every six months. Determine the outstanding balance after five years.

SOLUTION

$PV_n = PV = 25\,000.00;\ P/Y = 2;\ C/Y = 2;\ I/Y = 11;\ i = \frac{11\%}{2} = 5.5\% = 0.055$

The accumulated value of the original principal after five years

$$\begin{aligned} FV &= 25\,000.00(1.055^{10}) \\ &= 25\,000.00(1.708145) \\ &= \$42\,703.61 \end{aligned}$$

The accumulated value of the first ten payments

$$\begin{aligned} FV_{10} &= 2000.00\left(\frac{1.055^{10} - 1}{0.055}\right) \\ &= 2000.00(12.875354) \\ &= \$25\,750.71 \end{aligned}$$

Programmed Solution

("END" mode) (Set P/Y = 2; C/Y = 2) [2nd] (CLR TVM) 25 000 [±] [PV]

11 [I/Y] 10 [N] [CPT] [FV] [42 703.61146]

0 [PV] 2000 [±] [PMT] 11 [I/Y] 10 [N] [CPT] [FV] [25 750.70758]

The outstanding balance after five years is 42 703.61 − 25 750.71 = \$16 952.90.

D. Finding the interest paid and the principal repaid; constructing partial amortization schedules

Apart from computing the outstanding balance at any one time, all the other information contained in an amortization schedule, such as interest paid and principal repaid, can also be computed.

EXAMPLE 8.1I

For Example 8.1C, compute

(i) the interest paid in the fifth payment period;
(ii) the principal repaid in the fifth payment period.

SOLUTION

(i) The interest due for any given payment period is based on the outstanding balance at the beginning of the period. This balance is the same as the outstanding balance at the end of the previous payment period.

To find the interest paid by the fifth payment, we need to know the outstanding balance after the fourth payment.

$$\begin{aligned} PV_4 &= 682.55\left(\frac{1 - 1.02^{-4}}{0.02}\right) \\ &= 682.55(3.8077287) \\ &= \$2598.97 \end{aligned}$$

Excel

You can use Excel's ***Cumulative Interest Paid Between Two Periods (CUMIPMT)*** and ***Cumulative Principal Paid Between Two Periods (CUMPRINC)*** functions to answer questions like these. Refer to **CUMIPMT** and **CUMPRINC** on the Spreadsheet Template Disk to learn how to use these Excel functions.

Programmed Solution

("END" mode)(Set P/Y = 4; C/Y = 4) 0 [FV] 682.55 [±] [PMT] 8 [I/Y] 4 [N] [CPT] [PV] [2598.965223]

Interest for Payment Period 5 is 2598.97(0.02) = $51.98.

(ii) Principal repaid = Amount paid − Interest paid
= 682.55 − 51.98
= $630.57

EXAMPLE 8.1J

Jackie Kim borrowed $6000.00 from her trust company at 9% compounded monthly. She was to repay the loan with monthly payments over five years.

(i) What is the interest included in the 20th payment?

(ii) What is the principal repaid in the 36th payment period?

(iii) Construct a partial amortization schedule showing the details of the first three payments, the 20th payment, the 36th payment, and the last three payments, and determine the totals of amount paid, interest paid, and principal repaid.

SOLUTION

$PV_n = 6000.00$; $n = 5(12) = 60$; P/Y = 12; C/Y = 12; I/Y = 9;

$$i = \frac{9\%}{12} = 0.75\% = 0.0075$$

$$6000.00 = PMT\left(\frac{1 - 1.0075^{-60}}{0.0075}\right)$$

$$6000.00 = PMT(48.173374)$$

$$PMT = \$124.55$$

Programmed Solution

("END" mode) (Set P/Y = 12; C/Y = 12) 0 [FV] 6000 [±] [PV] 9 [I/Y] 60 [N] [CPT] [PMT] [124.550131]

(i) The outstanding balance after the 19th payment

$$PV_{19} = 124.55\left(\frac{1 - 1.0075^{-41}}{0.0075}\right)$$

$$= 124.55(35.183064)$$

$$= \$4382.05$$

Programmed Solution

0 [FV] 124.55 [±] [PMT] 9 [I/Y] 41 [N] [CPT] [PV] [4382.050802]

The interest paid by the 20th payment is 4382.05(0.0075) = \$32.87.

(ii) The outstanding balance after the 35th payment

$$
\begin{aligned}
PV_{35} &= 124.55\left(\frac{1 - 1.0075^{-25}}{0.0075}\right) \\
&= 124.55(22.718755) \\
&= \$2829.62
\end{aligned}
$$

Programmed Solution

0 [FV] 124.55 [±] [PMT] 9 [I/Y] 25 [N] [CPT] [PV] [2829.620994]

The interest included in Payment 36 is 2829.62(0.0075) 5 \$21.22.
The principal repaid by Payment 36 is 124.55 2 21.22 5 \$103.33.

(iii) We can develop the first three payments of the amortization schedule in the usual way. To show the details of the 20th payment, we need to know the outstanding balance after 19 payments (computed in part (i)). For the 36th payment, we need to know the outstanding balance after 35 payments (computed in part (ii)). Since there are 60 payments, the last three payments are Payments 58, 59, and 60. To show the details of these payments, we must determine the outstanding balance after Payment 57.

$$
\begin{aligned}
PV_{57} &= 124.55\left(\frac{1 - 1.0075^{-3}}{0.0075}\right) \\
&= 124.55(2.955556) \\
&= \$368.11
\end{aligned}
$$

Programmed Solution

0 [FV] 124.55 [±] [PMT] 9 [I/Y] 3 [N] [CPT] [PV] [368.1145294]

Partial amortization schedule

Payment Number	Amount Paid	Interest Paid $i = 0.0075$	Principal Repaid	Outstanding Principal Balance
0				6000.00
1	124.55	45.00	79.55	5920.45
2	124.55	44.40	80.15	5840.30
3	124.55	43.80	80.75	5759.55
•	•	•	•	•
•	•	•	•	•
19	•	•	•	4382.05
20	124.55	32.87	91.68	4290.37
•	•	•	•	•
•	•	•	•	•
35	•	•	•	2829.62
36	124.55	21.22	103.33	2726.29
•	•	•	•	•
•	•	•	•	•
57	•	•	•	368.11
58	124.55	2.76	121.79	246.32
59	124.55	1.85	122.70	123.62
60	124.55	0.93	123.62	0.00
TOTAL	7473.00	1473.00	6000.00	

Note: The total principal repaid must be $6000.00; the total amount paid is 124.55(60) = $7473.00; the total interest paid is 7473.00 − 6000.00 = $1473.00.

EXAMPLE 8.1K

The Erin Construction Company borrowed $75 000.00 at 14% compounded quarterly to buy construction equipment. Payments of $3500.00 are to be made at the end of every three months.

(i) Determine the principal repaid in the 16th payment.
(ii) Construct a partial amortization schedule showing the details of the first three payments, the 16th payment, the last three payments, and the totals.

SOLUTION

(i) Since the quarterly payments are not computed blended payments, the last payment will probably be different from the preceding equal payments. Use the retrospective method for finding the outstanding balance.

PV = 75 000.00; PMT = 3500.00; P/Y = 4; C/Y = 4; I/Y = 14;

$$i = \frac{14\%}{4} = 3.5\% = 0.035$$

The accumulated value of the original principal after the 15th payment

$$\begin{aligned} FV &= 75\,000.00(1.035^{15}) \\ &= 75\,000.00(1.675349) \\ &= \$125\,651.16 \end{aligned}$$

Programmed Solution

("END" mode) (Set P/Y = 4; C/Y = 4) 2nd (CLR TVM) 75 000 ± PV

14 I/Y 15 N CPT FV 125 651.1623

The accumulated value of the first 15 payments

$$FV_{15} = 3500.00\left(\frac{1.035^{15} - 1}{0.035}\right)$$
$$= 3500.00(19.295681)$$
$$= \$67\,534.88$$

Programmed Solution

0 PV 3500 ± PMT 14 I/Y 15 N CPT FV 67 534.88308

The outstanding principal after the 15th payment
= 125 651.16 − 67 534.88
= $58 116.28

The interest included in the 16th payment is 58 116.28(0.035) = $2034.07.
The principal repaid by the 16th payment is 3500.00 − 2034.07 = $1465.93.

(ii) To show details of the last three payments, we need to know the number of payments required to amortize the loan principal.

$PV_n = 75\,000.00$; PMT = 3500.00; $i = 3.5\%$

$$75\,000.00 = 3500.00\left(\frac{1 - 1.035^{-n}}{0.035}\right)$$
$$0.75 = 1 - 1.035^{-n}$$
$$1.035^{-n} = 0.25$$
$$-n \ln 1.035 = \ln 0.25$$
$$-n(0.034401) = -1.386294$$
$$n = 40.297584$$

Programmed Solution

0 FV 75 000 ± PV 3500 PMT 14 I/Y CPT N 40.29758337

Forty-one payments (40 payments of $3500.00 each plus a final payment) are required. This means the amortization schedule should show details of Payments 39, 40, and 41. To do so, we need to know the outstanding balance after 38 payments.

The accumulated value of the original loan principal after 38 payments

$$FV = 75\,000.00(1.035^{38})$$
$$= 75\,000.00(3.696011)$$
$$= \$277\,200.85$$

Programmed Solution

2nd (CLR TVM) 75 000 ± PV 14 I/Y 38 N CPT FV 277 200.8486

The accumulated value of the first 38 payments

$$FV_{38} = 3500.00\left(\frac{1.035^{38} - 1}{0.035}\right)$$
$$= 3500.00(77.028895)$$
$$= \$269\ 601.13$$

Programmed Solution

0 PV 3500 ± PMT 14 I/Y 38 N CPT FV 269 601.1315

The outstanding balance after the 38th payment
= 277 200.85 − 269 601.13
= $7599.72

Partial amortization schedule

Payment Number	Amount Paid	Interest Paid $i = 0.035$	Principal Repaid	Outstanding Principal Balance
0				75 000.00
1	3 500.00	2 625.00	875.00	74 125.00
2	3 500.00	2 594.38	905.62	73 219.38
3	3 500.00	2 562.68	937.32	72 282.06
•	•	•	•	•
•	•	•	•	•
•	•	•	•	•
15	•	•	•	58 116.28
16	3 500.00	2 034.07	1 465.93	56 650.35
•	•	•	•	•
•	•	•	•	•
•	•	•	•	•
38	•	•	•	7 599.72
39	3 500.00	265.99	3 234.01	4 365.71
40	3 500.00	152.80	3 347.20	1 018.51
41	1 054.16	35.65	1 018.51	0.00
TOTAL	141 054.16	66 054.16	75 000.00	

E. Computer application—amortization schedule

The amortization schedule in Example 8.1C displays the manual calculations for the repayment of $5000.00 with quarterly payments and interest at 8% compounded quarterly. Microsoft Excel and other spreadsheet programs can be used to create a file that will immediately display the results of a change in the principal, interest rate, or amount paid.

Following are general instructions for creating a file to calculate the amortization schedule in Example 8.1C. The formulas in the spreadsheet file were created using Excel; however, most spreadsheets work in a similar manner.

This exercise assumes a basic understanding of spreadsheet applications, but an individual who has no previous experience with spreadsheets will be able to complete it.

STEP 1 Enter the labels shown in Figure 8.3 in row 1 and in column A.

FIGURE 8.3

Workbook 1

	A	B	C	D	E	F
1	Payment Number	Amount Paid	Interest Paid	Principal Repaid	Outstanding Principal Balance	
2	0				5000.00	682.55
3	1	= F2	= F4*E2	= B3−C3	= E2−D3	
4	2	= F2	= F4*E3	= B4−C4	= E3−D4	= 0.08/4
5	3	= F2	= F4*E4	= B5−C5	= E4−D5	
6	4	= F2	= F4*E5	= B6−C6	= E5−D6	
7	5	= F2	= F4*E6	= B7−C7	= E6−D7	
8	6	= F2	= F4*E7	= B8−C8	= E7−D8	
9	7	= F2	= F4*E8	= B9−C9	= E8−D9	
10	8	= C10+D10	= F4*E9	= E9	= E9−D10	
11	Totals	= SUM(B3:B10)	= SUM(C3:C10)	= SUM(D3:D10)		

STEP 2 Enter the principal in cell E2. Do not type in the dollar sign or a comma.

STEP 3 Enter only the formulas shown in cells B3, C3, D3, and E3. Make sure that the formula entry includes the dollar ($) sign as shown in the figure.

STEP 4 The formulas that were entered in Step 3 can be copied through the remaining cells.

(a) Select and Copy the formulas in cells B3, C3, D3, and E3.
(b) Select cells B4 to B9 and Paste.
(c) The formulas are now active in all the cells.

Alternatively, to copy formulas in Excel you can use the Fill handle, dragging it down the desired range through consecutive cells.

STEP 5 Enter the formulas shown in cells B10, C10, D10, and E10.

STEP 6 Enter the formula shown in cell B11, and then use Copy and Paste to enter the formulas in cells C11 and D11.

STEP 7 To ensure readability of the spreadsheet, format the numbers to display with two decimal places, and widen the columns to display the full labels.

This spreadsheet can now be used to reflect changes in aspects of the loan and create a new amortization schedule. Use cell E2 for new principal amounts and cell F4 for new interest rates.

EXERCISE 8.1

If you choose, you can use Excel's **CUMIPMT**, **CUMPRINC**, **NPER**, **PMT**, or **PV** functions or **Template4** to answer the questions indicated below. Refer to the Spreadsheet Template Disk to find **Template4** or to learn how to use these Excel functions.

For each of the following four debts amortized by equal payments made at the end of each payment interval, compute (a) the size of the periodic payments; (b) the outstanding principal at the time indicated; (c) the interest paid; and (d) the principal repaid by the payment following the time indicated for finding the outstanding principal.

	Debt Principal	Repayment Period	Payment Interval	Interest Rate	Conversion Period	Outstanding Principal Required After:
1.	$12 000.00	8 years	3 months	10%	quarterly	20th payment
2.	$8 000.00	5 years	1 month	12%	monthly	30th payment
3.	$15 000.00	10 years	6 months	8%	semi-annually	15th payment
4.	$9 600.00	7 years	3 months	6%	quarterly	12th payment

For each of the following four debts repaid by periodic payments as shown, compute (a) the number of payments required to amortize the debts; (b) the outstanding principal at the time indicated.

	Debt Principal	Debt Payment	Payment Interval	Interest Rate	Conversion Period	Outstanding Principal Required After:
1.	$12 000.00	$750.00	3 months	8%	quarterly	16th payment
2.	$7 800.00	$175.00	1 month	12%	monthly	24th payment
3.	$21 000.00	$2000.00	6 months	9%	semi-annually	10th payment
4.	$15 000.00	$800.00	3 months	6%	quarterly	12th payment

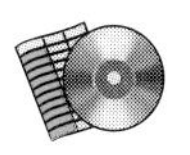

Answer each of the following questions.

1. Mr. and Mrs. Good purchased a ski chalet for $36 000.00. They paid $4000.00 down and agreed to make equal payments at the end of every three months for fifteen years. Interest is 8% compounded quarterly.

(a) What size payment are the Goods making every three months?
(b) How much will they owe after ten years?
(c) How much will they have paid in total after fifteen years?
(d) How much interest will they pay in total?

2. A contractor's price for a new building was $96 000.00. Slade Inc., the buyers of the building, paid $12 000.00 down and financed the balance by making equal payments at the end of every six months for twelve years. Interest is 12% compounded semi-annually.
   (a) What is the size of the semi-annual payment?
   (b) How much will Slade Inc. owe after eight years?
   (c) What is the total cost of the building for Slade Inc.?
   (d) What is the total interest included in the payments?
3. Sam's Auto Repairs Inc. borrowed $5500.00 to be repaid by monthly payments over four years. Interest on the loan is 9% compounded monthly.
   (a) What is the size of the periodic payment?
   (b) What is the outstanding principal after the thirteenth payment?
   (c) What is the interest paid in the fourteenth payment?
   (d) How much principal is repaid in the fourteenth payment?
4. To start their business, Ming and Ling borrowed $24 000.00 to be repaid by semi-annual payments over twelve years. Interest on the loan is 7% compounded semi-annually.
   (a) What is the size of the periodic payment?
   (b) What is the outstanding principal after the seventh payment?
   (c) What is the interest paid in the eighth payment?
   (d) How much principal is repaid in the eighth payment?
5. A loan of $10 000.00 with interest at 10% compounded annually is to be amortized by equal payments at the end of each year for seven years. Find the size of the annual payments and construct an amortization schedule showing the total paid and the cost of financing.
6. A loan of $8000.00 is repaid by equal payments made at the end of every three months for two years. If interest is 7% compounded quarterly, find the size of the quarterly payments and construct an amortization schedule showing the total paid and the total cost of the loan.
7. Hansco borrowed $9200.00 paying interest at 13% compounded annually. If the loan is repaid by payments of $2000.00 made at the end of each year, construct an amortization schedule showing the total paid and the total interest paid.
8. Pinto Bros. are repaying a loan of $14 500.00 by making payments of $2600.00 at the end of every six months. If interest is 7% compounded semi-annually, construct an amortization schedule showing the total paid and the total cost of the loan.
9. For Question 5, calculate the interest included in the fourth payment. Verify your answer by checking the amortization schedule.
10. For Question 6, calculate the principal repaid in the fifth payment period. Verify your answer by checking the amortization schedule.

11. For Question 7, calculate the principal repaid in the fourth payment period. Verify your answer by checking the amortization schedule.

12. For Question 8, calculate the interest included in the fifth payment. Verify your answer by checking the amortization schedule.

13. Apex Corporation borrowed $85 000.00 at 8% compounded quarterly for eight years to buy a warehouse. Equal payments are made at the end of every three months.

(a) Determine the size of the quarterly payments.
(b) Compute the interest included in the 16th payment.
(c) Determine the principal repaid in the 20th payment period.
(d) Construct a partial amortization schedule showing details of the first three payments, the last three payments, and totals.

14. Mr. Brabham borrowed $7500.00 at 15% compounded monthly. He agreed to repay the loan in equal monthly payments over five years.

(a) What is the size of the monthly payment?
(b) How much of the 25th payment is interest?
(c) What is the principal repaid in the 40th payment period?
(d) Prepare a partial amortization schedule showing details of the first three payments, the last three payments, and totals.

15. Thornhill Equipment Co. borrowed $24 000.00 at 11% compounded semi-annually. It is to repay the loan by payments of $2500.00 at the end of every six months.

(a) How many payments are required to repay the loan?
(b) How much of the sixth payment is interest?
(c) How much of the principal will be repaid in the tenth payment period?
(d) Construct a partial amortization schedule showing details of the first three payments, the last three payments, and totals.

16. Locust Inc. owes $16 000.00 to be repaid by monthly payments of $475.00. Interest is 6% compounded monthly.

(a) How many payments will Locust Inc. have to make?
(b) How much interest is included in the 18th payment?
(c) How much of the principal will be repaid in the 30th payment period?
(d) Construct a partial amortization schedule showing details of the first three payments, the last three payments, and totals.

8.2 Amortization Involving General Annuities

A. Finding the periodic payment and constructing amortization schedules

If the length of the payment interval is different from the length of the interest conversion period, the equal debt payments form a general annuity. The amortization of such debts involves the same principles and methods discussed in Section 8.1 except that general annuity formulas are applicable. Provided that the payments are made at the end of the payment intervals, use Formula 6.5.

$$PV_{nc} = PMT\left[\frac{1-(1+p)^{-n}}{p}\right]$$

where $p = (1+i)^c - 1$ — Formula 6.5

EXAMPLE 8.2A A debt of $30 000.00 with interest at 12% compounded quarterly is to be repaid by equal payments at the end of each year for seven years.

(i) Compute the size of the yearly payments.
(ii) Construct an amortization schedule.

SOLUTION (i) $PV_{nc} = 30\,000.00$; $n = 7$; $P/Y = 1$; $C/Y = 4$; $c = 4$; $I/Y = 12$;

$$i = \frac{12\%}{4} = 3\% = 0.03$$

$$p = 1.03^4 - 1 = 1.125509 - 1 = 0.125509 = 12.5509\%$$

$$30\,000.00 = PMT\left(\frac{1-1.125509^{-7}}{0.125509}\right)$$

$$30\,000.00 = PMT(4.485129)$$

$$PMT = \$6688.77$$

Programmed Solution

("END" mode) (Set P/Y = 1; C/Y = 4) 0 FV 30 000 ± PV 12 I/Y 7 N CPT PMT 6688.77031

(ii) ***Amortization schedule***

Payment Number	Amount Paid	Interest Paid $p = 0.1255088$	Principal Repaid	Outstanding Principal Balance
0				30 000.00
1	6 688.77	3 765.26	2 923.51	27 076.49
2	6 688.77	3 398.34	3 290.43	23 786.06
3	6 688.77	2 985.36	3 703.41	20 082.65
4	6 688.77	2 520.55	4 168.22	15 914.43
5	6 688.77	1 997.40	4 691.37	11 223.06
6	6 688.77	1 408.59	5 280.18	5 942.88
7	6 688.76	745.88	5 942.88	0.00
TOTAL	46 821.38	16 821.38	30 000.00	

B. Finding the outstanding principal

1. Finding the Outstanding Principal When All Payments Are Equal
When all payments are equal, the prospective method used in Section 8.1 is the more direct method.

EXAMPLE 8.2B

For Example 8.2A, compute the outstanding balance after three payments.

SOLUTION

The outstanding balance after three payments is the present value of the remaining four payments.

PMT = 6688.77; $n = 4$; $c = 4$; $i = 3\%$; $p = 12.5509\%$

$$\begin{aligned} PV_{nc} &= 6688.77\left(\frac{1 - 1.125509^{-4}}{0.125509}\right) \\ &= 6688.77(3.0024431) \\ &= \$20\,082.65 \end{aligned}$$

Programmed Solution

("END" mode) (Set P/Y = 1; C/Y = 4) 0 [FV] 6688.77 [±] [PMT] 12 [I/Y] 4 [N] [CPT] [PV] [20 082.65134]

The outstanding balance after three payments is $20 082.65.

EXAMPLE 8.2C

A $25 000.00 mortgage amortized by monthly payments over twenty years is renewable after five years.

(i) If interest is 8.5% compounded semi-annually, what is the outstanding balance at the end of the five-year term?
(ii) If the mortgage is renewed for a further three-year term at 8% compounded semi-annually, what is the size of the new monthly payment?
(iii) What is the payout figure at the end of the three-year term?

SOLUTION

(i) $PV_{nc} = 25\,000.00$; $n = 20(12) = 240$; P/Y = 12; C/Y = 2; I/Y = 8.5;

$c = \frac{2}{12} = \frac{1}{6}$; $i = \frac{8.5\%}{2} = 4.25\% = 0.0425$;

$p = 1.0425^{\frac{1}{6}} - 1 = 1.006961 - 1 = 0.006961 = 0.6961\%$

$$\begin{aligned} 25\,000.00 &= \text{PMT}\left(\frac{1 - 1.006961^{-240}}{0.006961}\right) \\ 25\,000.00 &= \text{PMT}(116.47420) \\ \text{PMT} &= \$214.64 \end{aligned}$$

The number of outstanding payments after five years is 15(12) = 180.

$$\begin{aligned} PV_{nc} &= 214.64\left(\frac{1 - 1.006961^{-180}}{0.006961}\right) \\ &= 214.64(102.44245) \\ &= \$21\,988.25 \end{aligned}$$

Programmed Solution

("END" mode) (Set P/Y = 12; C/Y = 2) 0 FV 25 000 ± PV 8.5 I/Y

240 N CPT CPT 214.639800

0 FV 214.64 ± PMT 8.5 I/Y 180 N CPT PV 21 988.24658

The outstanding balance after five years is $21 988.25.

(ii) After the five-year term is up, the outstanding balance of $21 988.25 is to be amortized over the remaining 15 years.

$PV_{nc} = 21\,988.25$; $n = 15(12) = 180$; P/Y = 12; C/Y = 2; I/Y = 8; $c = \frac{1}{6}$;
$i = \frac{8\%}{2} = 4\% = 0.04$;
$p = 1.04^{\frac{1}{6}} - 1 = 1.006558 - 1 = 0.006558 = 0.6558\%$

$$21\,988.25 = PMT\left(\frac{1 - 1.006558^{-180}}{0.006558}\right)$$
$$21\,988.25 = PMT(105.46822)$$
$$PMT = \$208.48$$

Programmed Solution

("END" mode) (Set P/Y = 12; C/Y = 2) 0 FV 21 988.25 ± PV 8 I/Y

180 N CPT PMT 208.482240

The monthly payment for the three-year term will be $208.48.

(iii) At the end of the three-year term, the number of outstanding payments is 144.

$$PV_{nc} = 208.48\left(\frac{1 - 1.006558^{-144}}{0.006558}\right)$$
$$= 208.48(92.99485)$$
$$= \$19\,387.57$$

Programmed Solution

("END" mode) (Set P/Y = 12; C/Y = 2) 0 FV 208.48 ± PMT 8 I/Y

144 N CPT PV 19 387.56586

The outstanding balance at the end of the three-year term will be $19 387.57.

2. Finding the Outstanding Balance When All Payments Except the Final Payment Are Equal

When the final payment is different from the preceding payments, use the retrospective method for finding the outstanding balance. Since this method requires finding the future value of an ordinary general annuity, Formula 6.2 applies.

$$FV_{nc} = PMT\left[\frac{(1+p)^n - 1}{p}\right] \quad \text{where } p = (1+i)^c - 1$$ — Formula 6.2

EXAMPLE 8.2D

A loan of $12 000.00 with interest at 12% compounded monthly and amortized by payments of $700.00 at the end of every three months is repaid in full after three years. What is the payout figure just after the last regular payment?

SOLUTION

Accumulate the value of the original principal after three years.

$PV = 12\,000.00; \quad n = 3(12) = 36; \quad i = \frac{12\%}{12} = 1\% = 0.01$

$$\begin{aligned} FV &= 12\,000.00(1.01^{36}) \\ &= 12\,000.00(1.430769) \\ &= \$17\,169.23 \end{aligned}$$

Programmed Solution

("END" mode) (Set P/Y = 12; C/Y = 12) [2nd] (CLR TVM) 12 000 [±] [PV] 12 [I/Y] 36 [N] [CPT] [FV] [17 169.2254]

Accumulate the value of the twelve payments made.

$PMT = 700.00; \; n = 3(4) = 12; \; P/Y = 4; C/Y = 12; I/Y = 12; c = \frac{12}{4} = 3;$

$i = 1\%$

$p = 1.01^3 - 1 = 1.030301 - 1 = 0.030301 = 3.0301\%$

$$\begin{aligned} FV_{nc} &= 700.00\left(\frac{1.030301^{12} - 1}{0.030301}\right) \\ &= 700.00(14.216322) \\ &= \$9951.43 \end{aligned}$$

Programmed Solution

("END" mode) (Set P/Y = 4; C/Y = 12) 0 [PV] 700 [±] [PMT] 12 [I/Y] 12 [N] [CPT] [FV] [9 951.425646]

The outstanding balance after three years is 17 169.23 − 9951.43 = $7217.80. The payout figure after three years is $7217.80.

Alternative Programmed Solution

(Set P/Y = 4; C/Y = 12) 12000 [±] [PV] 700 [PMT] 12 [I/Y]

12 [N] [CPT] [FV] [7217.799757]

C. Finding the interest paid and the principal repaid; constructing partial amortization schedules

EXAMPLE 8.2E

Mr. and Mrs. Poh took out a $40 000.00, 25-year mortgage renewable after five years. The mortgage bears interest at 9.5% compounded semi-annually and is amortized by equal monthly payments.

(i) What is the interest included in the 13th payment?
(ii) How much of the principal is repaid by the 13th payment?
(iii) What is the total interest cost during the first year?
(iv) What is the total interest cost during the fifth year?
(v) What will be the total interest paid by the Pohs during the initial five-year term?

SOLUTION

(i) First find the size of the monthly payment.

$PV_{nc} = 40\,000.00$; $n = 300$; P/Y = 12; C/Y = 2; I/Y = 9.5;

$c = \frac{1}{6}$; $i = 4.75\%$

$p = 1.0475^{\frac{1}{6}} - 1 = 1.007764 - 1 = 0.007764 = 0.7764\%$

$$40\,000.00 = \text{PMT}\left(\frac{1 - 1.007764^{-300}}{0.007764}\right)$$

$$40\,000.00 = \text{PMT}(116.140291)$$

$$\text{PMT} = \$344.41$$

Programmed Solution

("END" mode) (Set P/Y = 12; C/Y = 2) 0 [FV] 40 000 [±] [PV] 9.5 [I/Y]

300 [N] [CPT] [PMT] [344.411053]

Now find the outstanding balance after one year.

PMT = 344.41; $p = 0.7764\%$

The number of outstanding payments after one year $n = 288$.

$$PV_{nc} = \text{PMT}\left(\frac{1 - 1.007764^{-288}}{0.007764}\right)$$

$$= 344.41(114.90953)$$

$$= \$39\,576.05$$

Programmed Solution

0 [FV] 344.41 [±] [PMT] 9.5 [I/Y] 288 [N] [CPT] [PV] [39 576.05454]

The resulting difference in the present value amount is due to rounding of the payment.

The interest in the 13th payment is 39 576.05(0.007764) = \$307.28.

(ii) The principal repaid in the 13th payment is 344.41 − 307.28 = \$37.13.

(iii) The total amount paid during the first year is 344.41(12) = \$4132.92
The total principal repaid is 40 000.00 − 39 576.05 = 423.95

The total cost of interest for the first year = \$3708.97

(iv) After four years, $n = 300 - 48 = 252$.

$$PV_{nc} = 344.41\left(\frac{1 - 1.007764^{-252}}{0.007764}\right)$$
$$= 344.41(110.45204)$$
$$= \$38\,040.79$$

Programmed Solution

("END" mode) (Set P/Y = 12; C/Y = 2) 0 [FV] 344.41 [±] [PMT] 9.5 [I/Y] 252 [N] [CPT] [PV] [38 040.84291]

After five years, $n = 300 - 60 = 240$.

$$PV_{nc} = 344.41\left(\frac{1 - 1.007764^{-240}}{0.007764}\right)$$
$$= 344.41(108.66827)$$
$$= \$37\,426.44$$

Programmed Solution

("END" mode) (Set P/Y = 12; C/Y = 2) 0 [FV] 344.41 [±] [PMT] 9.5 [I/Y] 240 [N] [CPT] [PV] [37 426.49132]

The total amount paid during the fifth year is 344.41(12) = \$4132.92
The total principal repaid in Year 5 is 38 040.84 − 37 426.49 = 614.35
The total cost of interest in Year 5 = \$3518.57

(v) The total amount paid during the first five years is 344.41(60) = \$20 664.60
The total principal repaid is 40 000.00 − 37 426.49 = 2573.51
The total cost of interest for the first five years = \$18 091.09

EXAMPLE 8.2F

Confederated Venture Company financed a project by borrowing \$120 000.00 at 7% compounded annually and is repaying the loan at the rate of \$7000.00 due at the end of every three months.

(i) Compute the interest paid and the principal repaid by the tenth payment.
(ii) Construct a partial amortization schedule showing the first three payments, the tenth payment, the last three payments, and the totals.

SOLUTION

(i) Accumulate the value of the original loan after nine payments.

$PV = 120\ 000.00;\ n = 9;\ P/Y = 4;\ C/Y = 1;\ I/Y = 7;\ c = \frac{1}{4} = 0.25;$

$i = 7\% = 0.07;$

$p = 1.07^{0.25} - 1 = 1.0170585 - 1 = 0.0170585 = 1.70585\%$

$$\begin{aligned} FV &= 120\,000.00(1.017059^9) \\ &= 120\ 000.00(1.16443) \\ &= \$139\ 731.61 \end{aligned}$$

Programmed Solution

("END" mode) (Set P/Y = 4; C/Y = 1) [2nd] (CLR TVM) 120 000 [±] [PV]
7 [I/Y] 9 [N] [CPT] [FV] [139 731.6366]

Accumulate the value of the first nine payments.

$PMT = 7000.00;\ n = 9$

$$\begin{aligned} FV_{nc} &= 7000.00\left(\frac{1.017059^9 - 1}{0.017059}\right) \\ &= 7000.00(9.639186) \\ &= \$67\,474.30 \end{aligned}$$

Programmed Solution

("END" mode) (Set P/Y = 4; C/Y = 1) 0 [PV] 7000 [±] [PMT] 7 [I/Y]
9 [N] [CPT] [FV] [67 474.30606]

The outstanding balance after nine payments is 139 731.64 − 67 474.31 = \$72 257.33.

The interest included in the tenth payment is 72 257.33(0.017059) = \$1232.60.

The principal repaid by the tenth payment is 7000.00 − 1232.60 = \$5767.40.

(ii) To show the details of the last three payments, we need to know the number of payments required to amortize the loan.

$PV_{nc} = 120\ 000.00;\ PMT = 7000.00;\ p = 1.70585\%$

$$\begin{aligned}
120\,000.00 &= 7000.00\left(\frac{1-1.017059^{-n}}{0.017059}\right)\\
0.2924314 &= 1-1.017059^{-n}\\
1.017059^{-n} &= 0.707569\\
-n\ln 1.017059 &= \ln 0.707569\\
-n(0.016915) &= -0.345921\\
n &= 20.451013
\end{aligned}$$

Programmed Solution

("END" mode) (Set P/Y = 4; C/Y = 1) 0 [FV] 120 000 [±] [PV]

7000 [PMT] 7 [I/Y] [CPT] [N] [20.450975]

Twenty-one payments (20 payments of \$7000.00 plus a final payment) are required. This means the amortization schedule needs to show details of Payments 19, 20, and 21. To do so, we need to know the outstanding balance after 18 payments. Accumulate the value of the original principal after 18 payments.

$$\begin{aligned}
FV &= 120\,000.00(1.017059^{18})\\
&= 120\,000.00(1.355897)\\
&= \$162\,707.68
\end{aligned}$$

Programmed Solution

("END" mode) (Set P/Y = 4; C/Y = 1) [2nd] (CLR TVM) 120 000 [±] [PV]

7 [I/Y] 18 [N] [CPT] [FV] [162 707.7523]

Accumulate the value of the first 18 payments.

$$\begin{aligned}
FV_{nc} &= 7000.00\left(\frac{1.017059^{18}-1}{0.017059}\right)\\
&= 7000.00(20.863343)\\
&= \$146\,043.40
\end{aligned}$$

Programmed Solution

("END" mode) (Set P/Y = 4; C/Y = 1) 0 [PV] 7000 [±] [PMT]

7 [I/Y] 18 [N] [CPT] [FV] [146 043.4329]

The outstanding balance after 18 payments is 162 707.75 − 146 043.43 = \$16 664.32.

Partial amortization schedule

Payment Number	Amount Paid	Interest Paid $p = 0.0170585$	Principal Repaid	Outstanding Principal Balance
0				120 000.00
1	7 000.00	2 047.02	4 952.98	115 047.02
2	7 000.00	1 962.53	5 037.47	110 009.55
3	7 000.00	1 876.60	5 123.40	104 886.15
•	•	•	•	•
•	•	•	•	•
9	•	•	•	72 257.33
10	7 000.00	1 232.60	5 767.40	66 489.93
•	•	•	•	•
•	•	•	•	•
•	•	•	•	•
18	•	•	•	16 664.32
19	7 000.00	284.27	6 715.73	9 948.59
20	7 000.00	169.71	6 830.29	3 118.30
21	3 171.49	53.19	3 118.30	0.00
TOTAL	143 171.49	23 171.49	120 000.00	

As an alternative, use the calculator's Amortization function. Instructions on its use are available on the CD-ROM accompanying this book.

D. Computer application—Amortization schedule

Excel

As shown in Section 8.1E, amortization schedules can be created using spreadsheet programs like Excel. Refer to Section 8.1E for instructions for creating amortization schedules in spreadsheets. Using those instructions and the formulas shown in Figure 8.4 below, you can create the amortization schedule for the debt in Example 8.2A.

FIGURE 8.4

Workbook 1

	A	B	C	D	E	F
1	Payment Number	Amount Paid	Interest Paid	Principal Repaid	Outstanding Principal Balance	
2	0				30000.00	6688.77
3	1	= F2	= F4*E2	= B3−C3	= E2−D3	
4	2	= F2	= F4*E3	= B4−C4	= E3−D4	0.1255088
5	3	= F2	= F4*E4	= B5−C5	= E4−D5	
6	4	= F2	= F4*E5	= B6−C6	= E5−D6	
7	5	= F2	= F4*E6	= B7−C7	= E6−D7	
8	6	= F2	= F4*E7	= B8−C8	= E7−D8	
9	7	= C9+D9	= F4*E8	= E8	= E8−D9	
10	Totals	= SUM(B3:B9)	= SUM(C3:C9)	= SUM(D3:D9)		

EXERCISE 8.2

Excel If you choose, you can use Excel's **CUMIPMT**, **CUMPRINC**, **EFFECT**, **NPER**, **PMT**, or **PV** functions or **Template4** to answer the questions indicated below. Refer to the Spreadsheet Template Disk to find **Template4** or to learn how to use these Excel functions.

For each of the following four debts amortized by equal payments made at the end of each payment interval, compute: (a) the size of the periodic payments; (b) the outstanding principal at the time indicated; (c) the interest paid; and (d) the principal repaid by the payment following the time indicated for finding the outstanding principal.

	Debt Principal	Repayment Period	Payment Interval	Interest Rate	Conversion Period	Outstanding Principal Required After:
1.	$36 000.00	20 years	6 months	8%	quarterly	25th payment
2.	$15 000.00	10 years	3 months	12%	monthly	15th payment
3.	$8 500.00	5 years	1 month	6%	semi-annually	30th payment
4.	$9 600.00	7 years	3 months	9%	semi-annually	10th payment

For each of the following four debts repaid by periodic payments as indicated, compute: (a) the number of payments required to amortize the debts; and (b) the outstanding principal at the time indicated.

	Debt Principal	Debt Payment	Payment Interval	Interest Rate	Conversion Period	Outstanding Principal Required After:
1.	$6 000.00	$400.00	3 months	6%	monthly	10th payment
2.	$8 400.00	$1200.00	6 months	10%	quarterly	5th payment
3.	$23 500.00	$1800.00	3 months	7%	annually	14th payment
4.	$18 200.00	$430.00	1 month	8%	semi-annually	48th payment

Answer each of the following questions.

1. A debt of $45 000.00 is repaid over fifteen years with semi-annual payments. Interest is 9% compounded monthly.

(a) What is the size of the periodic payments?
(b) What is the outstanding principal after the eleventh payment?
(c) What is the interest paid in the twelfth payment?
(d) How much principal is repaid in the twelfth payment?

2. A debt of $60 000.00 is repaid over twenty-five years with monthly payments. Interest is 7% compounded semi-annually.

(a) What is the size of the periodic payments?
(b) What is the outstanding principal after the 119th payment?
(c) What is the interest paid in the 120th payment?
(d) How much principal is repaid in the 120th payment?

3. A $36 000.00 mortgage amortized by monthly payments over twenty-five years is renewable after three years.
   (a) If interest is 7% compounded semi-annually, what is the size of each monthly payment?
   (b) What is the mortgage balance at the end of the three-year term?
   (c) How much interest will have been paid during the first three years?
   (d) If the mortgage is renewed for a further three-year term at 9% compounded semi-annually, what will be the size of the monthly payments for the renewal period?

4. Fink and Associates bought a property valued at $80 000.00 for $15 000.00 down and a mortgage amortized over fifteen years. The firm makes equal pay ments due at the end of every three months. Interest on the mortgage is 6.5% compounded annually and the mortgage is renewable after five years.
   (a) What is the size of each quarterly payment?
   (b) What is the outstanding principal at the end of the five-year term?
   (c) What is the cost of the mortgage for the first five years?
   (d) If the mortgage is renewed for a further five years at 9% compounded semi-annually, what will be the size of each quarterly payment?

5. A loan of $10 000.00 with interest at 10% compounded quarterly is repaid by payments of $950.00 made at the end of every six months.
   (a) How many payments will be required to amortize the loan?
   (b) If the loan is repaid in full after six years, what is the payout figure?
   (c) If paid out, what is the total cost of the loan?

6. The owner of the Blue Goose Motel borrowed $12 500.00 at 12% compounded semi-annually and agreed to repay the loan by making payments of $700.00 at the end of every three months.
   (a) How many payments will be needed to repay the loan?
   (b) How much will be owed at the end of five years?
   (c) How much of the payments made at the end of five years will be interest?

7. A loan of $16 000.00 with interest at 9% compounded quarterly is repaid in seven years by equal payments made at the end of each year. Find the size of the annual payments and construct an amortization schedule showing the total paid and the total interest.

8. A debt of $12 500.00 with interest at 7% compounded semi-annually is repaid by payments of $1900.00 made at the end of every three months. Construct an amortization schedule showing the total paid and the total cost of the debt.

9. For Question 7, calculate the interest included in the fifth payment period. Verify your answer by checking the amortization schedule.

10. For Question 8, compute the principal repaid in the sixth payment period. Verify your answer by checking the amortization schedule.

11. A \$40 000.00 mortgage amortized by monthly payments over 25 years is renewable after five years.

(**a**) If interest is 8.5% compounded semi-annually, what is the size of each monthly payment?
(**b**) Find the total interest paid during the first year.
(**c**) Compute the interest included in the 48th payment.
(**d**) If the mortgage is renewed after five years at 10.5% compounded semi-annually, what is the size of the monthly payment for the renewal period?
(**e**) Construct a partial amortization schedule showing details of the first three payments for each of the two five-year terms.

12. A debt of \$32 000.00 is repaid by payments of \$2950.00 made at the end of every six months. Interest is 12% compounded quarterly.

(**a**) What is the number of payments needed to retire the debt?
(**b**) What is the cost of the debt for the first five years?
(**c**) What is the interest paid in the tenth payment period?
(**d**) Construct a partial amortization schedule showing details of the first three payments, the last three payments, and totals.

8.3 Finding the Size of the Final Payment

A. Three methods for computing the final payment

When all payments except the final payment are equal, three methods are available to compute the size of the final payment.

METHOD 1 Compute the value of the term n and determine the present value of the outstanding fractional payment. If the final payment is made at the end of the payment interval, add interest for one payment interval to the present value.

METHOD 2 Use the retrospective method to compute the outstanding principal after the last of the equal payments. If the final payment is made at the end of the payment interval, add to the outstanding principal the interest for one payment interval.

METHOD 3 Assume all payments to be equal, compute the overpayment, and subtract the overpayment from the size of the equal payments. This method must not be used when the payments are made at the beginning of the payment interval.

EXAMPLE 8.3A For Example 8.1D, compute the size of the final payment using each of the three methods. Compare the results with the size of the payment shown in the amortization schedule.

SOLUTION

METHOD 1 $PV_n = 15\,000.00$; PMT = 2500.00; P/Y = 4; C/Y = 4; I/Y = 10; $i = 2.5\%$

$$15\,000.00 = 2500.00\left(\frac{1 - 1.025^{-n}}{0.025}\right)$$

$$\begin{aligned} 6.00 &= \frac{1 - 1.025^{-n}}{0.025} \\ 0.15 &= 1 - 1.025^{-n} \\ 1.025^{-n} &= 0.85 \\ -n \ln 1.025 &= \ln 0.85 \\ -n(0.024693) &= -0.162519 \\ n &= 6.5816844 \end{aligned}$$

Programmed Solution

("END" mode) (Set P/Y = 4; C/Y = 4) 0 [FV] 15 000 [±] [PV] 2500 [PMT] 10 [I/Y] [CPT] [N] [6.58168223]

PMT = 2500.00; $i = 2.5\%$; $n = 0.581684$

$$\begin{aligned} PV_n &= 2500.00\left(\frac{1 - 1.025^{-0.5816844}}{0.025}\right) \\ &= 2500.00(0.570426) \\ &= \$1426.06 \end{aligned}$$

Programmed Solution

0 [FV] 2500 [±] [PMT] 10 [I/Y] 0.581684 [N] [CPT] [PV] [1426.058892]

Interest for one interval is 1426.06(0.025) = \$35.65.
Final payment is 1426.06 + 35.65 = \$1461.71.

METHOD 2

Compute the value of n as done in Method 1.
Since $n = 6.5816844$, the number of equal payments is 6.
The accumulated value of the original principal after six payments

$$\begin{aligned} FV &= 15\,000.00(1.025^{6}) \\ &= 15\,000.00(1.159693) \\ &= \$17\,395.40 \end{aligned}$$

Programmed Solution

("END" mode) (Set P/Y = 4; C/Y = 4) [2nd] (CLR TVM) 15 000 [±] [PV] 10 [I/Y] 6 [N] [CPT] [FV] [17 395.40127]

The accumulated value of the first six payments

$$\begin{aligned} FV_6 &= 2500.00\left(\frac{1.025^{6} - 1}{0.025}\right) \\ &= 2500.00(6.387737) \\ &= \$15\,969.34 \end{aligned}$$

Programmed Solution

0 [PV] 2500 [±] [PMT] 10 [I/Y] 6 [N] [CPT] [FV] [15 969.34182]

The outstanding balance after six payments is 17 395.40 − 15 969.34 = \$1426.06.

The final payment = outstanding balance + interest for one period
= the accumulated value of \$1426.06 for one year
= 1426.06(1.025)
= \$1461.71

METHOD 3

Compute the value of n as done in Method 1.
Since $n = 6.581684$, the number of assumed full payments is 7.
The accumulated value of the original principal after seven payments

$$\begin{aligned} FV &= 15\,000.00(1.025^7) \\ &= 15\,000.00(1.188686) \\ &= \$17\,830.29 \end{aligned}$$

Programmed Solution

("END" mode) (Set P/Y = 4; C/Y = 4) [2nd] (CLR TVM) 15 000 [±] [PV] 10 [I/Y]
7 [N] [CPT] [FV] [17 830.28631]

The accumulated value of seven payments

$$\begin{aligned} FV_7 &= 2500.00\left(\frac{1.025^7 - 1}{0.025}\right) \\ &= 2500.00(7.547430) \\ &= \$18\,868.58 \end{aligned}$$

Programmed Solution

0 [PV] 2500 [±] [PMT] 10 [I/Y] 7 [N] [CPT] [FV] [18 868.57537]

Since the accumulated value of seven payments is greater than the accumulated value of the original principal, there is an overpayment.

18 868.58 − 17 830.29 = \$1038.29

The size of the final payment is 2500.00 − 1038.29 = \$1461.71.

Note:

1. For all three methods, you must determine the term n using the methods shown in Chapters 5 and 6.
2. When using a scientific calculator, Method 1 is preferable because it is the most direct method. It is the method used in Subsection B below.

POINTERS AND PITFALLS

When dealing with loan repayment problems in which the size of the final payment must be determined, once n has been calculated, the size of the final payment *cannot* be calculated by simply multiplying the periodic pay (PMT) by the decimal portion of n (i.e., the non-integral part of n). For example, the *correct* value for the size of the final rent payment in Example 8.3A, Method 1 is $1461.71. If you had done it *incorrectly*, the final rent payment would have been $1454.21, the product of 0.586488 × $2500.00.

B. Applications

EXAMPLE 8.3B

On his retirement, Art received a bonus of $8000.00 from his employer. Taking advantage of the existing tax legislation, he invested his money in an annuity that provides for semi-annual payments of $1200.00 at the end of every six months. If interest is 6.25% compounded semi-annually, determine the size of the final payment (see Chapter 5, Example 5.5D).

SOLUTION

$PV_n = 8000.00$; PMT = 1200.00; P/Y = 2; C/Y = 2; I/Y = 6.25;

$$i = \frac{6.25\%}{2} = 3.125\%$$

$$\begin{aligned}
8000.00 &= 1200.00\left(\frac{1 - 1.03125^{-n}}{0.03125}\right)\\
0.208333 &= 1 - 1.03125^{-n}\\
1.03125^{-n} &= 0.791667\\
-n \ln 1.03125 &= \ln 0.791667\\
-n(0.030772) &= -0.233615\\
n &= 7.59188 \text{ (half-year periods)}
\end{aligned}$$

The annuity will be in existence for four years. Art will receive seven payments of $1200.00 and a final payment that will be less than $1200.00.

PMT = 1200.00; $i = 3.125\%$; $n = 0.59188$

$$\begin{aligned}
PV_n &= 1200.00\left(\frac{1 - 1.03125^{-0.59188}}{0.03125}\right)\\
&= 1200.00(0.577545)\\
&= \$693.05
\end{aligned}$$

Programmed Solution

("END" mode) (Set P/Y = 2; C/Y = 2) 0 [FV] 8000 [±] [PV] 1200 [PMT] 6.25 [I/Y] [CPT] [N] [7.591883613]

0 [FV] 1200 [±] [PMT] 6.25 [I/Y] 0.591883613 [N] [CPT] [PV] [693.0578625]

Final payment including the interest for one payment interval is 693.06(1.03125) = $714.72.

EXAMPLE 8.3C

A lease contract valued at $7800.00 is to be fulfilled by rental payments of $180.00 due at the beginning of each month. If money is worth 9% compounded monthly, determine the size of the final lease payment (see Chapter 7, Example 7.1O).

SOLUTION

$PV_n(\text{due}) = 7800.00$; $PMT = 180.00$; $P/Y = 12$; $C/Y = 12$; $I/Y = 9$;

$i = \frac{9\%}{12} = 0.75\%$

$n = 52.123125$ (see solution to Example 7.1O)

Present value of the final payment

$PMT = 180.00; \quad i = 0.75\%; \quad n = 0.123125$

$$\begin{aligned} PV_n(\text{due}) &= 180.00(1.0075)\left(\frac{1 - 1.0075^{-0.123125}}{0.0075}\right) \quad \text{Formula 6.2} \\ &= 180.00(1.0075)(0.122609) \\ &= 180.00(0.123529) \\ &= \$22.24 \end{aligned}$$

Programmed Solution

("BGN" mode) (Set P/Y = 12; C/Y = 12) 0 [FV] 180 [PMT] 9 [I/Y] 7800 [±] [PV] [CPT] [N] 52.12312544

("BGN" mode) 0 [FV] 180 [±] [PMT] 9 [I/Y] 0.12312544 [N] [CPT] [PV] 22.23525214

Since the payment is made at the beginning of the last payment interval, no interest is added. The final payment is $22.24.

EXAMPLE 8.3D

Payments of $500.00 deferred for nine years are received at the end of each month from a fund of $10 000.00 deposited at 10.5% compounded monthly. Determine the size of the final payment (see Chapter 7, Example 7.3E).

SOLUTION

$PV_n(\text{defer}) = 10\,000.00$; $PMT = 500.00$; $d = 9(12) = 108$; $P/Y = 12$; $C/Y = 12$;

$I/Y = 10.5$; $i = \frac{10.5\%}{12} = 0.875\%$

$n = 68.288334$ (see solution to Example 7.3E)

Present value of the final payment

$PMT = 500.00; \quad i = 0.875\%; \quad n = 0.288334$

$$\begin{aligned} PV_n &= 500.00\left(\frac{1 - 1.00875^{-0.288334}}{0.00875}\right) \\ &= 500.00(0.286720) \\ &= \$143.36 \end{aligned}$$

Programmed Solution

("END" mode) (Set P/Y = 12; C/Y = 12) 0 [FV] 500 [±] [PMT] 10.5 [I/Y] 0.288334 [N] [CPT] [PV] [143.359793]

The size of the final payment is 143.36(1.00875) = \$144.61.

EXAMPLE 8.3E

A business valued at \$96 000.00 is purchased for a down payment of 25% and payments of \$4000.00 at the end of every three months. If interest is 9% compounded monthly, what is the size of the final payment? (See Chapter 6, Example 6.4B.)

SOLUTION

$PV_{nc} = 96\ 000.00(0.75) = 72\ 000.00$; PMT = 4000.00; P/Y = 4; C/Y = 12;
I/Y = 9; $c = \frac{12}{4} = 3$; $i = \frac{9\%}{12} = 0.75\% = 0.0075$; $p = 1.0075^3 - 1 = 2.26692\%$;
$n = 23.390585$ (see solution to Example 6.4B)

Present value of the final payment

PMT = 4000.00; $p = 2.26692\%$; $n = 0.390585$

$$PV_n = 4000.00\left(\frac{1 - 1.022669^{-0.390585}}{0.022669}\right)$$
$$= 4000.00(0.384538)$$
$$= \$1538.15$$

Programmed Solution

("END" mode) (Set P/Y = 4; C/Y = 12) 0 [FV] 4000 [±] [PMT] 9 [I/Y] 0.390585 [N] [CPT] [PV] [1538.151278]

The final payment is 1538.15(1.0226692) = \$1573.02.

EXAMPLE 8.3F

Ted Davis, having reached his goal of a \$140 000.00 balance in his RRSP, converts it into an RRIF and withdraws from it \$1650.00 at the beginning of each month. If interest is 5.75% compounded quarterly, what is the size of the final withdrawal? (See Chapter 7, Example 7.2G.)

SOLUTION

$PV_{nc}(\text{due}) = 140\ 000.00$; PMT = 1650.00; P/Y = 12; C/Y = 4; I/Y = 5.75;

$c = \frac{4}{12} = \frac{1}{3}$; $i = \frac{5.75\%}{4} = 1.4375\% = 0.014375$;

$p = 1.014375^{\frac{1}{3}} - 1 = 1.004769 - 1 = 0.4769\%$

$n = 108.32388$ (see solution to Example 7.2G)

Present value of final payment

PMT = 1650.00; $p = 0.47689\%$; $n = 0.32388$

$$PV_n = 1650.00(1.004769)\left(\frac{1 - 1.004769^{-0.32395}}{0.004769}\right)$$

$$= 1650.00(1.004769)(0.322861)$$

$$= 1650.00(0.324401)$$

$$= \$535.26$$

Programmed Solution

("BGN" mode) (Set P/Y = 12; C/Y = 4) 0 [PV] 1650 [±] [PMT] 5.75 [I/Y]

0.32388 [N] [CPT] [PV] [535.261739]

Since the payment is at the beginning of the last payment interval, it is $535.26.

EXERCISE 8.3

Excel If you choose, you can use Excel's **NPER** or **PV** functions to answer the questions indicated below. Refer to the Spreadsheet Template Disk to learn how to use these Excel functions.

For each of the following six loans, compute the size of the final payment.

	Principal	Periodic Payment	Payment Interval	Payment Made At:	Interest Rate	Conversion Period
1.	$17 500.00	$1100.00	3 months	end	9%	quarterly
2.	$7 800.00	$775.00	6 months	beginning	7%	semi-annually
3.	$9 300.00	$580.00	3 months	beginning	7%	quarterly
4.	$15 400.00	$1600.00	6 months	end	8%	quarterly
5.	$29 500.00	$1650.00	3 months	end	9%	monthly
6.	$17 300.00	$425.00	1 month	beginning	6%	quarterly

Answer each of the following questions.

1. A loan of $7200.00 is repaid by payments of $360.00 at the end of every three months. Interest is 11% compounded quarterly.

(**a**) How many payments are required to repay the debt?
(**b**) What is the size of the final payment?

2. Seanna O'Brien receives pension payments of $3200.00 at the end of every six months from a retirement fund of $50 000.00. The fund earns 7% compounded semi-annually.

(**a**) How many payments will Seanna receive?
(**b**) What is the size of the final pension payment?

3. A loan of $35 000.00 is repaid by payments of $925.00 at the end of every month. Interest is 12% compounded monthly.
   (a) How many payments are required to repay the debt?
   (b) What is the size of the final payment?

4. An annuity with a cash value of $10 500.00 pays $900.00 at the beginning of every three months. The investment earns 11% semi-annually.
   (a) How many payments will be paid?
   (b) What is the size of the final annuity payment?

5. Payments of $1200.00 are made out of a fund of $25 000.00 at the end of every three months. If interest is 6% compounded monthly, what is the size of the final payment?

6. A debt of $30 000.00 is repaid in monthly installments of $550.00. If interest is 8% compounded quarterly, what is the size of the final payment?

7. A lease valued at $20 000.00 requires payments of $1000.00 every three months due in advance. If money is worth 7% compounded quarterly, what is the size of the final lease payment?

8. Eduardo Martinez has saved $125 000.00. If he withdraws $1250.00 at the beginning of every month and interest is 10.5% compounded monthly, what is the size of the last withdrawal?

9. Equipment priced at $42 000.00 was purchased on a contract requiring payments of $5000.00 at the beginning of every six months. If interest is 9% compounded quarterly, what is the size of the final payment?

10. Noreen Leung has agreed to purchase her partner's share in the business by making payments of $1100.00 every three months. The agreed transfer value is $16 500.00 and interest is 10% compounded annually. If the first payment is due at the date of the agreement, what is the size of the final payment?

11. David Jones has paid $16 000.00 for a retirement annuity from which he will receive $1375.00 at the end of every three months. The payments are deferred for ten years and interest is 10% compounded quarterly.
    (a) How many payments will David receive?
    (b) What is the size of the final payment?
    (c) How much will David receive in total?
    (d) How much of what he receives will be interest?

12. A contract valued at $27 500.00 requires payments of $6000.00 every six months. The first payment is due in four years and interest is 11% compounded semi-annually.
    (a) How many payments are required?
    (b) What is the size of the last payment?
    (c) How much will be paid in total?
    (d) How much of what is paid is interest?

8.4 Residential Mortgages in Canada

A. Basic concepts and definitions

A **residential mortgage** is a claim to a residential property given by a borrower to a lender as security for the repayment of a loan. It is often the largest amount of money ever borrowed by an individual. The borrower is called the **mortgagor**; the lender is called the **mortgagee**. The **mortgage contract** spells out the obligations of the borrower and the rights of the lender, including the lender's rights in case of default in payment by the borrower. If the borrower is unable to make the mortgage payments, the lender ultimately has the right to dispose of the property under *power of sale* provisions.

To secure legal claim against a residential property, the lender must register the mortgage against the property at the provincial government's land titles office. A **first mortgage** is the first legal claim against a residential property if the mortgage payments cannot be made and the property must be sold. **Equity** in a property is the difference between the property's market value and the total debts, or mortgages, registered against the property. It is possible to have a **second mortgage** on a residential property that is backed by equity in the property, even if there is a first mortgage already registered against the property. If the borrower defaults on the mortgage payments and the property must be sold, the first mortgagee gets paid before the second mortgagee. For this reason, second mortgages are considered riskier investments than first mortgages. They command higher interest rates than first mortgages to compensate for this risk. Home improvement loans and home equity loans are often secured by second mortgages. It is even possible to obtain *third mortgages* against residential properties. Third mortgages rank behind first and second mortgages. Thus, they command much higher interest rates.

Financial institutions offer two types of mortgages—fixed-rate mortgages and demand (or variable-rate) mortgages. A **fixed-rate mortgage** is a mortgage for which the rate of interest is fixed for a specific period of time. A **demand** (or **variable-rate**) **mortgage** is a mortgage for which the rate of interest changes as money market conditions change. The interest rate change is usually related to the change in a bank's prime lending rate. Both types of mortgage are usually repaid by equal payments that blend principal and interest. Payments are often required to be made monthly, but some lenders are more flexible and allow semi-monthly, bi-weekly, and even weekly payments.

For all types of mortgages, the amortization period is a part of the mortgage agreement. The amortization period is used to calculate the amount of the blended payments. The most common amortization period is twenty-five years for fixed-rate mortgages and twenty years for demand mortgages. Shorter amortization periods may be used at the discretion of the lender or the borrower.

The *term* of the mortgage specifies the period of time for which the interest rate is fixed. By definition, only fixed-rate mortgages have terms. The term ranges from six months to five years. Longer terms, such as seven and ten years, are becoming more frequently available.

For fixed-rate mortgages, Canadian law dictates that interest must be calculated semi-annually or annually, not in advance. In this context, "not in advance" means that interest is calculated at the end of each six-month or twelve-month period, not at the beginning. It is Canadian practice to calculate fixed-rate mortgage interest semi-annually, not in advance.

Fixed-rate mortgages can be either open or closed. Closed mortgages restrict the borrower's ability to increase payments, make lump-sum payments, change the term of the mortgage, or transfer the mortgage to another lender without penalty. Most closed mortgages contain some prepayment privileges. For example, some mortgages allow a lump-sum payment each year of up to 10% or 15% of the original mortgage principal, usually on the anniversary date of the mortgage. Some lenders permit increases in the periodic payments up to 10% or 15% once each calendar year. Changes in the term or transfers are usually subject to prohibitive penalties.

Open mortgages allow prepayment or repayment of the mortgage at any time without penalty. They are available from most lenders for terms of up to two years. However, interest rates on open mortgages are significantly higher than interest rates on closed mortgages. For a six-month term, the usual charge (i.e., the premium) for an open mortgage is an interest rate at least 0.5% (i.e., 50 basis points) higher than for a closed mortgage. For one-year and two-year terms, the interest rate is usually at least 1.0% (i.e., 100 basis points) higher.

As we stated above, a demand (or variable-rate) mortgage is a mortgage for which the rate of interest changes over time. Interest is calculated on a daily basis. Therefore, demand mortgages are not subject to the legal restriction of compounding semi-annually, not in advance. Demand mortgages also do not have a fixed term, because the interest rate may fluctuate.

B. CMHC mortgages

Fixed-rate mortgages and demand mortgages are usually available from financial institutions for up to 75% of the value of a property (calculated as the lesser of the purchase price and the appraised value of a property). This means the borrower needs to have at least a 25% down payment for the purchase of a residential property or at least 25% equity in a property. To borrow beyond the 75% level, the mortgage must be insured by the Canada Mortgage and Housing Corporation (CMHC).

Canada Mortgage and Housing Corporation (**CMHC**) is the corporation of the federal government that administers the National Housing Act (NHA) and provides mortgage insurance to lenders. CMHC acts as the insurer for the lender in the event that the borrower defaults on the mortgage payments. Anyone buying a home as a principal residence is an eligible borrower.

All CMHC borrowers must meet the following conditions:

- The maximum **Gross Debt Service** (**GDS**) **ratio**, including heating costs, is 32%. The GDS ratio is the percent of your gross annual income required to cover housing costs such as mortgage payments, property taxes, and heating costs.)

- The maximum **Total Debt Service (TDS) ratio** is 40%. The TDS ratio is the percent of your gross annual income required to cover housing costs *and* all other debts and obligations, such as a car loan.)
- The maximum amortization period is twenty-five years.
- Borrowers are required to demonstrate, at the time of their application, their ability to cover closing costs equal to at least 1.5% of the purchase price.
- If a borrower is using a financial gift to pay a part of the down payment, the borrower must be in possession of the funds thirty days before making an offer to purchase.

The maximum mortgage loan available depends on the home's geographic location in Canada. Currently, maximum insurable house prices range from $300 000 in Vancouver, Victoria, and Toronto, to $125 000 in other areas of Canada.

If the mortgage is insured by CMHC, home buyers can borrow up to 95% of the purchase price. The borrower must pay to CMHC a fixed administration fee (currently $75.00 for basic service; $235.00 for full service), as well as an insurance premium. The borrower can pay these fees separately or have them added to the mortgage balance.

CMHC insurance premiums depend on the loan-to-lending-value ratio, defined as Loan Principal ÷ Purchase Price. For example, a home buyer borrowing $160 000 to purchase a $200 000 home will have a loan-to-lending-value ratio of 160 000 ÷ 200 000 = 80%.

For single advance borrowers, the insurance premium for ratios exceeding 75% are currently as follows:

more than 75% to 80% inclusive	1.25% of loan principal
more than 80% to 85% inclusive	2.00% of loan principal
more than 85% to 90% inclusive	2.50% of loan principal
more than 90% to 95% inclusive	3.75% of loan principal

EXAMPLE 14.4A

The Wongs want to purchase a home in Vancouver with a CMHC-approved mortgage. According to the CMHC terms and conditions, determine the maximum initial mortgage balance if the Wongs opt for full service and want the administration fee and the insurance premium added to the maximum allowable loan balance.

SOLUTION

The maximum house price in Vancouver qualifying for a CMHC mortgage is $300 000.

The maximum CMHC loan balance = 95% of $300 000 = $285 000.

The full service fee = $235.00.

The insurance premium = 3.75% of $285 000 = $10 687.50.

The maximum initial mortgage balance
= 285 000.00 + 235.00 + 10 687.50 = $295 922.50.

DID YOU KNOW?

There are a number of different strategies that home buyers might consider to reduce the amount of interest they pay over the life of a mortgage:

(i) *Comparison shop* at a number of local banks, trust companies, credit unions, reputable mortgage brokers, and other lenders to find the lowest mortgage interest rates available.

(ii) Reduce the *amortization period.* For example, suppose you have arranged a $100 000 mortgage at 6% p.a. compounded semi-annually with monthly payments of $712.19. By cutting the amortization period from 20 years to 15 years, total interest savings amount to $19 747.20.

(iii) Consider *bi-weekly payments* rather than monthly payments. For example, on a $100 000 mortgage at 6% p.a. compounded semi-annually, bi-weekly payments of $356.10 (rather than monthly payments of $712.19) would result in (a) total interest savings of $11 251.09 and (b) an amortization period of 17 years, 3 months, and 2 weeks, instead of 20 years.

(iv) Avoid paying the additional costs associated with a CMHC-insured mortgage by *making a down payment of at least 25% of the purchase price* of the home. By doing this, a homeowner qualifies for a conventional (non-CMHC-insured) mortgage, thereby avoiding the insurance premium that would be added to the mortgage amount. Total savings on a 20-year, $100 000 mortgage at 6% compounded semi-annually range between $854.40 (for a 0.5% CMHC insurance-premium rate) and $4272.00 (for a 2.5% CMHC insurance-premium rate).

C. Computing the effective rate of interest for fixed-rate mortgages

For residential mortgages, Canadian legislation requires the rate of interest charged by the lender to be calculated annually or semi-annually, not in advance. The fixed rates advertised, posted, or quoted by lenders are usually nominal annual rates. To meet the legislated requirements, the applicable nominal annual rate of interest must be converted into the equivalent effective rate of interest per payment period.

This is done as explained in Section 6.1A by using Formula 6.1.

$$p = (1 + i)^c - 1$$

where p = the effective rate of interest per payment period
i = the rate per conversion period

$$c = \frac{\text{THE NUMBER OF INTEREST CONVERSION PERIODS PER YEAR}}{\text{THE NUMBER OF PAYMENT PERIODS PER YEAR}}$$

The prevailing practice is semi-annual compounding and monthly payment for most mortgages. For most mortgages (with semi-annual compounding and monthly payment),

$$c = \frac{2 \text{ (compounding periods per year)}}{12 \text{ (payments per year)}} = \frac{1}{6}$$

EXAMPLE 8.4B

Suppose a financial institution posted the interest rates for closed mortgages shown below. The interest is compounded semi-annually and the mortgages require monthly payments.

Term	Interest Rate
6 months	5.25%
1 year	5.50%
2 years	6.50%
3 years	7.00%
4 years	7.25%
5 years	7.50%

Compute the effective rate of interest per payment period for each term.

SOLUTION

$c = \frac{1}{6}; \quad p = (1 + i)^{\frac{1}{6}} - 1$

For the six-month term, $i = \frac{5.25\%}{2} = 2.625\% = 0.02625$

$$\begin{aligned} p &= (1 + 0.02625)^{\frac{1}{6}} - 1 \\ &= 1.004328 - 1 \\ &= 0.004328, \text{ or } 0.4328\% \end{aligned}$$

For the one-year term, $i = \frac{5.50\%}{2} = 2.75\% = 0.0275$

$p = (1.0275^{\frac{1}{6}}) - 1 = 0.004532 = 0.4532\%$

You can obtain the effective rates using a preprogrammed financial calculator by using the function keys as follows:

("END" mode) (Set P/Y = 1; C/Y = 1) [2nd] (CLR TVM) 1 [±] [PV]

$\frac{1}{6} = 0.166667$ [N] (enter applicable interest rate)

[I/Y] [CPT] [FV] [Display Shows $(1 + p)$]

For the two-year term, $i = 3.25\% = 0.0325$; $p = (1.0325^{\frac{1}{6}}) - 1 = 0.005345 = 0.5345\%$

[2nd] (CLR TVM) 1 [±] [PV] 0.166667 [N]

3.25 [I/Y] [CPT] [FV] [1.005344741]

$p = 0.534474\%$

(Note that if you perform these calculations in succession, you do not have to key in PV and N each time.)

For the three-year term, $i = 3.5\% = 0.035$; $p = (1.035^{\frac{1}{6}}) - 1 = 0.00575 = 0.575\%$

2nd (CLR TVM) 1 ± PV 0.166667 N

3.5 I/Y CPT FV 1.005750041

$p = 0.5750041\%$

For the four-year term, $i = 3.625\% = 0.03625$; $p = (1.03625^{\frac{1}{6}}) - 1 = 0.005952 = 0.5952\%$

2nd (CLR TVM) 1 ± PV 0.1666667 N

3.625 I/Y CPT FV 1.005952385

$p = 0.5952385\%$

For the five-year term, $i = 3.75\% = 0.0375$; $p = (1.0375^{\frac{1}{6}}) - 1 = 0.006155 = 0.6155\%$

2nd (CLR TVM) 1 ± PV 0.1666667 N

3.75 I/Y CPT FV 1.006154525

$p = 0.6154525\%$

D. Computing mortgage payments and balances

Blended residential mortgage payments are ordinary general annuities. Therefore, Formula 12.3 applies.

$$PV_{nc} = PMT\left[\frac{1-(1+p)^{-n}}{p}\right], \text{ where } p = (1+i)^c - 1$$ ——————Formula 5.5

The periodic payment PMT is calculated using the method shown in Section 6.3.

EXAMPLE 8.4C

A mortgage for $120 000.00 is amortized over 25 years. Interest is 7.5% p.a., compounded semi-annually, for a five-year term and payments are monthly.

(i) Compute the monthly payment.
(ii) Compute the balance at the end of the five-year term.
(iii) Compute the monthly payment if the mortgage is renewed for a four-year term at 7.0% compounded semi-annually.

SOLUTION

(i) When computing the monthly payment, n is the total number of payments in the amortization period. The term for which the rate of interest is fixed (in this case, five years) does *not* enter the calculation.

$PV_{nc} = 120\,000.00$; $n = 12(25) = 300$; $P/Y = 12$; $C/Y = 2$;

$I/Y = 7.5$; $i = \frac{7.5\%}{2} = 3.75\%$; $c = \frac{1}{6}$

We must first compute the effective monthly rate of interest.

$p = (1.0375)^{\frac{1}{6}} - 1 = 1.0061545 - 1 = 0.0061545 = 0.061545\%$

$$120\,000.00 = \text{PMT}\left(\frac{1 - 1.0061545^{-300}}{0.0061545}\right)$$ ——— using Formula 6.3

$$120\,000.00 = \text{PMT}(136.695485)$$
$$\text{PMT} = \$877.86$$

Programmed Solution

("END" mode) (Set P/Y = 12; C/Y = 2) 0 [FV] 120 000 [±] [PV] 7.5 [I/Y]
300 [N] [CPT] [PMT] [877.865900]

The monthly payment for the original five-year term is $877.87.

(ii) The balance at the end of the five-year term is the present value of the outstanding payments. After five years, 60 of the required 300 payments have been made; 240 payments remain outstanding.

PMT = 877.87; $n = 240$; $p = 0.615453\%$

$$PV_{nc} = 877.87\left(\frac{1 - 1.0061545^{-240}}{0.0061545}\right)$$ ——— using Formula 5.5

$$= 877.87(125.219078)$$
$$= \$109\,926.07$$

Programmed Solution

0 [FV] 877.87 [±] [PMT] 240 [N] 7.5 [I/Y] [CPT] [PV] [109 925.8315]

The mortgage balance at the end of the first five-year term is $109 925.83.

(iii) For the renewed term, the starting principal is the balance at the end of the five-year term. The amortization period is the number of years remaining after the initial term. We must recalculate p for the new interest rate.

$PV_{nc} = 109\,925.83$; $n = 12(20) = 240$; I/Y = 7; $i = 3.5\%$

$$p = (1.035)^{\frac{1}{6}} - 1 = 1.00575 - 1 = 0.00575 = 0.575\%$$

$$109\,925.83 = \text{PMT}\left(\frac{1 - 1.00575^{-240}}{0.00575}\right)$$
$$109\,925.83 = \text{PMT}(129.986983)$$
$$\text{PMT} = 845.67$$

Programmed Solution

0 [FV] 109 925.83 [±] [PV] 7 [I/Y]
240 [N] [CPT] [PMT] [845.671113]

The monthly payment for the renewed four-year term is $845.67.

E. Rounded payments

Mortgage payments are sometimes rounded up to an exact dollar value (such as to the next dollar, the next five dollars, or the next ten dollars). The payment of \$877.87 might be rounded up to \$878 or \$880 or even \$900. Rounded payments up to a higher value will result in a lower balance at the end of the term. In the final renewal term, rounding will affect the size of the final payment.

EXAMPLE 8.4D

A mortgage balance of \$17 322.35 is renewed for the remaining amortization period of three years at 8% compounded semi-annually.

(i) Compute the size of the monthly payments.
(ii) Determine the size of the last payment if the payments computed in part (i) have been rounded up to the next ten dollars.

SOLUTION

(i) $PV_{nc} = 17\,322.35$; $n = 12(3) = 36$; $P/Y = 12$; $C/Y = 2$; $I/Y = 8$;

$i = 4\%$; $c = \frac{1}{6}$

$p = (1.04)^{\frac{1}{6}} - 1 = 1.006558 - 1 = 0.65582\%$

$$17\,322.35 = PMT\left(\frac{1 - 1.0065582^{-36}}{0.0065582}\right)$$

$$17\,322.35 = PMT(31.973035)$$

$$PMT = \$541.78$$

Programmed Solution

("END" mode) (Set P/Y = 12; C/Y = 2) 0 FV 17 322.35 ± PV 8 I/Y

36 N CPT PMT 541.779935

The monthly payment is \$541.78.

(ii) If payments are rounded to \$550.00, the last payment will be less than \$550.00. To calculate the size of the last payment we need to determine the number of payments of \$550.00 that are required to amortize the loan balance.

$PV_{nc} = 17\,322.35$; $PMT = 550.00$; $p = 0.65582\%$

$$17\,322.35 = 550.00\left(\frac{1 - 1.0065582^{-n}}{0.0065582}\right)$$

$$0.2065517 = 1 - 1.006558^{-n}$$

$$1.0065582^{-n} = 0.7934483$$

$$-n(\ln 1.0065582) = \ln 0.7934483$$

$$-n(0.0065368) = -0.2313669$$

$$n = 35.394520$$

Programmed Solution

0 [FV] 17 322.35 [±] [PV] 8 [I/Y]

550 [PMT] [CPT] [N] 35.394580

There will be 35 payments of \$550.00 and a final payment smaller than \$550.00. We need to determine the balance after the 35th payment.

$\text{PMT} = 550.00; \quad n = 0.394580; \quad p = 0.65582\%$

$$\text{PV}_{nc} = 550.00\left(\frac{1 - 1.0065582^{-0.39458}}{0.0065582}\right)$$

$$= 550.00(0.392785)$$

$$= \$216.03$$

Programmed Solution

0 [FV] 550.00 [±] [PMT] 8 [I/Y] 0.394580 [N]

[CPT] [PV] 216.031533

The final payment includes interest on the balance of \$216.03.

Final payment = \$216.03(1.0065582) = \$217.45.

BUSINESS MATH NEWS BOX

A Cheap Source of Funds—Mortgages

Royal Bank of Canada has a wide assortment of residential mortgages and options attached to those mortgages that allow mortgages to grow with the needs of homeowners and their families. The Mortgage Add-On Option is one such option that allows homeowners to use the equity in their home as a source of affordable funds. These funds could be used for a major renovation to the home or for other purposes, such as to pay for a major purchase or to finance a child's education.

The Mortgage Add-On Option lets you access cash at mortgage interest rates, which are usually much lower than other borrowing rates. When you use the Mortgage Add-On Option to borrow an additional amount during your mortgage term, your existing mortgage rate and the current market rate for the remaining term of your mortgage are blended together to give you one "weighted" annual interest rate on your new mortgage balance for the remainder of your term. Your regular payment is adjusted to reflect your new mortgage principal amount and interest rate.

With the Mortgage Add-On Option, you may generally borrow up to 75% of the appraised value of your home at the time of the Add-On, minus the amount of your outstanding mortgage. You don't have to wait until your mortgage is up for renewal. You can use the Mortgage Add-On Option as often as you wish during the life of your mortgage, provided your outstanding balance never exceeds 75% of the appraised value of your home at the time of the Add-On.

QUESTIONS

Chris and Faye Bartley have a home that is currently appraised at $200 000 and the present value of their existing mortgage is $60 000. The interest rate is 6% compounded semi-annually, payments are made monthly, and the remaining amortization term is ten years. Using the Mortgage Add-On Option, the Bartleys want to borrow the maximum, which is 75% of the current appraised value of their home, for extensive renovations. The current interest rate is 8% compounded semi-annually. They will continue to make monthly mortgage payments.

1. What is the amount of extra borrowing the Bartleys are seeking from the bank by using the Mortgage Add-On Option?
2. What is the weighted annual interest rate on the new mortgage balance?
3. What is the amount of the Bartleys' new monthly payment if the new mortgage balance is amortized over the remaining term?

Source: "Residential Mortgages. Mortgage Add-On Option." Royal Bank of Canada, www.royalbank.com/products/mortgages, June 2003. Reproduced with permission.

F. Mortgage statement

Currently, when financial institutions record monthly mortgage payments, they calculate interest for the exact number of days that have elapsed since the last payment. This is done by multiplying the effective monthly rate of interest by 12 to convert it into a simple annual rate of interest. The annual rate is then multiplied by the number of days expressed as a fraction of 365.

For example, $p = 0.65582\%$ becomes the simple annual interest rate $12(0.65582\%) = 7.86984\%$.

This approach takes into account that the number of days elapsed between payments fluctuates depending on the number of days in a particular month. It also allows for fluctuations in receiving payments, and permits semi-monthly, bi-weekly, or weekly payments for mortgages requiring contractual monthly payments.

EXAMPLE 8.4E

A credit union member made the contractual mortgage payment of $725.00 on May 31, leaving a mortgage loan balance of $75 411.79. The effective monthly fixed rate was 0.53447%. The member made the contractual payments on June 28, July 31, August 30, September 30, October 29, November 28, and December 30. The credit union agreed to convert the fixed-rate mortgage to a demand mortgage on October 29 at 5.25% compounded annually. This rate was changed to 4.75% on December 2. Determine the mortgage balance on December 30.

SOLUTION

The monthly effective rate $p = 0.53447\%$ is equivalent to the simple annual rate $12(0.53447\%) = 6.41364\%$.

Payment Date	Number of Days	Amount Paid	Interest Paid	Principal	Balance Repaid
May 31	rate is 6.41364%				75 411.79
Jun 28	28	725.00	371.03	353.97	75 057.82
Jul 31	33	725.00	435.23	289.77	74 768.05
Aug 30	30	725.00	394.14	330.86	74 437.19
Sept 30	31	725.00	405.47	319.53	74 117.66
Oct 29	29	725.00	377.69	347.31	73 770.35
Oct 29	rate becomes 5.25%				
Nov 28	30	725.00	318.32	406.68	73 363.67
Dec 02	rate becomes 4.75%				
Dec 30	32	725.00	309.54*	415.46	72 948.21

******Note:*** Interest calculation for December is

4 days at 5.25% on $73 363.67	$ 42.21
28 days at 4.75% on $73 363.67	267.33
TOTAL	$309.54

EXAMPLE 8.4F

A mortgage of $55 000.00 closed on April 12, amortized over 15 years at 8.50% compounded semi-annually for a five-year term. It requires contractual monthly payments rounded up to the nearest $10. The lender agreed to accept bi-weekly payments of half the contractual monthly amount starting April 24. The mortgagor's second June payment was three days late.

(i) Determine the size of the contractual monthly payment.
(ii) Produce a mortgage statement to June 30.
(iii) Compute the accrued interest on June 30.

SOLUTION

(i) $PV_{nc} = 55\,000.00; \quad n = 12(15) = 180; \quad i = 4.25; \quad c = \frac{1}{6}$

$$p = (1.0425)^{\frac{1}{6}} - 1 = 1.006961 - 1 = 0.006961\%$$

$$55\,000.00 = PMT\left(\frac{1 - 1.006961^{-180}}{0.006961}\right)$$

$$55\,000.00 = PMT(102.44)$$

$$PMT = 536.89$$

The contractual monthly payment is $540.00.

(ii) The bi-weekly payment is $270.00.
The annual rate of interest = 12(0.69611) = 8.35332%.

Payment Date	Number of Days	Amount Paid	Interest Paid	Principal	Balance Repaid
April 12					55 000.00
April 24	12	270.00	151.05	118.95	54 881.05
May 8	14	270.00	175.84	94.16	54 786.89
May 22	14	270.00	175.54	94.46	54 692.43
June 5	14	270.00	175.24	94.76	54 597.67
June 22	17	270.00	212.42	57.58	54 540.09

The mortgage balance on June 30 is \$54 540.09.

(iii) On June 30 interest has accrued for 8 days.

The amount of accrued interest $= 54\,540.09(0.083533)\left(\frac{8}{365}\right) = \$99.86.$

EXERCISE 8.4

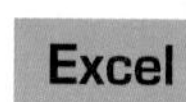

Excel

If you choose, you can use Excel's **NPER**, **PMT**, **PV**, or **RATE** functions or **Template4** to answer the questions indicated below. Refer to the Spreadsheet Template Disk to find **Template4** or to learn how to use these Excel functions.

A. Answer each of the following questions.

1. A \$90 000.00 mortgage is to be amortized by making monthly payments for 25 years. Interest is 8.5% compounded semi-annually for a five-year term.

(a) Compute the size of the monthly payment.
(b) Determine the balance at the end of the five-year term.
(c) If the mortgage is renewed for a three-year term at 7% compounded semi-annually, what is the size of the monthly payment for the renewal term?

2. A demand (variable-rate) mortgage of \$150 000.00 is amortized over 20 years by equal monthly payments. After 18 months the original interest rate of 6% p.a. was raised to 6.6% p.a. Two years after the mortgage was taken out, it was renewed at the request of the mortgagor at a fixed rate of 7.5% compounded semi-annually for a four-year term.

(a) Calculate the mortgage balance after 18 months.
(b) Compute the size of the new monthly payment at the 6.6% rate of interest.
(c) Determine the mortgage balance at the end of the four-year term.

3. A \$40 000.00 mortgage is to be repaid over a ten-year period by monthly payments rounded up to the next higher \$50.00. Interest is 9% compounded semi-annually.

(a) Determine the number of rounded payments required to repay the mortgage.
(b) Determine the size of the last payment.
(c) Calculate the amount of interest saved by rounding the payments up to the next higher \$50.

4. A mortgage balance of \$23 960.00 is to be repaid over a seven-year term by equal monthly payments at 11% compounded semi-annually. At the request of the mortgagor, the monthly payments were set at \$440.00.

(a) How many payments will the mortgagor have to make?
(b) What is the size of the last payment?
(c) Determine the difference between the total actual amount paid and the total amount required to amortize the mortgage by the contractual monthly payments.

5. A mortgage of $80 000.00 is amortized over 15 years by monthly payments of $826.58. What is the nominal annual rate of interest compounded semi-annually?

6. At what nominal annual rate of interest will a $195 000.00 demand (variable-rate) mortgage be amortized by monthly payments of $1606.87 over 20 years?

7. Interest for the initial four-year term of a $105 000.00 mortgage is 7.25% compounded semi-annually. The mortgage is to be repaid by equal monthly payments over 20 years. The mortgage contract permits lump-sum payments at each anniversary date up to 10% of the original principal.

(a) What is the balance at the end of the four-year term if a lump-sum payment of $7000.00 is made at the end of the third year?

(b) How many more payments will be required after the four-year term if there is no change in the interest rate?

(c) What is the difference in the cost of the mortgage if no lump-sum payment is made?

8. The Berezins agreed to monthly payments rounded up to the nearest $100.00 on a mortgage of $36 000.00 amortized over ten years. Interest for the first five years was 8.75% compounded semi-annually. After 30 months, as permitted by the mortgage agreement, the Berezins increased the rounded monthly payment by 10%.

(a) Determine the mortgage balance at the end of the five-year term.

(b) If the interest rate remains unchanged over the remaining term, how many more of the increased payments will amortize the mortgage balance?

(c) How much did the Berezins save by exercising the increase-in-payment option?

9. A $40 000.00 mortgage taken out on June 1 is to be repaid by monthly payments rounded up to the nearest $10.00. The payments are due on the first day of each month starting July 1. The amortization period is 12 years and interest is 5.5% compounded semi-annually for a six-month term. Construct an amortization schedule for the six-month term.

10. For Question 9, produce the mortgage statement for the six-month term. Assume all payments have been made on time. Compare the balance to the balance in Question 9 and explain why there may be a difference.

11. For the mortgage in Question 9, develop a mortgage statement for the six-month term if semi-monthly payments equal to one-half of the monthly payment are made on the first day and the 16th day of each month. The first payment is due June 16. Compare the balance to the balances in Question 9 and Question 10. Explain why there are differences.

12. For the mortgage in Question 9, develop a mortgage statement for the six-month term if bi-weekly payments equal to one-half of the rounded monthly payments are made starting June 16. Compare the balance to the balances in Questions 9, 10, and 11. Explain why it differs significantly from the other three balances.

Review Exercise

1. Sylvie Cardinal bought a business for \$45 000.00. She made a down payment of \$10 000.00 and agreed to repay the balance by equal payments at the end of every three months for eight years. Interest is 8% compounded quarterly.
   (a) What is the size of the quarterly payments?
   (b) What will be the total cost of financing?
   (c) How much will Sylvie owe after five years?
   (d) How much interest will be included in the 20th payment?
   (e) How much of the principal will be repaid by the 24th payment?
   (f) Construct a partial amortization schedule showing details of the first three payments, Payments 10, 11, 12, the last three payments, and totals.

2. Angelo Lemay borrowed \$8000.00 from his credit union. He agreed to repay the loan by making equal monthly payments for five years. Interest is 9% compounded monthly.
   (a) What is the size of the monthly payments?
   (b) How much will the loan cost him?
   (c) How much will Angelo owe after eighteen months?
   (d) How much interest will he pay in his 36th payment?
   (e) How much of the principal will be repaid by the 48th payment?
   (f) Prepare a partial amortization schedule showing details of the first three payments, Payments 24, 25, 26, the last three payments, and totals.

3. Comfort Swim Limited borrowed \$40 000.00 for replacement of equipment. The debt is repaid in installments of \$2000.00 made at the end of every three months.
   (a) If interest is 7% compounded quarterly, how many payments are needed?
   (b) How much will Comfort Swim owe after two years?
   (c) How much of the 12th payment is interest?
   (d) How much of the principal will be repaid by the 20th payment?
   (e) Construct a partial amortization schedule showing details of the first three payments, the last three payments, and totals.

4. A \$48 000.00 mortgage amortized by monthly payments over 35 years is renewable after five years. Interest is 9% compounded semi-annually.
   (a) What is the size of the monthly payments?
   (b) How much interest is paid during the first year?
   (c) How much of the principal is repaid during the first five-year term?
   (d) If the mortgage is renewed for a further five-year term at 8% compounded semi-annually, what will be the size of the monthly payments?
   (e) Construct a partial amortization schedule showing details of the first three payments for each of the two five-year terms, the last three payments for the second five-year term, and totals at the end of the second five-year term.

5. Pelican Recreational Services owes \$27 500.00 secured by a collateral mortgage. The mortgage is amortized over fifteen years by equal payments made at the end of every three months and is renewable after three years.
   (a) If interest is 7% compounded annually, what is the size of the payments?
   (b) How much of the principal is repaid by the fourth payment?
   (c) What is the balance at the end of the three-year term?
   (d) If the mortgage is renewed for a further four years but amortized over eight years and interest is 7.5% compounded semi-annually, what is the size of the quarterly payments for the renewal period?

(e) Construct a partial amortization schedule showing details of the first three payments for each of the two terms, the last three payments in the four-year term, and totals at the end of the four-year term.

6. A debt of $17 500.00 is repaid by payments of $2850.00 made at the end of each year. Interest is 8% compounded semi-annually.
   (a) How many payments are needed to repay the debt?
   (b) What is the cost of the debt for the first three years?
   (c) What is the principal repaid in the fifth year?
   (d) Construct an amortization schedule showing details of the first three payments, the last three payments, and totals.

7. A debt of $25 000.00 is repaid by payments of $3500.00 made at the end of every six months. Interest is 11% compounded semi-annually.
   (a) How many payments are needed to repay the debt?
   (b) What is the size of the final payment?

8. Jane Evans receives payments of $900.00 at the beginning of each month from a pension fund of $72 500.00. Interest earned by the fund is 6.30% compounded monthly.
   (a) What is the number of payments Jane will receive?
   (b) What is the size of the final payment?

9. A lease agreement valued at $33 000.00 requires payment of $4300.00 every three months in advance. The payments are deferred for three years and money is worth 10% compounded quarterly.
   (a) How many lease payments are to be made under the contract?
   (b) What is the size of the final lease payment?

10. A contract worth $52 000.00 provides benefits of $20 000.00 at the end of each year. The benefits are deferred for ten years and interest is 11% compounded quarterly.
   (a) How many payments are to be made under the contract?
   (b) What is the size of the last benefit payment?

11. A mortgage for $135 000.00 is amortized over 25 years. Interest is 8.7% p.a., compounded semi-annually, for a five-year term and payments are monthly.
   (a) Compute the monthly payment.
   (b) Compute the balance at the end of the five-year term.
   (c) Compute the monthly payment if the mortgage is renewed for a three-year term at 7.8% compounded semi-annually.

12. A $180 000.00 mortgage is to be amortized by making monthly payments for 25 years. Interest is 7.9% compounded semi-annually for a four-year term.
   (a) Compute the size of the monthly payment.
   (b) Determine the balance at the end of the four-year term.
   (c) If the mortgage is renewed for a five-year term at 8.8% compounded semi-annually, what is the size of the monthly payment for the renewal term?

13. An $80 000.00 mortgage is to be repaid over a 10-year period by monthly payments rounded up to the next higher $50.00. Interest is 8.5% compounded semi-annually.
   (a) What is the number of rounded payments required to repay the mortgage?
   (b) What is the size of the last payment?
   (c) How much interest was saved by rounding the payments up to the next higher $50.00?

14. A $160 000.00 mortgage is to be repaid over a 20-year period by monthly payments rounded up to the next higher $100.00. Interest is 9.2% compounded semi-annually.
   (a) Determine the number of rounded payments required to repay the mortgage.
   (b) Determine the size of the last payment.
   (c) Calculate the amount of interest saved by rounding the payments up to the next higher $100.00.

15. A debt of $6500.00 is repaid in equal monthly installments over four years. Interest is 9% compounded monthly.
    (a) What is the size of the monthly payments?
    (b) What will be the total cost of borrowing?
    (c) What is the outstanding balance after one year?
    (d) How much of the 30th payment is interest?
    (e) Construct a partial amortization schedule showing details of the first three payments, the last three payments, and totals.

16. Milton Investments borrowed $32 000.00 at 11% compounded semi-annually. The loan is repaid by payments of $4500.00 due at the end of every six months.
    (a) How many payments are needed?
    (b) How much of the principal will be repaid by the fifth payment?
    (c) Prepare a partial amortization schedule showing the details of the last three payments and totals.

17. A mortgage of $95 000.00 is amortized over 25 years by monthly payments of $748.06. What is the nominal annual rate of interest compounded semi-annually?

18. At what nominal annual rate of interest will a $135 000.00 mortgage be amortized by monthly payments of $1370.69 over 15 years?

19. A $28 000.00 mortgage is amortized by quaterly payments over twenty years. The mortgage is renewable after three years and interest is 6% compounded semi-annually.
    (a) What is the size of the quarterly payments?
    (b) How much interest will be paid during the first year?
    (c) What is the balance at the end of the three-year term?
    (d) If the mortgage is renewed for another three years at 7% compounded annually, what will be the size of the quarterly payments for the renewal period?

20. The Superior Tool Company is repaying a debt of $16 000.00 by payments of $1000.00 made at the end of every three months. Interest is 7.5% compounded monthly.
    (a) How many payments are needed to repay the debt?
    (b) What is the size of the final payment?

Self-Test

1. A $9000.00 loan is repaid by equal monthly payments over five years. What is the outstanding balance after two years if interest is 12% compounded monthly?
2. A loan of $15 000.00 is repaid by quarterly payments of $700.00 each at 8% compounded quarterly. What is the principal repaid by the 25th payment?
3. A $50 000.00 mortgage is amortized by monthly payments over twenty years. If interest is 9% compounded semi-annually, how much interest will be paid during the first three years?
4. A debt of $24 000.00 is repaid by quarterly payments of $1100.00. If interest is 6% compounded quarterly, what is the size of the final payment?

5. A $190 000.00 mortgage is to be amortized by making monthly payments for 20 years. Interest is 6.5% compounded semi-annually for a three-year term.
   (a) Compute the size of the monthly payment.
   (b) Determine the balance at the end of the three-year term.
   (c) If the mortgage is renewed for a five-year term at 7.25% compounded semi-annually, what is the size of the monthly payment for the renewal term?

6. A $140 000.00 mortgage is to be repaid over a 15-year period by monthly payments rounded up to the next higher $50.00. Interest is 8.25% compounded semi-annually.
   (a) Determine the number of rounded payments required to repay the mortgage.
   (b) Determine the size of the last payment.
   (c) Calculate the amount of interest saved by rounding the payments up to the next higher $50.00.

7. A mortgage of $145 000 is amortized over 25 years by monthly payments of $1297.00. What is the nominal annual rate of interest compounded semi-annually?

8. A loan of $12 000.00 is amortized over ten years by equal monthly payments at 7.5% compounded monthly. Construct an amortization schedule showing details of the first three payments, the fortieth payment, the last three payments, and totals.

Challenge Problems

1. A debt is amortized by monthly payments of $250.00. Interest is 8% compounded monthly. If the outstanding balance is $3225.68 just after a particular payment (say, the xth payment), what was the balance just after the previous payment (i.e., the $(x - 1)$th payment)?

2. Captain Sinclair has been posted to Cold Lake, Alberta. He prefers to purchase a condo rather than live on the base. He knows that in four years he will be posted overseas. The condo he wishes to purchase will require a mortgage of $130 000, and he has narrowed his choices to two lenders. Trust Company A is offering a five-year mortgage at 6.75% compounded semi-annually. This mortgage can be paid off at any time but there is a penalty clause in the agreement requiring two months' interest on the remaining principal. Trust Company B is offering a five-year mortgage for 7% compounded semi-annually. It can be paid off at any time without penalty. Both mortgages are amortized over 25 years and require monthly payments. Captain Sinclair will have to sell his condo in four years and pay off the mortgage at that time before moving overseas. Given that he expects to earn 3% compounded annually on his money over the next five years, which mortgage offer is cheaper? By how much is it cheaper?

Case Study 8.1 Managing a Mortgage

» Chris and Briana purchased their first home, a condominium. When they made the purchase, they assumed a $100 000 mortgage from the previous owner. The mortgage had a 9% semi-annually compounded interest rate, was amortized over 25 years, and had a five-year term. Payments were made monthly.

After three years, interest rates had fallen. Chris and Briana had been told that they should pay out the old mortgage, in spite of the interest penalties, and negotiate a new mortgage at a much more reasonable rate. They met with the loans officer at their bank, who laid out the options for them.

Interest on mortgages with a five-year term was 6% compounded semi-annually, the lowest rate in many years. The loans officer had informed Chris and Briana that there is a penalty for renegotiating a mortgage early, before the end of the current term. According to their mortgage contract, the penalty for renegotiating the mortgage before the end of the five-year term is the greater of:

A. Three months' interest at the original rate of interest. (Banks generally calculate this as one month's interest on the mortgage principal remaining to be paid, multiplied by three.)

B. The interest differential over the remainder of the original term. (Banks generally calculate this as the difference between the interest the bank would have earned over the remainder of the original term at the original [higher] mortgage rate and at the renegotiated [lower] mortgage rate.)

The loans officer also explained that there are two options for paying the penalty amount: (1) you can pay the full amount of the penalty at the beginning of the new mortgage period; or (2) the penalty amount can be added to the principal when the mortgage is renegotiated, allowing the penalty to be paid off over the term of the new mortgage.

Chris and Briana agreed to look at their options before giving the loans officer their final answer.

QUESTIONS

1. Suppose there was no penalty for refinancing the mortgage after three years. How much would Chris and Briana save per month by refinancing their mortgage for a five-year term at the new rate?
2. Suppose Chris and Briana decide to refinance their mortgage for a five-year term at the new interest rate.
   (a) What is the amount of penalty A?
   (b) What is the amount of penalty B?
   (c) What penalty would Chris and Briana have to pay in this situation?
3. If they pay the full amount of the penalty at the beginning of the new five-year term, what will Chris and Briana's new monthly payment be?
4. If the penalty amount is added to the principal when the mortgage is renegotiated, what will the new monthly payment be?

Case Study 8.2 Steering the Business

» On February 1, 2005, Sandra and her friend Francisco arranged a loan to purchase two used black limousines for $35 000.00 and $50 000.00 respectively, and launched their new business, Classy Limousine Services. The loan was for two years at 9.5% compounded semi-annually. Payments were made quarterly beginning on May 1, 2005.

Business was brisk, especially for weddings. To meet the demand, the partners bought a new white stretch limousine on April 1, 2006 for $110 000.00. They arranged a three-year loan for this amount at 8.8% interest compounded monthly. Payments were made monthly beginning on May 1, 2006.

On August 1, 2006, the partners were given an opportunity to buy a black super-stretch limousine for $140 000.00. They arranged a three-year loan for this amount at 7.8% interest compounded monthly. The monthly payments began on September 1, 2006.

Business continued to increase. Sandra and Francisco discussed the need for a new parking garage and office space they could own instead of rent. When an industrial warehouse large enough to house their expanding fleet became available, they decided to purchase it. The $280 000.00 mortgage had an interest rate of 8.9% compounded semi-annually for a five-year term. It was amortized over 20 years, and the monthly mortgage payments began on March 1, 2007.

Concerned about their cash flows, Sandra and Francisco decided to make no more large purchases for the next year.

QUESTIONS

1. What is the size of the quarterly payments for the original loan obtained on February 1, 2005?

2. What is the size of the monthly payments for the loan on the white stretch limousine?

3. What is the size of the monthly payments for the loan on the black super-stretch limousine?

4. **(a)** What is the size of the monthly payments required for the mortgage on the warehouse?
(b) What principal will remain to be paid at the end of the mortgage's five-year term?

5. On each of the following dates, what is the total of all the loan payments that must be paid?
(a) May 1, 2006
(b) November 1, 2006
(c) July 1, 2007

SUMMARY OF FORMULAS

No new formulas were introduced in this chapter. However, some of the formulas introduced in Chapters 3 to 6 have been used, namely Formulas 3.1A, 5.2, 5.5, 6.1, 6.2, 6.3, and 6.5.

GLOSSARY

Amortization repayment of both interest and principal of interest-bearing debts by a series of equal payments made at equal intervals of time *(p. 309)*

Amortization schedule a schedule showing in detail how a debt is repaid *(p. 310)*

Canada Mortgage and Housing Corporation (CMHC) the corporation of the federal government that administers the National Housing Act (NHA) and provides mortgage insurance to lenders *(p. 348)*

Demand mortgage a mortgage for which the rate of interest changes as money market conditions change *(p. 347)*

Equity the difference between the price for which a property could be sold and the total debts registered against the property *(p. 347)*

First mortgage the first legal claim registered against a property; in the event of default by the borrower, first mortgagees are paid before all other claimants *(p. 347)*

Fixed-rate mortgage a mortgage for which the rate of interest is fixed for a specific period of time; it can be open or closed *(p. 347)*

Gross Debt Service (GDS) ratio the percent of gross annual income required to cover such housing costs as mortgage payments, property taxes, and heating costs *(p. 348)*

Mortgage contract a document specifying the obligations of the borrowers and the rights of the lender *(p. 347)*

Mortgagee the lender *(p. 347)*

Mortgagor the borrower *(p. 347)*

Prospective method a method for finding the outstanding debt balance that considers the payments that remain outstanding *(p. 316)*

Residential mortgage a claim to a residential property given by a borrower to a lender as security for the repayment of a loan *(p. 347)*

Retrospective method a method of finding the outstanding balance of a debt that considers the payments that have been made *(p. 316)*

Second mortgage the second legal claim registered against a property; in the event of default by the borrower, second-mortgage holders are paid only after first-mortgage holders have been paid *(p. 347)*

Total Debt Service (TDS) ratio the percent of gross annual income required to cover such housing costs as mortgage payments, property taxes, and heat, and all other debts and obligations *(p. 349)*

Variable-rate mortgage *see* **Demand mortgage**

USEFUL INTERNET SITES

www.cmhc-schl.gc.ca

Canada Mortgage and Housing Corporation (CMHC) As the Government of Canada's national housing agency, CMHC plays a major role in Canada's housing industry. CMHC develops new ways to finance home purchases.

www.tdcanadatrust.com

Mortgage Calculator Click on Mortgages, then under Tools and Resources click on Mortgage Calculator. By entering values for the relevant variables in this tool on the Canada Trust site, you can calculate the payment, principal, or amortization period of a mortgage. You can also obtain mortgage rates and general mortgage information.

www.businessfinancemag.com

***Business Finance* Magazine** *Business Finance* presents information on business management and technology issues for accountants and finance professionals, controllers, chief financial officers, and treasurers.

9 Bond Valuation and Sinking Funds

OBJECTIVES

Upon completing this chapter, you will be able to do the following:

1. Determine the purchase price of bonds, redeemable at par or otherwise, bought on or between interest dates.
2. Calculate the premium or discount on the purchase of a bond.
3. Construct bond schedules showing the amortization of premiums or accumulation of discounts.
4. Calculate the yield rate for bonds bought on the market by the method of averages.
5. Make sinking fund computations when payments form simple annuities, including the size of the periodic payments, accumulated balance, interest earned, and increase in the fund.
6. Construct complete or partial sinking fund schedules.

Many financial planners recommend bonds as a part of a balanced investment portfolio. Historically, bonds represent a less risky investment than stocks. Companies and governments issue bonds to raise large sums of money. To have the funds to pay for the bonds when they mature, companies often create a *sinking fund* by making regular equal investments from the outset of the bond issue. Both individuals and companies use annuity formulas to decide about possible investments in bonds and sinking funds.

INTRODUCTION

Bonds are contracts used to borrow sizeable sums of money, usually from a large group of investors. The indenture, or contract, for most bonds provides for the repayment of the principal at a specified future date plus periodic payment of interest at a specified percent of the face value. Bonds are negotiable; that is, they can be freely bought and sold. The mathematical issues arising from the trading of bonds are the topic of this chapter.

9.1 Purchase Price of Bonds

A. Basic concepts and terminology

Corporations and governments use bonds to borrow money, usually from a large group of lenders (investors). To deal with the expected large number of investors, the borrower prints up written contracts, called *bonds* or *debentures*, in advance. The printed bonds specify the terms of the contract including

(a) the **face value** (or **par value** or **denomination**), which is the amount owed to the holder of the bond, usually a multiple of $100 such as $100, $1000, $5000, $10 000, $25 000, $100 000;

(b) the **bond rate** (or **coupon rate** or **nominal rate**), which is the rate of interest paid, usually semi-annually, based on the face value of the bond;

(c) the **redemption date** (or **maturity date** or **due date**), which is the date on which the principal of the bond is to be repaid;

(d) the **redemption price**, which is the money paid by the issuer to the bond-holder at the date of surrender of the bonds.

Most bonds are **redeemable at par**; that is, they are redeemable at their *face* value. However, some bonds have a redemption feature either to make the bonds more attractive to the investor or because they are callable, that is, because they can be redeemed *before* maturity. In either case, the bonds will be **redeemable at a premium**, that is, at a price *greater* than their face price. The redemption price in such cases is stated as a percent of the face value. For example, the redemption price of a $5000 bond redeemable at 104 is 104% of $5000, or $5200.

Investors in bonds expect to receive periodic interest payments during the term of the bond from the date of issue to the date of maturity and they expect to receive the principal at the date of maturity. The bond rate is used to determine the periodic interest payments.

To facilitate the payment of interest, some bonds have dated interest **coupons** attached that can be cashed on or after the stated interest payment date at any bank. For example, a twenty-year, $1000.00 bond bearing interest at 10% payable semi-annually will have attached to it at the date of issue forty coupons of $50.00 each. Each coupon represents the semi-annual interest due on each of the forty interest payment dates.

The issuer may or may not offer security such as real estate, plant, or equipment as a guarantee for the repayment of the principal. Bonds for which no security is offered are called **debentures**.

Bonds are marketable and may be freely bought and sold. When investors acquire bonds, they buy two promises:

1. A promise to be paid the redemption price of the bond at maturity.
2. A promise to be paid the periodic interest payments according to the rate of interest stated on the bond.

Two basic problems arise for investors when buying bonds:

1. What should be the purchase price of a bond to provide the investor with a given rate of return?
2. What is the rate of interest that a bond will yield if purchased at a given price?

In this section we will deal with the first of these two problems.

B. Purchase price of a bond bought on an interest date

EXAMPLE 9.1A A \$1000 bond bearing interest at 6% payable semi-annually is due in four years. If money is worth 7% compounded semi-annually, what is the value of the bond if purchased today?

SOLUTION The buyer of the bond acquires two promises:

1. A promise of \$1000.00 four years from now.
2. A promise of \$30.00 interest due at the end of every six months (the annual interest is 6% of \$1000.00 = \$60.00, half of which is paid after the first six months, the other half at the end of the year).

The two promises can be represented on a time graph as Figure 9.1 shows.

FIGURE 9.1 **Graphical Representation of Method and Data**

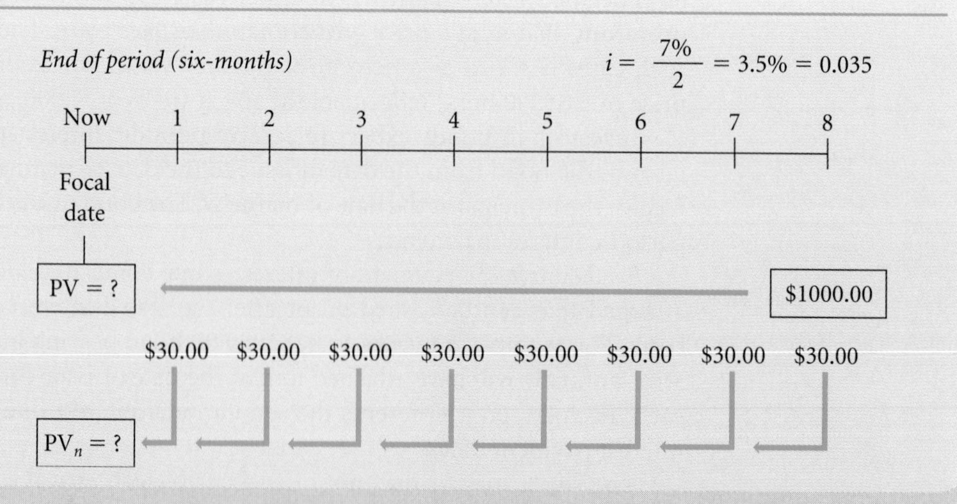

The focal date for evaluating the two promises is "now"and the rate of interest to be used for the valuation is 7% compounded semi-annually.

The value at the focal date of the redemption price of \$1000.00 is its present value.

$FV = 1000.00;\ n = 4(2) = 8;\ P/Y = 2; C/Y = 2; I/Y = 7;\ i = \frac{7\%}{2} = 3.5\% = 0.035$

$PV = 1000.00(1.035)^{-8}$ ——— using Formula 3.1C

$= 1000.00(0.759412)$

$= \$759.41$

Programmed Solution

("END" mode) (Set P/Y = 2; C/Y = 2) 2nd (CLR TVM) 1000 FV 7 I/Y 8 N CPT PV -759.4115562

The value "now" of the semi-annual interest payments is the present value of an ordinary annuity.

$$PMT = \frac{6\% \text{ of } 1000.00}{2} = 30.00; \; n = 8; \; i = 3.5\%$$

$$\begin{aligned} PV_n &= 30.00\left(\frac{1 - 1.035^{-8}}{0.035}\right) \quad \text{using Formula 5.2} \\ &= 30.00(6.873956) \\ &= \$206.22 \end{aligned}$$

Programmed Solution

0 FV 30 PMT 7 I/Y 8 N CPT PV 206.2186661

The purchase price of the bond is the sum of the present values of the two promises

$= PV + PV_n$
$= 759.41 + 206.22 = \$965.63$

THE PURCHASE PRICE OF A BOND BOUGHT ON AN INTEREST PAYMENT DATE = THE PRESENT VALUE OF THE REDEMPTION PRICE + THE PRESENT VALUE OF THE INTEREST PAYMENTS

The two steps involved in using the above relationship can be combined.

$$\text{PURCHASE PRICE} = P + A_n$$
$$= S(1+i)^{-n} + R\left[\frac{1-(1+i)^{-n}}{i}\right]$$

restated as

$$\text{PURCHASE PRICE} = PV + PV_n$$
$$= FV(1+i)^{-n} + PMT\left[\frac{1-(1+i)^{-n}}{i}\right]$$

Formula 9.1

where FV = the redemption price of the bond;
PMT = the periodic interest payment (coupon);
n = the number of outstanding interest payments (or compounding periods);
i = the yield rate per payment interval.

Note: It is important to recognize that two rates of interest are used in determining the purchase price:

1. The bond rate, which determines the size of the periodic interest payments (coupons);
2. The **yield rate**, which is used to determine the present values of the two promises.

EXAMPLE 9.1B

A \$5000 bond bearing interest at 10.5% payable semi-annually is redeemable at par in ten years. If it is bought to yield 9% compounded semi-annually, what is the purchase price of the bond?

SOLUTION

The redemption price FV = 5000.00;

$$\text{the coupon PMT} = \frac{5000.00(0.105)}{2} = 262.50;$$

$$n = 10(2) = 20;\ \text{P/Y} = 2;\ \text{C/Y} = 2;\ \text{I/Y} = 9;\ i = \frac{9\%}{2} = 4.5\% = 0.045$$

$$\begin{aligned}
\text{Purchase price} &= \text{PRESENT VALUE OF THE REDEMPTION PRICE} + \text{PRESENT VALUE OF THE COUPONS} \\
&= \text{PV} + \text{PV}_n \\
&= 5000.00(1.045^{-20}) + 262.50\left(\frac{1-1.045^{-20}}{0.045}\right) \\
&= 5000.00(0.414643) + 262.50(13.007936) \\
&= 2073.21 + 3414.58 \\
&= \$5487.79
\end{aligned}$$

Programmed Solution

("END" mode) (Set P/Y = 2; C/Y = 2) [2nd] (CLR TVM) 5000 [FV] 9 [I/Y]

20 [N] [CPT] [PV] [-2073.214298]

0 [FV] 262.50 [PMT] 9 [I/Y] 20 [N] [CPT] [PV] [-3414.583319]

(these keystrokes can be eliminated since they are still programmed in the calculator from the previous step)

The purchase price is 2073.21 + 3414.58 = \$5487.79.

EXAMPLE 9.1C

A bond with a par value of \$10 000 is redeemable at 106 in 25 years. The coupon rate is 8% payable semi-annually. What is the purchase price of the bond to yield 10% compounded semi-annually?

SOLUTION

The redemption price FV = 10 000.00(1.06) = 10 600.00;

$$\text{the coupon PMT} = \frac{10\,000.00(0.08)}{2} = 400.00;\ n = 25(2) = 50;$$

$$i = \frac{10\%}{2} = 5\% = 0.05$$

$$\begin{aligned}
\text{Purchase price} &= \text{PV} + \text{PV}_n \\
&= 10\,600.00(1.05^{-50}) + 400.00\left(\frac{1-1.05^{-50}}{0.05}\right) \\
&= 10\,600.00(0.087204) + 400.00(18.255925) \\
&= 924.36 + 7302.37 \\
&= \$8226.73
\end{aligned}$$

Programmed Solution

("END" mode) (Set P/Y = 2; C/Y = 2) 2nd (CLR TVM) 10 600 FV 10 I/Y 50 N CPT PV -924.3595059

0 FV 400 PMT 10 I/Y 50 N CPT PV −7302.370184

The purchase price is 924.36 + 7302.37 = $8226.73

EXAMPLE 9.1D

A $25 000 bond bearing interest at 11% payable quarterly is redeemable at 107 in twelve years. Find the purchase price of the bond to yield 10% compounded quarterly.

SOLUTION

The redemption price FV = 25 000.00(1.07) = 26 750.00;

the coupon PMT $= \dfrac{25\,000.00(0.11)}{4} = 687.50$;

$n = 12(4) = 48$; P/Y = 4; C/Y = 4; I/Y = 10; $i = \dfrac{10\%}{4} = 2.5\% = 0.025$

$$\begin{aligned}\text{Purchase price} &= 26\,750.00(1.025^{-48}) + 687.50\left(\frac{1 - 1.025^{-48}}{0.025}\right)\\ &= 26\,750.00(0.3056712) + 687.50(27.773154)\\ &= 8176.70 + 19\,094.04\\ &= \$27\,270.74\end{aligned}$$

Programmed Solution

("END" mode) (Set P/Y = 4; C/Y = 4) 2nd (CLR TVM) 26 750 FV 10 I/Y 48 N CPT PV -8176.703457

0 FV 687.50 PMT 10 I/Y 48 N CPT PV −19 094.04318

The purchase price is 8176.70 + 19 094.04 = $27 270.74.

In the preceding problems, the bond interest payment period and the yield rate conversion period were equal in length. This permitted the use of simple annuity formulas. However, when the bond interest payment period and the yield rate conversion period are not equal in length, general annuity formulas must be used.

EXAMPLE 9.1E

A municipality issues ten-year bonds in the amount of $1 000 000. Interest on the bonds is 10% payable annually. What is the issue price of the bonds if the bonds are sold to yield 11% compounded quarterly?

SOLUTION

The redemption price of the bonds FV = 1 000 000.00;

the annual interest payment PMT = 1 000 000.00(0.10) = $100 000.00.

Since the interest payment period (annual) is not equal in length to the yield rate conversion period (quarterly), the interest payments form an ordinary general annuity.

$n = 10$; P/Y = 1; C/Y = 4; $c = \frac{4}{1} = 4$; I/Y = 11; $i = \frac{11\%}{4} = 2.75\% = 0.0275$;

$p = 1.0275^4 - 1 = 1.114621 - 1 = 0.114621 = 11.46213\%$

The present value of the redemption price

$$\begin{aligned} PV &= 1\,000\,000.00(1.114621^{-10}) \\ &= 1\,000\,000.00(0.337852) \\ &= \$337\,852.10 \end{aligned}$$

The present value of the annual interest payments

$$\begin{aligned} PV_{nc} &= 100\,000.00\left[\frac{(1 - 1.114621^{-10})}{0.114621}\right] \\ &= 100\,000.00(5.776831) \\ &= \$577\,683.12 \end{aligned}$$

——— using Formula 6.3

Programmed Solution

("END" mode) (Set P/Y = 1; C/Y = 4) 2nd (CLR TVM) 1 000 000 FV 11 I/Y 10 N CPT PV -337 852.2208

0 FV 100 000 PMT 11 I/Y 10 N CPT PV -577 683.2174

The issue price is 337 852.22 + 577 683.22 = $915 535.44.

We can modify Formula 9.1 to allow for the general annuity case by using Formula 6.3.

$$\text{PURCHASE PRICE} = P + A_{nc} = S(1 + p)^{-n} + R\left[\frac{1 - (1 + p)^{-n}}{p}\right]$$

restated as

$$\text{PURCHASE PRICE} = PV + PV_{nc} = FV(1 + p)^{-n} + PMT\left[\frac{1 - (1 + p)^{-n}}{p}\right]$$

— Formula 9.2

where $p = (1 + i)^c - 1$

EXAMPLE 9.1F

A \$100 000 bond redeemable at 103 bearing interest at 7.5% payable semi-annually is bought eight years before maturity to yield 8% compounded quarterly. What is the purchase price of the bond?

SOLUTION

The redemption price FV = 100 000.00(1.03) = \$103 000.00;

the size of the semi-annual coupon PMT = $100\,000.00\left(\frac{0.075}{2}\right) = \$3750.00$

$n = 8(2) = 16$; P/Y = 2; C/Y = 4; $c = \frac{4}{2} = 2$; I/Y = 8; $i = \frac{8\%}{4} = 2\% = 0.02$;

$p = 1.02^2 - 1 = 1.0404 - 1 = 0.0404 = 4.04\%$

The purchase price of the bond using Formula 15.2

$$= 103\,000.00(1.0404^{-16}) + 3750.00\left(\frac{1 - 1.0404^{-16}}{0.0404}\right)$$
$$= 103\,000.00(0.530633) + 3750.00(11.617988)$$
$$= 54\,655.23 + 43\,567.45$$
$$= \$98\,222.68$$

Programmed Solution

("END" mode) (Set P/Y = 2; C/Y = 4) 2nd (CLR TVM) 103 000 FV 8 I/Y 16 N CPT PV -54 655.23026

0 FV 3750 PMT 8 I/Y 16 N CPT PV -43 567.45326

The purchase price is 54 655.23 + 43 567.45 = \$98 222.68.

C. Purchase price of bonds between interest dates

The trading of bonds is, of course, not restricted to interest dates. In practice, most bonds are bought and sold between interest dates.

In such cases, we can compute the price of the bond on the date of purchase by first finding the purchase price on the interest date immediately preceding the date of purchase. The resulting value can then be accumulated using the future value formula for simple interest at the nominal yield rate for the number of days elapsed between the interest payment date and the purchase date.

DID YOU KNOW?

Every year, during the month of October, the Government of Canada offers for sale a new series of Canada Savings Bonds (CSBs). CSBs are different from other corporate and government bonds because they cannot be freely traded after they have been purchased. CSBs also have a number of other unique features.

Two types of CSBs are available: Regular Interest Bonds (or R-Bonds) and Compound Interest Bonds (or C-Bonds). R-Bonds pay simple interest every year, which you can receive by cheque or direct deposit to your bank account. C-Bonds pay compound interest, which is reinvested each year by the government. CSBs are available in denominations of \$100 (C-Bonds only), \$300, \$500, \$1000, \$5000, and \$10 000. The government sets a purchase limit for each year's series of CSBs. (In 2003, the limit was \$200 000.00.)

The Government of Canada fully guarantees the principal and interest of CSBs. When CSBs are issued, interest rates are set and guaranteed for each year of the CSB's life. If market interest rates rise, CSB interest rates usually rise, but they never fall below the guaranteed rates. Also, you can cash CSBs at any time. If you cash CSBs within the first three months, you receive the full face value of the CSBs. After three months, you receive the full face value plus all interest earned for each full month elapsed since November 1 of the year you purchased the CSBs. Canadian residents can buy CSBs at any bank, credit union, caisse populaire, or trust company, and most investment dealers.

EXAMPLE 9.1G

A bond with a face value of \$1000 bearing interest at 10% payable semi-annually matures on August 1, 2008. What is the purchase price of the bond on April 18, 2006 to yield 9% compounded semi-annually?

SOLUTION

STEP 1

Find the purchase price on the preceding interest date.
The redemption price FV = \$1000.00;

the semi-annual coupon $PMT = 1000.00\left(\frac{0.10}{2}\right) = \$50.00$.

Since the maturity date is August 1, the semi-annual interest dates are February 1 and August 1. The interest date preceding the date of purchase is February 1, 2006. The time period from February 1, 2006 to the date of maturity is 2.5 years.

$n = 2.5(2) = 5;\ P/Y = 2;\ C/Y = 2;\ I/Y = 9;\ i = \frac{9\%}{2} = 4.5\% = 0.045$

The purchase price of the bond on February 1, 2006

$$= 1000.00(1.045^{-5}) + 50.00\left(\frac{1 - 1.045^{-5}}{0.045}\right)$$
$$= 1000.00(0.802451) + 50.00(4.389977)$$
$$= 802.45 + 219.50$$
$$= \$1021.95$$

Programmed Solution

("END" mode) (Set P/Y = 2; C/Y = 2) 2nd (CLR TVM) 1000 FV 9 I/Y

5 N CPT PV −802.4510465

0 FV 50 PMT 9 I/Y 5 N CPT PV −219.4988372

The purchase price is 802.45 + 219.50 = $1021.95.

STEP 2 Accumulate the purchase price on February 1, 2006 to the purchase date at simple interest.

The number of days from February 1, 2006 to April 18, 2006 is 76; the number of days in the interest interval February 1, 2006 to August 1, 2006 is 181.

$$\text{PV} = 1021.95; \quad r = i = 0.045; \quad t = \frac{76}{181}$$

$$\text{FV} = 1021.95\left[1 + 0.045\left(\frac{76}{181}\right)\right] = \$1041.26.$$

The purchase price of the bond on April 18, 2006 is $1041.26.

D. Flat price and quoted price

In Example 15.1F, the total purchase price of $1041.26 is called the **flat price**. This price includes interest that has accrued from February 1, 2006 to April 18, 2006 but will not be paid until August 1, 2006.

The actual accrued interest is $1000.00(0.05)\left(\frac{76}{181}\right) = \$20.99$.

As far as the seller of the bond is concerned, the net price of the bond is 1041.26 − 20.99 = $1020.27. This price is called the **quoted price** or **market price.**

QUOTED PRICE = FLAT PRICE − ACCRUED INTEREST

FLAT PRICE = QUOTED PRICE + ACCRUED INTEREST

In a stable market, the flat price of bonds increases as the accrued interest increases and drops suddenly by the amount of the interest paid on the interest date. To avoid this fluctuation in price, which is entirely due to the accrued interest, bonds are offered for sale at the quoted price. The accrued interest is added to obtain the total price or flat price.

EXAMPLE 9.1H A $5000 bond redeemable at par in seven years and four months bearing interest at 6.5% payable semi-annually is bought to yield 7.5% compounded semi-annually. Determine

(i) the purchase price (flat price);
(ii) the accrued interest;
(iii) the market price (quoted price).

SOLUTION (i) The redemption price FV = $5000.00;

the semi-annual coupon $\text{PMT} = 5000.00\left(\frac{0.065}{2}\right) = \$162.50$.

The interest date preceding the purchase date is 7.5 years before maturity.

$n = 7.5(2) = 15$; P/Y = 2; C/Y = 2; I/Y = 7.5; $i = \frac{7.5\%}{2} = 3.75\% = 0.0375$

The purchase price on the interest date preceding the date of purchase

$= 5000.00(1.0375^{-15}) + 162.50\left(\frac{1 - 1.0375^{-15}}{0.0375}\right)$

$= 5000.00(0.575676) + 162.50(11.315296)$

$= 2878.38 + 1838.74$

$= \$4717.12$

Programmed Solution

("END" mode) (Set P/Y = 2; C/Y = 2) [2nd] (CLR TVM) 5000 [FV] 7.5 [I/Y]

15 [N] [CPT] [PV] [-2878.381956]

0 [FV] 162.50 [PMT] 7.5 [I/Y] 15 [N] [CPT] [PV] [-1838.735638]

The purchase price is 2878.38 + 1838.74 = \$4717.12.

The accumulated value two months later (flat price)

$PV = 4717.12; \quad r = i = 0.0375; \quad t = \frac{2}{6};$

$FV = 4717.12\left[1 + 0.0375\left(\frac{2}{6}\right)\right]$

$= \$4776.08$

The purchase price (flat price) is \$4776.08.

(ii) The accrued interest is $5000.00(0.0325)\left(\frac{2}{6}\right) = \$54.17$.

(iii) The quoted price (market price) is 4776.08 − 54.17 = \$4721.91.

EXAMPLE 9.1I

A \$10 000, 11% bond redeemable at 108 matures on May 1, 2017. Interest is payable semi-annually. The bond is purchased on March 21, 2005 to yield 10% compounded semi-annually.

(i) What is the purchase price of the bond?
(ii) What is the accrued interest?
(iii) What is the quoted price?

SOLUTION

(i) The redemption price FV = 10 000.00(1.08) = \$10 800.00;

the semi-annual coupon $PMT = 10\,000.00\left(\frac{0.11}{2}\right) = \$550.00$.

The interest dates on the bond are May 1 and November 1.
The interest date preceding the date of purchase is November 1, 2004.
The time period November 1, 2004 to May 1, 2017 is 12.5 years.

$n = 12.5(2) = 25; \ P/Y = 2; \ C/Y = 2; \ I/Y = 10; \ i = \frac{10\%}{2} = 5\% = 0.05$

The purchase price on November 1, 2004,

$= 10\,800.00(1.05^{-25}) + 550.00\left(\frac{1 - 1.05^{-25}}{0.05}\right)$

$= 10\,800.00(0.295303) + 550.00(14.093945)$
$= 3189.27 + 7751.67$
$= \$10\,940.94$

Programmed Solution

("END" mode) (Set P/Y = 2; C/Y = 2) 2nd (CLR TVM) 10 800 FV 10 I/Y 25 N CPT PV -3189.269934

0 FV 550 PMT 10 I/Y 25 N CPT PV -7751.669511

The purchase price is 3189.27 + 7751.67 = \$10 940.94.

The time period November 1, 2004 to March 21, 2005 contains 140 days; the number of days in the interest payment interval November 1, 2004 to May 1, 2005 is 181.

$PV = 10\,940.94; \quad r = i = 0.05; \quad t = \frac{140}{181}$

The accumulated value (flat price) on March 21, 2005,

$$= 10\,940.94\left[1 + 0.05\left(\frac{140}{181}\right)\right] = \$11\,364.07$$

(ii) The actual accrued interest is $10\,000.00(0.055)\left(\frac{140}{181}\right) = \$425.41$.

(iii) The quoted price is 11 364.07 − 425.41 = \$10 938.66.

Most financial calculators have a Bond function. To view instructions on their use, see the CD-ROM accompanying this book.

EXERCISE 9.1

Excel

Excel has a ***Bond Purchase Price (PRICE)*** function you can use to find the price per \$100 face value of a security that pays periodic interest. If you choose, you can use this function to answer the questions in Part A and Part B below. Refer to **PRICE** on the Spreadsheet Template Disk to learn how to use this Excel function.

A. Determine the purchase price at the indicated time before redemption of each of the bonds shown in the table on the next page.

	Par Value	Redeemed At:	Bond Rate Payable Semi-annually	Time Before Redemption	Yield Rate	Conversion Period
1.	$100 000	par	7%	5.5 years	7.5%	semi-annually
2.	$5 000	par	10%	12 years	9%	semi-annually
3.	$25 000	103	6%	7 years	7%	semi-annually
4.	$1 000	110	8.5%	12.5 years	8%	semi-annually
5.	$50 000	par	6.5%	10 years	6%	annually
6.	$20 000	par	7.5%	6.5 years	8%	quarterly
7.	$8 000	104	8%	18.5 years	9%	monthly
8.	$3 000	107	10%	20 years	12%	quarterly

Excel Excel has a number of functions you can choose to answer the questions in Part B below: ***Days Since Last Interest Date (COUPDAYBS)***, ***Total Number of Days in the Coupon Period (COUPDAYS)***, and ***Total Number of Remaining Coupon Periods (COUPNUM)***. Refer to **COUPDAYBS**, **COUPDAYS**, and **COUPNUM** on the Spreadsheet Template Disk to learn how to use these Excel functions.

B. Answer each of the following questions.

1. A $500 bond is redeemable at par on March 1, 2009. Interest is 6% payable semi-annually. Find the purchase price of the bond on September 1, 2003 to yield 7.5% compounded semi-annually.

2. A $25 000, 10% bond redeemable at par is purchased twelve years before maturity to yield 7% compounded semi-annually. If the bond interest is payable semi-annually, what is the purchase price of the bond?

3. A $15 000, 9.5% bond redeemable at par is purchased six years and four months before maturity to yield 10% semi-annually. If the bond interest is payable semi-annually, what is the purchase price of the bond?

4. A $5000, 6% bond redeemable at par is purchased twelve years and nine months before maturity to yield 6.5% semi-annually. If the bond interest is payable semi-annually, what is the purchase price of the bond?

5. A $10 000, 9% bond redeemable at 108 is purchased nine years and five months before maturity to yield 8.5% semi-annually. If the bond interest is payable semi-annually, what is the purchase price of the bond?

6. A $2000, 7.5% bond redeemable at 105 is purchased five years and ten months before maturity to yield 9.5% semi-annually. If the bond interest is payable semi-annually, what is the purchase price of the bond?

7. A $1000, 9% bond redeemable at 104 is purchased 8.5 years before maturity to yield 6.5% compounded semi-annually. If the bond interest is payable semi-annually, what is the purchase price of the bond?

8. A $100 000 bond is redeemable at 108 in fifteen years. If interest on the bond is 7.5% payable semi-annually, what is the purchase price to yield 8% compounded semi-annually?

9. A 25-year bond issue of $5 000 000 redeemable at par and bearing interest at 7.25% payable annually is sold to yield 8.5% compounded semi-annually. What is the issue price of the bonds?

10. A $100 000 bond bearing interest at 6.75% payable semi-annually is bought eight years before maturity to yield 7.35% compounded annually. If the bond is redeemable at par, what is the purchase price?

11. Bonds with a par value of $40 000 redeemable at 103 in 7.5 years bearing interest at 8% payable quarterly are sold to yield 6.8% compounded semi-annually. Determine the purchase price of the bonds.

12. Six $1000 bonds with 11.5% coupons payable semi-annually are bought to yield 8.4% compounded monthly. If the bonds are redeemable at 109 in eight years, what is the purchase price?

13. A $25 000, 10% bond redeemable at par on December 1, 2011 is purchased on September 25, 2000 to yield 7.6% compounded semi-annually. Bond interest is payable semi-annually.

(**a**) What is the flat price of the bond?
(**b**) What is the accrued interest?
(**c**) What is the quoted price?

14. A $100 000 bond redeemable at par on October 1, 2024 is purchased on January 15, 2003. Interest is 5.9% payable semi-annually and the yield is 9% compounded semi-annually.

(**a**) What is the purchase price of the bond?
(**b**) How much interest has accrued?
(**c**) What is the market price?

15. A $5000, 9.5% bond redeemable at 104 matures on August 1, 2010. If the coupons are payable semi-annually, what is the quoted price on May 10, 2001 to yield 8.5% compounded semi-annually?

16. Bonds in denominations of $1000 redeemable at 107 are offered for sale. If the bonds mature in six years and ten months and the coupon rate is 9.5% payable quarterly, what is the market price of the bonds to yield 8% compounded quarterly?

9.2 Premium and Discount

A. Basic concepts—bond rate versus yield (or market) rate

Comparing the redemption values with the purchase prices obtained in Examples 9.1A to 9.1I (see Table 9.1) shows that the purchase price is sometimes less than and sometimes more than the redemption value.

TABLE 9.1 Comparison of Redemption Values with Purchase Prices for Examples 15.1A to 15.1I

Example	Redemption Value FV	Purchase (Flat) Price PP	Comparison of FV and PP	Premium or Discount	Amount of Premium or Discount	Bond Rate b	Yield Rate i	b Versus i
15.1A	\$1 000.00	\$965.63	FV > PP	discount	\$34.37	3%	3.5%	$b < i$
15.1B	\$5 000.00	\$5 487.79	PP > FV	premium	\$487.79	5.25%	4.5%	$b > i$
15.1C	\$10 600.00	\$8 226.73	FV > PP	discount	\$2 373.27	4%	5%	$b < i$
15.1D	\$26 750.00	\$27 270.74	PP > FV	premium	\$520.74	2.75%	2.5%	$b > i$
15.1E	\$1 000 000.00	\$915 535.44	FV > PP	discount	\$84 464.56	10%	11.46%	$b < i$
15.1F	\$103 000.00	\$98 222.68	FV > PP	discount	\$4 777.32	3.75%	4.04%	$b < i$
15.1G	\$1 000.00	\$1 041.26	PP > FV	premium	\$41.26	5%	4.5%	$b > i$
15.1H	\$5 000.00	\$4 776.08	FV > PP	discount	\$223.92	3.25%	3.75%	$b < i$
15.1I	\$10 800.00	\$11 364.07	PP > FV	premium	\$564.07	5.5%	5%	$b > i$

If the purchase price of a bond is greater than the redemption price, the bond is said to be bought at a premium and the difference between the purchase price and the redemption price is called the **premium**.

PREMIUM = PURCHASE PRICE − REDEMPTION PRICE
where purchase price > redemption price

If the purchase price of a bond is less than the redemption price, the bond is said to be bought at a discount and the difference between the redemption price and the purchase price is called the **discount.**

DISCOUNT = REDEMPTION PRICE − PURCHASE PRICE
where redemption price > purchase price

An examination of the size of the bond rate b relative to the size of the market rate i shows that this relationship determines whether there is a premium or discount.

The bond rate (or coupon rate) stated on the bond is the percent of the *face* value of the bond that will be paid at the end of each interest period to the bondholder. This rate is established at the time of issue of the bonds and remains the same throughout the term of the bond.

On the other hand, the rate at which lenders are willing to provide money fluctuates in response to economic conditions. The combination of factors at work in the capital market at any given time in conjunction with the perceived risk associated with a particular bond determines the yield rate (or market rate) for a bond and thus the price at which a bond will be bought or sold.

The bond rate and the market rate are usually *not* equal. However, if the two rates happen to be equal, then bonds that are redeemable at par will sell at their face value. If the bond rate is *less* than the market rate, the bond will sell at a price less than the face value, that is, at a *discount*. If the bond rate is *greater* than the market rate, the bond will sell at a price above its face value, that is, at a *premium*.

Conversely, if a bond is redeemable at par (that is, at 100), purchasers will realize the bond rate if they pay 100. They will realize less than the bond rate if they buy at a premium and more than the bond rate if they buy at a discount.

At any time, one of three possible situations exists for any given bond:

(1) Bond rate = Market rate ($b = i$) The bond sells at *par*.
(2) Bond rate < Market rate ($b < i$) The bond sells at a *discount*.
(3) Bond rate > Market rate ($b > i$) The bond sells at a *premium*.

EXAMPLE 9.2A

A $10 000 bond is redeemable at par and bears interest at 10% compounded semi-annually.

(i) What is the purchase price ten years before maturity if the market rate compounded semi-annually is
(a) 10%; (b) 12%; (c) 8%?

(ii) What is the purchase price five years before maturity if the market rate compounded semi-annually is
(a) 10%; (b) 12%; (c) 8%?

SOLUTION

(i) FV = 10 000.00; PMT = 10 000.00(0.05) = 500.00; $n = 10(2) = 20$; $b = 5\%$

(a) $i = \frac{10\%}{2} = 5\% = 0.05$; $(b = i)$

$$\text{Purchase price} = 10\,000.00(1.05^{-20}) + 500.00\left(\frac{1 - 1.05^{-20}}{0.05}\right)$$
$$= 10\,000.00(0.376890) + 500.00(12.46221)$$
$$= 3768.90 + 6231.11$$
$$= \$10\,000.01$$

Programmed Solution

("END" mode) (Set P/Y = 2; C/Y = 2) [2nd] (CLR TVM) 10 000 [FV] 10 [I/Y] 20 [N] [CPT] [PV] [-3768.894829]

0 [FV] 500 [PMT] 10 [I/Y] 20 [N] [CPT] [PV] [-6231.105171]

The purchase price is 3768.89 + 6231.11 = $10 000.00.
The bond sells at par.

(b) $i = \frac{12\%}{2} = 6\% = 0.06; \quad (b < i)$

$$\begin{aligned}\text{Purchase price} &= 10\,000.00(1.06^{-20}) + 500.00\left(\frac{1 - 1.06^{-20}}{0.06}\right)\\ &= 10\,000.00(0.311805) + 500.00(11.469921)\\ &= 3118.05 + 5734.96\\ &= \$8853.01\end{aligned}$$

Programmed Solution

("END" mode) (Set P/Y = 2; C/Y = 2) [2nd] (CLR TVM) 10 000 [FV] 12 [I/Y]

20 [N] [CPT] [PV] [-3118.047269]

0 [FV] 500 [PMT] 12 [I/Y] 20 [N] [CPT] [PV] [-5734.960609]

The purchase price is 3118.05 + 5734.96 = $8853.01.

The bond sells below par.

The discount is 10 000.00 − 8853.01 = $1146.99.

(c) $i = \frac{8\%}{2} = 4\% = 0.04; \quad (b > i)$

$$\begin{aligned}\text{Purchase price} &= 10\,000.00(1.04^{-20}) + 500.00\left(\frac{1 - 1.04^{-20}}{0.04}\right)\\ &= 10\,000.00(0.456387) + 500.00(13.590326)\\ &= 4563.87 + 6795.16\\ &= \$11\,359.03\end{aligned}$$

Programmed Solution

("END" mode) (Set P/Y = 2; C/Y = 2) [2nd] (CLR TVM) 10 000 [FV] 8 [I/Y]

20 [N] [CPT] [PV] [-4563.869462]

0 [FV] 500 [PMT] 8 [I/Y] 20 [N] [CPT] [PV] [-6795.163172]

The purchase price is 4563.87 + 6795.16 = $11 359.03.

The bond sells above par.

The premium is 11 359.03 − 10 000.00 = $1359.03.

(ii) FV = 10 000.00; PMT = 500.00; $n = 5(2) = 10$; $b = 5\%$

(a) $i = 5\%; \ (b = i)$

$$\begin{aligned}\text{Purchase price} &= 10\,000.00(1.05^{-10}) + 500.00\left(\frac{1 - 1.05^{-10}}{0.05}\right)\\ &= 6139.13 + 3860.87\\ &= \$10\,000.00\end{aligned}$$

Programmed Solution

("END" mode) (Set P/Y = 2; C/Y = 2) 2nd (CLR TVM) 10 000 FV 10 I/Y

10 N CPT PV -6139.132535

0 FV 500 PMT 10 I/Y 10 N CPT PV −3860.867465

The purchase price is 6139.13 + 3860.87 = $10 000.00.

The bond sells at par.

(b) $i = 6\%$; $(b < i)$

$$\begin{aligned}\text{Purchase price} &= 10\,000.00(1.06^{-10}) + 500.00\left(\frac{1 - 1.06^{-10}}{0.06}\right)\\ &= 5583.95 + 3680.04\\ &= \$9263.99\end{aligned}$$

Programmed Solution

("END" mode) (Set P/Y = 2; C/Y = 2) 2nd (CLR TVM) 10 000 FV 12 I/Y

10 N CPT PV -5583.947769

0 FV 500 PMT 12 I/Y 10 N CPT PV −3680.043526

The purchase price is 5583.95 + 3680.04 = $9263.99.

The bond sells below par.

The discount is 10 000.00 − 9263.99 = $736.01. It is smaller than in part (i) because the time to maturity is shorter.

(c) $i = 4\%$; $(b > i)$

$$\begin{aligned}\text{Purchase price} &= 10\,000.00(1.04^{-10}) + 500.00\left(\frac{1 - 1.04^{-10}}{0.04}\right)\\ &= 6755.64 + 4055.45\\ &= \$10\,811.09\end{aligned}$$

Programmed Solution

("END" mode) (Set P/Y = 2; C/Y = 2) 2nd (CLR TVM) 10 000 FV 8 I/Y

10 N CPT PV -6755.641688

0 FV 500 PMT 8 I/Y 10 N CPT PV −4055.44789

The purchase price is 6755.64 + 4055.45 = $10 811.09.

The bond sells above par.

The premium is 10 811.09 − 10 000.00 = $811.09. It is smaller than in part (i) because the time to maturity is shorter.

B. Direct method of computing the premium or discount—alternative method for finding the purchase price

EXAMPLE 9.2B

A \$5000, 12% bond with semi-annual coupons is bought six years before maturity to yield 10% compounded semi-annually. Determine the premium.

SOLUTION

FV = 5000.00; P/Y = 2; C/Y = 2; I/Y = 10; $b = \frac{12\%}{2} = 6\% = 0.06$;

PMT = 5000.00(0.06) = 300.00; $i = \frac{10\%}{2} = 5\% = 0.05$; $n = 6(2) = 12$

Since $b > i$, the bond will sell at a premium.

$$\begin{aligned}\text{Purchase price} &= 5000.00(1.05^{-12}) + 300.00\left(\frac{1 - 1.05^{-12}}{0.05}\right)\\ &= 5000.00(0.556837) + 300.00(8.863252)\\ &= 2784.19 + 2658.97\\ &= \$5443.16\end{aligned}$$

Programmed Solution

("END" mode) (Set P/Y = 2; C/Y = 2) [2nd] (CLR TVM) 5000 [FV] 10 [I/Y]

12 [N] [CPT] [PV] [-2784.187091]

0 [FV] 300 [PMT] 10 [I/Y] 12 [N] [CPT] [PV] [−2658.975491]

The purchase price is 2784.19 + 2658.97 = \$5443.16.
The premium is 5443.16 − 5000.00 = \$443.16.

While you can always determine the premium by the basic method using Formula 9.1, it is more convenient to determine the premium directly by considering the relationship between the bond rate b and the yield rate i.

As previously discussed, a premium results when $b > i$. When this is the case, the premium is paid because the periodic interest payments received exceed the periodic interest required according to the yield rate.

In Example 9.2B

The semi-annual interest payment	5000.00(0.06) = \$300.00
The required semi-annual interest based on the yield rate	5000.00(0.05) = \$250.00
The excess of the actual interest received over the required interest to make the yield rate	= \$ 50.00

This excess is received at the end of every payment interval; thus it forms an ordinary annuity whose present value can be computed at the yield rate i.

PMT (the excess interest) = 50.00; $i = 5\%$; $n = 12$

$$PV_n = 50.00\left(\frac{1 - 1.05^{-12}}{0.05}\right)$$
$$= 50.00(8.863252)$$
$$= \$443.16$$

Programmed Solution

("END" mode) (Set P/Y = 2; C/Y = 2) 0 [FV] 50 [PMT] 10 [I/Y]

12 [N] [CPT] [PV] [−443.1625818]

The premium is $443.16.

The purchase price is 5000.00 + 443.16 = $5443.16.

The premium is the present value of the ordinary annuity formed by the excess of the actual bond interest over the required interest based on the yield rate. We can obtain the purchase price by adding the premium to the redemption price.

$$\text{PREMIUM} = (\text{PERIODIC BOND INTEREST} - \text{REQUIRED INTEREST})\left[\frac{1 - (1 + i)^{-n}}{i}\right]$$
$$= (\text{FACE VALUE} \times b - \text{REDEMPTION PRICE} \times i)\left[\frac{1 - (1 + i)^{-n}}{i}\right]$$

EXAMPLE 9.2C A $5000, 6% bond with semi-annual coupons is bought six years before maturity to yield 8% compounded semi-annually. Determine the discount.

SOLUTION $FV = 5000.00$; $P/Y = 2$; $C/Y = 2$; $b = \frac{6\%}{2} = 3\% = 0.03$;

$PMT = 5000.00(0.03) = 150.00$; $I/Y = 8$; $i = \frac{8\%}{2} = 4\% = 0.04$; $n = 6(2) = 12$

Since $b < i$, the bond will sell at a discount.

$$\text{Purchase price} = 5000.00(1.04^{-12}) + 150.00\left(\frac{1 - 1.04^{-12}}{0.04}\right)$$
$$= 5000.00(0.624597) + 150.00(9.385074)$$
$$= 3122.99 + 1407.76$$
$$= \$4530.75$$

Programmed Solution

("END" mode)(Set P/Y = 2; C/Y = 2) [2nd] (CLR TVM) 5000 [FV]

8 [I/Y] 12 [N] [CPT] [PV] [-3122.985248]

0 [FV] 150 [PMT] 8 [I/Y] 12 [N] [CPT] [PV] [–1407.761064]

The purchase price is 3122.99 + 1407.76 = $4530.75.

The discount is 5000.00 − 4530.75 = $469.25.

As in the case of a premium, while you can always determine the discount by the basic method using Formula 9.1, it is more convenient to determine the discount directly.

When $b < i$, a discount results. The discount on a bond is received because the periodic interest payments are less than the periodic interest required to earn the yield rate.

In Example 9.2C

The semi-annual interest payment	$5000.00(0.03) = \$150.00$
The required semi-annual interest based on the yield rate	$5000.00(0.04) = \$200.00$
The shortage of the actual interest received compared to the required interest based on the yield rate	$= \$50.00$

This shortage occurs at the end of every interest payment interval; it forms an ordinary annuity whose present value can be computed at the yield rate i.

$\text{PMT} = 50.00; \quad i = 4\%; \quad n = 12$

$$\begin{aligned} PV_n &= 50.00\left(\frac{1 - 1.04^{-12}}{0.04}\right) \\ &= 50.00(9.385074) \\ &= \$469.25 \end{aligned}$$

Programmed Solution

("END" mode) (Set P/Y = 2; C/Y = 2) 0 [FV] 50 [PMT] 8 [I/Y]

12 [N] [CPT] [PV] [-469.253688]

The discount is \$469.25.

The purchase price is $5000.00 - 469.25 = \$4530.75$.

The discount is the present value of the ordinary annuity formed by the shortage of the actual bond interest received as compared to the required interest based on the yield rate. We can obtain the purchase price by subtracting the discount from the redemption price.

$$\begin{aligned} \text{DISCOUNT} &= (\text{REQUIRED INTEREST} - \text{PERIODIC BOND INTEREST})\left[\frac{1 - (1 + i)^{-n}}{i}\right] \\ &= -(\text{PERIODIC BOND INTEREST} - \text{REQUIRED INTEREST})\left[\frac{1 - (1 + i)^{-n}}{i}\right] \\ &= -(\text{FACE VALUE} \times b - \text{REDEMPTION PRICE} \times i)\left[\frac{1 - (1 + i)^{-n}}{i}\right] \end{aligned}$$

Since in both cases the difference between the periodic bond interest and the required interest is involved, the premium or discount on the purchase of a bond can be obtained using the same relationship.

$$\begin{matrix}\text{PREMIUM} \\ \text{or} \\ \text{DISCOUNT}\end{matrix} = (b \times \text{FACE VALUE} - i \times \text{REDEMPTION PRICE})\left[\frac{1-(1+i)^{-n}}{i}\right]$$ —— Formula 9.3

EXAMPLE 9.2D

A \$1000, 8.5% bond with semi-annual coupons redeemable at par in fifteen years is bought to yield 7% compounded semi-annually. Determine

(i) the premium or discount;
(ii) the purchase price.

SOLUTION

(i) FV = 1000.00; P/Y = 2; C/Y = 2; $b = \frac{8.5\%}{2} = 4.25\% = 0.0425$;
PMT = 1000.00(0.0425) = 42.50; I/Y = 7; $i = \frac{7\%}{2} = 3.5\% = 0.035$;
$n = 15(2) = 30$

Since $b > i$, the bond will sell at a premium.

The required interest based on the yield rate is 1000.00(0.035) = 35.00; the excess interest is 42.50 − 35.00 = 7.50.

The premium is $7.50\left(\frac{1-1.035^{-30}}{0.035}\right) = 7.50(18.392045) = \$137.94$.

Programmed Solution

("END" mode) (Set P/Y = 2; C/Y = 2) 0 [FV] 7.50 [PMT] 7 [I/Y]
30 [N] [CPT] [PV] [−137.9403406]

(ii) The purchase price is 1000.00 + 137.94 = \$1137.94.

EXAMPLE 9.2E

A \$50 000, 10% bond with quarterly coupons redeemable at par in ten years is purchased to yield 11% compounded quarterly.

(i) What is the premium or discount?
(ii) What is the purchase price?

SOLUTION

(i) FV = 50 000.00; P/Y = 4; C/Y = 4; $b = \frac{10\%}{4} = 2.5\% = 0.025$;
$n = 10(4) = 40$; I/Y = 11; $i = \frac{11\%}{4} = 2.75\% = 0.0275$

Since $b < i$, the bond will sell at a discount.

$$\begin{aligned}\text{Discount} &= (0.025 \times 50\,000.00 - 0.0275 \times 50\,000.00)\left(\frac{1-1.0275^{-40}}{0.0275}\right) \\ &= (1250.00 - 1375.00)(24.078101) \\ &= (-125.00)(24.078101) \\ &= -\$3009.76\end{aligned}$$

← using Formula 9.3

← the negative sign indicates a discount

Programmed Solution

("END" mode)

First compute PMT = $(0.025 \times 50\,000 - 0.0275 \times 50\,000) = -125.00$

(Set P/Y = 4; C/Y = 4) 0 [FV] 125 [PMT] 11 [I/Y]

40 [N] [CPT] [PV] [−3009.762633]

(ii) The purchase price is 50 000.00 − 3009.76 = $46 990.24.

EXAMPLE 9.2F

Bonds with a face value of $15 000 redeemable at 108 with interest at 9% payable semi-annually are bought twelve years before maturity to yield 11% compounded semi-annually.

(i) What is the premium or discount?
(ii) What is the purchase price?

SOLUTION

(i) FV = 15 000.00(1.08) = 16 200.00; P/Y = 2; C/Y = 2;

$b = \frac{9\%}{2} = 4.5\% = 0.045;\ n = 12(2) = 24;$ I/Y = 11;

$i = \frac{11\%}{2} = 5.5\% = 0.055$

Since $b < i$ and the redemption price is greater than par, the bond will sell at a discount.

$$\begin{aligned} \text{Discount} &= (0.045 \times 15\,000.00 - 0.055 \times 16\,200.00)\left(\frac{1 - 1.055^{-24}}{0.055}\right) \\ &= (675.00 - 891.00)(13.151699) \\ &= (-216.00)(13.151699) \\ &= -\$2840.77 \end{aligned}$$

Programmed Solution

("END" mode)

PMT = $(0.045 \times 15\,000 - 0.055 \times 16\,200) = -216.00$

(Set P/Y = 2; C/Y = 2) 0 [FV] 216 [PMT] 11 [I/Y]

24 [N] [CPT] [PV] [−2840.766974]

(ii) The purchase price is 16 200.00 − 2840.77 = $13 359.23.

EXAMPLE 9.2G

A $10 000, 8% bond with quarterly coupons redeemable at 106 in seven years is purchased to yield 6% compounded quarterly.

(i) What is the premium or discount?
(ii) What is the purchase price?

SOLUTION

(i) FV= 10 000.00(1.06) = 10 600.00; P/Y = 4; C/Y = 4; I/Y = 6; $b = \frac{8\%}{4} = 2\% = 0.02;$

$n = 7(4) = 28;\ i = \frac{6\%}{4} = 1.5\% = 0.015$

Since $b > i$, the bond is expected to sell at a premium.

$$\text{Premium} = (0.02 \times 10\,000.00 - 0.015 \times 10\,600.00)\left(\frac{1 - 0.015^{-28}}{0.015}\right)$$
$$= (200.00 - 159.00)(22.726717)$$
$$= (41.00)(22.726717)$$
$$= \$931.80$$

Programmed Solution

("END" mode)

PMT = (0.02 × 10 000 − 0.015 × 10 600) = 200.00 − 159.00 = 41.00

(Set P/Y = 4; C/Y = 4) 0 FV 41 PMT 6 I/Y

28 N CPT PV −931.795385

(ii) The purchase price is 10 600.00 + 931.80 = $11 531.80.

EXAMPLE 9.2H

A $1000, 7% bond redeemable at 110 with interest payable annually is bought nine years before maturity to yield 6.5% compounded annually. Determine

(i) the premium or discount;
(ii) the purchase price.

SOLUTION

(i) FV = 1000.00(1.10) = 1100.00; P/Y = 1; C/Y = 1; $b = 7\% = 0.07$; $n = 9$; I/Y = 6.5; $i = 6.5\% = 0.065$

Since $b > i$, the bond is expected to be sold at a premium. While this is always true for bonds redeemable at par, it does not necessarily follow for bonds redeemable above par.

In this particular case:

the actual bond interest per year	1000.00(0.07)	= $70.00
the interest required to make the yield rate	1100.00(0.065)	= $71.50

Because of the redemption premium, the required interest exceeds the actual interest: the bond will, in fact, sell at a discount. This conclusion is borne out when using Formula 9.3.

$$\text{Premium/Discount} = (0.07 \times 1000.00 - 0.065 \times 1100.00)\left(\frac{1 - 1.065^{-9}}{0.065}\right)$$
$$= (70.00 - 71.50)(6.656104)$$
$$= (-1.50)(6.656104)$$
$$= -\$9.98$$

Programmed Solution

("END" mode)

PMT = (0.07 × 1000 − 0.065 × 1100) = −1.50

(Set P/Y = 1; C/Y = 1) 0 FV 1.50 PMT 6.5 I/Y

9 N CPT PV −9.984156281

(ii) Since the answer (PMT) is negative, the bond sells at a discount of \$9.98. The purchase price is $1100.00 - 9.98 = \$1090.02$.

EXAMPLE 9.21

A \$25 000 bond, redeemable at 104 on July 1, 2015 with 11% coupons payable quarterly, is bought on May 20, 2006 to yield 10% compounded quarterly. What is

(i) the premium or discount?
(ii) the purchase price?
(iii) the quoted price?

SOLUTION

$FV = 25\,000.00(1.04) = 26\,000.00$; $P/Y = 4$; $C/Y = 4$; $I/Y = 10$

$b = \frac{11\%}{4} = 2.75\% = 0.0275$; $I/Y = 10$; $i = \frac{10\%}{4} = 2.5\% = 0.025$

The interest payment dates are October 1, January 1, April 1, and July 1. The interest payment date preceding the date of purchase is April 1, 2006. The time period April 1, 2006 to July 1, 2015 contains 9 years and 3 months; $n = 9.25(4) = 37$.

The premium on April 1, 2006,

$$= (25\,000.00 \times 0.0275 - 26\,000.00 \times 0.025)\left(\frac{1 - 1.025^{-37}}{0.025}\right)$$
$$= (687.50 - 650.00)(23.957318)$$
$$= (37.50)(23.957318)$$
$$= \$898.40$$

Programmed Solution

("END" mode)

$PMT = (25\,000 \times 0.0275 - 26\,000 \times 0.025) = 37.50$

(Set P/Y = 4; C/Y = 4) 0 [FV] 37.50 [PMT] 10 [I/Y]

37 [N] [CPT] [PV] [−898.3994293]

The purchase price on April 1, 2006 is $26\,000.00 + 898.40 = \$26\,898.40$.

The time period April 1 to May 20 contains 49 days; the number of days in the interest payment interval April 1 to July 1 is 91.

$PV = 26\,898.40$; $r = i = 0.025$; $t = \frac{49}{91}$

The accumulated value (flat price) on May 20, 2006,

$$= 26\,898.40\left[1 + 0.025\left(\frac{49}{91}\right)\right]$$
$$= 26\,898.40(1.013462)$$
$$= \$27\,260.49$$

The accrued interest to May 20 is $25\,000.00(0.0275)\left(\frac{49}{91}\right) = \$370.19$.

The quoted price is $27\,260.49 - 370.19 = \$26\,890.30$.

Thus, on May 20, 2006,

(i) the premium is 26 890.30 − 26 000.00 = \$890.30;
(ii) the purchase price is \$27 260.49;
(iii) the quoted price is \$26 890.30.

EXERCISE 9.2

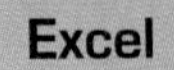

Excel If you choose, you can use these Excel functions to answer the questions indicated below: ***Bond Purchase Price (PRICE)***, ***Days Since Last Interest Date (COUPDAYBS)***, ***Total Number of Days in the Coupon Period (COUPDAYS)***, and ***Total Number of Remaining Coupon Periods (COUPNUM)***. Refer to **PRICE**, **COUPDAYBS**, **COUPDAYS**, and **COUPNUM** on the Spreadsheet Template Disk to learn how to use these Excel functions.

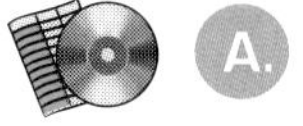

A. For each of the six bonds in the table below, use Formula 9.3 to determine

(a) the premium or discount;

(b) the purchase price.

	Par Value	Redeemed At:	Bond Rate Payable Semi-annually	Time Before Redemption	Yield Rate Compounded Semi-annually
1.	\$25 000	par	6%	10 years	9%
2.	\$5 000	par	8.5%	8 years	7%
3.	\$10 000	104	9%	15 years	11%
4.	\$7 000	110	9%	5 years	8.5%
5.	\$1 000	105	6.5%	6 years, 10 months	8%
6.	\$50 000	108	12%	4 years, 5 months	11.5%

B. Answer each of the following questions.

1. A \$100 000, 8% bond redeemable at par with quarterly coupons is purchased to yield 6.5% compounded quarterly. Find the premium or discount and the purchase price if the bond is purchased

(a) fifteen years before maturity; **(b)** five years before maturity.

2. A \$5000, 7.5% bond redeemable at 104 with semi-annual coupons is purchased to yield 6% compounded semi-annually. What is the premium or discount and the purchase price if the bond is bought

(a) ten years before maturity? **(b)** six years before maturity?

3. A \$25 000, 9% bond redeemable at par with interest payable annually is bought six years before maturity. Determine the premium or discount and the purchase price if the bond is purchased to yield

(a) 13.5% compounded annually; **(b)** 6% compounded annually.

4. A \$1000, 8% bond redeemable at 108 in seven years bears coupons payable annually. Compute the premium or discount and the purchase price if the yield, compounded annually, is

(a) 6.5%; **(b)** 7.5%; **(c)** 8.5%.

5. Twelve $1000 bonds redeemable at par bearing interest at 10% payable semi-annually and maturing on September 1, 2006 are bought on June 18, 2001 to yield 7% compounded semi-annually. Determine

   (a) the premium or discount on the preceding interest payment date;
   (b) the purchase price;
   (c) the quoted price.

6. Bonds with a face value of $30 000 redeemable at 107 on June 1, 2012 are offered for sale to yield 9.2% compounded quarterly. If interest is 7% payable quarterly and the bonds are bought on January 24, 2003, what is

   (a) the premium or discount on the interest payment date preceding the date of sale?
   (b) the purchase price?
   (c) the quoted price?

7. A $5 000 000 issue of ten-year bonds redeemable at par offers 7.25% coupons payable semi-annually. What is the issue price of the bonds to yield 8.4% compounded monthly?

8. A $3000 issue of nine-year bonds redeemable at 107 offers 7.5% coupons paid semi-annually. What is the issue price of the bonds to yield 10.5% semi-annually?

9. Twenty $5000 bonds redeemable at 110 bearing 12% coupons payable quarterly are sold eight years before maturity to yield 11.5% compounded annually. What is the purchase price of the bonds?

10. Sixty $1000 bonds redeemable at 108 bearing 7% coupons payable semi-annually are sold seven years before maturity to yield 9.5% compounded semi-annually. What is the purchase price of the bonds?

BUSINESS MATH NEWS BOX

The Price of Liquidity

Investors now have a choice when they invest in the Government of Canada. The traditional Canada Savings Bonds that many Canadians have invested in for many years have been complemented by a new series of bonds called Canada Premium Bonds. The Government of Canada guarantees the principal and interest on both series of bonds. The difference lies in the flexibility to withdraw your money for other investments or uses. Canada Savings Bonds are cashable at any time, with interest being paid up to the first day of the month in which redemption takes place. Canada Premium Bonds, however, are cashable without penalty only on the anniversary of their issue date or during the 30 days thereafter. Canada Premium Bonds, however, carry a higher rate of interest. The interest rate is fixed and interest is compounded if the Canada Premium Bonds are held for more than one year. Canada Savings Bonds have a lower minimum guaranteed interest rate than Canada Premium Bonds for series on sale at the same time with the same issue date, and interest is also compounded if Canada Savings Bonds are held for more than one year. Both Canada Premium Bonds and Canada Savings Bonds are available in Regular, non-compounding-interest versions as well to which interest is paid annually on the anniversary of issue.

QUESTIONS

1. Assume the annual rate of interest on Canada Savings Bonds is set at 2.00% on April 1, 2004, and 2.50% on April 1, 2005 and 3.00% on April 1, 2006. What is the accumulated value of a $1000 Canada Savings Bond purchased on April 1, 2004, on April 1, 2007?
2. Assume the annual rate of interest on Canada Premium Bonds is set at 2.50% on April 1, 2004, and 3.00% on April 1, 2005 and 3.00% on April 1, 2006. What is the accumulated value of a $1000 Canada Premium Bond purchased on April 1, 2004, on April 1, 2007?
3. What is the rate of interest compounded annually for a Canada Premium Bond purchased on April 1, 2004, on April 1, 2007?
4. Why would an investor purchase Canada Savings Bonds instead of Canada Premium Bonds? Why would an investor purchase Canada Premium Bonds instead of Canada Savings Bonds?

9.3 Bond Schedules

A. Amortization of premium

If a bond is purchased for more than the redemption price, the resulting premium is not recovered when the bond is redeemed at maturity, so it becomes a capital *loss*. To avoid the capital loss at maturity, the premium is written down gradually over the period from the date of purchase to the maturity date. The writing down of the premium gradually reduces the bond's book value until it equals the redemption price at the date of maturity.

The process of writing down the premium is called **amortization of the premium**. The most direct method of amortizing a premium assigns the difference between the interest received (coupon) and the interest required according to the yield rate to write down the premium. The details of writing down the premium are often shown in a tabulation referred to as a *schedule of amortization of premium*.

EXAMPLE 9.3A

A $1000, 12% bond redeemable at par matures in three years. The coupons are payable semi-annually and the bond is bought to yield 10% compounded semi-annually.

(i) Compute the purchase price.
(ii) Construct a schedule of amortization of premium.

SOLUTION

(i) FV = 1000.00; $n = 3(2) = 6$; P/Y = 2; C/Y = 2;

$b = \frac{12\%}{2} = 6\% = 0.06$; I/Y = 10; $i = \frac{10\%}{2} = 5\% = 0.05$

Since $b > i$, the bond sells at a premium.

$$\text{Premium} = (0.06 \times 1000.00 - 0.05 \times 1000.00)\left(\frac{1 - 1.05^{-6}}{0.05}\right)$$
$$= (60.00 - 50.00)(5.075692)$$
$$= (10.00)(5.075692)$$
$$= \$50.76$$

Programmed Solution

PMT = (0.06 × 1000.00 − 0.05 × 1000.00) = 10.00

("END" mode)

(Set P/Y = 2; C/Y = 2) 0 [FV] 10 [PMT] 10 [I/Y]

6 [N] [CPT] [PV] [−50.75692067]

The purchase price is 1000.00 + 50.76 = $1050.76.

(ii) ***Schedule of amortization of premium***

End of Interest Payment Interval	Bond Interest Received (Coupon) $b = 6\%$	Interest on Book Value at Yield Rate $i = 5\%$	Amount of Premium Amortized	Book Value of Bond	Remaining Premium
0				1050.76	50.76
1	60.00	52.54	7.46	1043.30	43.30
2	60.00	52.17	7.83	1035.47	35.47
3	60.00	51.77	8.23	1027.24	27.24
4	60.00	51.36	8.64	1018.60	18.60
5	60.00	50.93	9.07	1009.53	9.53
6	60.00	50.47	9.53	1000.00	0.00
TOTAL	360.00	309.24	50.76		

Explanations of schedule

1. The original book value shown is the purchase price of $1050.76.
2. At the end of the first interest payment interval, the interest received (coupon) is 1000.00(0.06) = $60.00; the interest required according to the yield rate is 1050.76(0.05) = $52.54; the difference 60.00 − 52.54 = 7.46 is used to write down the premium to $43.30 and reduces the book value from $1050.76 to $1043.30.
3. The coupon at the end of the second interest payment interval is again $60.00. The interest required according to the yield rate is 1043.30(0.05) = $52.17; the difference 60.00 − 52.17 = 7.83 reduces the premium to $35.47 and the book value of the bond to $1035.47.
4. Continue in a similar manner until the maturity date when the redemption price is reached. If a rounding error becomes apparent at the end of the final interest payment interval, adjust the final interest on the book value at the yield rate to make the premium zero and to obtain the exact redemption price as the book value.
5. The totals provide useful accounting information showing the total interest received ($360.00) and the net income realized ($309.24).

EXAMPLE 9.3B

A $25 000, 7.5% bond redeemable at 106 with coupons payable annually matures in seven years. The bond is bought to yield 6% compounded annually.

(i) Compute the premium and the purchase price.
(ii) Construct a schedule of amortization of premium.

SOLUTION

(i) FV = 25 000.00(1.06) = 26 500.00; $n = 7$; P/Y = 1; C/Y = 1;

$b = 7.5\% = 0.075$; 1/Y = 6; $i = 6\% = 0.06$

Since $b > i$, the bond is expected to sell at a premium.

$$\begin{aligned}\text{Premium} &= (0.075 \times 25\,000.00 - 0.06 \times 26\,500.00)\left(\frac{1 - 1.06^{-7}}{0.06}\right)\\ &= (1875.00 - 1590.00)(5.582381)\\ &= (285.00)(5.582381)\\ &= \$1590.98\end{aligned}$$

Programmed Solution

PMT = $(0.075 \times 25\,000 - 0.06 \times 26\,500) = 285$

("END" mode)

(Set P/Y = 1; C/Y = 1) 0 [FV] 285 [PMT] 6 [I/Y] 7 [N] [CPT] [PV] [−1590.97871]

The purchase price is 26 500.00 + 1590.98 = $28 090.98.

(ii) ***Schedule of amortization of premium***

End of Interest Payment Interval	Coupon $b = 7.5\%$	Interest on Book Value at Yield Rate $i = 6\%$	Amount of Premium Amortized	Book Value of Bond	Remaining Premium
0				28 090.98	1590.98
1	1 875.00	1 685.46	189.54	27 901.44	1401.44
2	1 875.00	1 674.09	200.91	27 700.53	1200.53
3	1 875.00	1 662.03	212.97	27 487.56	987.56
4	1 875.00	1 649.25	225.75	27 261.81	761.81
5	1 875.00	1 635.71	239.29	27 022.52	522.52
6	1 875.00	1 621.35	253.65	26 768.87	268.87
7	1 875.00	1 606.13	268.87	26 500.00	0.00
TOTAL	13 125.00	11 534.02	1590.98		

B. Accumulation of discount

If a bond is bought at less than the redemption price, there will be a gain at the time of redemption equal to the amount of discount. It is generally accepted accounting practice that this gain does not accrue in total to the accounting period in which the bond is redeemed. Instead, some of the gain accrues to each of the accounting periods from the date of purchase to the date of redemption.

To adhere to this practice, the discount is decreased gradually so that the book value of the bond increases gradually until, at the date of redemption, the discount is reduced to zero while the book value equals the redemption price. The process of reducing the discount so as to increase the book value is called **accumulation of discount.**

In the case of discount, the interest required according to the yield rate is greater than the actual interest received (the coupon). Similar to amortization of a premium, the most direct method of accumulating a discount assigns the difference between the interest required by the yield rate and the coupon to reduce the discount. The details of decreasing the discount while increasing the book value of a bond are often shown in a tabulation called a *schedule of accumulation of discount.*

EXAMPLE 9.3C

A $10 000 bond, redeemable at par in four years with 5.5% coupons payable semi-annually, is bought to yield 7% compounded semi-annually.

(i) Determine the discount and the purchase price.
(ii) Construct a schedule of accumulation of discount.

SOLUTION

(i) FV = 10 000.00; $n = 4(2) = 8$; P/Y = 2; C/Y = 2;

$b = \frac{5.5\%}{2} = 2.75\% = 0.0275$; I/Y = 7; $i = \frac{7\%}{2} = 3.5\% = 0.035$

Since $b < i$, the bond sells at a discount.

$$\text{Discount} = (0.0275 \times 10\,000.00 - 0.035 \times 10\,000.00)\left(\frac{1 - 1.035^{-8}}{0.035}\right)$$
$$= (275.00 - 350.00)(6.873956)$$
$$= (-75.00)(6.873956)$$
$$= -\$515.55$$

Programmed Solution

PMT = $(0.0275 \times 10\,000 - 0.035 \times 10\,000) = -75.00$

("END" mode)

(Set P/Y = 2; C/Y = 2) 0 [FV] 75 [PMT] 7 [I/Y]

8 [N] [CPT] [PV] [−515.5466653]

The purchase price is 10 000.00 − 515.55 = $9484.45.

(ii) ***Schedule of accumulation of discount***

End of Interest Payment Interval	Coupon $b = 2.75\%$	Interest on Book Value at Yield Rate $i = 3.5\%$	Amount of Discount Accumulated	Book Value of Bond	Remaining Discount
0				9484.55	515.55
1	275.00	331.96	56.96	9541.41	458.59
2	275.00	333.95	58.95	9600.36	399.64
3	275.00	336.01	61.01	9661.37	338.63
4	275.00	338.15	63.15	9724.52	275.48
5	275.00	340.36	65.36	9789.88	210.12
6	275.00	342.65	67.65	9857.53	142.47
7	275.00	345.01	70.01	9927.54	72.46
8	275.00	347.46	72.46	10 000.00	0.00
TOTAL	2200.00	2715.55	515.55		

Explanations of schedule

1. The original book value shown is the purchase price of $9484.55.
2. At the end of the first interest payment interval, the coupon is 10 000.00(0.0275) = $275.00; the interest required according to the yield rate is 9484.55(0.035) = $331.96; the difference used to reduce the discount and to increase the book value is 331.96 − 275.00 = $56.96; the book value is 9484.55+ 56.96 = $9541.41; and the remaining discount is 515.55 − 56.96 = $458.59.
3. The coupon at the end of the second interest payment interval is again $275.00; the interest required on the book value is 9541.41(0.035) = $333.95; the difference is 333.95 − 275.00 = $58.95; the book value is 9541.41 + 58.95 = $9600.36; and the remaining discount is 458.59 − 58.95 = $399.64.
4. Continue in a similar manner until the maturity date when the redemption price is reached. If a rounding error becomes apparent at the end of the final interest payment interval, adjust the final interest on the book value at the yield rate to make the remaining discount equal to zero and obtain the exact redemption price as the book value.
5. The totals provide useful accounting information showing the total interest received ($2200.00) and the net income realized ($2715.55).

EXAMPLE 9.3D

A $5000, 10% bond redeemable at 102 on April 1, 2008 with coupons payable quarterly is bought on October 1, 2006 to yield 13% compounded quarterly.

(i) Compute the discount and the purchase price.
(ii) Construct a schedule showing the accumulation of the discount.

SOLUTION

(i) FV = 5000.00(1.02) = 5100.00; P/Y = 4; C/Y = 4;
the time period October 1, 2006 to April 1, 2008 contains 18 months

$$n = \frac{18}{3} = 6 \text{ (quarters)};$$

$$b = \frac{10\%}{4} = 2.5\% = 0.025; \quad i = \frac{13\%}{4} = 3.25\% = 0.0325$$

Since $b < i$, the bond sells at a discount.

$$\text{Discount} = (0.025 \times 5000.00 - 0.0325 \times 5100.00)\left(\frac{1 - 1.0325^{-6}}{0.0325}\right)$$
$$= (125.00 - 165.75)(5.372590)$$
$$= (-40.75)(5.372590)$$
$$= -\$218.93$$

Programmed Solution

PMT = (0.025× 5000 − 0.0325 × 5100) = −40.75

("END" mode)

(Set P/Y = 4; C/Y = 4) 0 [FV] 40.75 [PMT] 13 [I/Y]

6 [N] [CPT] [PV] [−218.93304]

The purchase price is 5100.00 − 218.93 = $4881.07.

(ii) ***Schedule of accumulation of discount***

End of Interest Payment Interval	Coupon $b = 2.5\%$	Interest on Book Value at Yield Rate $i = 3.25\%$	Amount of Discount Accumulated	Book Value of Bond	Remaining Discount
Oct. 1, 2006				4881.07	218.93
Jan. 1, 2007	125.00	158.63	33.63	4914.70	185.30
Apr. 1, 2007	125.00	159.73	34.73	4949.43	150.57
July 1, 2007	125.00	160.86	35.86	4985.29	114.71
Oct. 1, 2007	125.00	162.02	37.02	5022.31	77.69
Jan. 1, 2008	125.00	163.23	38.23	5060.54	39.46
Apr. 1, 2008	125.00	164.46	39.46	5100.00	0.00
TOTAL	750.00	968.93	218.93		

C. Book value of a bond—finding the gain or loss on the sale of a bond

EXAMPLE 9.3E

A $10 000, 8% bond redeemable at par with semi-annual coupons was purchased fifteen years before maturity to yield 6% compounded semi-annually. The bond was sold three years later at 101¼. Find the gain or loss on the sale of the bond.

SOLUTION

The market quotation of 101¼ indicates that the bond was sold at 101.25% of its face value. The proceeds from the sale of the bond are 10 000.00(1.0125)

= \$10 125.00. To find the gain or loss on the sale of the bond, we need to know the book value of the bond at the date of sale. This we can do by determining the original purchase price, constructing a bond schedule, and reading the book value at the time of sale from the schedule.

$FV = 10\,000.00; \quad n = 15(2) = 30$

$P/Y = 2; C/Y = 2; \quad I/Y = 6; \quad b = \frac{8\%}{2} = 4\% = 0.04; \quad i = \frac{6\%}{2} = 3\% = 0.03$

Since $b > i$, the bond was bought at a premium.

$$\begin{aligned}\text{Premium} &= (0.04 \times 10\,000.00 - 0.03 \times 10\,000.00)\left(\frac{1 - 1.03^{-30}}{0.03}\right)\\ &= (400.00 - 300.00)(19.600441)\\ &= (100.00)(19.600441)\\ &= \$1960.04\end{aligned}$$

Programmed Solution

$PMT = (0.04 \times 10\,000 - 0.03 \times 10\,000) = 100.00$

("END" mode)

(Set P/Y = 2; C/Y = 2) 0 FV 100 PMT 6 I/Y 30 N

CPT PV −1960.044135

The purchase price is 10 000.00 + 1960.04 = \$11 960.04.

Schedule of amortization of premium

End of Interest Payment Interval	Coupon $b = 4\%$	Interest on Book Value at Yield Rate $i = 3\%$	Amount of Premium Amortized	Book Value of Bond	Remaining Premium
0				11 960.04	1960.04
1	400.00	358.80	41.20	11 918.84	1918.84
2	400.00	357.57	42.43	11 876.41	1876.41
3	400.00	356.29	43.71	11 832.70	1832.70
4	400.00	354.98	45.02	11 787.68	1787.68
5	400.00	353.63	46.37	11 741.31	1741.31
6	400.00	352.24	47.76	11 693.55	1693.55
	etc.				

The book value after three years (six semi-annual periods) is \$11 693.55. Since the book value is greater than the proceeds, the loss on the sale of the bond is 11 693.55 − 10 125.00 = \$1568.55.

We can solve this problem more quickly by computing the book value directly. The book value of a bond at a given time is the purchase price of the bond on that date. We can determine the book value of a bond without constructing a bond schedule by using Formula 9.1 or 9.3. This approach can also be used to verify book values in a bond schedule.

$FV = 10\,000.00; \quad n = (15 - 3)(2) = 24; \quad b = 4\%; \quad i = 3\%$

$$\begin{aligned}\text{Premium} &= (0.04 \times 10\,000.00 - 0.03 \times 10\,000.00)\left(\frac{1 - 1.03^{-24}}{0.03}\right)\\ &= (100.00)(16.935542)\\ &= \$1693.55\end{aligned}$$

Programmed Solution

PMT = (0.04 × 10 000 − 0.03 × 10 000) = 100.00

("END" mode)

(Set P/Y = 2; C/Y = 2) 0 [FV] 100 [PMT] 6 [I/Y]

24 [N] [CPT] [PV] −1693.554212

The purchase price is 10 000.00 + 1693.55 = \$11 693.55.

The loss on the sale is 11 693.55 − 10 125.00 = \$1568.55.

EXAMPLE 9.3F

A \$5000, 11% bond redeemable at 106 with semi-annual coupons was purchased twelve years before maturity to yield 10.5% compounded semi-annually. The bond is sold five years later at 98 7/8. Find the gain or loss on the sale of the bond.

SOLUTION

Proceeds from the sale of the bond are 5000.00(0.98875) = \$4943.75.

$FV = 5000.00(1.06) = 5300.00; \; n = (12 - 5)(2) = 14; \; P/Y = 2;$

$C/Y = 2; \; b = \frac{11\%}{2} = 5.5\% = 0.055; \; I/Y = 10.5; \; i = \frac{10.5\%}{2} = 5.25\% = 0.0525$

$$\begin{aligned}\text{Premium/Discount} &= (0.055 \times 5000.00 - 0.0525 \times 5300.00)\left(\frac{1 - 1.0525^{-14}}{0.0525}\right)\\ &= (275.00 - 278.25)(9.742301)\\ &= (-3.25)(9.742301)\\ &= -\$31.66 \quad \text{discount}\end{aligned}$$

Programmed Solution

PMT = (0.055 × 5000 − 0.0525 × 5300) = −3.25

("END" mode)

(Set P/Y = 2; C/Y = 2) 0 [FV] 3.25 [PMT] 10.5 [I/Y]

14 [N] [CPT] [PV] −31.66247742

The purchase price or book value is 5300.00 − 31.66 = \$5268.34.
The loss on the sale is 5268.34 − 4943.75 = \$324.59.

EXAMPLE 9.3G

A \$1000, 10% bond with quarterly coupons redeemable at 104 on May 1, 2012 was purchased on August 1, 2002 to yield 12% compounded quarterly. If the bond is sold at 95 ½ on December 11, 2005, what is the gain or loss on the sale of the bond?

SOLUTION

The interest payment dates are August 1, November 1, February 1, and May 1. The interest date preceding the date of sale is November 1, 2005.

The proceeds from the sale of the bond on December 11, 2005,

$= 1000.00(0.955) +$ accrued interest from November 1 to December 11

$$= 955.00 + 1000.00(0.025)\left(\frac{40}{92}\right)$$
$$= 955.00 + 10.87$$
$$= \$965.87$$

$FV = 1000.00(1.04) = 1040.00;$

$P/Y = 4;\ C/Y = 4;\ b = \dfrac{10\%}{4} = 2.5\% = 0.025;\ I/Y = 12;\ i = \dfrac{12\%}{4} = 3\% = 0.03$

The time interval November 1, 2005 to May 1, 2012 contains six years and six months: $n = 6.5(4) = 26$.

Since $b < i$, the bond will sell at a discount.

$$\text{Discount} = (0.025 \times 1000.00 - 0.03 \times 1040.00)\left(\frac{1 - 1.03^{-26}}{0.03}\right)$$
$$= (25.00 - 31.20)(17.876842)$$
$$= (-6.20)(17.876842)$$
$$= -\$110.84$$

Programmed Solution

$PMT = (0.025 \times 1000 - 0.03 \times 1040) = -6.20$

("END" mode)

(Set P/Y = 4; C/Y = 4) 0 [FV] 6.20 [PMT] 12 [I/Y]

26 [N] [CPT] [PV] [−110.836423]

The purchase price on November 1, 2005 is $1040.00 - 110.84 = \$929.16$.

The accumulated value on December 11, 2005,

$$= 929.16\left[1 + 0.03\left(\frac{40}{92}\right)\right]$$
$$= 929.16(1.0130435)$$
$$= \$941.28$$

The gain from the sale of the bond is $965.87 - 941.28 = \$24.59$.

EXERCISE 9.3

Excel If you choose, you can use these Excel functions to answer the questions indicated below: ***Bond Purchase Price (PRICE)***, ***Days Since Last Interest Date (COUPDAYBS)***, ***Total Number of Days in the Coupon Period (COUPDAYS)***, and ***Total Number of Remaining Coupon Periods (COUPNUM)***. Refer to **PRICE**, **COUPDAYBS**, **COUPDAYS**, and **COUPNUM** on the Spreadsheet Template Disk to learn how to use these Excel functions.

A. For each of the following bonds, compute the premium or discount and the purchase price, and construct the appropriate bond schedule.

1. A $5000, 6% bond redeemable at par in three-and-a-half years with semi-annual coupons is purchased to yield 6.5% compounded semi-annually.
2. A $25 000 bond with interest at 12.5% payable quarterly redeemable at par is bought two years before maturity to yield 11% compounded quarterly.
3. A $1000, 12% bond with semi-annual coupons redeemable at 103 on September 1, 2005 is bought on March 1, 2002 to yield 10% compounded semi-annually.
4. A $10 000, 7.75% bond with annual coupons redeemable at 110 in seven years is bought to yield 7.25% compounded annually.

B. Find the gain or loss on the sale of each of the following bonds without constructing a bond schedule.

1. A $25 000, 10.5% bond redeemable at par with semi-annual coupons bought ten years before maturity to yield 12% compounded semi-annually is sold four years before maturity at 99¼.
2. Four $5000, 8.5% bonds with interest payable semi-annually redeemable at par were bought twenty years before maturity to yield 7.5% compounded semi-annually. The bonds were sold three years later at 103⅝.
3. Seven $1000, 9.25% bonds with annual coupons redeemable at 107 were bought nine years before maturity to yield 8.25% compounded annually. The bonds are sold three years before maturity at 94½.
4. A $100 000, 7% bond with semi-annual coupons redeemable at 102 was purchased eleven-and-a-half years before maturity to yield 6% compounded semi-annually. The bond was sold five years later at 99⅛.

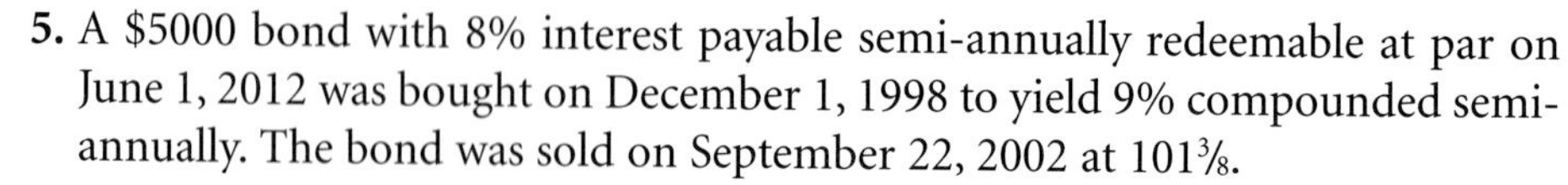

5. A $5000 bond with 8% interest payable semi-annually redeemable at par on June 1, 2012 was bought on December 1, 1998 to yield 9% compounded semi-annually. The bond was sold on September 22, 2002 at 101⅜.

6. Three $10 000, 10.5% bonds with quarterly coupons redeemable at 109 on August 1, 2009 were bought on May 1, 1995 to yield 12% compounded quarterly. The bonds were sold on January 16, 2003 at 93½.

9.4 Finding the Yield Rate

A. Quoted price of a bond—buying bonds on the market

Bonds are usually bought or sold through a bond exchange where agents trade bonds on behalf of their clients. To allow for the different denominations, bonds are offered at a quoted price stated as a percent of their face value.

It is understood that, if the bond is bought between interest dates, such a quoted price does not include any accrued interest. As explained in Section 9.1, the seller of a bond is entitled to the interest earned by the bond to the date of sale and the interest is added to the quoted price to obtain the purchase price (flat price).

EXAMPLE 9.4A

A $5000, 8% bond with semi-annual coupons payable April 1 and October 1 is purchased on August 25 at $104\frac{3}{4}$. What is the purchase price of the bond?

SOLUTION

The quoted price is 5000.00(1.0475) = $5237.50.

The time period April 1 to August 25 contains 146 days; the number of days in the interest payment interval April 1 to October 1 is 183.

$$PV = 5000.00; \quad r = i = \frac{8\%}{2} = 4\% = 0.04; \quad t = \frac{146}{183}$$

The accrued interest is $5000.00(0.04)\frac{146}{183} = \$159.56$.

The purchase price (flat price) is 5237.50 + 159.56 = $5397.06.

B. Finding the yield rate—the average investment method

When bonds are bought on the market, the yield rate is not directly available; it needs to be determined. The simplest method in use is the so-called **method of averages,** which gives a reasonable approximation of the yield rate as the ratio of the average income per interest payment interval to the average book value.

$$\text{APPROXIMATE VALUE OF } i = \frac{\text{AVERAGE INCOME PER INTEREST PAYMENT INTERVAL}}{\text{AVERAGE BOOK VALUE}}$$

e of where

$$\text{AVERAGE BOOK VALUE} = \frac{1}{2}(\text{QUOTED PRICE} + \text{REDEMPTION PRICE})$$

and

$$\text{AVERAGE INCOME PER INTEREST PAYMENT INTERVAL} = \frac{\text{TOTAL INTEREST PAYMENTS} \begin{array}{l} - \text{PREMIUM} \\ + \text{DISCOUNT} \end{array}}{\text{NUMBER OF INTEREST PAYMENT INTERVALS}}$$

Formula 9.4

EXAMPLE 9.4B

A \$25 000, 7.5% bond with semi-annual coupons redeemable at par in ten years is purchased at 103½. What is the approximate yield rate?

SOLUTION

The quoted price (initial book value) is 25 000.00(1.035) = \$25 875.00; the redemption price is \$25 000.00.

The average book value is $\frac{1}{2}(25\,875.00 + 25\,000.00) = \$25\,437.50$.

The semi-annual interest payment is $25\,000.00\left(\frac{0.075}{2}\right) = \$937.50$;

the number of interest payments to maturity is 10(2) = 20;
the total interest payments are 20(937.50) = \$18 750.00;
the premium is 25 875.00 − 25 000.00 = \$875.00.

$$\text{Average income per interest payment interval} = \frac{18\,750.00 - 875.00}{20} = \$893.75$$

$$\text{Approximate value of } i = \frac{893.75}{25\,437.50} = 0.035135 = 3.51\%$$

The yield rate is 2(3.51) = 7.02%.

EXAMPLE 9.4C

Eight \$1000, 6% bonds with semi-annual coupons redeemable at 105 in seventeen years are purchased at 97⅜. What is the approximate yield rate?

SOLUTION

The quoted price is 8000.00(0.97375) = \$7790.00;
the redemption price is 8000.00(1.05) = \$8400.00;

the average book value is $\frac{(7790.00 + 8400.00)}{2} = \$8095.00$.

The semi-annual interest payment is $8000.00\left(\frac{0.06}{2}\right) = \$240.00$;
the number of interest payments to maturity is 17(2) = 34;
the total interest payments are 34(240.00) = \$8160.00;
the bond discount is 8400.00 − 7790.00 = \$610.00.

$$\text{Average income per interest payment interval} = \frac{(8160.00 + 610.00)}{34} = \$257.94$$

The approximate value of i is $\frac{257.94}{8095.00} = 0.031864 = 3.19\%$.

The approximate yield rate is 2(3.19%) = 6.38%.

EXAMPLE 9.4D

A \$5000, 10% bond with semi-annual coupons redeemable at par on July 15, 2018 is quoted on December 2, 2006 at 103¾. What is the approximate yield rate?

SOLUTION

To find the approximate yield rate for a bond purchased between interest dates, assume that the price was quoted on the nearest interest date. Since the interest dates are January 15 and July 15, the nearest interest date is January 15, 2007, which is 11.5 years before maturity.

The quoted price is $5000.00(1.0375) = \$5187.50$;
the redemption price is 5000.00;

the average book value is $\frac{(5187.50 + 5000.00)}{2} = \$5093.75.$

The semi-annual interest is $5000.00\left(\frac{0.10}{2}\right) = \$250.00;$

the number of interest payments to maturity is $11.5(2) = 23$;
the total interest payments are $23(250.00) = \$5750.00$;
the premium is $5187.50 - 5000.00 = \$187.50$.

The average income per interest payment interval $= \frac{(5750.00 - 187.50)}{23}$

$= \$241.85$

The approximate value of i is $\frac{241.85}{5093.75} = 0.047480 = 4.75\%.$

The approximate yield rate is $2(4.75\%) = 9.50\%$.

C. Finding the accurate yield rate by trial and error

A method of trial and error similar to the one used to find the nominal rate of interest may be used to obtain as precise an approximation to the yield rate as desired. The method is illustrated in Appendix A on the CD-ROM to this book.

EXERCISE 9.4

Use the method of averages to find the approximate yield rate for each of the six bonds shown in the table below.

	Face Value	Bond Rate Payable Semi-annually	Time Before Redemption	Redeemed At:	Market Quotation
1.	\$10 000	6%	15 years	par	$101\frac{3}{8}$
2.	\$5 000	10.5%	7 years	par	$94\frac{3}{4}$
3.	\$25 000	7.5%	10 years	104	$97\frac{1}{8}$
4.	\$1 000	8.5%	8 years	109	101
5.	\$50 000	9%	5 years, 4 months	par	$98\frac{7}{8}$
6.	\$20 000	7%	9 years, 8 months	106	$109\frac{1}{4}$

9.5 Sinking Funds

A. Finding the size of the periodic payment

Sinking funds are interest-bearing accounts into which payments are made at periodic intervals to provide a desired sum of money at a specified future time. Such funds usually involve large sums of money used by both the private and the public sector to repay loans, redeem bonds, finance future capital acquisitions, provide for the replacement of depreciable plant and equipment, and recover investments in depletable natural resources.

The basic problem in dealing with sinking funds is to determine the *size of the periodic payments* that will accumulate to a known future amount. These payments form an annuity in which the accumulated value is known.

Depending on whether the periodic payments are made at the end or at the beginning of each payment period, the annuity formed is an ordinary annuity or an annuity due. Depending on whether or not the payment interval is equal in length to the interest conversion period, the annuity formed is a simple annuity or a general annuity. However, since sinking funds are normally set up so that the payment interval and the interest conversion period are equal in length, only the simple annuity cases are considered in this text.

(a) For sinking funds with payments at the end of each payment interval,

$$FV_n = PMT\left[\frac{(1+i)^n - 1}{i}\right]$$ ——— Formula 5.1

(b) For sinking funds with payments at the beginning of each payment interval,

$$FV_n(\text{due}) = PMT(1+i)\left[\frac{(1+i)^n - 1}{i}\right]$$ ——— Formula 1.1

EXAMPLE 9.5A

Western Oil plans to create a sinking fund of \$20 000.00 by making equal deposits at the end of every six months for four years. Interest is 6% compounded semi-annually.

(i) What is the size of the semi-annual deposit into the fund?
(ii) What is the total amount deposited into the fund?
(iii) How much of the fund will be interest?

SOLUTION

(i) $FV_n = 20\,000.00$; P/Y = 2; C/Y = 2; $n = 4(2) = 8$; I/Y = 6; $i = \frac{6\%}{2} = 3\%$

$$20\,000.00 = PMT\left(\frac{1.03^8 - 1}{0.03}\right)$$

$$20\,000.00 = PMT(8.892336)$$

$$PMT = \$2249.13$$

Programmed Solution

("END" mode)

(Set P/Y = 2; C/Y = 2) 0 [PV] 20 000 [FV] 6 [I/Y]

8 [N] [CPT] [PMT] [-2249.127777]

The size of the semi-annual payment is $2249.13.

(ii) The total deposited into the sinking fund is 8(2249.13) = $17 993.04.

(iii) The amount of interest in the fund is 20 000.00 − 17 993.04 = $2006.96.

EXAMPLE 9.5B

Ace Machinery wants to provide for replacement of equipment seven years from now estimated to cost $60 000.00. To do so, the company set up a sinking fund into which it will pay equal sums of money at the beginning of each of the next seven years. Interest paid by the fund is 11.5% compounded annually.

(i) What is the size of the annual payment into the fund?
(ii) What is the total paid into the fund by Ace Machinery?
(iii) How much of the fund will be interest?

SOLUTION

(i) FV_n(due) = 60 000.00; P/Y = 1; C/Y = 1; I/Y = 11.5; $n = 7$; $i = 11.5\%$

$$60\,000.00 = PMT(1.115)\left(\frac{1.115^7 - 1}{0.115}\right)$$

$$60\,000.00 = PMT(1.115)(9.934922)$$

$$60\,000.00 = PMT(11.077438)$$

$$PMT = \$5416.42$$

Programmed Solution

("BGN" mode)

(Set P/Y = 1; C/Y = 1) 0 [PV] 60 000 [FV] 11.5 [I/Y]

7 [N] [CPT] [PMT] [-5416.415057]

The size of the annual payment is $5416.42.

(ii) The total paid into the fund by Ace Machinery will be 7(5416.42) = $37 914.94.

(iii) The interest earned by the fund will be 60 000.00 − 37 914.94 = $22 085.06.

B. Constructing sinking fund schedules

The details of a sinking fund can be presented in the form of a schedule. Sinking fund schedules normally show the payment number (or payment date), the periodic payment into the fund, the interest earned by the fund, the increase in the fund, and the accumulated balance.

EXAMPLE 9.5C

Construct a sinking fund schedule for Example 9.5A.

SOLUTION

PMT = 2249.13; $n = 8$; $i = 3\% = 0.03$

Sinking fund schedule

Payment Interval Number	Periodic Payment	Interest for Payment Interval $i = 0.03$	Increase in Fund	Balance in Fund at End of Payment Interval
0				0.00
1	2 249.13	0.00	2 249.13	2 249.13
2	2 249.13	67.47	2 316.60	4 565.73
3	2 249.13	136.97	2 386.10	6 951.83
4	2 249.13	208.55	2 457.68	9 409.51
5	2 249.13	282.29	2 531.42	11 940.93
6	2 249.13	358.23	2 607.36	14 548.29
7	2 249.13	436.45	2 685.58	17 233.87
8	2 249.13	517.02	2 766.15	20 000.02
TOTAL	17 993.04	2006.98	20 000.02	

Explanations regarding the construction of the sinking fund schedule

1. The payment number 0 is used to introduce the beginning balance.
2. The first deposit is made at the end of the first payment interval. The interest earned by the fund during the first payment interval is \$0, the increase in the fund is \$2249.13 and the balance is \$2249.13.
3. The second deposit is added at the end of the second payment interval. The interest for the interval is 0.03(2249.13) = \$67.47. The increase in the fund is 2249.13 + 67.47 = \$2316.60 and the new balance in the fund is 2249.13 + 2316.60 = \$4565.73.
4. The third deposit is made at the end of the third payment interval. The interest for the interval is 0.03(4565.73) = \$136.97, the increase in the fund is 2249.13 + 136.97 = \$2386.10, and the new balance in the fund is 2386.10 + 4565.73 = \$6951.83.
5. Calculations for the remaining payment intervals are made in a similar manner.
6. The final balance in the sinking fund will probably be slightly different from the expected value. This difference is a result of rounding. The balance may be left as shown (\$20 000.02) or the exact balance of \$20 000.00 may be obtained by adjusting the last payment to \$2249.11.
7. The three totals shown are useful and should be obtained for each schedule. The total increase in the fund must be the same as the final balance. The total periodic payments are 8(2249.13) = 17 993.04. The total interest is the difference: 20 000.02 − 17 993.04 = \$2006.98.

EXAMPLE 9.5D

Construct a sinking fund schedule for Example 9.5B (an annuity due with payments at the beginning of each payment interval).

SOLUTION

PMT = 5416.42 (made at the beginning); $n = 7$; $i = 11.5\% = 0.115$

Sinking fund schedule

Payment Interval Number	Periodic Payment	Interest for Payment Interval $i = 0.115$	Increase in Fund	Balance in Fund at End of Payment Interval
0				0.00
1	5 416.42	622.89	6 039.31	6 039.31
2	5 416.42	1 317.41	6 733.83	12 773.14
3	5 416.42	2 091.80	7 508.22	20 281.36
4	5 416.42	2 955.24	8 371.66	28 653.02
5	5 416.42	3 917.99	9 334.41	37 987.43
6	5 416.42	4 991.44	10 407.86	48 395.29
7	5 416.42	6 188.35	11 604.77	60 000.06
TOTAL	37 914.94	22 085.12	60 000.06	

Explanations regarding the construction of the sinking fund schedule

1. The starting balance is $0.00.
2. The first deposit is made at the beginning of the first payment interval and the interest earned by the fund during the first payment interval is 0.115(5416.42) = $622.89. The increase in the fund is 5416.42 + 622.89 = $6039.31 and the balance is $6039.31.
3. The second deposit is made at the beginning of the second payment interval, the interest earned is 0.115(6039.31 + 5416.42) = $1317.41, the increase is 5416.42 + 1317.41 = $6733.83, and the balance is 6039.31 + 6733.83 = $12 773.14.
4. The third deposit is made at the beginning of the third payment interval, the interest earned is 0.115(12 773.14 + 5416.42) = $2091.80, the increase is 5416.42 + 2091.80 = $7508.22, and the balance is 12 773.14 + 7508.22 = $20 281.36.
5. Calculations for the remaining payment intervals are made in a similar manner. Be careful to add the deposit to the previous balance when computing the interest earned.
6. The final balance of $60 000.06 is slightly different from the expected balance of $60 000.00 due to rounding. The exact balance may be obtained by adjusting the last payment to $5416.36.
7. The total increase in the fund must equal the final balance of $60 000.06. The total periodic payments are 7(5416.42) = $37 914.94. The total interest is 60 000.06 − 37 914.94 = $22 085.12.

C. Finding the accumulated balance and interest earned or increase in a sinking fund for a payment interval; constructing partial sinking fund schedules

EXAMPLE 9.5E For Examples 9.5A and 9.5B, compute

(i) the accumulated value in the fund at the end of the third payment interval;
(ii) the interest earned by the fund in the fifth payment interval;
(iii) the increase in the fund in the fifth interval.

SOLUTION

(i) The balance in a sinking fund at any time is the accumulated value of the payments made into the fund.

For Example 9.5A (when payments are made at the end of each payment interval)

PMT = 2249.13; $n = 3$; $i = 3\%$

$$FV_n = 2249.13\left(\frac{1.03^3 - 1}{0.03}\right)$$
$$= 2249.13(3.0909)$$
$$= \$6951.83$$

see sinking fund schedule, Example 9.5C

Programmed Solution

("END" mode)

(Set P/Y = 2; C/Y = 2) 0 [PV] 2249.13 [±] [PMT] 6 [I/Y] 3 [N] [CPT] [FV] 6951.835917

For Example 9.5B (when payments are made at the beginning of each payment interval)

PMT = 5416.42; $n = 3$; $i = 11.5\%$

$$FV_n(\text{due}) = 5416.42(1.115)\left(\frac{1.115^3 - 1}{0.115}\right)$$
$$= 5416.42(1.115)(3.358225)$$
$$= \$20\,281.36$$

see sinking fund schedule, Example 9.5D

Programmed Solution

("BGN" mode)

(Set P/Y = 1; C/Y = 1) 0 [PV] 5416.42 [±] [PMT] 11.5 [I/Y] 3 [N] [CPT] [FV] 20 281.35612

(ii) The interest earned during any given payment interval is based on the balance in the fund at the beginning of the interval. This figure is the same as the balance at the end of the previous payment interval.

For Example 9.5A
The balance at the end of the fourth payment interval

$$\begin{aligned} FV_4 &= 2249.13\left(\frac{1.03^4 - 1}{0.03}\right) \\ &= 2249.13(4.183627) \\ &= \$9409.52 \end{aligned}$$

Programmed Solution

("END" mode)

(Set P/Y = 2; C/Y = 2) 0 [PV] 2249.13 [±] [PMT]

6 [I/Y] 4 [N] [CPT] [FV] [9409.520995]

The interest earned by the fund in the fifth payment interval is 0.03(9409.52) = \$282.29.

For Example 9.5B
The balance in the fund at the end of the fourth payment interval

$$\begin{aligned} FV_4 &= 5416.42(1.115)\left(\frac{1.115^4 - 1}{0.115}\right) \\ &= 5416.42(1.115)(4.744421) \\ &= \$28\,653.02 \end{aligned}$$

Programmed Solution

("BGN" mode)

(Set P/Y = 1; C/Y = 1) 0 [PV] 5416.42 [±] [PMT]

11.5 [I/Y] 4 [N] [CPT] [FV] [28 653.02037]

The interest earned by the fund in the fifth payment interval is 0.115(28 653.02 + 5416.42) = 0.115(34 069.44) = \$3917.99.

(iii) The increase in the sinking fund in any given payment interval is the interest earned by the fund during the payment interval plus the periodic payment.

For Example 9.5A
The increase in the fund during the fifth payment interval is 282.29 + 2249.13 = \$2531.42.

For Example 9.5B
The increase in the fund during the fifth payment interval is 3917.99 + 5416.42 = \$9334.41.

EXAMPLE 9.5F

The board of directors of National Credit Union decided to establish a building fund of $130 000.00 by making equal deposits into a sinking fund at the end of every three months for seven years. Interest is 12% compounded quarterly.

(i) Compute the increase in the fund during the twelfth payment interval.
(ii) Construct a partial sinking fund schedule showing details of the first three deposits, the twelfth deposit, the last three deposits, and totals.

SOLUTION

Size of the quarterly deposit:

$FV_n = 130\,000.00$; P/Y = 4; C/Y = 4; I/Y = 12; $n = 7(4) = 28$; $i = \frac{12\%}{4} = 3\%$

$$130\,000.00 = PMT\left(\frac{1.03^{28} - 1}{0.03}\right)$$
$$130\,000.00 = PMT(42.930922)$$
$$PMT = 3028.12$$

Programmed Solution

("END" mode)

(Set P/Y = 4; C/Y = 4) 0 [PV] 130 000 [FV] 12 [I/Y]
28 [N] [CPT] [PMT] [-3028.12034]

(i) Balance in the fund at the end of the eleventh payment interval

$$FV_n = 3028.12\left(\frac{1.03^{11} - 1}{0.03}\right)$$
$$= 3028.12(12.807796)$$
$$= \$38\,783.54$$

Programmed Solution

("END" mode)

(Set P/Y = 4; C/Y = 4) 0 [PV] 3028.12 [±] [PMT]
12 [I/Y] 11 [N] [CPT] [FV] [38 783.54229]

The interest earned by the fund during the twelfth payment interval is 0.03(38 783.54) = $1163.51.

The increase in the fund during the twelfth payment interval is 1163.51 + 3028.12 = $4191.63.

(ii) The last three payments are Payments 26, 27, and 28. To show details of these, we must know the accumulated value after 25 payment intervals.

$$FV_{25} = 3028.12\left(\frac{1.03^{25} - 1}{0.03}\right) = 3028.12(36.459264) = \$110\,403.03$$

Programmed Solution

("END" mode)

(Set P/Y = 4; C/Y = 4) 0 [PV] 3028.12 [±] [PMT]

12 [I/Y] 25 [N] [CPT] [FV] [110 403.0275]

Partial sinking fund schedule

Payment Interval Number	Periodic Payment Made at End	Interest for Payment Interval $i = 0.03$	Increase in Fund	Balance in Fund at End of Payment Interval
0				0.00
1	3 028.12	0.00	3 028.12	3 028.12
2	3 028.12	90.84	3 118.96	6 147.08
3	3 028.12	184.41	3 212.53	9 359.61
•	•	•	•	•
•	•	•	•	•
11	•	•	•	38 783.54
12	3 028.12	1 163.51	4 191.63	42 975.17
•	•	•	•	•
•	•	•	•	•
25	•	•	•	110 403.03
26	3 028.12	3 312.09	6 340.21	116 743.24
27	3 028.12	3 502.30	6 530.42	123 273.66
28	3 028.12	3 698.21	6 726.33	129 999.99
TOTAL	84 787.36	45 212.63	129 999.99	

EXAMPLE 9.5G

Laurin and Company want to build up a fund of $75 000.00 by making payments of $2000.00 at the beginning of every six months into a sinking fund earning 11% compounded semi-annually. Construct a partial sinking fund schedule showing details of the first three payments, the last three payments, and totals.

SOLUTION

To show details of the last three payments, we need to know the number of payments.

$FV_n(\text{due}) = 75\,000.00$; $PMT = 2000.00$; $P/Y = 2$; $C/Y = 2$; $I/Y = 11$;

$$i = \frac{11\%}{2} = 5.5\%$$

$$75\,000.00 = 2000.00(1.055)\left(\frac{1.055^n - 1}{0.055}\right)$$

$$1.055^n = 2.954976$$

$$n \ln 1.055 = \ln 2.954976$$

$$n(0.053541) = 1.0834906$$

$$n = 20.236741$$

Programmed Solution

("BGN" mode)

(Set P/Y = 2; C/Y = 2) 0 [PV] 75 000 [FV] 2000 [±] [PMT]

11 [I/Y] [CPT] [N] 20.23674097

Twenty-one payments are needed. The last three payments are Payments 19, 20, and 21. The balance in the fund at the end of the 18th payment interval

$$\begin{aligned} FV_{18}(\text{due}) &= 2000.00(1.055)\left(\frac{1.055^{18} - 1}{0.055}\right) \\ &= 2000.00(1.055)(29.481205) \\ &= \$62\,205.34 \end{aligned}$$

Programmed Solution

("BGN" mode)

(Set P/Y = 2; C/Y = 2) 0 [PV] 2000 [±] [PMT]

11 [I/Y] 18 [N] [CPT] [FV] 62 205.3422

Partial sinking fund schedule

Payment Interval Number	Periodic Payment Made at Beginning	Interest for Payment Interval $i = 0.055$	Increase in Fund	Balance in Fund at End of Payment Interval
0				0.00
1	2 000.00	110.00	2 110.00	2 110.00
2	2 000.00	226.05	2 226.05	4 336.05
3	2 000.00	348.48	2 348.48	6 684.53
•	•	•	•	•
•	•	•	•	•
18	•	•	•	62 205.34
19	2 000.00	3 531.29	5 531.29	67 736.63
20	2 000.00	3 835.51	5 835.51	73 572.14
21	1 427.86	0.00	1 427.86	75 000.00
TOTAL	41 427.86	33 572.14	75 000.00	

Note: The desired balance in the sinking fund will be reached at the beginning of the 21st payment interval by depositing $1427.86.

D. Computer application—sinking fund schedule

The schedule in Example 9.5C displays the manual calculations for a sinking fund with an interest rate of 3% and eight periodic payments. Microsoft Excel and other spreadsheet programs can be used to create a file that will immediately display the results of a change in the interest rate, the increase in the fund and the accumulated balance.

Excel

The following steps are general instructions for creating a file to calculate the sinking fund schedule in Example 9.5C. The formulas in the spreadsheet file were created using Excel; however, most spreadsheets work similarly.

Although this exercise assumes a basic understanding of spreadsheet applications, someone without previous experience with spreadsheets will be able to complete it.

STEP 1 Enter the labels shown in Figure 9.2 in row 1 and in column A.

STEP 2 Enter the numbers shown in cells E2, F2 and F4. Do not type in a dollar sign or a comma.

STEP 3 Enter only the formulas shown in cells B3, C3, D3, and E3. Make sure that the formula entry includes the dollar ($) sign as shown in the figure.

STEP 4 The formulas entered in Step 3 can be copied through the remaining cells.

(a) Select and Copy the formulas in cells B3, C3, D3, and E3.
(b) Select cells B4 to B10 and then Paste.
(c) The formulas are now active in all the cells.

STEP 5 Enter the formula shown in cell B11, and then use Copy and Paste to enter the formula in cells C11 and D11.

STEP 6 To ensure readability of the spreadsheet, format the numbers to display in two decimal places, and widen the columns to display the full labels.

This spreadsheet can now be used to reflect changes in aspects of the sinking fund and create a new schedule. Use cell F2 for new payment amounts and cell F4 for new interest rates.

FIGURE 9.2

Workbook 1

	A	B	C	D	E	F
1	Payment Interval	Periodic Payment	Interest	Increase in Fund	Balance at End	
2	0				0	2249.13
3	1	= F2	= F4*E2	= B3+C3	= E2+D3	
4	2	= F2	= F4*E3	= B4+C4	= E3+D4	= 0.03
5	3	= F2	= F4*E4	= B5+C5	= E4+D5	
6	4	= F2	= F4*E5	= B6+C6	= E5+D6	
7	5	= F2	= F4*E6	= B7+C7	= E6+D7	
8	6	= F2	= F4*E7	= B8+C8	= E7+D8	
9	7	= F2	= F4*E8	= B9+C9	= E8+D9	
10	8	= F2	= F4*E9	= B10+C10	= E9+D10	
11	Totals	= SUM(B3:B10)	= SUM(C3:C10)	= SUM(D3:D10)		

E. Debt retirement by the sinking fund method

When a sinking fund is created to retire a debt, the debt principal is repaid in total at the due date from the proceeds of the sinking fund while interest on the principal is paid periodically. The payments into the sinking fund are usually made at the same time as the interest payments are made. The sum of the two payments (debt interest payment plus payment into the sinking fund) is called the **periodic cost of the debt**. The difference between the debt principal and the sinking fund balance at any point is called the **book value of the debt**.

POINTERS AND PITFALLS

When a debt is retired by the sinking fund method, the borrower is, in effect, paying two separate annuities. Sinking fund installments are made to the fund's trustee, while periodic interest payments are made to the lender. Because of these two payment streams, two interest rates must be quoted in questions involving sinking fund debt retirement. One rate determines the *interest revenue* generated by the sinking fund, and the other rate determines the *interest penalty* paid by the borrower to the lender.

EXAMPLE 9.5H

The City Board of Education borrowed \$750 000.00 for twenty years at 13% compounded annually to finance construction of Hillview Elementary School. The board created a sinking fund to repay the debt at the end of twenty years. Equal payments are made into the sinking fund at the end of each year and interest earned by the fund is 10.5% compounded annually. Rounding all computations to the nearest dollar,

(i) determine the annual cost of the debt;
(ii) compute the book value of the debt at the end of ten years;
(iii) construct a partial sinking fund schedule showing the book value of the debt, the three first payments, the three last payments, and the totals.

SOLUTION

(i) The annual interest cost on the principal:

PV = 750 000; P/Y = 1; C/Y = 1; I/Y = 13; $i = 13\% = 0.13$;
I = 750 000(0.13) = \$97 500

The annual payment into the sinking fund:

$FV_n = 750\,000$; $n = 20$; I/Y = 10.5; $i = 10.5\%$

$$750\,000 = \text{PMT}\left(\frac{1.105^{20} - 1}{0.105}\right)$$

$$750\,000 = \text{PMT}(60.630808)$$

$$\text{PMT} = \$12\,370$$

Programmed Solution

("END" mode)

(Set P/Y = 1; C/Y = 1) 0 [PV] 750 000 [FV] 10.5 [I/Y]

20 [N] [CPT] [PMT] [−12 369.94895]

The annual cost of the debt is 97 500 + 12 370 = \$109 870.

(ii) The balance in the sinking fund after the tenth payment

$$FV_{10} = 12\,370\left(\frac{1.105^{10} - 1}{0.105}\right)$$
$$= 12\,370(16.324579)$$
$$= \$201\,935$$

Programmed Solution

("END" mode)

(Set P/Y = 1; C/Y = 1) 0 PV 12 370 ± PMT 10.5 I/Y
10 N CPT FV 201 935.0483

The book value of the debt at the end of the tenth year is
750 000 − 201 935 = $548 065.

(iii) The last three payments are Payments 18, 19, and 20. The balance in the sinking fund at the end of Year 17

$$FV_{17} = 12\,370\left(\frac{1.105^{17} - 1}{0.105}\right) = 12\,370(42.47213) = \$525\,380$$

Programmed Solution

("END" mode)

(Set P/Y = 1; C/Y = 1) 0 PV 12 370 ± PMT
10.5 I/Y 17 N CPT FV 525 380.2441

Partial sinking fund schedule

Payment Interval Number	Periodic Payment Made at End	Interest for Payment Interval $i = 0.105$	Increase in Fund	Balance in Fund at End of Payment Interval	Book Value of Debt
0				0	750 000
1	12 370	0	12 370	12 370	737 630
2	12 370	1 299	13 669	26 039	723 961
3	12 370	2 734	15 104	41 143	708 857
•	•	•	•	•	•
•	•	•	•	•	•
17	•	•	•	525 380	224 620
18	12 370	55 165	67 535	592 915	157 085
19	12 370	62 256	74 626	667 541	82 459
20	12 367	70 092	82 459	750 000	0
TOTAL	247 397	502 603	750 000		

Note: The last payment has been adjusted to create a fund of exactly $750 000.

EXERCISE 9.5

Excel

If you choose, you can use Excel's ***Periodic Payment Size (PMT)*** functions to answer the questions indicated below. Refer to **PMT** on the Spreadsheet Template Disk to learn how to use these Excel functions.

A. For each of the four sinking funds listed in the table below, compute (a) the size of the periodic payment; (b) the accumulated balance at the time indicated.

	Amount of Sinking Fund	Payment Interval	Payments Made At:	Term	Interest Rate	Conversion Period	Accumulated Balance Required After:
1.	$15 000.00	6 months	end	10 years	6%	semi-annually	10th payment
2.	$9 600.00	1 month	end	8 years	12%	monthly	36th payment
3.	$8 400.00	1 month	beginning	15 years	9%	monthly	96th payment
4.	$21 000.00	3 months	beginning	20 years	8%	quarterly	28th payment

B. Each of the four debts listed in the table below is retired by the sinking fund method. Interest payments on the debt are made at the end of each payment interval and the payments into the sinking fund are made at the same time. Determine

(a) the size of the periodic interest expense of the debt;
(b) the size of the periodic payment into the sinking fund;
(c) the periodic cost of the debt;
(d) the book value of the debt at the time indicated.

	Debt Principal	Term of Debt	Payment Interval	Interest Rate On Debt	Interest Rate On Fund	Conversion Period	Book Value Required After:
1.	$20 000.00	10 years	3 months	10%	12%	quarterly	6 years
2.	$14 500.00	8 years	6 months	7%	8.5%	semi-annually	5 years
3.	$10 000.00	5 years	1 month	7.5%	6.0%	monthly	4 years
4.	$40 000.00	15 years	3 months	8.0%	7.0%	quarterly	10 years

C. Answer each of the following questions.

1. Hein Engineering expects to expand its plant facilities in six years at an estimated cost of $75 000.00. To provide for the expansion, a sinking fund has been established into which equal payments are made at the end of every three months. Interest is 5% compounded quarterly.

(a) What is the size of the quarterly payments?
(b) How much of the maturity value will be payments?
(c) How much interest will the fund contain?

2. To redeem a $100 000.00 promissory note due in ten years, Cobblestone Enterprises has set up a sinking fund earning 7.5% compounded semi-annually. Equal deposits are made at the beginning of every six months.

(a) What is the size of the semi-annual deposits?
(b) How much of the maturity value of the fund is deposits?
(c) How much is interest?

3. Equal deposits are made into a sinking fund at the end of each year for seven years. Interest is 5.5% compounded annually and the maturity value of the fund is $20 000.00. Find the size of the annual deposits and construct a sinking fund schedule showing totals.

4. A sinking fund amounting to $15 000.00 is to be created by making payments at the beginning of every six months for four years. Interest earned by the fund is 12.5% compounded semi-annually. Determine the size of the semi-annual payments and prepare a sinking fund schedule showing totals.

5. For Question 3, calculate the increase in the fund for the fourth year. Verify your answer by checking the sinking fund schedule.

6. For Question 4, compute the interest earned during the fifth payment interval. Verify your answer by checking the sinking fund schedule.

7. HY Industries Ltd. plans to replace a warehouse in twelve years at an anticipated cost of $45 000.00. To pay for the replacement, a sinking fund has been established into which equal payments are made at the end of every quarter. Interest is 10% compounded quarterly.

   (a) What is the size of the quarterly payments?
   (b) What is the accumulated balance just after the sixteenth payment?

8. To provide for expansion, Champlain Company has established a sinking fund earning 7% semi-annually. The fund is anticipated to reach a balance of $72 000.00 in fifteen years. Payments are made at the beginning of every six months.

   (a) What is the size of the semi-annual payment?
   (b) What is the accumulated balance at the end of the twentieth payment period?

9. Winooski & Co. has borrowed $95 000.00 for capital expansion. The company must pay the interest on the loan at the end of every six months and make equal payments at the time of the interest payments into a sinking fund until the loan is retired in twenty years. Interest on the loan is 9% compounded semi-annually and interest on the sinking fund is 7% compounded semi-annually. (Round all answers to the nearest dollar.)

   (a) Determine the size of the periodic interest expense of the debt.
   (b) Determine the size of the periodic payment into the sinking fund.
   (c) What is the periodic cost of the debt?
   (d) What is the book value of the debt after fifteen years?

10. The City of Chatham has borrowed $80 000.00 to expand a community centre. The city must pay the interest on the loan at the end of every month and make equal payments at the time of the interest payments into a sinking fund until the loan is retired in twelve years. Interest on the loan is 6% compounded monthly and interest on the sinking fund is 7.5% compounded monthly. (Round all answers to the nearest dollar.)

    (a) Determine the size of the periodic interest expense of the debt.
    (b) Determine the size of the periodic payment into the sinking fund.
    (c) What is the periodic cost of the debt?
    (d) What is the book value of the debt after eight years?

11. Kirk, Klein & Co. requires $100 000.00 fifteen years from now to retire a debt. A sinking fund is established into which equal payments are made at the end of every month. Interest is 7.5% compounded monthly.

(a) What is the size of the monthly payment?
(b) What is the balance in the sinking fund after five years?
(c) How much interest will be earned by the fund in the 100th payment interval?
(d) By how much will the fund increase during the 150th payment interval?
(e) Construct a partial sinking fund schedule showing details of the first three payments, the last three payments, and totals.

12. The Town of Keewatin issued debentures worth $120 000.00 maturing in ten years to finance construction of water and sewer facilities. To redeem the debentures, the town council decided to make equal deposits into a sinking fund at the beginning of every three months. Interest earned by the sinking fund is 6% compounded quarterly.

(a) What is the size of the quarterly payment into the sinking fund?
(b) What is the balance in the fund after six years?
(c) How much interest is earned by the fund in the 28th payment interval?
(d) By how much will the fund increase in the 33rd payment interval?
(e) Prepare a partial sinking fund schedule showing details of the first three payments, the last three payments, and totals.

13. The Township of Langley borrowed $300 000.00 for road improvements. The debt agreement requires that the township pay the interest on the loan at the end of each year and make equal deposits at the time of the interest payments into a sinking fund until the loan is retired in twenty years. Interest on the loan is 8.25% compounded annually and interest earned by the sinking fund is 5.5% compounded annually. (Round all answers to the nearest dollar.)

(a) What is the annual interest expense?
(b) What is the size of the annual deposit into the sinking fund?
(c) What is the total annual cost of the debt?
(d) How much is the increase in the sinking fund in the tenth year?
(e) What is the book value of the debt after fifteen years?
(f) Construct a partial sinking fund schedule showing details, including the book value of the debt, for the first three years, the last three years, and totals.

14. Ontario Credit Union borrowed $225 000.00 at 13% compounded semi-annually from League Central to build an office complex. The loan agreement requires payment of interest at the end of every six months. In addition, the credit union is to make equal payments into a sinking fund so that the principal can be retired in total after fifteen years. Interest earned by the fund is 11% compounded semi-annually. (Round all answers to the nearest dollar.)

(a) What is the semi-annual interest payment on the debt?
(b) What is the size of the semi-annual deposits into the sinking fund?
(c) What is the total annual cost of the debt?
(d) What is the interest earned by the fund in the twentieth payment interval?
(e) What is the book value of the debt after twelve years?
(f) Prepare a partial sinking fund schedule showing details, including the book value of the debt, for the first three years, the last three years, and totals.

Review Exercise

1. A $5000, 11.5% bond with interest payable semi-annually is redeemable at par in twelve years. What is the purchase price to yield
   (a) 10.5% compounded semi-annually?
   (b) 13% compounded semi-annually?

2. A $10 000, 6% bond with semi-annual coupons is redeemable at 108. What is the purchase price to yield 7.5% compounded semi-annually
   (a) nine years before maturity?
   (b) fifteen years before maturity?

3. A $25 000, 9% bond with interest payable quarterly is redeemable at 104 in six years. What is the purchase price to yield 8.25% compounded annually?

4. A $1000, 9.5% bond with semi-annual coupons redeemable at par on March 1, 2010 was purchased on September 19, 2001 to yield 7% compounded semi-annually. What was the purchase price?

5. Four $5000, 7% bonds with semi-annual coupons are bought seven years before maturity to yield 6% compounded semi-annually. Find the premium or discount and the purchase price if the bonds are redeemable
   (a) at par; (b) at 107.

6. Nine $1000, 8% bonds with interest payable semi-annually and redeemable at par are purchased ten years before maturity. Find the premium or discount and the purchase price if the bonds are bought to yield
   (a) 6%; (b) 8%; (c) 10%.

7. A $100 000, 5% bond with interest payable semi-annually redeemable at par on July 15, 2012 was purchased on April 18, 2001 to yield 7% compounded semi-annually. Determine
   (a) the premium or discount;
   (b) the purchase price;
   (c) the quoted price.

8. Four $10 000 bonds bearing interest at 6% payable quarterly and redeemable at 106 on September 1, 2014 were purchased on January 23, 2002 to yield 5% compounded quarterly. Determine
   (a) the premium or discount;
   (b) the purchase price;
   (c) the quoted price.

9. A $5000, 8% bond with semi-annual coupons redeemable at 108 in ten years is purchased to yield 10% compounded semi-annually. What is the purchase price?

10. A $1000 bond bearing interest at 8% payable semi-annually redeemable at par on February 1, 2009 was purchased on October 12, 2002 to yield 7% compounded semi-annually. Determine the purchase price.

11. A $25 000, 13% bond with semi-annual coupons redeemable at 107 on June 15, 2014 was purchased on May 9, 2003 to yield 14.5% compounded semi-annually. Determine
    (a) the premium or discount;
    (b) the purchase price;
    (c) the quoted price.

12. A $50 000, 11% bond with semi-annual coupons redeemable at par on April 15, 2007 was purchased on June 25, 2000 at 92⅜. What was the approximate yield rate?

13. A $1000, 8.5% bond with interest payable annually is purchased six years before maturity to yield 10.5% compounded annually. Compute the premium or discount and the purchase price and construct the appropriate bond schedule.

14. A $5000, 12.25% bond with interest payable annually redeemable at par in seven years is purchased to yield 13.5% compounded annually. Find the premium or discount and the purchase price and construct the appropriate bond schedule.

15. A \$20 000, 9.5% bond with semi-annual coupons redeemable at 105 in three years is purchased to yield 8% compounded semi-annually. Find the premium or discount and purchase price and construct the appropriate bond schedule.

16. Three \$25 000, 11% bonds with semi-annual coupons redeemable at par were bought eight years before maturity to yield 12% compounded semi-annually. Determine the gain or loss if the bonds are sold at 89 3/8 five years later.

17. A \$10 000 bond with 5% interest payable quarterly redeemable at 106 on November 15, 2012 was bought on July 2, 1996 to yield 9% compounded quarterly. If the bond was sold at 92 3/4 on September 10, 2002, what was the gain or loss on the sale?

18. A \$25 000, 9.5% bond with semi-annual coupons redeemable at par is bought sixteen years before maturity at 78 1/4. What was the approximate yield rate?

19. A \$10 000, 7.5% bond with quarterly coupons redeemable at 102 on October 15, 2013 was purchased on May 5, 2001 at 98 3/4. What is the approximate yield rate?

20. What is the approximate yield realized if the bond in Question 19 was sold on August 7, 2006 at 92?

21. A 6.5% bond of \$50 000 with interest payable quarterly is to be redeemable at par in twelve years.

(a) What is the purchase price to yield 8% compounded quarterly?

(b) What is the book value after nine years?

(c) What is the gain or loss if the bond is sold nine years after the date of purchase at 99 5/8?

22. A \$100 000, 10.75% bond with interest payable annually is redeemable at 103 in eight years. What is the purchase price to yield 12% compounded quarterly?

23. A \$5000, 14.5% bond with semi-annual coupons redeemable at par on August 1, 2014 was purchased on March 5, 2003 at 95 1/2. What was the approximate yield rate?

24. A \$25 000, 8% bond with semi-annual coupons, redeemable at 104 in fifteen years, is purchased to yield 6% compounded semi-annually. Determine the gain or loss if the bond is sold three years later at 107 1/4.

25. To provide for the purchase of heavy construction equipment estimated to cost \$110 000.00, Valmar Construction is paying equal sums of money at the end of every six months for five years into a sinking fund earning 7.5% compounded semi-annually.

(a) What is the size of the semi-annual payment into the sinking fund?

(b) Compute the balance in the fund after the third payment.

(c) Compute the amount of interest earned during the sixth payment interval.

(d) Construct a sinking fund schedule showing totals. Check your answers to parts (b) and (c) with the values in the schedule.

26. Alpha Corporation is depositing equal sums of money at the beginning of every three months into a sinking fund to redeem a \$65 000.00 promissory note due eight years from now. Interest earned by the fund is 12% compounded quarterly.

(a) Determine the size of the quarterly payments into the sinking fund.

(b) Compute the balance in the fund after three years.

(c) Compute the increase in the fund during the 24th payment interval.

(d) Construct a partial sinking fund schedule showing details of the first three deposits, the last three deposits, and totals.

27. The municipality of Kirkfield borrowed \$100 000.00 to build a recreation centre. The debt principal is to be repaid in eight years and interest at 13.75% compounded annually is to be paid annually. To provide for the retirement of the debt, the municipal council set up a sinking fund into which equal payments are made at the time of the annual interest payments. Interest earned by the fund is 11.5% compounded annually.

(a) What is the annual interest payment?

(b) What is the size of the annual payment into the sinking fund?

(c) What is the total annual cost of the debt?

(d) Compute the book value of the debt after three years.

(e) Compute the interest earned by the fund in Year 6.

(f) Construct a sinking fund schedule showing the book value of the debt and totals. Verify your computations in parts (d) and (e) against the schedule.

28. The Harrow Board of Education financed the acquisition of a building site through a $300 000.00 long-term promissory note due in fifteen years. Interest on the promissory note is 9.25% compounded semi-annually and is payable at the end of every six months. To provide for the redemption of the note, the board agreed to make equal payments at the end of every six months into a sinking fund paying 8% compounded semi-annually. (Round all answers to the nearest dollar.)

(a) What is the semi-annual interest payment?

(b) What is the size of the semi-annual payment into the sinking fund?

(c) What is the annual cost of the debt?

(d) Compute the book value of the debt after five years.

(e) Compute the increase in the sinking fund in the 20th payment interval.

(f) Construct a partial sinking fund schedule showing details, including the book value of the debt, for the first three years, the last three years, and totals.

29. Northern Flying Service is preparing to buy an aircraft estimated to cost $60 000.00 by making equal payments at the end of every three months into a sinking fund for five years. Interest earned by the fund is 8% compounded quarterly.

(a) What is the size of the quarterly payment into the sinking fund?

(b) How much of the maturity value of the fund will be interest?

(c) What is the accumulated value of the fund after two years?

(d) How much interest will the fund earn in the 15th payment interval?

30. A sinking fund of $10 000.00 is to be created by equal annual payments at the beginning of each year for seven years. Interest earned by the fund is 7.5% compounded annually.

(a) Compute the annual deposit into the fund.

(b) Construct a sinking fund schedule showing totals.

31. Joe Ngosa bought a retirement fund for $15 000.00. Beginning twenty-five years from the date of purchase, he will receive payments of $17 500.00 at the beginning of every six months. Interest earned by the fund is 12% compounded semi-annually.

(a) How many payments will Joe receive?

(b) What is the size of the last payment?

32. The town of Kildare bought firefighting equipment for $96 000.00. The financing agreement provides for annual interest payments and equal payments into a sinking fund for ten years. After ten years the proceeds of the sinking fund will be used to retire the principal. Interest on the debt is 14.5% compounded annually and interest earned by the sinking fund is 13% compounded annually.

(a) What is the annual interest payment?

(b) What is the size of the annual payment into the sinking fund?

(c) What is the total annual cost of the debt?

(d) What is the book value of the debt after four years?

(e) Construct a partial sinking fund schedule showing details, including the book value of the debt, for the last three years and totals.

Self-Test

1. A $10 000, 10% bond with quarterly coupons redeemable at par in fifteen years is purchased to yield 11% compounded quarterly. Determine the purchase price of the bond.
2. What is the purchase price of a $1000, 7.5% bond with semi-annual coupons redeemable at 108 in ten years if the bond is bought to yield 6% compounded semi-annually?
3. A $5000, 8% bond with semi-annual coupons redeemable at 104 is bought six years before maturity to yield 6.5% compounded semi-annually. Determine the premium or discount.
4. A $20 000, 10% bond with semi-annual coupons redeemable at par March 1, 2009 was purchased on November 15, 2002 to yield 9% compounded semi-annually. What was the purchase price of the bond?
5. A $5000, 7% bond with semi-annual coupons redeemable at 102 on December 15, 2012 was purchased on November 9, 2001 to yield 8.5% compounded semi-annually. Determine the quoted price.
6. A $5000, 11.5% bond with semi-annual coupons redeemable at 105 is bought four years before maturity to yield 13% compounded semi-annually. Construct a bond schedule.
7. A $100 000, 13% bond with semi-annual interest payments redeemable at par on July 15, 2010 is bought on September 10, 2003 at 102⅝. What was the approximate yield rate?
8. A $25 000, 6% bond with semi-annual coupons redeemable at 106 in twenty years is purchased to yield 8% compounded semi-annually. Determine the gain or loss if the bond is sold seven years after the date of purchase at 98¼.
9. A $10 000, 12% bond with semi-annual coupons redeemable at par on December 1, 2012 was purchased on July 20, 2001 at 93⅞. Compute the approximate yield rate.
10. Cottingham Pies made semi-annual payments into a sinking fund for ten years. If the fund had a balance of $100 000.00 after ten years and interest is 11% compounded semi-annually, what was the accumulated balance in the fund after seven years?
11. A fund of $165 000.00 is to be accumulated in six years by making equal payments at the beginning of each month. If interest is 7.5% compounded monthly, how much interest is earned by the fund in the twentieth payment interval?
12. Gillian Armes invested $10 000.00 in an income fund at 13% compounded semi-annually for twenty years. After twenty years, she is to receive semi-annual payments of $10 000.00 at the end of every six-month period until the fund is exhausted. What is the size of the final payment?
13. A company financed a plant expansion of $750 000.00 at 9% compounded annually. The financing agreement requires annual payments of interest and the funding of the debt through equal annual payments for fifteen years into a sinking fund earning 7% compounded annually. What is the book value of the debt after five years?

14. Annual sinking fund payments made at the beginning of every year for six years earning 11.5% compounded annually amount to $25 000.00 at the end of six years. Construct a sinking fund schedule showing totals.

Challenge Problems

1. A $2000 bond with annual coupons is redeemable at par in five years. If the first coupon is $400, and subsequent annual coupons are worth 75% of the previous year's coupon, find the purchase price of the bond that would yield an interest rate of 10% compounded annually.
2. An issue of bonds, redeemable at par in *n* years, is to bear coupons at 9% compounded semi-annually. An investor offers to buy the entire issue at a premium of 15%. At the same time, the investor advises that if the coupon rate were raised to 10% compounded semi-annually, he would offer to buy the whole issue at a premium of 25%. At what yield rate compounded semi-annually are these two offers equivalent?

Case Study 9.1 Investing in Bonds

» Recently Ruja attended a personal financial planning seminar. The speaker mentioned that bonds should be a part of everyone's balanced investment portfolio, even if they are only a small part. Ruja's RRSP contains mutual funds and a guaranteed investment certificate (GIC), but no bonds. She has decided to invest up to $4500.00 of her RRSP funds in bonds and has narrowed her choices to three.

Bond A is a $1000, 6.8% bond with semi-annual coupons redeemable in five years. Ruja can purchase up to four of these bonds at 104.25.

Bond B is a $1000, 7.1% bond with semi-annual coupons redeemable in four years. She can purchase up to four of these bonds at 103.10.

Bond C is a $1000, 6.2% bond with semi-annual coupons redeemable in seven years. Ruja can purchase up to four of these bonds at 101.85.

QUESTIONS

1. Suppose Ruja wants to invest in only one bond. Use the average investment method to answer the following questions.
   (a) What is the approximate yield rate of Bond A?
   (b) What is the approximate yield rate of Bond B?
   (c) What is the approximate yield rate of Bond C?
   (d) Assume Ruja is willing to hold the bond she chooses until it matures. Which bond has the highest yield?

2. Suppose Ruja decides to buy two $1000 denominations of Bond A. Bond A's semi-annual coupons are payable on January 1 and July 1. Suppose Ruja purchases these bonds on February 27.

   (a) What is the accrued interest on these two bonds up to the date of Ruja's purchase?
   (b) What is Ruja's purchase price (or flat price) for these two bonds?

3. Suppose Ruja decides to buy two $1000 denominations of Bond B on February 27. Bond B's semi-annual coupons are payable on February 1 and August 1.

   (a) What is the accrued interest on these two bonds up to the date of Ruja's purchase?
   (b) What is Ruja's purchase price (or flat price) for these two bonds?

Case Study 9.2 The Business of Bonds

» Beaucage Development Company is developing a new process to manufacture compact discs. The development costs were higher than expected, so Beaucage required an immediate cash inflow of $4 800 000.00. To raise this money, the company decided to issue bonds. Since Beaucage had no expertise in issuing and selling bonds, the company decided to work with an investment dealer. The investment dealer bought the company's entire bond issue at a discount, then sold the bonds to the public at face value or the current market value. To ensure it would raise the $4 800 000.00 it required, Beaucage issued 5000 bonds with a face value of $1000 each on January 20, 2004. Interest is paid semi-annually on July 20 and January 20, beginning July 20, 2004. The bonds pay interest at 7.5% compounded semi-annually.

Beaucage directors realize that when the bonds mature on January 20, 2024, there must be $5 000 000.00 available to repay the bondholders. To have enough money on hand to meet this obligation, the directors set up a sinking fund using a specially designated savings account. The company earns interest of 5.5% compounded semi-annually on this sinking fund account. The directors began making semi-annual payments to the sinking fund on July 20, 2004.

Beaucage Development Company issued the bonds, sold them all to the investment dealer, and used the money raised to continue its research and development.

QUESTIONS

1. How much would an investor have to pay for one of these bonds to earn 8% compounded semi-annually?
2. (a) What is the size of the sinking fund payment?
   (b) What will be the total amount deposited into the sinking fund account?
   (c) How much of the sinking fund will be interest?
3. Suppose Beaucage discovers on January 20, 2014 that it can earn 8% interest compounded semi-annually on its sinking fund account.
   (a) What is the balance in the sinking fund after the January 20, 2014 sinking fund payment?
   (b) What is the new sinking fund payment if the fund begins to earn 8% on January 21, 2014?
   (c) What will be the total amount deposited into the sinking fund account over the life of the bonds?
   (d) How much of the sinking fund will then be interest?
   (e) How does the amount of sinking fund interest calculated in part (d) compare to the amount of interest calculated in Question 2(c)?

SUMMARY OF FORMULAS

Formula 9.1

$$PP = FV(1 + i)^{-n} + PMT\left[\frac{1 - (1 + i)^{-n}}{i}\right]$$

Basic formula for finding the purchase price of a bond when the interest payment interval and the yield rate conversion period are equal

Formula 9.2

$$PP = FV(1 + p)^{-n} + PMT\left[\frac{1 - (1 + p)^{-n}}{p}\right]$$

where $p = (1 + i)^c - 1$

Basic formula for finding the purchase price of a bond when the interest payment interval and the yield rate conversion period are different

Formula 9.3

Direct formula for finding the premium or discount of a bond (a negative answer indicates a discount)

$$\text{PREMIUM OR DISCOUNT} = (b \times \text{FACE VALUE} - i \times \text{REDEMPTION PRICE})\left[\frac{1 - (1 + i)^{-n}}{i}\right]$$

Formula 9.4

Basic formula for finding the yield rate using the method of averages.

$$\text{APPROXIMATE VALUE OF } i = \frac{\text{AVERAGE INCOME PER INTEREST PAYMENT INTERVAL}}{\text{AVERAGE BOOK VALUE}}$$

where

$$\text{AVERAGE BOOK VALUE} = \frac{1}{2}(\text{QUOTED PRICE} + \text{REDEMPTION PRICE})$$

and

$$\text{AVERAGE INCOME PER INTEREST PAYMENT INTERVAL} = \frac{\text{TOTAL INTEREST PAYMENTS} \begin{array}{l} - \text{PREMIUM} \\ + \text{DISCOUNT} \end{array}}{\text{NUMBER OF INTEREST PAYMENT INTERVALS}}$$

In addition, Formulas 3.1C, 5.1, 5.2, 6.3, and 7.1 were used in this chapter.

GLOSSARY

Accumulation of discount the process of reducing a bond discount *(p. 398)*

Amortization of premium the process of writing down a bond premium *(p. 395)*

Bond rate the rate of interest paid by a bond, stated as a percent of the face value *(p. 369)*

Book value of a debt the difference at any time between the debt principal and the associated sinking fund balance *(p. 418)*

Coupon a voucher attached to a bond to facilitate the collection of interest by the bondholder *(p. 369)*

Coupon rate *see* **Bond rate**

Debentures bonds for which no security is offered *(p. 369)*

Denomination *see* **Face value**

Discount the difference between the purchase price of a bond and its redemption price when the purchase price is less than the redemption price *(p. 382)*

Due date *see* **Redemption date**

Face value the amount owed by the issuer of the bond to the bondholder *(p. 369)*

Flat price the total purchase price of a bond (including any accrued interest) *(p. 377)*

Market price *see* **Quoted price**

Maturity date *see* **Redemption date**

Method of averages a method for finding the approximate yield rate *(p. 405)*

Nominal rate *see* **Bond rate**

Par value *see* **Face value**

Periodic cost of a debt the sum of the interest paid and the payment into the sinking fund when a debt is retired by the sinking fund method *(p. 418)*

Premium the difference between the purchase price of a bond and its redemption price when the purchase price is greater than the redemption price *(p. 382)*

Quoted price the net price of a bond (without accrued interest) at which a bond is offered for sale *(p. 377)*

Redeemable at a premium bonds whose redemption price is greater than the face value *(p. 369)*

Redeemable at par bonds that are redeemed at their face value *(p. 369)*

Redemption date the date at which the bond principal is repaid *(p. 369)*

Redemption price the amount that the issuer of the bond pays to the bondholder upon surrender of the bond on or after the date of maturity *(p. 369)*

Sinking fund a fund into which payments are made to provide a specific sum of money at a future time; usually set up for the purpose of meeting some future obligation *(p. 408)*

Yield rate the rate of interest that an investor earns on his or her investment in a bond *(p. 371)*

USEFUL INTERNET SITES

www.cis-pec.gc.ca
Canada Investment and Savings (CIS) CIS is a special operating agency of the Department of Finance that markets and manages savings and investment products for Canadians. This Website provides information on products, including Canada Savings Bonds and Canada Premium Bonds.

www.bankofcanada.ca/en
Bond Market Rates Click on Bond Securities. The current Government of Canada bond yields and marketable bond average yields are found at this site. The site also provides links to selected historical interest rates.

www.carswell.com/payroll/index.asp
The Payroll Community Hosted by Carswell publishers, this site provides information about payroll, FAQs, new products, payroll publications, and links to relevant sites.

www.benefitscanada.com
Benefits Canada This magazine deals with employee benefits and pension investments.

10 Investment Decision Applications

OBJECTIVES

Upon completing this chapter, you will be able to do the following:

1. Determine the discounted value of cash flows and choose between alternative investments on the basis of the discounted cash flow criterion.
2. Determine the net present value of a capital investment project and infer from the net present value whether a project is feasible or not.
3. Compute the rate of return on investment.

Choices among different investment opportunities must be made often by both individuals and companies. Whether one is deciding between buying and leasing a car or how to increase plant capacity, an understanding of the time value of money is critical. When comparing different ways of achieving the same goal, we should always examine cash flows at the same point in the time—usually at the beginning when a decision must be made. Only then can we know whether it is better to buy or lease that car, or to expand the plant now or to wait.

INTRODUCTION

When making investment decisions, all decision makers must consider the comparative effects of alternative courses of action on the cash flows of a business or of an individual. Since cash flow analysis needs to take into account the time value of money (interest), present-value concepts are useful.

When only cash inflows are considered, the value of the discounted cash flows is helpful in guiding management toward a rational decision. If outlays as well as inflows are considered, the net present value concept is applicable in evaluating projects.

The net present value method indicates whether or not a project will yield a specified rate of return. To be a worthwhile investment, the rate of return on a capital project must be attractive. Required rates of return tend to be high and may even reach or exceed the 20% level. Knowing the actual rate of return provides useful information to the decision maker. It may be computed using the net present value concept.

10.1 Discounted Cash Flow

A. Evaluation of capital expenditures—basic concepts

Projects expected to generate benefits over a period of time longer than one year are called *capital investment projects,* and they result in *capital expenditures.* The benefits resulting from these projects may be in either monetary or non-monetary form. The methods of analysis considered in this chapter will deal only with investment projects generating cash flows in monetary form.

While capital expenditures normally result in the acquisition of assets, the primary purpose of investing in capital expenditures is to acquire a future stream of benefits in the form of an inflow of cash. When an investment is being considered, analysis of the anticipated future cash flows aids in the decision to acquire or replace assets, and whether to buy or lease them.

The analysis generally uses the technique of **discounted cash flow**. This technique involves estimating all anticipated future cash flows flowing from an investment or project, projecting an interest rate, and calculating the present value of these cash flows. It is important that the present value be calculated, not the future value, due to the need to make a decision on the investment in the beginning, or in the present, before any cash or other resources are invested. The decision to be made might involve determining whether to invest in a project, or choosing which project to invest in.

In some situations, the cash flows estimated may not be certain, or there may be other, non-financial concerns. The analysis techniques considered in this text are concerned only with the amount and the timing of cash receipts and cash payments under the assumption that the amount and timing of the cash flow are certain.

From the mathematical point of view, the major issue in evaluating capital expenditure projects is the time value of money. This value prevents direct comparison of cash received and cash payments made at different times. The concept of present value, as introduced in Chapter 3 and subsequent chapters, provides the vehicle for making sums of money received or paid at different times comparable at a given time.

B. Discounted cash flow

Discounted cash flow is the present value of all cash payments. When using the discounting technique to evaluate alternatives, two fundamental principles serve as decision criteria.

1. The *bird-in-the-hand principle*—Given that all other factors are equal, earlier benefits are preferable to later benefits.
2. *The-bigger-the-better principle*—Given that all other factors are equal, bigger benefits are preferable to smaller benefits.

EXAMPLE 10.1A

Suppose you are offered a choice of receiving \$1000.00 today or receiving \$1000.00 three years from now. What is the preferred choice?

SOLUTION

Accepting the bird-in-the-hand principle, you should prefer to receive \$1000.00 today rather than three years from now. The rationale is that \$1000.00 can be invested to earn interest and will accumulate in three years to a sum of money greater than \$1000.00. Stated another way, the present value of \$1000.00 to be received in three years is less than \$1000.00 today.

EXAMPLE 10.1B

Consider a choice of \$2000.00 today or \$3221.00 five years from now. Which alternative is preferable?

SOLUTION

No definite answer is possible without considering interest. A rational choice must consider the time value of money; that is, we need to know the rate of interest. Once a rate of interest is established, we can make the proper choice by considering the present value of the two sums of money and applying the the-bigger-the-better principle.

If you choose "now" as the focal date, three outcomes are possible.

1. The present value of \$3221.00 is greater than \$2000.00. In this case, the preferred choice is \$3221.00 five years from now.
2. The present value of \$3221.00 is less than \$2000.00. In this case, the preferred choice is \$2000.00 now.
3. The present value of \$3221.00 equals \$2000.00. In this case, either choice is equally acceptable.

(a) *Suppose the rate of interest is 8%.*
$FV = 3221.00; \quad i = 8\% = 0.08; \quad n = 5$
$PV = 3221.00(1.08^{-5}) = 3221.00(0.680583) = \$2192.16$
Since at 8% the discounted value of \$3221.00 is greater than \$2000.00, the preferred choice at 8% is \$3221.00 five years from now.

(b) *Suppose the rate of interest is 12%.*
$FV = 3221.00; \quad i = 12\% = 0.12; \quad n = 5$
$PV = 3221.00(1.12^{-5}) = 3221.00(0.567427) = \$1827.68$
Since at 12% the discounted value is less than \$2000.00, the preferred choice is \$2000.00 now.

(c) *Suppose the rate of interest is 10%.*
$FV = 3221.00; \quad i = 10\% = 0.10; \quad n = 5$
$PV = 3221.00(1.10^{-5}) = 3221.00(0.620921) = \$1999.99$
Since at 10% the discounted value is equal to \$2000.00, the two choices are equally acceptable.

Programmed Solution

("END" mode)

(Set P/Y = 1; C/Y = 1)

(a) 2nd (CLR TVM) 3221 FV 8 I/Y 5 N CPT PV -2192.158478

(b) 2nd (CLR TVM) 3221 FV 12 I/Y 5 N CPT PV -1827.681902

(c) 2nd (CLR TVM) 3221 FV 10 I/Y 5 N CPT PV -1999.987582

EXAMPLE 10.1C

Two investment alternatives are available. Alternative A yields a return of $6000 in two years and $10 000 in five years. Alternative B yields a return of $7000 now and $7000 in seven years. Which alternative is preferable if money is worth
(i) 11%? (ii) 15%?

SOLUTION

To determine which alternative is preferable, we need to compute the present value of each alternative and choose the alternative with the higher present value.

Since the decision is to be made immediately, choose a focal point of "now."

(i) For $i = 11\%$

Alternative A

The present value of Alternative A is the sum of the present values of $6000 in two years and $10 000 in five years.

Present value of $6000 in two years $= 6000(1.11^{-2}) = 6000(0.811622)$	=	$ 4 870
Present value of $10 000 in five years $= 10\,000(1.11^{-5}) = 10\,000(0.593451)$	=	5 935
The present value of Alternative A	=	$10 805

Alternative B

The present value of Alternative B is the sum of the present values of $7000 now and $7000 in seven years.

Present value of $7000 now	=	$ 7 000
Present value of $7000 in seven years $= 7000(1.11^{-7}) = 7000(0.481658)$	=	3 372
The present value of Alternative B	=	$10 372

Programmed Solution

Alternative A

("END" mode)

(Set P/Y = 1; C/Y = 1)

2nd (CLR TVM) 6000 FV 11 I/Y 2 N CPT PV -4869.734599

2nd (CLR TVM) 10 000 FV 11 I/Y 5 N CPT PV -5934.513281

$4870 + 5935 = $10 805

Alternative B

2nd (CLR TVM) 7000 FV 11 I/Y 7 N CPT PV −3371.608876

$3372 + 7000 = $10 372

Since at 11% the present value of Alternative A is greater than the present value of Alternative B, Alternative A is preferable.

(ii) For $i = 15\%$

Alternative A

Present value of $6000 in two years		
$= 6000(1.15^{-2}) = 6000(0.756144)$	=	$4537
Present value of $10 000 in five years		
$= 10\ 000(1.15^{-5}) = 10\ 000(0.497177)$	=	4972
The present value of Alternative A	=	$9509

Alternative B

Present value of $7000 now	=	$7000
Present value of $7000 in seven years		
$= 7000(1.15^{-7}) = 7000(0.375937)$	=	2632
The present value of Alternative B	=	$9632

Programmed Solution

Alternative A

("END" mode)

(Set P/Y = 1; C/Y = 1)

2nd (CLR TVM) 6000 FV 15 I/Y 2 N CPT PV −4536.862004

2nd (CLR TVM) 10 000 FV 15 I/Y 5 N CPT PV −4971.767353

$4537 + 4972 = $9509

Alternative B

2nd (CLR TVM) 7000 FV 15 I/Y 7 N CPT PV −2631.559279

$2632 + 7000 = $9632

Since at 15% the present value of Alternative B is greater than the present value of Alternative A, Alternative B is preferable.

Note: Applying present value techniques to capital investment problems usually involves estimates. For this reason, dollar amounts in the preceding example and all following examples may be rounded to the nearest dollar. We suggest that you do the same when working on problems of this nature.

EXAMPLE 10.1D

An insurance company offers to settle a claim either by making a payment of \$50 000 immediately or by making payments of \$8000 at the end of each year for ten years. What offer is preferable if interest is 8% compounded annually?

SOLUTION

Present value of \$8000 at the end of each year for ten years is the present value of an ordinary annuity in which PMT = 8000, $n = 10$, and $i = 8\%$.

$$PV_n = 8000\left(\frac{1 - 1.08^{-10}}{0.08}\right) = 8000(6.710081) = \$53\,681$$

Programmed Solution

("END" mode)

(Set P/Y = 1; C/Y = 1)

0 [FV] 8000 [PMT] 10 [N] 8 [I/Y] [CPT] [PV] [-53680.65119]

Since the immediate payment is smaller than the present value of the annual payments of \$8000, the annual payments of \$8000 are preferable.

DID YOU KNOW?

There are several numerical measures that are used by corporations, financial analysts, and investors to determine whether a particular common stock's trading price is realistic:

(i) Book Value: Book value is calculated as (Total assets − Total liabilities) ÷ Total number of shares of common stock. Because the dollar amount of assets is sometimes overstated or understated in the financial records of a company compared to their market value, the book value may be misleading.

(ii) Earnings per Share: Earnings per share is calculated as After-tax earnings ÷ Total number of shares of common stock. As you might expect, any increase in earnings per share is generally viewed as a healthy sign for a corporation and its shareholders.

(iii) Price-Earnings (P/E) Ratio: The price-earnings (P/E) ratio is calculated as the Price per common share ÷ Earnings per share. The P/E ratio is a key factor used by serious investors to evaluate stocks. A low P/E ratio indicates that a particular stock may be a good investment because its stock price is lower than it could be given its level of earnings. A high P/E ratio suggests that it could be a poor investment. For most corporations, P/E ratios range between 5 and 25. Some dot-com or Internet-related companies have P/E ratios of 60 and higher or have negative P/E ratios because they have not yet earned income. Negative P/E ratios are not normally reported.

(iv) Beta: Beta is an index reported in many financial publications that compares the risk associated with a particular stock to the risk of the stock market in general. The beta for the stock market in general is 1.0, while the majority of stocks have beta values between 0.5 and 2.0. Typically, low-risk stocks have low beta values and high-risk stocks have high beta values.

EXAMPLE 10.1E

National Credit Union needs to decide whether to buy a high-speed scanner for \$6000 and enter a service contract requiring the payment of \$45 at the end of every three months for five years, or to enter a five-year lease requiring the payment of \$435 at the beginning of every three months. If leased, the scanner can be bought after five years for \$600. At 9% compounded quarterly, should the credit union buy or lease?

SOLUTION

To make a rational decision, the credit union should compare the present value of the cash outlays if buying the scanner with the present value of the cash outlays if leasing the scanner.

Present value of the decision to buy

Present value of cash payment for the scanner = \$6000

Present value of the service contract involves an ordinary annuity in which PMT = 45, $n = 20$, P/Y = 4; C/Y = 4; I/Y = 9; $i = 2.25\%$

$$= 45\left(\frac{1 - 1.0225^{-20}}{0.0225}\right) = 45(15.963712) \quad = \quad 718$$

Present value of decision to buy = \$6718

Present value of the decision to lease

Present value of the quarterly lease payments involves an annuity due: PMT = 435, $n = 20$, $i = 2.25\%$

$$= 435(1.0225)\left(\frac{1 - 1.0225^{-20}}{0.0225}\right) = 435(1.0225)(15.963712) \quad = \quad \$7100$$

Present value of purchase price after five years

$= 600(1.0225^{-20}) = 600(0.640816)$ = 384

Present value of decision to lease = \$7484

Programmed Solution

Present value of the decision to buy

Present value of cash payment for the scanner = \$6000

Present value of the service contract

("END" mode)

(Set P/Y = 4; C/Y = 4)

0 [FV] 45 [PMT] 20 [N] 9 [I/Y] [CPT] [PV] [−718.3670566]

Present value of decision to buy is \$6000 + 718 = \$6718.

Present value of the decision to lease

Present value of the quarterly lease payments

("BGN" mode)

0 [FV] 435 [PMT] 20 [N] 9 [I/Y] [CPT] [PV] [−7100.459716]

Present value of purchase price after five years

0 [PMT] 600 [FV] 20 [N] 9 [I/Y] [CPT] [PV] [−384.489883]

Present value of the decision to lease is \$7100 + 384 = \$7484.

In the case of costs, the selection criterion follows the the-smaller-the-better principle. Since the present value of the decision to buy is smaller than the present value of the decision to lease, the credit union should buy the scanner.

EXAMPLE 10.1F

Hans Machine Service needs a brake machine. The machine can be purchased for $4600 and after five years will have a salvage value of $490, or the machine can be leased for five years by making monthly payments of $111 at the beginning of each month. If money is worth 10% compounded annually, should Hans Machine Service buy or lease?

SOLUTION

Alternative 1: Buy machine

Present value of cash price	=	$4600
Less: Present value of salvage value		
$= 490(1.10^{-5}) = 490(0.620921)$	=	304
Present value of decision to buy	=	$4296

Alternative 2: Lease machine

The monthly lease payments form a general annuity due in which

PMT = 111; P/Y = 12; C/Y = 1; I/Y = 10; $c = \frac{1}{12}$; $n = 60$; $i = 10\%$;

$p = 1.10^{\frac{1}{12}} - 1 = 1.007974 - 1 = 0.7974\%$

Present value of the monthly lease payments

$$= 111(1.007974)\left(\frac{1 - 1.007974^{-60}}{0.007974}\right)$$
$$= 111(1.007974)(47.538500)$$
$$= \$5319$$

The present value of the decision to lease is $5319.

Programmed Solution

Alternative 1: Buy machine

Present value of cash price = $4600

Less: Present value of salvage value

("END" mode)

(Set P/Y = 12; C/Y = 1)

[2nd] (CLR TVM) 490 [FV] 60 [N] 10 [I/Y] [CPT] [PV] [-304.2514483]

Present value of decision to buy is $4600 − 304 = $4296.

Alternative 2: Lease machine

Present value of the monthly lease payments

("BGN" mode) (Set P/Y = 12; C/Y = 1) 0 [FV] 111 [PMT]

60 [N] 10 [I/Y] [CPT] [PV] [-5318.851263]

Since the present value of the decision to buy is smaller than the present value of the decision to lease, Hans Machine Service should buy the machine.

EXERCISE 10.1

Excel You can use Excel's ***Net Present Value (NPV)*** function to answer the questions below. Refer to **NPV** on the Spreadsheet Template Disk to learn how to use this function.

For each of the following, compute the present value of each alternative and determine the preferred alternative according to the discounted cash flow criterion.

1. The D Company must make a choice between two investment alternatives. Alternative 1 will return the company $20 000 at the end of three years and $60 000 at the end of six years. Alternative 2 will return the company $13 000 at the end of each of the next six years. The D Company normally expects to earn a rate of return of 12% on funds invested.
2. The B Company has a policy of requiring a rate of return on investment of 16%. Two investment alternatives are available but the company may choose only one. Alternative 1 offers a return of $50 000 after four years, $40 000 after seven years, and $30 000 after ten years. Alternative 2 will return the company $750 at the end of each month for ten years.
3. An obligation can be settled by making a payment of $10 000 now and a final payment of $20 000 in five years. Alternatively, the obligation can be settled by payments of $1500 at the end of every three months for five years. Interest is 10% compounded quarterly.
4. An unavoidable cost may be met by outlays of $10 000 now and $2000 at the end of every six months for seven years or by making monthly payments of $500 in advance for seven years. Interest is 7% compounded annually.
5. A company must purchase new equipment costing $2000. The company can pay cash on the basis of the purchase price or make payments of $108 per month for 24 months. Interest is 7.8% compounded monthly. Should the company purchase the new equipment with cash or make payments on the installment plan?
6. For less than a dollar a day, Jerri can join a fitness club. She would have to pay $24.99 per month for 30 months, or she can pay a lump sum of $549 at the beginning. Interest is 16.2% compounded monthly. Should Jerri pay a lump sum or use the monthly payment feature?

B. Answer each of the following questions.

1. A contract offers $25 000 immediately and $50 000 in five years or $10 000 at the end of each year for ten years. If money is worth 6%, which offer is preferable?
2. A professional sports contract offers $400 000 per year paid at the end of each of six years or $100 000 paid now, $200 000 paid at the end of each of the second and third years and 800 000 paid at the end of each of the last three years. If money is worth 7.3%, which offer is preferable?
3. Bruce and Carol want to sell their business. They have received two offers. If they accept Offer A, they will receive $15 000 immediately and $20 000 in three years. If they accept Offer B, they will receive $3000 now and $3000 at the end of every six months for six years. If interest is 10%, which offer is preferable?

4. When Peter decided to sell his farm, he received two offers. If he accepts the first offer, he would receive \$250 000 now, \$750 000 one year from now, and \$500 000 two years from now. If he accepts the second offer, he would receive \$600 000 now, \$300 000 one year from now, and \$600 000 two years from now. If money is worth 9.8%, which offer should he accept?

5. A warehouse can be purchased for \$90 000. After twenty years the property will have a residual value of \$30 000. Alternatively, the warehouse can be leased for twenty years at an annual rent of \$10 000 payable in advance. If money is worth 8%, should the warehouse be purchased or leased?

6. A car costs \$9500. Alternatively, the car can be leased for three years by making payments of \$240 at the beginning of each month and can be bought at the end of the lease for \$4750. If interest is 9% compounded semi-annually, which alternative is preferable?

10.2 Net Present Value Method

A. Introductory examples

EXAMPLE 10.2A Net cash inflows from two ventures are as follows:

End of Year	1	2	3	4	5	Total
Venture A	12 000	14 400	17 280	20 736	24 883	89 299
Venture B	17 000	17 000	17 000	17 000	17 000	85 000

Which venture is preferable if the required yield is 20%?

SOLUTION

Present value of Venture A

$= 12\,000(1.20^{-1}) + 14\,400(1.20^{-2}) + 17\,280(1.20^{-3})$
$+ 20\,736(1.20^{-4}) + 24\,883(1.20^{-5})$
$= 12\,000(0.833333) + 14\,400(0.694444) + 17\,280(0.578704)$
$+ 20\,736(0.482253) + 24\,883(0.401878)$
$= 10\,000 + 10\,000 + 10\,000 + 10\,000 + 10\,000$
$= \$50\,000$

Present value of Venture B

$$= 17\,000\left(\frac{1 - 1.20^{-5}}{0.20}\right) = 17\,000(2.990612) = \$50\,840$$

Programmed Solution

Present value of Venture A

("END" mode)

(Set P/Y = 1; C/Y = 1)

2nd (CLR TVM) 12 000 FV 20 I/Y 1 N CPT PV -10000

2nd (CLR TVM) 14 400 FV 20 I/Y 2 N CPT PV −10000

2nd (CLR TVM) 17 280 FV 20 I/Y 3 N CPT PV −10000

2nd (CLR TVM) 20 736 FV 20 I/Y 4 N CPT PV −10000

2nd (CLR TVM) 24 883 FV 20 I/Y 5 N CPT PV −10000

The present value of Venture A is $10 000 + 10 000 + 10 000 + 10 000 + 10 000 = $50 000.

Present value of Venture B

0 FV 17 000 ± PMT 20 I/Y 5 N CPT PV 50840.40638

The present value of Venture B is $50 840.

Since at 20% the present value of Venture B is greater than the present value of Venture A, Venture B is preferable to Venture A.

EXAMPLE 10.2B

Assume for Example 10.2A that Venture A requires an immediate non-recoverable outlay of $9000 while Venture B requires a non-recoverable outlay of $11 000. At 20%, which venture is preferable?

SOLUTION

	Venture A	*Venture B*
Present value of cash inflows	$50 000	$50 840
Present value of immediate outlay	9000	11 000
Net present value	$41 000	$39 840

Since the net present value of Venture A is greater than the net present value of Venture B, Venture A is preferable.

B. The net present value concept

When the present value of the cash outlays is subtracted from the present value of cash inflows, the resulting difference is called the **net present value**. In Example 10.2A, Venture B was preferable, where only the cash inflows were considered. In Example 10.2B, the present value of the outlays as well as the present value of the cash inflows resulted in the net present value being calculated. This approach is necessary when the cash outlays are different.

NET PRESENT VALUE (NPV) = PRESENT VALUE OF INFLOWS − PRESENT VALUE OF OUTLAYS ——— Formula 10.1

Since the net present value involves the difference between the present value of the inflows and the present value of the outlays, three outcomes are possible:

1. If the present value of the inflows is greater than the present value of the outlays, then the net present value is greater than zero.
2. If the present value of the inflows is smaller than the present value of the outlays, then the net present value is smaller than zero.

3. If the present value of the inflows equals the present value of the outlays, then the net present value is zero.

$$PV_{IN} > PV_{OUT} \longrightarrow NPV > 0 \text{ (positive)}$$
$$PV_{IN} = PV_{OUT} \longrightarrow NPV = 0$$
$$PV_{IN} < PV_{OUT} \longrightarrow NPV < 0 \text{ (negative)}$$

Criterion rule

At the organization's required rate of return, accept those capital investment projects that have a positive or zero net present value and reject those projects that have a negative net present value.

For a given rate of return:
ACCEPT if $NPV > 0$ or $NPV = 0$;
REJECT if $NPV < 0$.

To distinguish between a negative and a positive net present value, use
$NPV = PV_{IN} - PV_{OUT}$.

If a company is considering more than one project but can choose only one, the project with the greatest positive net present value is preferable.

Assumptions about the timing of inflows and outlays

The net present value method of evaluating capital investment projects is particularly useful when cash outlays are made and cash inflows received at various times. Since the *timing* of the cash flows is of prime importance, follow these assumptions regarding the timing of cash inflows and cash outlays.

Unless otherwise stated:

1. All cash inflows (benefits) are assumed to be received at the end of a period.
2. All cash outlays (costs) are assumed to be made at the beginning of a period.

C. Applications

EXAMPLE 10.2C A company is offered a contract promising annual net returns of \$36 000 for seven years. If it accepts the contract, the company must spend \$150 000 immediately to expand its plant. After seven years, no further benefits are available from the contract and the plant expansion undertaken will have no residual value. Should the company accept the contract if the required rate of return is

(i) 12%? (ii) 18%? (iii) 15%?

SOLUTION The net inflows and outlays can be represented on a time graph.

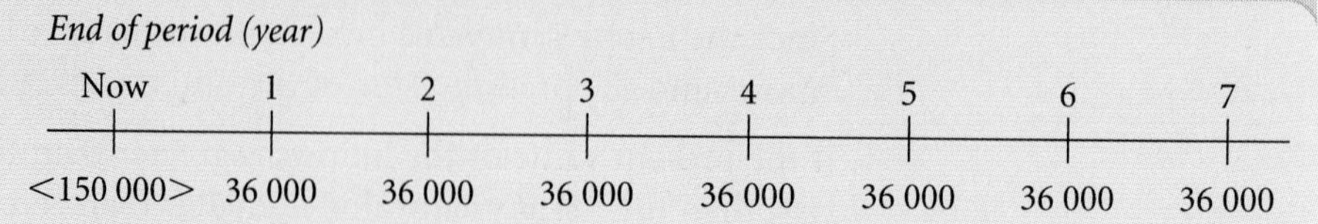

Note: Cash outlays (costs) are identified in such diagrams by a minus sign or by using accounting brackets.

(i) For $i = 12\%$

Since we assume the annual net returns (benefits) are received at the end of a period unless otherwise stated, they form an ordinary annuity in which

PMT = 36 000; $n = 7$; $i = 12\%$

$PV_{IN} = 36\,000\left(\frac{1 - 1.12^{-7}}{0.12}\right) = 36\,000(4.563756)$	=	\$164 295
PV_{OUT} = Present value of 150 000 now	=	150 000
The net present value (NPV) = 164 295 − 150 000	=	\$ 14 295

Since at 12% the net present value is greater than zero, the contract should be accepted. The fact that the net present value at 12% is positive means that the contract offers a return on investment of more than 12%.

(ii) For $i = 18\%$

$PV_{IN} = 36\,000\left(\frac{1 - 1.18^{-7}}{0.18}\right) = 36\,000(3.811528)$	=	\$137 215
PV_{OUT}	=	150 000
NPV = 137 215 − 150 000	=	−\$12 785

Since at 18% the net present value is less than zero, the contract should not be accepted. The contract does not offer the required rate of return on investment of 18%.

(iii) For $i = 15\%$

$PV_{IN} = 36\,000\left(\frac{1 - 1.15^{-7}}{0.15}\right) = 36\,000(4.16042)$	=	\$149 775
PV_{OUT}	=	\$150 000
NPV = 149 775 − 150 000	=	−\$225

The net present value is slightly negative, which means that the net present value method does not provide a clear signal as to whether to accept or reject the contract. The rate of return offered by the contract is almost 15%.

Programmed Solution

(i) For $i = 12\%$

("END" mode)

PV_{IN}: (Set P/Y = 1; C/Y = 1) 0 [FV] 36 000 [PMT] 7 [N]

12 [I/Y] [CPT] [PV] [−164295.2354]

PV_{OUT}: = Present value of 150 000 now = \$150 000

The net present value (NPV) = \$164 295 − 150 000 = \$14 295.

(ii) For $i = 18\%$

("END" mode)

PV_{IN}: (Set P/Y = 1; C/Y = 1) 0 [FV] 36 000 [PMT] 7 [N]

18 [I/Y] [CPT] [PV] [−137214.9934]

PV_{OUT}: = \$150 000

NPV = \$137 215 − 150 000 = −\$12 785

(iii) For i = 15%

("END" mode)

PV_{IN}: (Set P/Y = 1; C/Y = 1) 0 FV 36 000 PMT 7 N

15 I/Y CPT PV −149775.1104

PV_{OUT}: = \$150 000

NPV = \$149 775 − 150 000 = −\$225

EXAMPLE 10.2D

A project requires an initial investment of \$80 000 with a residual value of \$15 000 after six years. It is estimated to yield annual net returns of \$21 000 for six years. Should the project be undertaken at 16%?

SOLUTION

The cash flows are represented in the diagram below.

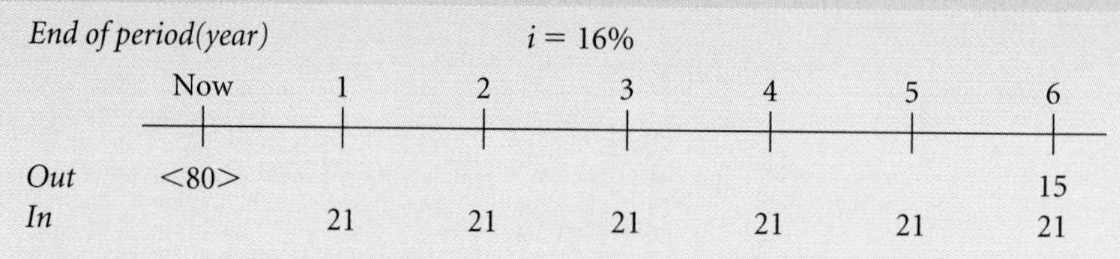

Note: The residual value of \$15 000 is considered to be a reduction in outlays. Its present value should be subtracted from the present value of other outlays.

$$PV_{IN} = 21\,000\left(\frac{1 - 1.16^{-6}}{0.16}\right) = 21\,000(3.684736) \qquad = \$77\,379$$

$$PV_{OUT} = 80\,000 - 15\,000(1.16^{-6})$$

$$= 80\,000 - 15\,000(0.410442) = 80\,000 - 6157 \qquad = 73\,843$$

$$= \text{Net present value (NPV)} \qquad = \$\ 3\,536$$

Programmed Solution

("END" mode)

PV_{IN}: (Set P/Y = 1; C/Y = 1) 0 FV 21 000 PMT 6 N

16 I/Y CPT PV −77379.45407

PV_{OUT}: 0 PMT 15 000 FV 6 N 16 I/Y CPT PV -6156.63382

NPV = \$77 379 − (80 000 − 6157) = \$3536

Since the net present value is positive (the present value of the benefits is greater than the present value of the costs), the rate of return on the investment is greater than 16%. The project should be undertaken.

EXAMPLE 10.2E

The UBA Corporation is considering developing a new product. If undertaken, the project requires the outlay of \$100 000 per year for three years. Net returns beginning in Year 4 are estimated at \$65 000 per year for twelve years. The residual value of the outlays after fifteen years is \$30 000. If the corporation requires a return on investment of 14%, should it develop the new product?

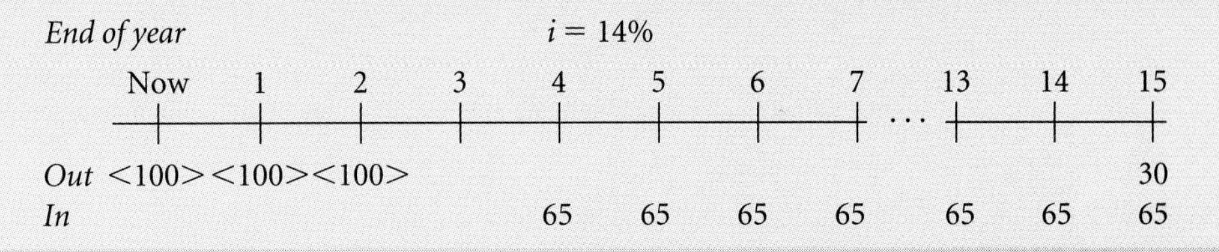

SOLUTION

The net returns, due at the end of Year 4 to Year 15 respectively, form an ordinary annuity deferred for *three* years in which PMT = 65 000, $n = 12$, $d = 3$, $i = 14\%$.

$$
\begin{aligned}
PV_{IN} &= 65\,000\left(\frac{1 - 1.14^{-12}}{0.14}\right)(1.14^{-3}) \\
&= 65\,000(5.660292)(0.674972) \\
&= \$248\,335
\end{aligned}
$$

The outlays, assumed to be made at the beginning of each year, form an annuity due in which PMT = 100 000, $n = 3$, $i = 14\%$.

$$
\begin{aligned}
PV_{OUT} &= 100\,000(1.14)\left(\frac{1 - 1.14^{-3}}{0.14}\right) - 30\,000(1.14^{-15}) \\
&= 100\,000(1.14)(2.321632) - 30\,000(0.140096) \\
&= 264\,666 - 4203 \\
&= \$260\,463 \\
NPV &= 248\,335 - 260\,463 = <\$12\,128>
\end{aligned}
$$

Programmed Solution

("END" mode)

PV_{IN}: (Set P/Y = 1; C/Y = 1) 0 [FV] 65 000 [PMT] 12 [N]

14 [I/Y] [CPT] [PV] [−367918.9882]

0 [PMT] 367 919 [FV] 3 [N] 14 [I/Y] [CPT] [PV] [−248334.8453]

("BGN" mode)

PV_{OUT}: 0 [FV] 100 000 [PMT] 3 [N] 14 [I/Y] [CPT] [PV] [−264666.0511]

0 [PMT] 30 000 [FV] 15 [N] 14 [I/Y] [CPT] [PV] [−4202.894462]

NPV = \$248 335 − (264 666 − 4203) = −\$12 128

Since the net present value is negative, the investment does not offer a 14% return. The corporation should not develop the product.

EXAMPLE 10.2F A feasibility study concerning a contemplated venture yielded the following estimates:

Initial cost outlay: $1 300 000;
further outlays in Years 2 to 5: $225 000 per year;
residual value after 20 years: $625 000;
net returns: Years 5 to 10: $600 000 per year;
Years 11 to 20: $500 000 per year.

Should the venture be undertaken if the required return on investment is 15%?

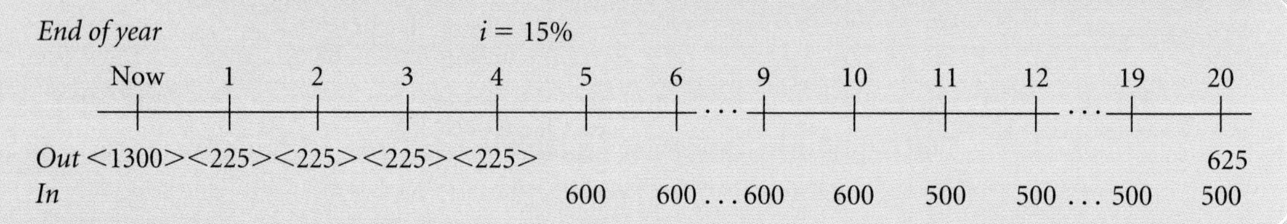

SOLUTION

$$\begin{aligned} PV_{IN} &= 600\,000\left(\frac{1-1.15^{-6}}{0.15}\right)(1.15^{-4}) + 500\,000\left(\frac{1-1.15^{-10}}{0.15}\right)(1.15^{-10}) \\ &= 600\,000(3.784483)(0.571753) + 500\,000(5.018769)(0.247185) \\ &= 1\,298\,274 + 620\,281 \\ &= \$1\,918\,555 \end{aligned}$$

$$\begin{aligned} PV_{OUT} &= 1\,300\,000 + 225\,000\left(\frac{1-1.15^{-4}}{0.15}\right) - 625\,000(1.15^{-20}) \\ &= 1\,300\,000 + 225\,000(2.854978) - 625\,000(0.061100) \\ &= 1\,300\,000 + 642\,370 - 38\,188 \\ &= \$1\,904\,182 \end{aligned}$$

$$NPV = 1\,918\,555 - 1\,904\,182 = \$14\,373$$

Programmed Solution

("END" mode)

PV_{IN}: (Set P/Y = 1; C/Y = 1)

0 FV 600 000 PMT 15 I/Y 6 N CPT PV −2270689.616

0 PMT 2 270 689 FV 15 I/Y 4 N CPT PV -1298273.805

0 FV 500 000 PMT 15 I/Y 10 N CPT PV −2509384.313

0 PMT 2 509 384 FV 15 I/Y 10 N CPT PV -620281.3466

PV_{IN} = $1 298 274 + 620 281 = $1 918 555

PV_{OUT} 0 FV 225 000 PMT 15 I/Y 4 N CPT PV −642 370.1316

0 PMT 625 000 FV 15 I/Y 20 N CPT PV -38187.67434

PV_{OUT} = $1 300 000 + 642 370 − 38 188 = $1 904 182
NPV = $1 918 555 − 1 904 182 = $14 373

Since the net present value is positive, the rate of return on investment is greater than 15%. The venture should be undertaken.

Net present value can be determined by using a preprogrammed financial calculator. For instructions on using this function, refer to the CD-ROM accompanying this book.

EXERCISE 10.2

If you choose, you can use Excel's ***Net Present Value (NPV)*** function to answer the questions in Part A and Part B below. Refer to **NPV** on the Spreadsheet Template Disk to learn how to use this Excel function.

For each of the following six investment choices, compute the net present value. Determine which investment should be accepted or rejected according to the net present value criterion.

1. A contract is estimated to yield net returns of $3500 quarterly for seven years. To secure the contract, an immediate outlay of $50 000 and a further outlay of $30 000 three years from now are required. Interest is 12% compounded quarterly.
2. Replacing old equipment at an immediate cost of $50 000 and an additional outlay of $30 000 six years from now will result in savings of $3000 per quarter for twelve years. The required rate of return is 10% compounded annually.
3. A business has two investment choices. Alternative 1 requires an immediate outlay of $2000 and offers a return of $7000 after seven years. Alternative 2 requires an immediate outlay of $1800 in return for which $250 will be received at the end of every six months for the next seven years. The required rate of return on investment is 17% compounded semi-annually.
4. Suppose you are offered two investment alternatives. If you choose Alternative 1, you will have to make an immediate outlay of $9000. In return, you will receive $500 at the end of every three months for the next ten years. If you choose Alternative 2, you will have to make an outlay of $4000 now and $5000 in two years. In return, you will receive $30 000 ten years from now. Interest is 12% compounded semi-annually.
5. You have two investment alternatives. Alternative 1 requires an immediate outlay of $8000. In return, you will receive $900 at the end of every quarter for the next three years. Alternative 2 requires an immediate outlay of $2000, and an outlay of $1000 in two years. In return, you will receive $300 at the end of every quarter for the next three years. Interest is 7% compounded quarterly. Which alternative would you choose? Why?

6. Your old car cost you $300 per month in gas and repairs. If you replace it, you could sell the old car immediately for $2000. To buy a new car that would last five years, you need to pay out $10 000 immediately. Gas and repairs would cost you only $120 per month on the new car. Interest is 9% compounded monthly. Should you buy a new car? Why?

B. Answer each of the following questions.

1. Teck Engineering normally expects a rate of return of 12% on investments. Two projects are available but only one can be chosen. Project A requires an immediate investment of $4000. In return, a revenue payment of $4000 will be received in four years and a payment of $9000 in nine years. Project B requires an investment of $4000 now and another $2000 in three years. In return, revenue payments will be received in the amount of $1500 per year for nine years. Which project is preferable?

2. The owner of a business is presented with two alternative projects. The first project involves the investment of $5000 now. In return the business will receive a payment of $8000 in four years and a payment of $8000 in ten years. The second project involves an investment of $5000 now and another $5000 three years from now. The returns will be semi-annual payments of $950 for ten years. Which project is preferable if the required rate of return is 14% compounded annually?

3. Northern Track is developing a special vehicle for Arctic exploration. The development requires investments of $60 000, $50 000, and $40 000 for the next three years respectively. Net returns beginning in Year 4 are expected to be $33 000 per year for twelve years. If the company requires a rate of return of 14%, compute the net present value of the project and determine whether the company should undertake the project.

4. The Kellog Company has to make a decision about expanding its production facilities. Research indicates that the desired expansion would require an immediate outlay of $60 000 and an outlay of a further $60 000 in five years. Net returns are estimated to be $15 000 per year for the first five years and $10 000 per year for the following ten years. Find the net present value of the project. Should the expansion project be undertaken if the required rate of return is 12%?

5. Agate Marketing Inc. intends to distribute a new product. It is expected to produce net returns of $15 000 per year for the first four years and $10 000 per year for the following three years. The facilities required to distribute the product will cost $36 000 with a disposal value of $9000 after seven years. The facilities will require a major facelift costing $10 000 each after three and after five years respectively. If Agate requires a return on investment of 20%, should the company distribute the new product?

6. A company is considering a project that will require a cost outlay of $15 000 per year for four years. At the end of the project the salvage value will be $10 000. The project will yield returns of $60 000 in Year 4 and $20 000 in Year 5. There are no returns after Year 5. Alternative investments are available that will yield a return of 16%. Should the company undertake the project?

7. Demand for a product manufactured by Eagle Company is expected to be 15 000 units per year during the next ten years. The net return per unit is $2. The manufacturing process requires the purchase of a machine costing $140 000. The machine has an economic life of ten years and a salvage value of $20 000 after ten years. Major overhauls of the machine require outlays of

$20 000 after four years and $40 000 after seven years. Should Eagle invest in the machine if it requires a return of 12% on its investments?

8. Magnum Electronics Company expects a demand of 20 000 units per year for a special purpose component during the next six years. Net return per unit is $4.00. To produce the component, Magnum must buy a machine costing $250 000 with a life of six years and a salvage value of $40 000 after six years. The company estimates that repair costs will be $20 000 per year during Years 2 to 6. If Magnum requires a return on investment of 18%, should it market the component?

BUSINESS MATH NEWS BOX

Yamaha recently ran the following advertisement:

Lease a New	
Vstar 1100 Classic	Kodiak Ultramatic 4 × 4
for $222.95*	for $176.88*
per month for	per month for
36 months with	36 months with
$1500.00 down.	$1200.00 down.

*Plus applicable taxes, freight, and PDI. Does not include insurance, licence, or registration fees. Limited time offer. Consumer may be required to purchase leased goods at the end of the lease term for $5399.50 for Vstar™ and $4274.50 for Kodiak™ plus applicable taxes. See your dealer for return or refinancing options.

Source: Used with permission from Yamaha Motor Canada Ltd.

QUESTIONS

To answer these questions, assume the combined PST and GST tax rate is 15%. Assume freight, PDI, insurance, licence, and registration costs are the same whether you lease or buy a vehicle.

1. Suppose the Vstar 1100 Classic has a manufacturer's suggested retail price (MSRP) of $10 799.00 plus taxes.

(a) If you can earn 10% on your money, is it cheaper to lease or buy this motorcycle?

(b) If you can earn 20% on your money, is it cheaper to lease or buy this motorcycle?

2. Suppose the Kodiak Ultramatic 4 × 4 has an MSRP of $8549.00 plus taxes.

(a) If you can earn 10% on your money, is it cheaper to lease or buy this motorcycle?

(b) If you can earn 20% on your money, is it cheaper to lease or buy this motorcycle?

(c) Based on your calculations in parts (a) and (b), about how much interest would you have to earn on your money to make leasing a Kodiak Ultramatic 4 × 4 cost the same as buying it?

10.3 Finding the Rate of Return on Investment

A. Net present value, profitability index, rate of return

The **rate of return** on investment (R.O.I.) is widely used to measure the value of an investment. Since it takes interest into account, knowing the rate of return that results from a capital investment project provides useful information when evaluating a project.

The method of finding the rate of return that is explained and illustrated in this section uses the net present value concept introduced in Section 10.2. However, instead of being primarily concerned with a specific discount rate and with comparing the present value of the cash inflows and the present value of the cash outlays, this method is designed to determine the rate of return on the investment.

As explained in Section 10.2, three outcomes are possible when using Formula 16.1. These three outcomes indicate whether the rate of return is greater than, less than, or equal to the discount rate used in finding the net present value.

1. If the net present value is greater than zero (positive), then the rate of return is greater than the discount rate used to determine the net present value.
2. If the net present value is less than zero (negative), then the rate of return is less than the discount rate used.
3. If the net present value is equal to zero, then the rate of return is equal to the rate of discount used.

If NPV > 0 (POSITIVE)	R.O.I. $> i$
If NPV < 0 (NEGATIVE)	R.O.I. $< i$
If NPV $= 0$	R.O.I. $= i$

It follows, then, that the rate of return on investment (R.O.I.) is that rate of discount for which the NPV $= 0$, that is, for which $PV_{IN} = PV_{OUT}$.

The above definition of the rate of return and the relationship between the net present value, the rate of discount used to compute the net present value, and the rate of return are useful in developing a method of finding the rate of return.

However, before computing the rate of return, it is useful to consider a ratio known as the **profitability index** or **discounted benefit-cost ratio.** It is defined as the ratio that results when comparing the present value of the cash inflows with the present value of the cash outlays.

$$\text{PROFITABILITY INDEX (or DISCOUNTED BENEFIT-COST RATIO)} = \frac{PV_{IN}}{PV_{OUT}}$$ Formula 10.2

Since a division is involved, three outcomes are possible when computing this ratio.

1. If the numerator (PV_{IN}) is greater than the denominator (PV_{OUT}), then the profitability index is greater than one.

2. If the numerator (PV_{IN}) is less than the denominator (PV_{OUT}), then the profitability index is less than one.
3. If the numerator (PV_{IN}) is equal to the denominator (PV_{OUT}), then the profitability index is equal to one.

The three outcomes give an indication of the rate of return.

1. If the profitability index is greater than 1, then the rate of return is greater than the discount rate used.
2. If the profitability index is less than 1, then the rate of return is less than the discount rate used.
3. If the profitability index is equal to 1, then the rate of return equals the discount rate used.

The relationship between the present value of the inflows, the present value of the outlays, the net present value, the profitability index, and the rate of return at a given rate of discount i is summarized below.

PV_{IN} Versus PV_{OUT}	Net Present Value (NPV)	Profitability Index	Rate of Return (R.O.I.)
$PV_{IN} > PV_{OUT}$	$NPV > 0$	> 1	$> i$
$PV_{IN} = PV_{OUT}$	$NPV = 0$	$= 1$	$= i$
$PV_{IN} < PV_{OUT}$	$NPV < 0$	< 1	$< i$

B. Procedure for finding the rate of return by trial and error

From the relationships noted above, the rate of return on investment can be defined as the rate of discount for which the present value of the inflows (benefits) equals the present value of the outlays (costs). This definition implies that the rate of return is the rate of discount for which the net present value equals zero or for which the profitability index (benefit-cost ratio) equals 1. This conclusion permits us to determine the rate of return by trial and error.

STEP 1 Arbitrarily select a discount rate and compute the net present value at that rate.

STEP 2 From the outcome of Step 1, draw one of the three conclusions.

(a) If NPV = 0, infer that the R.O.I. = i.

(b) If NPV > 0, infer that the R.O.I. > i.

(c) If NPV < 0, infer that the R.O.I. < i.

STEP 3 (a) If, in Step 1, NPV = 0, then R.O.I. = i and the problem is solved.

(b) If, in Step 1, NPV > 0 (positive), then we know that R.O.I. > i. A second attempt is needed. This second try requires choosing a discount rate greater than the rate used in Step 1 and computing the net present value using the higher rate.

If the resulting net present value is still positive, choose a still higher rate of discount and compute the net present value for that rate. Repeat this procedure until the selected rate of discount yields a negative net present value.

(c) If, in Step 1, NPV < 0 (negative), then we know that R.O.I. $< i$. The second try requires choosing a discount rate less than the rate used in Step 1 and computing the net present value using the lower rate.

If the resulting net present value is still negative, choose a still lower rate of discount and compute the net present value for that rate. Repeat this procedure until the selected rate of discount yields a positive net present value.

STEP 4 The basic aim of Step 3 is to find one rate of discount for which the net present value is positive and a second rate for which the net present value is negative. Once this has been accomplished, the rate of return must be a rate between the two rates used to generate a positive and a negative net present value.

You can now obtain a reasonably accurate value of the rate of return by using linear interpolation. To ensure sufficient accuracy in the answer, we recommend that the two rates of discount used when interpolating be no more than two percentage points apart. The worked examples in this section have been solved using successive even rates of discounts when interpolating.

STEP 5 (Optional) You can check the accuracy of the method of interpolation when using an electronic calculator by computing the net present value. Use as the discount rate the rate of return determined in Step 4. Expect the rate in Step 4 to be slightly too high. You can obtain a still more precise answer by further trials.

C. Selecting the rate of discount—using the profitability index

While the selection of a discount rate in Step 1 of the procedure is arbitrary, a sensible choice is one that is neither too high nor too low. Since the negative net present value immediately establishes a range between zero and the rate used, it is preferable to be on the high side. Choosing a rate of discount within the range 12% to 24% usually leads to quick solutions.

While the initial choice of rate is a shot in the dark, the resulting knowledge about the size of the rate of return combined with the use of the profitability index should ensure the selection of a second rate that is fairly close to the actual rate of return.

In making the second choice, use the profitability index.

1. Compute the index for the first rate chosen and convert the index into a percent.
2. Deduct 100% from the index and divide the difference by 4.
3. If the index is greater than 1, add the above result to obtain the rate that you should use for the second attempt. If, however, the index is smaller than 1, deduct the above result from the rate of discount initially used.

Assume that the rate of discount initially selected is 16%. The resulting $PV_{IN} = 150$ and the $PV_{OUT} = 120$.

1. The profitability index is $\frac{150}{120} = 1.25 = 125\%$.
2. The difference (125% − 100%) divided by 4 = 6.25%.
3. Since the index is greater than 1, add 6.25% to the initial rate of 16%; the recommended choice is 22%.

Assume that the rate of discount initially selected is 20%. The resulting $PV_{IN} = 200$ and the $PV_{OUT} = 250$.

1. The profitability index is $\frac{200}{250} = 0.80 = 80\%$.
2. The difference (80% − 100%) divided by 4 = −5%.
3. Since the index is less than 1, subtract 5% from the initial rate of 20%; the recommended choice is 15%. (If, as in this text, you are using only even rates, try either 14% or 16%.)

D. Using linear interpolation

The method of linear interpolation used in Step 4 of the suggested procedure is illustrated in Example 10.3A below.

EXAMPLE 10.3A Assume that the net present value of a project is \$420 at 14% and −\$280 at 16%. Use linear interpolation to compute the rate of return correct to the nearest tenth of a percent.

SOLUTION The data can be represented on a line diagram.

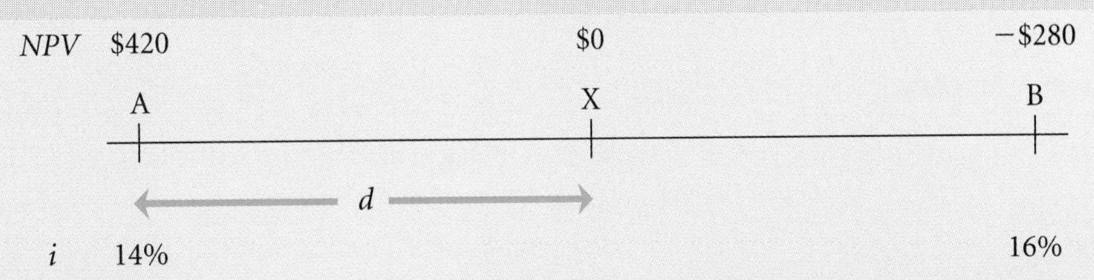

The line segment AB represents the distance between the two rates of discount that are associated with a positive and a negative net present value respectively.

At Point A, where $i = 14\%$, the NPV = 420;
at Point B, where $i = 16\%$, the NPV = −280;
at Point X, where i is unknown, the NPV = 0.

By definition, the rate of return is that rate of discount for which the net present value is zero. Since 0 is a number between 420 and −280, the NPV = 0 is located at a point on AB. This point is marked X.

We can obtain two useful ratios by considering the line segment from the two points of view shown in the diagram.

(i) In terms of the discount rate i,

AB = 2% ———— 16% − 14%

AX = d% ———— the unknown percent that must be added to 14% to obtain the rate of discount at which the NPV = 0

$$\frac{AX}{AB} = \frac{d\%}{2\%}$$

(ii) In terms of the net present value figures,

AB = 700 ———— 420 + 280

AX = 420

$$\frac{AX}{AB} = \frac{420}{700}$$

Since the ratio AX:AB is written twice, we can derive a proportion statement.

$$\frac{d\%}{2\%} = \frac{420}{700}$$

$$d\% = \frac{420}{700} \times 2\%$$

$$d\% = 1.2\%$$

Therefore, the rate at which the net present value is equal to zero is 14% + 1.2% = 15.2%. The rate of return on investment is 15.2%.

E. Computing the rate of return

EXAMPLE 10.3B A project requires an initial outlay of $25 000. The estimated returns are $7000 per year for seven years. Compute the rate of return (correct to the nearest tenth of a percent).

SOLUTION The cash flows (in thousands) are represented in the diagram below.

The inflows form an ordinary annuity since inflows are assumed to be received at the end of each year.

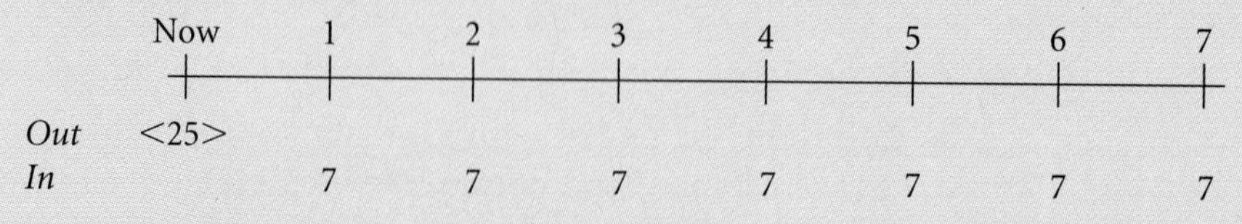

$$PV_{IN} = 7000\left[\frac{1-(1+i)^{-7}}{i}\right]$$

The outlays consist of an immediate payment.

$PV_{OUT} = 25\ 000$

To determine the rate of return, we will choose a rate of discount, compute the net present value, and try further rates until we find two successive even rates. For one, the NPV > 0 (positive) and, for the other, NPV < 0 (negative).

STEP 1 Try $i = 12\%$.

$$PV_{IN} = 7000\left(\frac{1 - 1.12^{-7}}{0.12}\right) = 7000(4.563757) = \$31\ 946$$

$$PV_{OUT} = 25\ 000$$

$$\text{NPV at } 12\% = \$\ 6\ 946$$

Since the NPV > 0, R.O.I. > 12%.

STEP 2 Compute the profitability index to estimate what rate should be used next.

$$\text{INDEX} = \frac{PV_{IN}}{PV_{OUT}} = \frac{31\ 946}{25\ 000} = 1.278 = 127.8\%$$

Since at $i = 12\%$, the profitability index is 27.8% more than 100%, the rate of discount should be increased by $\frac{27.8\%}{4} = 7\%$ approximately. To obtain another even rate, the increase should be either 6% or 8%. In line with the suggestion that it is better to go too high, increase the previous rate by 8% and try $i = 20\%$.

STEP 3 Try $i = 20\%$.

$$PV_{IN} = 7000\left(\frac{1 - 1.20^{-7}}{0.20}\right) = 7000(3.604592) = \$25\ 232$$

$$PV_{OUT} = 25\ 000$$

$$\text{NPV at } 20\% = \$\ 232$$

Since the NPV > 0, R.O.I. > 20%.

STEP 4 Since the net present value is still positive, a rate higher than 20% is needed. The profitability index at 20% is $\frac{25\ 232}{25\ 000} = 1.009 = 100.9\%$. The index exceeds 100% by 0.9%; division by 4 suggests an increase of 0.2%. For interpolation, the recommended minimum increase or decrease is 2%. The next try should use $i = 22\%$.

STEP 5 Try $i = 22\%$.

$$PV_{IN} = 7000\left(\frac{1 - 1.22^{-7}}{0.22}\right) = 7000(3.415506) = \$23\ 909$$

$$PV_{OUT} = -\$25\ 000$$

$$\text{NPV at } 22\% = -\$\ 1091$$

Since the NPV < 0, R.O.I. < 22%.
Therefore, 20% < R.O.I. < 22%.

STEP 6 Now that the rate of return has been located between two sufficiently close rates of discount, linear interpolation can be used as illustrated in Example 16.3A.

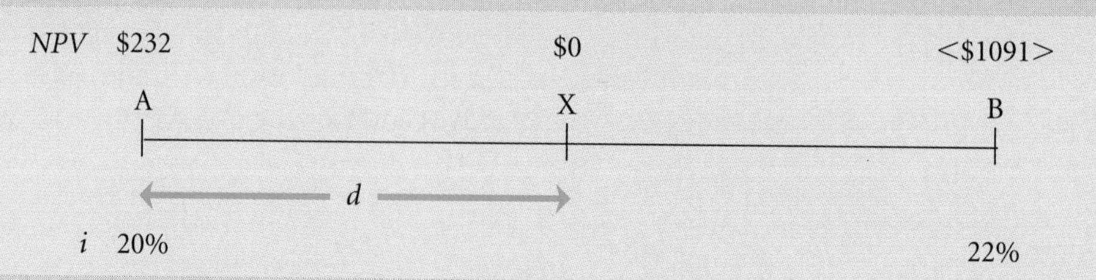

$$\frac{d}{2} = \frac{232}{232 + 1091}$$

$$d = \frac{232(2)}{1323} = 0.35$$

The rate of discount for which the NPV = 0 is approximately 20% + 0.35% = 20.35%. The rate of return is approximately 20.3%. (A more precisely computed value is 20.3382%.)

Note: Three attempts were needed to locate the R.O.I. between 20% and 22%. Three is the usual number of tries necessary. The minimum number is two attempts. Occasionally four attempts may be needed. To produce a more concise solution, organize the computation as shown below. Since estimates are involved, it is sufficient to use present value factors with only three decimal positions. In the following examples, all factors are rounded to three decimals.

Present Value of Amounts in General Form	Attempts					
	$i = 12\%$		$i = 20\%$		$i = 22\%$	
PV_{IN}	Factor	$	Factor	$	Factor	$
$7000\left[\frac{1-(1+i)^{-7}}{i}\right]$	4.563757	31 946	3.604592	25 232	3.415506	23 909
PV_{OUT} 25 000 now		25 000		25 000		25 000
NPV		6 946		232		<1 091>

Programmed Solution

(Set P/Y = 1; C/Y = 1)

Present Value of Amounts in General Form	Attempts					
	$i = 12\%$	$i = 20\%$	$i = 22\%$	$i = 20.5\%$	$i = 20.3\%$	$i = 20.34\%$
PV_{IN}	0 FV	0 FV	0 FV	0 FV	0 FV	0 FV
$7000\left[\frac{1-(1+i)^{-7}}{i}\right]$	−7000 PMT	−7000 PMT	−7000 PMT	−7000 PMT	−7000 PMT	−7000 PMT
	7 N	7 N	7 N	7 N	7 N	7 N
	12 I/Y	20 I/Y	22 I/Y	20.5 I/Y	20.3 I/Y	20.34 I/Y
	CPT	CPT	CPT	CPT	CPT	CPT
	PV	PV	PV	PV	PV	PV
PV_{IN}	31 946	25 232	23 909	24 890	25 026	24 999
PV_{OUT} 25 000 now	25 000	25 000	25 000	25 000	25 000	25 000
NPV	6946	232	−1091	−110	26	−1
R.O.I.	>12%	>20%	<22%	<20.5%	>20.3%	=20.34%

EXAMPLE 10.3C

A venture that requires an immediate outlay of $320 000 and an outlay of $96 000 after five years has a residual value of $70 000 after ten years. Net returns are estimated to be $64 000 per year for ten years. Compute the rate of return.

SOLUTION

The cash flow is represented in the diagram below (in thousands).

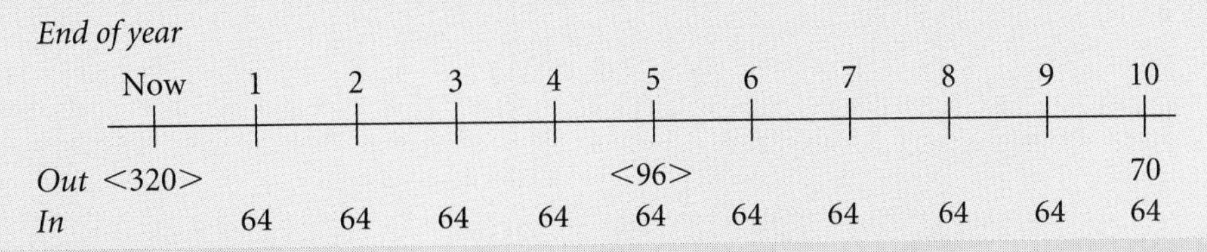

The computations are organized in a chart; explanations regarding the computations follow.

Present Value of Amounts in General Form	Attempts					
	i = 20%		i = 14%		i = 12%	
PV of benefits	**Factor**	**$**	**Factor**	**$**	**Factor**	**$**
$64\,000\left[\frac{1-(1+i)^{-10}}{i}\right]$	4.192	268 288	5.216	333 824	5.650	361 600
PV of costs:						
320 000 now		320 000		320 000		320 000
$96\,000(1+i)^{-5}$	0.402	38 592	0.519	49 824	0.567	54 432
<$70\,000(1+i)^{-10}$>	0.162	<11 340>	0.270	<18 900>	0.322	<22 540>
TOTAL		347 252		350 924		351 892
NPV		<78 964>		<17 100>		9 708

Explanations for computations

STEP 1

Try $i = 20\%$.
Since NPV < 0, R.O.I. $< 20\%$.

$$\text{Index at } 20\% = \frac{268\,288}{347\,252} = 0.773 = 77.3\%$$

$$\text{Reduction in rate} = \frac{22.7\%}{4} = 5.7\% \longrightarrow 6\%$$

STEP 2

Try $i = 14\%$.
NPV < 0; R.O.I. $< 14\%$

$$\text{Index} = \frac{333\,824}{350\,924} = 0.951 = 95.1\%$$

$$\text{Reduction in rate} = \frac{4.9\%}{4} = 1.2\% \longrightarrow 2\%$$

STEP 3

Try $i = 12\%$.
NPV > 0; R.O.I. $> 12\%$
$12\% <$ R.O.I. $< 14\%$

STEP 4

$$\frac{d}{2} = \frac{9708}{9708 + 17\,100} = \frac{9708}{26\,808} = 0.362131$$

$$d = 2(0.362131) = 0.724$$

The rate of discount at which the net present value is zero is 12% + 0.72% = 12.72%. The rate of return is 12.7%.

Programmed Solution

(Set P/Y = 1; C/Y = 1)

Present Value of Amounts in General Form	Attempts				
	$i = 20\%$	$i = 14\%$	$i = 12\%$	$i = 12.7\%$	$i = 12.68\%$
PV of benefits	0 FV	0 FV	0 FV	0 FV	0 FV
$64\,000\left[\frac{1-(1+i)^{-10}}{i}\right]$	−64 000 PMT	−64 000 PMT	−64 000 PMT	−64 000 PMT	−64 000 PMT
	10 N	10 N	10 N	10 N	10 N
	20 I/Y	14 I/Y	12 I/Y	12.7 I/Y	12.68 I/Y
	CPT	CPT	CPT	CPT	CPT
	PV	PV	PV	PV	PV
PV_{IN}	268 318	333 831	361 614	351 484	351 767
PV of costs					
320 000 now	320 000	320 000	320 000	320 000	320 000
$96\,000(1+i)^{-5}$	0 PMT	0 PMT	0 PMT	0 PMT	0 PMT
	96 000 FV	96 000 FV	96 000 FV	96 000 FV	96 000 FV
	5 N	5 N	5 N	5 N	5 N
	20 I/Y	14 I/Y	12 I/Y	12.7 I/Y	12.68 I/Y
	CPT	CPT	CPT	CPT	CPT
	PV	PV	PV	PV	PV
Outflow	−38 580	−49 859	−54 473	−52 802	−52 849
<70 000(1 + i)$^{-10}$>	70 000 FV	70 000 FV	70 000 FV	70 000 FV	70 000 FV
	10 N	10 N	10 N	10 N	10 N
	20 I/Y	14 I/Y	12 I/Y	12.7 I/Y	12.68 I/Y
	CPT	CPT	CPT	CPT	CPT
	PV	PV	PV	PV	PV
<inflow>	<11 305>	<18 882>	<22 538>	<21 177>	<21 214>
PV_{OUT}	347 275	350 977	351 935	351 625	351 635
NPV	<78 957>	<17 146>	9 679	<141>	132
R.O.I.	<20%	<14%	>12%	<12.7%	>12.68%

EXAMPLE 10.3D A project requires an immediate investment of $33 000 with a residual value of $7000 at the end of the project. It is expected to yield a net return of $7000 in Year 1, $8000 in Year 2, $11 000 per year for the following six years, and $9000 per year for the remaining four years. Find the rate of return.

SOLUTION The cash flows for the project (in thousands) are represented in the diagram below.

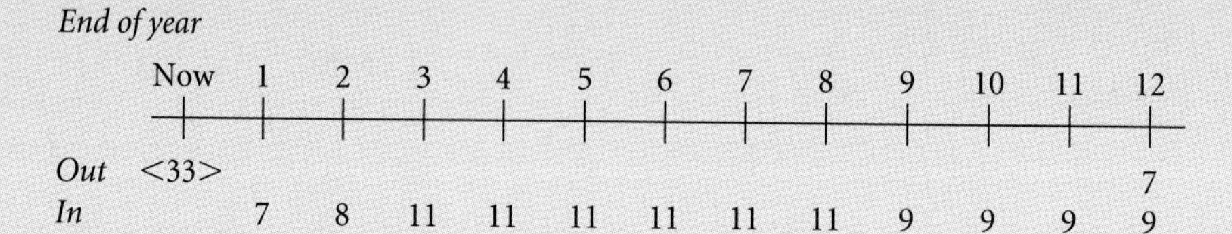

The computations are organized in the chart that follows.

Present Value of Amounts in General Form	Attempts					
	i = 20%		i = 28%		i = 26%	
PV of returns	Factor	$	Factor	$	Factor	$
$7000(1+i)^{-1}$	0.833	5 831	0.781	5 467	0.794	5 558
$8000(1+i)^{-2}$	0.694	5 552	0.610	4 880	0.630	5 040
$11\,000\left[\frac{1-(1+i)^{-6}}{i}\right]$	3.326		2.759		2.885	
	×	25 391	×	18 513	×	19 993
$\times(1+i)^{-2}$	0.694		0.610		0.630	
$9000\left[\frac{1-(1+i)^{-4}}{i}\right]$	2.589		2.241		2.320	
	×	5 429	×	2 803	×	3 278
$\times(1+i)^{-8}$	0.233		0.139		0.157	
TOTAL PV_{IN}		42 203		31 663		33 869
PV of costs						
33 000 now		33 000		33 000		33 000
$<7000(1+i)^{-12}>$	0.112	<784>	0.052	<364>	0.062	<434>
TOTAL PV_{OUT}		32 216		32 636		32 566
NPV		9 987		<973>		1 303

Explanations for computations

STEP 1 The present value of the returns consists of $7000 discounted for one year, $8000 discounted for two years, the present value of an ordinary annuity of six payments of $11 000 deferred for two years, and the present value of an ordinary annuity of four payments of $9000 deferred for eight years. The present value of the costs consists of the lump sum of $33 000 less the salvage value of $7000 discounted for twelve years.

STEP 2 The rate of discount chosen for the first attempt is 20%.

For $i = 20\%$, NPV > 0; R.O.I. > 20%

$$\text{Index} = \frac{42\,203}{32\,216} = 1.310 = 131.0\%$$

$$\text{Increase in rate} = \frac{31.0\%}{4} = 7.75\% \text{ or } 8\%$$

STEP 3 For $i = 28\%$, NPV < 0; R.O.I. < 28%

$$\text{Index} = \frac{31\,663}{32\,636} = 0.970 = 97.0\%$$

$$\text{Decrease in rate} = \frac{3\%}{4} = 0.75\% \text{ or } 2\% \text{ (rounded up)}$$

STEP 4 For $i = 26\%$, NPV > 0; R.O.I. > 26%

26% < R.O.I. < 28%

STEP 5

$$d = \frac{1303}{1303 + 973} \times 2 = \frac{2606}{2276} = 1.14499$$

The rate of discount for which the net present value is zero is approximately 26% + 1.14% = 27.14%. The rate of return, correct to the nearest tenth of a percent, is 27.1%.

POINTERS AND PITFALLS

Investors should always be aware of the fact and that higher rates of return (yield) are accompanied typically by higher levels of investment risk. For the most daring investors, high risk/high yield investments include derivatives, commodities, precious metals, gemstones, collectible items, common stocks, and growth stocks. For investors more comfortable with moderate risk/moderate yield, investment choices include mutual funds, real estate, corporate bonds, and preferred stocks. Low risk/low yield options such as savings accounts, term deposits, guaranteed investment certificates (GICs), and Canada Savings Bonds (CSBs) are designed to appeal to the most conservative investors.

Excel Internal rate of return can be determined by using a preprogrammed financial calculator. For instructions on using this function, refer to the CD-ROM accompanying this book.

EXERCISE 10.3

Excel If you choose, you can use Excel's ***Internal Rate of Return (IRR)*** function to answer the questions in Part B below. Refer to **IRR** on the Spreadsheet Template Disk to learn how to use this Excel function.

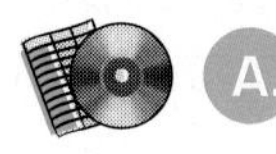

A. Use linear interpolation to find the approximate value of the rate of return for each of the projects below. State your answer correct to the nearest tenth of a percent.

	Positive NPV at *i*	Negative NPV at *i*
1.	\$2350 at 24%	−\$1270 at 26%
2.	\$850 at 8%	−\$370 at 10%
3.	\$135 at 20%	−\$240 at 22%
4.	\$56 at 16%	−\$70 at 18%

B. Find the rate of return for each of the six situations below (correct to the nearest tenth of a percent).

1. The proposed expansion of CIV Electronics' plant facilities requires the immediate outlay of \$100 000. Expected net returns are

Year 1: Nil	Year 2: \$30 000	Year 3: \$40 000
Year 4: \$60 000	Year 5: \$50 000	Year 6: \$20 000

2. The introduction of a new product requires an initial outlay of \$60 000. The anticipated net returns from the marketing of the product are expected to be \$12 000 per year for ten years.

3. Your firm is considering introducing a new product for which net returns are expected to be

Year 1 to Year 3 inclusive: \$2000 per year;
Year 4 to Year 8 inclusive: \$5000 per year;
Year 9 to Year 12 inclusive: \$3000 per year.

The introduction of the product requires an immediate outlay of \$15 000 for equipment estimated to have a salvage value of \$2000 after twelve years.

4. A project requiring an immediate investment of \$150 000 and a further outlay of \$40 000 after four years has a residual value of \$30 000 after nine years. The project yields a negative net return of \$10 000 in Year 1, a zero net return in Year 2, \$50 000 per year for the following four years, and \$70 000 per year for the last three years.

5. You are thinking of starting a hot dog business that requires an initial investment of \$16 000 and a major replacement of equipment after ten years amounting to \$8000. From competitive experience, you expect to have a net loss of \$2000 the first year, a net profit of \$2000 the second year, and, for the remaining years of the first fifteen years of operations, net returns of \$6000 per year. After fifteen years, the net returns will gradually decline and will be zero at the end of twenty-five years (assume returns of \$3000 per year for that period). After twenty-five years, your lease will expire. The salvage value of equipment at that time is expected to be just sufficient to cover the cost of closing the business.

6. The Blue Sky Ski Resort plans to install a new chair lift. Construction is estimated to require an immediate outlay of $220 000. The life of the lift is estimated to be fifteen years with a salvage value of $80 000. Cost of clearing and grooming the new area is expected to be $30 000 for each of the first three years of operation. Net cash inflows from the lift are expected to be $40 000 for each of the first five years and $70 000 for each of the following ten years.

Review Exercise

1. Wells Inc. has to choose between two investment alternatives. Alternative A will return the company $20 000 after three years, $60 000 after six years, and $40 000 after ten years. Alternative B will bring returns of $10 000 per year for ten years. If the company expects a return of 14% on investments, which alternative should it choose?
2. A piece of property may be acquired by making an immediate payment of $25 000 and payments of $37 500 and $50 000 three and five years from now respectively. Alternatively, the property may be purchased by making quarterly payments of $5150 in advance for five years. Which alternative is preferable if money is worth 15% compounded semi-annually?
3. An investor has two investment alternatives. If he chooses Alternative 1, he will have to make an immediate outlay of $7000 and will receive $500 every three months for the next nine years. If he chooses Alternative 2, he will have to make an immediate outlay of $6500 and will receive $26 000 after eight years. If interest is 12% compounded quarterly, which alternative should the investor choose on the basis of the net present value criterion?
4. Replacing old equipment at an immediate cost of $65 000 and $40 000 five years from now will result in a savings of $8000 semi-annually for ten years. At 14% compounded annually, should the old equipment be replaced?
5. A real estate development project requires annual outlays of $75 000 for eight years. Net cash inflows beginning in Year 9 are expected to be $250 000 per year for fifteen years. If the developer requires a rate of return of 18%, compute the net present value of the project.
6. A company is considering a project that will require a cost outlay of $30 000 per year for four years. At the end of the project, the company expects to salvage the physical assets for $30 000. The project is estimated to yield net returns of $60 000 in Year 4, $40 000 in Year 5, and $20 000 for each of the following five years. Alternative investments are available yielding a rate of return of 14%. Compute the net present value of the project.
7. An investment requires an initial outlay of $45 000. Net returns are estimated to be $14 000 per year for eight years. Determine the rate of return.
8. A project requires an initial outlay of $10 000 and promises net returns of $2000 per year over a twelve-year period. If the project has a residual value of $4000 after twelve years, what is the rate of return?
9. Compute the rate of return for Question 5.
10. Compute the rate of return for Question 6.
11. Superior Jig Co. has developed a new jig for which it expects net returns as follows.

    Year 1: $8000

    Year 2 to 6 inclusive: $12 000 per year

    Year 7 to 10 inclusive: $6000 per year

    The initial investment of $36 000 has a residual value of $9000 after ten years. Compute the rate of return.

12. The owner of a sporting goods store is considering remodelling the store in order to carry a larger inventory. The cost of remodelling and additional inventory is $60 000. The expected increase in net profit is $8000 per year for the next four years and $10 000 each year for the following six years. After ten years, the owner plans to retire and sell the business. She expects to recover the additional $40 000 invested in inventory but not the $20 000 invested in remodelling. Compute the rate of return.

13. Outway Ventures evaluates potential investment projects at 20%. Two alternative projects are available. Project A will return the company $5800 per year for eight years. Project B will return the company $13 600 after one year, $17 000 after five years, and $20 400 after eight years. Which alternative should the company choose according to the discounted cash flow criterion?

14. Project A requires an immediate investment of $8000 and another $6000 in three years. Net returns are $4000 after two years, $12 000 after four years, and $8000 after six years. Project B requires an immediate investment of $4000, another $6000 after two years, and $4000 after four years. Net returns are $3400 per year for seven years. Determine the net present value at 10%. Which project is preferable according to the net present value criterion?

15. Net returns from an investment are estimated to be $13 000 per year for twelve years. The investment involves an immediate outlay of $50 000 and a further outlay of $30 000 after six years. The investments are estimated to have a residual value of $10 000 after twelve years. Find the net present value at 20%.

16. The introduction of a new product requires an immediate outlay of $45 000. Anticipated net returns from the marketing of the product are expected to be $12 500 per year for ten years. What is the rate of return on the investment (correct to the nearest tenth of a percent)?

17. Games Inc. has developed a new electronic game and compiled the following product information.

	Production Cost	Promotion Cost	Sales Revenue
Year 1	$32 000	—	—
Year 2	32 000	$64 000	$ 64 000
Year 3	32 000	96 000	256 000
Year 4	32 000	32 000	128 000
Year 5	32 000		32 000

Should the product be marketed if the company requires a return of 16%?

18. Farmer Jones wants to convert his farm into a golf course. He asked you to determine his rate of return based on the following estimates.

Development cost for each of the first three years, $80 000.

Construction of a clubhouse in Year 4, $240 000.

Upon his retirement in fifteen years, improvements in the property will yield him $200 000.

Net returns from the operation of the golf course will be nil for the first three years and $100 000 per year afterwards until his retirement.

Self-Test

1. Opportunities Inc. requires a minimum rate of return of 15% on investment proposals. Two proposals are under consideration but only one may be chosen. Alternative A offers a net return of $2500 per year for twelve years. Alternative B offers a net return of $10 000 each year after four, eight, and twelve years respectively. Determine the preferred alternative according to the discounted cash flow criterion.

2. A natural resources development project requires an immediate outlay of $100 000 and $50 000 at the end of each year for four years. Net returns are nil for the first two years and $60 000 per year thereafter for fourteen years. What is the net present value of the project at 16%?

3. An investment of $100 000 yields annual net returns of $20 000 for ten years. If the residual value of the investment after ten years is $30 000, what is the rate of return on the investment (correct to the nearest tenth of a percent)?

4. A telephone system with a disposable value of $1200 after five years can be purchased for $6600. Alternatively, a leasing agreement is available that requires an immediate payment of $1500 plus payments of $100.00 at the beginning of each month for five years. If money is worth 12% compounded monthly, should the telephone system be leased or purchased?

5. A choice has to be made between two investment proposals. Proposal A requires an immediate outlay of $60 000 and a further outlay of $40 000 after three years. Net returns are $20 000 per year for ten years. The investment has no residual value after ten years. Proposal B requires outlays of $29 000 in each of the first four years. Net returns starting in Year 4 are $40 000 per year. The residual value of the investment after ten years is $50 000. Which proposal is preferable at 20%?

6. Introducing a new product requires an immediate investment in plant facilities of $180 000 with a disposal value of $45 000 after seven years. The facilities will require additional capital outlays of $50 000 each after three and five years respectively. Net returns on the investment are estimated to be $75 000 per year for each of the first four years and $50 000 per year for the remaining three years. Determine the rate of return on investment (correct to the nearest tenth of a percent).

Challenge Problems

1. The owners of a vegetable processing plant can buy a new conveyor system for $85 000. The owners estimate they can save $17 000 per year on labour and maintenance costs. They can purchase the same conveyor system with an automatic loader for $114 000. They estimate they can save $22 000 per year with this system. If the owners expect both systems to last ten years and they require at least 14% return per year, should the owners buy the system with the automatic loader?

2. CheeseWorks owns four dairies in your province and has planned upgrades for all locations. The owners are considering four projects, each of which is independent of the other three projects. The details of each project—A, B, C, and D—are shown below.

Project	Cost at Beginning of Year 1	Revenues, and Cost Savings at End of:				
		Year 1	Year 2	Year 3	Year 4	Year 5
A	300 000	150 000	120 000	120 000	0	0
B	360 000	0	40 000	200 000	200 000	200 000
C	210 000	10 000	10 000	100 000	120 000	120 000
D	125 000	30 000	40 000	40 000	40 000	40 000

The owners of CheeseWorks have $700 000 to invest in these projects. They expect at least 12% return on all of their projects. In which projects should the owners of a CheeseWorks invest to maximize the return on their investment?

Case Study 10.1 To Lease or Not to Lease?

» Jonathon was just hired for a new job. To travel to his new job, he needed a new car. Reading the newspaper, he noticed an ad for a 2003 Jeep Liberty Sport. It was just the car he wanted.

The ad quoted both a cash purchase price of $33 650 and a monthly lease payment option. Since he did not have enough money to pay for a car, he would have to borrow the money from the bank, paying interest of 5.3% compounded annually on the loan.

The lease option showed payments of $339 a month for 48 months with a $3300 down payment or equivalent trade. Freight and air tax was included. Jonathon did not currently have a vehicle to offer as a trade-in. If the vehicle was leased, after 48 months it could be purchased for $20 242. The lease was based on a finance interest rate of 5.3%. During the term of the lease, kilometres were limited to 81 600, with a charge of $0.15 per kilometre for excess kilometres. The costs included freight, but excluded taxes, registration, licence, and dealer administration charges. Jonathon was particularly impressed with the "seven-years or 115 000-kilometre" warranty on the engine and transmission. They also offered 24-hour roadside assistance.

Jonathon must decide whether to buy or lease this car. He lives in a province with a combined PST and GST tax rate of 15%. He realizes that the costs of licence and insurance must be paid, but he will ignore these in his calculations.

QUESTIONS

1. If Jonathon buys the car, what is the total purchase price, including taxes? Assume tax is charged on any additional costs.
2. Since Jonathon has no down payment, he must finance the car if he purchases it.
   (a) Is it cheaper to borrow the money from the bank or to lease?

(b) The dealer is offering a vehicle loan rate of 6% compounded annually. Is it better to buy or lease the car at this rate?
(c) Find the rate of interest at which the cost of buying is the same as the cost of leasing. Calculate your answer to two decimal places

3. Suppose Jonathon has a $2000 down payment for this car.
(a) What is the purchase price of the car if he pays cash for it? Assume the down payment is subtracted from the price of the car including tax.
(b) If the monthly lease payment is $279.00, is it cheaper to lease or buy the car if Jonathon can get the special dealer rate of 5%?
(c) Is there an advantage to having a $2000 down payment if you want to lease this car? Why or why not?

Case Study 10.2 Building a Business

» MAS Manufacturing has demolished an old warehouse to make room for additional manufacturing capacity. The company has decided to construct a new building, but must decide how to proceed. It has two alternatives for the new building, both of which will create a building with an expected life of fifty years. The residual value is unknown but will be the same for either alternative.

Alternative A is to construct a new building that would have 180 000 square feet. Construction costs will total $2 100 000 at the end of Year 1. Maintenance costs are expected to be $15 000 per year. The building will need to be repainted every ten years (starting in ten years) at an estimated cost of $10 000.

Alternative B is to construct the building in two stages: build 100 000 square feet now; and add 80 000 square feet in ten years. Construction costs for the first stage will be $1 600 000 at the end of Year 1. Construction costs for the second stage will be $900 000 when the addition is completed at the end of Year 10. Maintenance costs are expected to be $10 000 per year for the first ten years, then $17 000 per year after that. The building and the addition will have to be painted every ten years, beginning in Year 20, at an estimated cost of $10 000.

QUESTIONS

1. Suppose the company's required rate of return is 15%.
(a) What is the present value of Alternative A?
(b) What is the present value of Alternative B?
(c) Which alternative would you recommend on the basis of your discounted cash flow analysis?

2. Which alternative would you recommend if the company's required rate of return was 20%? Show all calculations.

3. Suppose the company could rent a portion of its building for $48 000 per year for the first ten years if it chose Alternative A.
(a) If the company's required rate of return is 15%, what is the net present value of Alternative A?
(b) On the basis of the new information, would you recommend Alternative A or Alternative B if the company's required rate of return is 15%?

SUMMARY OF FORMULAS

Formula 10.1

$$\text{NET PRESENT VALUE (NPV)} = \text{PRESENT VALUE OF INFLOWS} - \text{PRESENT VALUE OF OUTLAYS}$$

Formula for finding the difference between the present value of cash inflows and the present value of cash outflows, known as the net present value.

Formula 10.2

$$\text{PROFITABILITY INDEX} = \frac{\text{PRESENT VALUE OF INFLOWS}}{\text{PRESENT VALUE OF OUTLAYS}} = \frac{PV_{IN}}{PV_{OUT}}$$

Formula for finding the relationship by dividing the present value of cash inflows by the present value of cash outflows, known as the profitability index.

In addition, Formulas 3.1B, 3.1C, 5.2, 6.3, 7.2, and 7.4 were used in this chapter.

GLOSSARY

Discounted benefit-cost ratio *see* **Profitability index**

Discounted cash flow the present value of cash payments *(p. 432)*

Net present value the difference between the present value of the inflows (benefits) and the present value of the outlays (costs) of a capital investment project *(p. 441)*

Profitability index the ratio of the present value of the inflows (benefits) to the present value of the outlays (costs) of a capital investment project *(p. 450)*

Rate of return the rate of discount for which the net present value of a capital investment project is equal to zero *(p. 450)*

USEFUL INTERNET SITES

www.tse.com
TSX Group Visit this site for information on all the companies that are traded on the Toronto Stock Exchange (TSX).

www.investorwords.com
InvestorWords.com A comprehensive financial glossary with definitions and related terms

http://finance.yahoo.com
Yahoo! Finance A full education site that also contains quotes, charts, historical analysis, and reports.

Trade Discount, Cash Discount, Markup, and Markdown

OBJECTIVES

Upon completing this chapter, you will be able to do the following:

1. Solve problems involving trade discounts.
2. Calculate equivalent single rates of discount for discount series and solve problems involving discount series.
3. Apply the three most commonly used methods of cash discount.
4. Solve problems involving markup based on either cost or selling price.
5. Solve pricing problems involving markup, markdown, and discounts.

Suppose you were the owner of a furniture manufacturing company. You purchase your raw materials from suppliers, who offer their goods to you at a list, or catalogue, price. You might receive a trade discount if you pay for the materials promptly or if you purchase the materials in bulk. When you sell your furniture to furniture retailers, you offer your furniture at a list, or catalogue, price and you might offer a discount for prompt payment or bulk purchases. As you can see, we need to be careful, since the same terms are used by different companies to represent different dollar amounts.

INTRODUCTION

A product typically passes through a number of stages of the *merchandising chain* on its way from a raw material to a product purchased by the consumer.

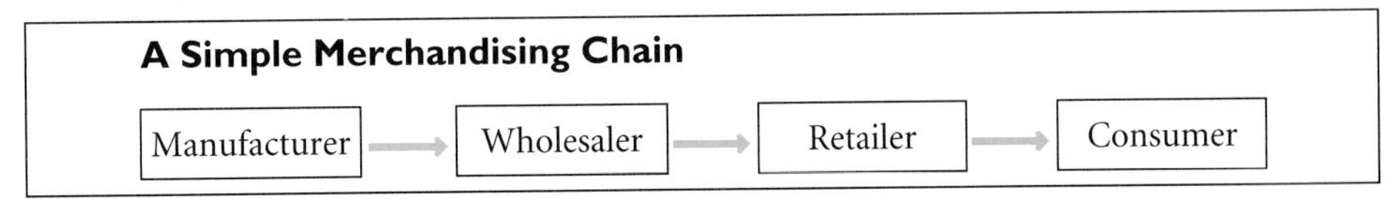

A simple merchandising chain may include manufacturers, wholesalers, and retailers, all of whom must make a profit on the product to remain in business. As the product is purchased and resold along the chain, each merchandiser adds a *markup* above the cost of buying the merchandise that increases the price of the product. Sometimes a merchandiser offers a discount in order to sell more product or to encourage prompt payment for the product. When the product is sold to the consumer, it may be *marked down* from the regular selling price in response to competitors' prices or other economic conditions.

This chapter deals with trade discount, cash discount, markup, and markdown.

I.1 Merchandising

As you make your way through the merchandising chain, you will find many of the same terms used to represent different things, depending on where you are in the chain. As shown in Figure I.1, the terms markup, list price, and trade discount are used throughout the merchandising chain.

If you are a manufacturer or supplier, you might mark up an item to create a list, or catalogue, price. If you want to sell the item to a wholesaler for less than list price, you might offer a trade discount. The wholesaler would add a markup to create a list price at which it offers the item to the retailer. The wholesaler could offer the retailer a trade discount to sell the item for less than the list price. The retailer would then add a markup and offer the item to the consumer at a regular selling price, or list price. The retailer might offer a markdown on the item to sell the item for less than the regular selling price.

For any particular situation, first identify where you are in the merchandising chain. You will then be able to understand and apply these terms correctly.

FIGURE I.1 Terminology Used in the Merchandising Chain

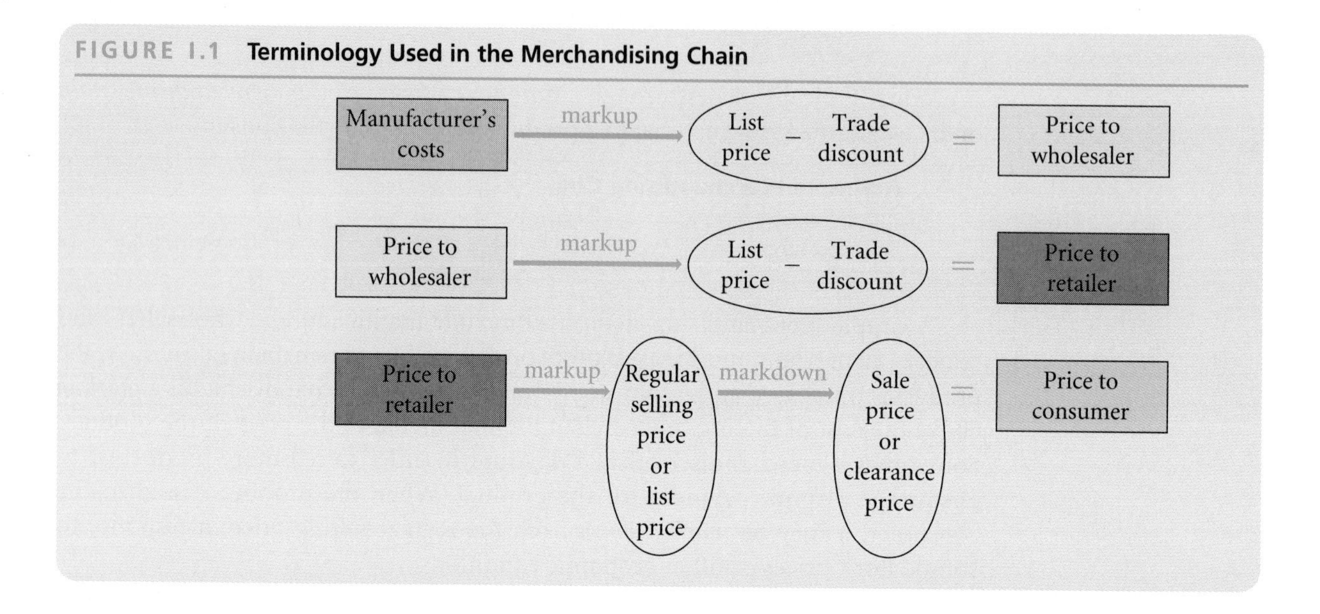

I.2 Trade Discount

A. Basic concepts and computations

The merchandising chain is made up of manufacturers, distributors, wholesalers, and retailers. Merchandise is usually bought and sold among these members of the chain on credit terms. The prices quoted to other members often involve *trade discounts.* A **trade discount** is a reduction of a catalogue or **list price** or **manufacturer's suggested retail price (MSRP)** and is usually stated as a percent of the list price or MSRP.

Trade discounts are used by manufacturers, wholesalers, and distributors as pricing tools for a number of reasons. The most important reasons are

(a) to facilitate the establishment of price differentials for different groups of customers;

(b) to facilitate the communication of changes in prices;

(c) to reduce the cost of making changes in prices in published catalogues.

When computing trade discounts, keep in mind that the rate of discount is based on the list price.

$$\text{AMOUNT OF DISCOUNT} = \text{RATE OF DISCOUNT} \times \text{LIST PRICE}$$ — **Formula I.1A**

The amount of discount is then subtracted from the list price. The remainder is the **net price**.

$$\text{NET PRICE} = \text{LIST PRICE} - \text{AMOUNT OF DISCOUNT}$$ — **Formula I.2**

B. Computing the amount of discount and net price

EXAMPLE I.2A

An item listed at \$80.00 is subject to a trade discount of 25%.
Compute (i) the amount of discount;
(ii) the net price.

SOLUTION

(i) Amount of trade discount = Rate of discount × List price
= (0.25)(80.00) = \$20.00

(ii) Net price = List price − Trade discount
= 80.00 − 20.00 = \$60.00

C. Computing the list price and net price when the discount is known

You can compute the list price and net price if the discount rate and discount amount are known. From Formula I.1A,

$$\text{LIST PRICE} = \frac{\text{AMOUNT OF DISCOUNT}}{\text{RATE OF DISCOUNT}}$$

Formula I.1B

EXAMPLE I.2B

The 30% discount on a tennis racket amounts to \$28.98.
Compute (i) the list price;
(ii) the net price.

SOLUTION

(i) $\text{List price} = \frac{\text{Amount of discount}}{\text{Rate of discount}} = \frac{28.98}{0.30} = \$96.60$

(ii) Net price = List price − Amount of discount
= 96.60 − 28.98 = \$67.62

D. Computing the rate of discount

Since the rate of trade discount is based on a list price, computing a rate of discount involves comparing the amount of discount to the list price.

$$\text{RATE OF TRADE DISCOUNT} = \frac{\text{AMOUNT OF DISCOUNT}}{\text{LIST PRICE}}$$

Formula I.1C

EXAMPLE I.2C

Find the rate of discount for
(i) skis listed at \$280.00 less a discount of \$67.20;
(ii) ski gloves listed at \$36.80 whose net price is \$23.92;
(iii) ski sweaters whose net price is \$55.68 after a discount of \$40.32.

SOLUTION

(i) $\text{Rate of discount} = \frac{\text{Amount of discount}}{\text{List price}} = \frac{67.20}{280.00} = 0.24 = 24\%$

(ii) Since Net price = List price − Amount of discount (Formula I.2),
Amount of discount = List price − Net price = 36.80 − 23.92 = $12.88

$$\text{Rate of discount} = \frac{\text{Amount of discount}}{\text{List price}} = \frac{12.88}{36.80} = 0.35 = 35\%$$

(iii) Since Net price = List price − Amount of discount (Formula I.2),
List price = Net price + Amount of discount = 55.68 + 40.32 = $96.00

$$\text{Rate of discount} = \frac{\text{Amount of discount}}{\text{List price}} = \frac{40.32}{96.00} = 0.42 = 42\%$$

E. The net factor approach to computing the net price

Instead of computing the amount of discount and then deducting this amount from the list price, the net price can be found by using the more efficient net factor approach developed in the following illustration.

Referring back to Example I.2A, the solution can be restated as follows.

List price	$80.00
Less trade discount 25% of 80.00	20.00
Net price	$60.00

Since the discount is given as a percent of the list price, the three dollar values may be stated as percents of list price.

List price	$80.00	⟶	100% of list price
Less trade discount	$20.00	⟶	25% of list price
Net price	$60.00	⟶	75% of list price

Note: The resulting "75%" is called the **net price factor** or **net factor** (in abbreviated form, **NPF**) and is obtained by deducting the 25% discount from 100%.

$$\text{NET PRICE FACTOR (NPF)} = 100\% - \%\ \text{DISCOUNT}$$ — **Formula I.3A**

The resulting relationship between net price and list price may be stated generally.

$$\text{NET PRICE} = \text{NET PRICE FACTOR (NPF)} \times \text{LIST PRICE}$$ — **Formula I.4A**

The two relationships represented by Formulae I.3A and I.4A can be restated in algebraic terms:

Convert the % discount into its decimal equivalent represented by d and express 100% by its decimal equivalent 1.

$$\text{NET PRICE FACTOR} = 1 - d$$ — **Formula I.3B**

Let the list price be represented by L and let the net price be represented by N.

$$N = (1 - d)L \quad \text{or} \quad N = L(1 - d)$$ — **Formula I.4B**

EXAMPLE I.2D

Find the net price for
(i) list price \$36.00 less 15%;
(ii) list price \$125.64 less 37.5%;
(iii) list price \$86.85 less $33\frac{1}{3}$%;
(iv) list price \$49.98 less $16\frac{2}{3}$%.

SOLUTION

(i) Net price = Net price factor × List price — using Formula I.4A
= (100% − 15%)(36.00) — using Formula I.3A
= (85%)(36) — subtract
= (0.85)(36) — convert the percent into a decimal
= \$30.60

(ii) Net price = (1 − *d*)L — using Formula I.4B
= (1 − 0.375)(125.64) — 37.5% = 0.375
= (0.625)(125.64)
= \$78.53

(iii) Net price = $(100\% - 33\frac{1}{3}\%)(86.85)$ — using Formula I.3A
= $(66\frac{2}{3}\%)(86.85)$
= (0.666667)(86.85) — use a sufficient number of decimals
= \$57.90

(iv) Net price = $(1 - 0.1\dot{6})(49.98)$ — using Formula I.4B
= $(0.8\dot{3})(49.98)$
= (0.833333)(49.98) — use a sufficient number of decimals
= \$41.65

EXAMPLE I.2E

A manufacturer can cover its cost and make a reasonable profit if it sells an article for \$63.70. At what price should the article be listed so that a discount of 30% can be allowed?

SOLUTION

Let the list price be represented by \$L.
The net price factor is 0.70 and the net price is \$63.70.

63.70 = 0.70L — using Formula I.4A

$$L = \frac{63.70}{0.70} = \$91.00$$

The article should be listed at \$91.00.

EXERCISE I.2

Find the missing values (represented by question marks) for each of the following questions.

	Rate of Discount	List Price	Amount of Discount	Net Price
1.	45%	$24.60	?	?
2.	$16\frac{2}{3}$%	$184.98	?	?
3.	?	$76.95	?	$51.30
4.	?	$724.80	?	$616.08
5.	?	?	$37.89	$214.71
6.	?	?	$19.93	$976.57
7.	62.5%	?	$83.35	?
8.	1.5%	?	$13.53	?
9.	37.5%	?	?	$84.35
10.	22%	?	?	$121.29

Answer each of the following questions.

1. A mountain bike listed for $975.00 is sold for $820.00. What rate of discount was allowed?
2. A washer-dryer combination listed at $1136.00 has a net price of $760.00. What is the rate of discount?
3. A 37.5% discount on a video recorder amounts to $913.50. What is the list price?
4. You can buy a set of golf clubs at Golf Liquidations for $762.50 below suggested retail price. Golf Liquidations claims that this represents a 62.5% discount. What is the suggested retail price (or list price)?
5. A $16\frac{2}{3}$% discount allowed on a silk shirt amounted to $14.82. What was the net price?
6. A store advertises a discount of $44.24 on winter boots. If the discount is 35%, for how much were the boots sold?
7. The net price of a freezer after a discount of $16\frac{2}{3}$% is $355.00. What is the list price?
8. The net price of an article is $63.31. What is the suggested retail price (the list price) if a discount of 35% was allowed?

I.3 Multiple Discounts

A. Discount series

If a list price is subject to two or more discounts, these discounts are called a **discount series.** A manufacturer may offer two or more **discounts** to different members of the merchandising chain. For example, a chain member closest to the consumer might be offered additional discounts, if there are fewer chain members who must make a profit on an item. If the manufacturer wants to encourage large-volume orders or early orders of seasonal items, it may offer additional discounts. For example, a manufacturer might offer a store a 5% discount on orders over 1000 items and an additional discount of 6% for ordering Christmas items in April. It may also offer additional discounts to compensate for advertising, promotion, and service costs handled by merchandising chain members.

DID YOU KNOW?

Today, discount series are becoming less common as more retailers and large buying groups buy directly from the manufacturer, bypassing chain members to keep prices low. These buyers demand the best price. They are not concerned with whether the best price is due to a discount series, a single large discount, or any other price-setting method.

When computing the net price, the discounts making up the discount series are applied to the list price successively. The net price resulting from the first discount becomes the list price for the second discount; the net price resulting from the second discount becomes the list price for the third discount; and so on. In fact, finding the net price when a list price is subject to a discount series consists of solving as many discount problems as there are discounts in the discount series.

EXAMPLE I.3A An item listed at $150.00 is subject to the discount series 20%, 10%, 5%. Determine the net price.

SOLUTION

List price	$150.00	Problem 1
Less first discount 20% of 150.00	30.00	
Net price after first discount	$120.00	Problem 2
Less second discount 10% of 120.00	12.00	
Net price after second discount	$108.00	Problem 3
Less third discount 5% of 108.00	5.40	
Net price	$102.60	

Because the solution to Example I.3A consists of three problems involving a simple discount, the net price factor approach can be used to solve it or any problem involving a series of discounts.

Problem 1 Net price after the first discount
= NPF for 20% discount × Original list price
= (1 − 0.20)(150.00)
= (0.80)(150.00)
= $120.00

Problem 2 Net price after the second discount
= NPF for 10% discount × Net price after the first discount
= (1 − 0.10)(120.00)
= (0.90)(120.00)
= (0.90)(0.80)(150.00)
= $108.00

Problem 3 Net price after the third discount

$$= (1 - 0.05)(108.00)$$
$$= (0.95)(108.00)$$
$$= (0.95)(0.90)(0.80)(150.00)$$
$$= \$102.60$$

The final net price of $102.60 is obtained from

(0.95)(0.90)(0.80)(150.00)
= (0.80)(0.90)(0.95)(150.00) ——— the order of the factors may be rearranged
= NPF for 20% × NPF for 10% × NPF for 5% × Original list price
= Product of the NPFs for the discounts in the discount series × Original list price
= Net price factor for the discount series × Original list price

This result may be generalized to find the net price for a list price subject to a discount series.

NPF FOR THE DISCOUNT SERIES = NPF FOR THE FIRST DISCOUNT × NPF FOR THE SECOND DISCOUNT × . . . × NPF FOR THE LAST DISCOUNT ——— **Formula I.5A**

NET PRICE = NPF FOR THE DISCOUNT SERIES × LIST PRICE ——— **Formula I.6A**

The two relationships represented by Formulas I.5A and I.6A can be restated in algebraic terms.

Let the net price be represented by N,
the original list price by L,
the first rate of discount by d_1,
the second rate of discount by d_2,
the third rate of discount by d_3, and
the last rate of discount by d_n.

Then Formula I.5A can be shown as

$$\text{NPF FOR A DISCOUNT SERIES} = (1 - d_1)(1 - d_2)(1 - d_3)\ldots(1 - d_n)$$ ——— **Formula I.5B**

and Formula I.6A can be shown as

$$\text{NET PRICE} = (1 - d_1)(1 - d_2)(1 - d_3)\ldots(1 - d_n)\text{L}$$ ——— **Formula I.6B**

EXAMPLE I.3B

Determine the net price for
(i) an office desk listed at $440.00 less 25%, 15%, 2%;
(ii) a power drill listed at $180.00 less 30%, 12.5%, 5%, 5%;

(iii) a home computer listed at $1260.00 less 33⅓%, 16⅔%, 2.5%;
(iv) an electronic chessboard listed at $1225.00 less 66⅔%, 8⅓%.

SOLUTION

(i) Net price = NPF for the discount series × List price — using Formula I.6A
= (1 − 0.25)(1 − 0.15)(1 − 0.02)(440.00)
= (0.75)(0.85)(0.98)(440.00)
= $274.89

(ii) Net price = $L(1 - d_1)(1 - d_2)(1 - d_3)(1 - d_4)$ — using Formula I.6B
= 180.00(1 − 0.30)(1 − 0.125)(1 − 0.05)(1 − 0.05)
= 180.00(0.70)(0.875)(0.95)(0.95)
= $99.50

(iii) Net price = $1260.00(1 - 0.3\dot{3})(1 - 0.1\dot{6})(1 - 0.025)$
= $1260.00(0.6\dot{6})(0.8\dot{3})(0.975)$
= 1260.00(0.666667)(0.833333)(0.975)
= $682.50

(iv) Net price = $(1225.00)(1 - 0.6\dot{6})(1 - 0.08\dot{3})$
= $(1225.00)(0.3\dot{3})(0.91\dot{6})$
= (1225.00)(0.333333)(0.916667)
= $374.31

B. Single equivalent rates of discount

For every discount series, a **single equivalent rate of discount** exists.

EXAMPLE I.3C

A manufacturer sells skidoos to dealers at a list price of $2100.00 less 40%, 10%, 5%. Determine
(i) the amount of discount;
(ii) the single rate of discount.

SOLUTION

(i) Net price = NPF × List price
= (1 − 0.40)(1 − 0.10)(1 − 0.05)(2100.00)
= (0.60)(0.90)(0.95)(2100.00)
= $1077.30

Amount of discount = List price − Net price
= 2100.00 − 1077.30 = $1022.70

(ii) $\text{Rate of discount} = \dfrac{\text{Amount of discount}}{\text{List price}} = \dfrac{1022.70}{2100.00} = 0.487 = 48.7\%$

Note: Taking off a single discount of 48.7% has the *same* effect as using the discount series 40%, 10%, 5%. That is, the single discount of 48.7% is equivalent to the discount series 40%, 10%, 5%. Caution: The sum of the discounts in the series, 40% + 10% + 5% or 55%, is *not* equivalent to the single discount.

You can find the single equivalent rate of discount by choosing a suitable list price and computing first the amount of discount and then the rate of discount.

EXAMPLE I.3D

Find the single equivalent rate of discount for the discount series 30%, 8%, 2%.

SOLUTION

Assume a list price of \$1000.00.

Net price = NPF for the series × List price
= (0.70)(0.92)(0.98)(1000.00)
= (0.63112)(1000.00)
= \$631.12

Amount of discount = 1000.00 − 631.12 = \$368.88

$$\text{Single equivalent rate of discount} = \frac{368.88}{1000.00} = 0.36888 = 36.888\%$$

Note: The net price factor for the series = (0.70)(0.92)(0.98) = 0.63112, and 1 − 0.63112 = 0.36888. Therefore, you can find the single equivalent rate of discount by subtracting the net price factor for the series from 1.

SINGLE EQUIVALENT RATE OF DISCOUNT FOR A DISCOUNT SERIES
= 1 − NPF FOR THE DISCOUNT SERIES
$= 1 - [(1 - d_1)(1 - d_2)(1 - d_3) \ldots (1 - d_n)]$ — **Formula I.7**

EXAMPLE I.3E

Determine the single equivalent rate of discount for each of the following discount series.

(i) 25%, 20%, 10%
(ii) 30%, 12.5%, 2.5%
(iii) $33\frac{1}{3}$%, 15%, 5%, 5%

SOLUTION

(i) Single equivalent rate of discount
= 1 − (1 − 0.25)(1 − 0.20)(1 − 0.10) — **using Formula I.7**
= 1 − (0.75)(0.80)(0.90)
= 1 − 0.540000
= 0.46
= 46%

(ii) Single equivalent rate of discount
= 1 − (1 − 0.30)(1 − 0.125)(1 − 0.025)
= 1 − (0.70)(0.875)(0.975)
= 1 − 0.5971887
= 0.402812
= 40.28125%

(iii) Single equivalent rate of discount
$= 1 - (1 - 0.3\dot{3})(1 - 0.15)(1 - 0.05)(1 - 0.05)$
$= 1 - (0.6\dot{6})(0.85)(0.95)(0.95)$
= 1 − 0.511417
= 0.488583
= 48.85833%

Note: When computing or using single equivalent rates of discount, use a sufficient number of decimals to ensure an acceptable degree of accuracy.

POINTERS AND PITFALLS

The single equivalent rate of discount is not simply the sum of the individual discounts. Proper application of Formula I.7 will always result in a single equivalent discount rate that is less than the sum of the individual discounts. You can use this fact to check whether the single equivalent discount rate you calculate is reasonable.

EXAMPLE I.3F

Determine the amount of discount for each of the following list prices subject to the discount series 40%, 12.5%, $8\frac{1}{3}$%, 2%.

(i) $625.00
(ii) $786.20
(iii) $1293.44

SOLUTION

Single equivalent rate of discount

$= 1 - (1 - 0.40)(1 - 0.125)(1 - 0.08\dot{3})(1 - 0.02)$
$= 1 - (0.60)(0.875)(0.91\dot{6})(0.98)$
$= 1 - 0.471625$
$= 0.528375$
$= 52.8375\%$

(i) Amount of discount = Rate of discount × List price
= (0.528375)(625.00)
= $330.23

(ii) Amount of discount = (0.528375)(786.20)
= $415.41

(iii) Amount of discount = (0.528375)(1293.44)
= $683.42

EXAMPLE I.3G

The local hardware store has listed a power saw for $136.00 less 30%. A department store in a nearby shopping mall lists the same model for $126.00 less 20%, less an additional 15%. What additional rate of discount must the hardware store give to meet the department store price?

SOLUTION

Hardware store net price = 136.00(0.70)	$95.20
Department store price = 126.00(0.80)(0.85)	85.68
Additional discount needed	$ 9.52

$$\text{Additional rate of discount needed} = \frac{9.52}{95.20} = 0.10 = 10\%$$

EXAMPLE I.3H

Redden Distributors bought a shipment of camcorders for $477.36 each. At what price were the camcorders listed if the list price was subject to discounts of 15%, 10%, 4%?

SOLUTION

Let the list price be $L.
The net price factor is (0.85)(0.90)(0.96) —— using Formula I.5A

The net price is \$477.36.
477.36 = L(0.85)(0.90)(0.96)

$$L = \frac{477.36}{(0.85)(0.90)(0.96)} = \$650.00$$

The camcorders were listed at \$650.00.

EXERCISE I.3

For each of the following six questions, find the missing values represented by the question marks.

	Rate of Discount	List Price	Net Price	Single Equivalent Rate of Discount
1.	25%, 10%	\$44.80	?	?
2.	$33\frac{1}{3}$%, 5%	\$126.90	?	?
3.	40%, 12.5%, 2%	\$268.00	?	?
4.	20%, $16\frac{2}{3}$%, 3%	\$72.78	?	?
5.	35%, $33\frac{1}{3}$%, 10%	?	\$617.50	?
6.	20%, 20%, 10%	?	\$53.28	?

Answer each of the following questions.

1. A patio chair is listed for \$240.00 less 30%, 20%, 5%.
  (a) What is the net price?
  (b) What is the total amount of discount allowed?
  (c) What is the exact single rate of discount that was allowed?

2. A power saw is listed for \$174.00 less $16\frac{2}{3}$%, 10%, 8% .
  (a) What is the net price?
  (b) How much is the amount of discount allowed?
  (c) What is the exact single rate of discount that was allowed?

3. Compute the equivalent single rate of discount for each of the following discount series.
  (a) 30%, 12.5% **(b)** $33\frac{1}{3}$%, 20%, 3%

4. Determine the equivalent single rate of discount for each of the following series of discounts.
  (a) $16\frac{2}{3}$%, 7.5% **(b)** 25%, $8\frac{1}{3}$%, 2%

5. An item listed by a wholesaler for \$750.00 less 20%, 5%, 2% is reduced at a clearance sale to \$474.81. What additional rate of discount was offered?

6. Crosstown Jewellers sells watches for \$340.00 less 25%. Its competitor across the street offer the same type of watch for \$360.00 less 30%, 15%. What additional rate of discount must Crosstown offer to meet the competitor's price?

7. Arrow Manufacturing offers discounts of 25%, 12.5%, 4% on a line of products. For how much should an item be listed if it is to be sold for $113.40?

8. What is the list price of an article that is subject to discounts of 33⅓%, 10%, 2% if the net price is $564.48?

9. A distributor lists an item for $85.00 less 20%. To improve lagging sales, the net price of the item is reduced to $57.80. What additional rate of discount does the distributor offer?

10. Polar Bay Wines advertises California Juice listed at $125.00 per bucket at a discount of 24%. A nearby competitor offers the same type of juice for $87.40 per bucket. What additional rate of discount must Polar Bay Wines give to meet the competitor's price?

I.4 Cash Discount

A. Basic concepts

Goods among manufacturers, wholesalers, distributors, and retailers are usually sold on credit rather than for cash. To encourage prompt payment, many businesses offer a reduction in the amount of the invoice. This reduction is called a **cash discount**.

Cash discounts are offered in a variety of ways. The three most commonly used methods are

1. **ordinary dating;**
2. **end-of-the-month** (or **proximo**) **dating;**
3. **receipt-of-goods dating.**

The invoice's terms of payment specify the method and size of the cash discount. Regardless of the method used, all **payment terms** have three things in common.

1. The **rate of discount** is stated as a percent of the net amount of the invoice. The net amount of the invoice is the amount after trade discounts are deducted.
2. The **discount period** is stipulated. The cash discount applies during this time period only.
3. The **credit period** is stipulated. The invoice must be paid during this time period.

If payment is not made during the stipulated discount period, the net amount of the invoice is to be paid by the end of the credit period. This date, called the *due date*, is either stipulated by the terms of payment or implied by the prevailing business practice. If payment is not made by the due date, the account is overdue and might be subject to late payment charges.

In dealing with cash discounts, the major problem is how to interpret the terms of payment. Otherwise, the mathematics of working with cash discounts is similar to that used in working with trade discounts.

B. Ordinary dating

The most frequently used method of offering a cash discount is ordinary dating, and the most commonly used payment terms are *2/10, n/30* (read *two ten, net thirty*).

This payment term means that if payment is made *within* ten days of the date of the invoice, a discount of 2% may be deducted from the net amount of the invoice. Otherwise, payment of the net amount of the invoice is due within thirty days. (See Figure I.2.)

FIGURE I.2 **Interpretation of Payment Terms**

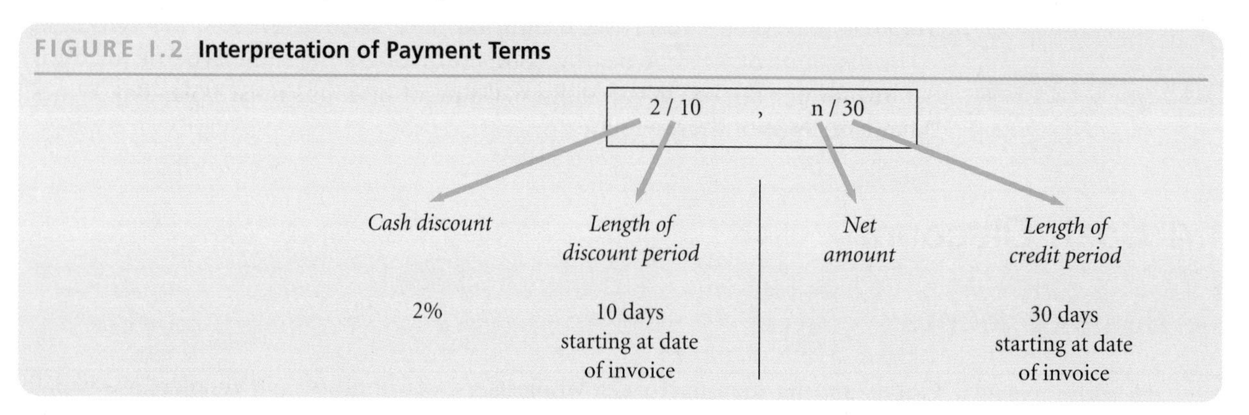

EXAMPLE I.4A Determine the payment needed to settle an invoice of $950.00 dated September 22, terms 2/10, n/30, if the invoice is paid

(i) on October 10; (ii) on October 1.

SOLUTION The terms of the invoice indicate a credit period of 30 days and state that a 2% discount may be deducted from the invoice amount of $950.00 if the invoice is paid within ten days of the invoice date of September 22. The applicable time periods and dates are shown in Figure I.3.

FIGURE I.3 **Discount and Credit Periods—Example I.3A, Ordinary Dating**

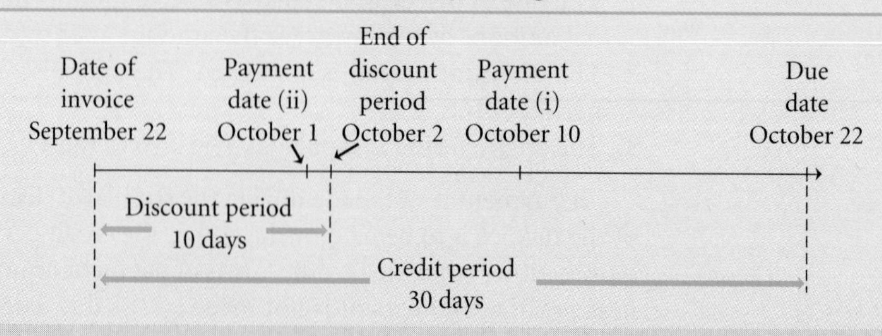

Ten days after September 22 is October 2. The discount period ends October 2.

(i) Payment on October 10 is beyond the last day for taking the discount. The discount cannot be taken. The full amount of the invoice of $950.00 must be paid.

(ii) October 1 is within the discount period; the 2% discount can be taken.

Amount paid = Net amount − 2% of the net amount
= 950.00 − 0.02(950.00)
= 950.00 − 19.00
= $931.00

Alternatively: using the net price factor approach,

Amount paid = NPF for a 2% discount × Net amount
= 0.98(950.00)
= $931.00

EXAMPLE I.4B

An invoice of $2185.65 dated May 31, terms 3/15, n/60, is paid in full on June 15. What is the size of the payment?

SOLUTION

The discount period ends June 15.

Since payment is made on June 15 (the last day for taking the discount), the 3% discount is allowed.

Amount paid = 0.97(2185.65) = $2120.08

EXAMPLE I.4C

An invoice for $752.84 dated March 25, terms 5/10, 2/30, n/60, is paid in full on April 20. What is the total amount paid to settle the account?

SOLUTION

The payment terms state that

(i) a 5% discount may be taken within ten days of the invoice date (up to April 4); or

(ii) a 2% discount may be taken within 30 days of the invoice date (after April 4 but no later than April 24); or

(iii) the net amount is due within 60 days of the invoice date if advantage is not taken of the cash discounts offered.

FIGURE I.4 **Discount and Credit Periods—Example I.4C, Ordinary Dating**

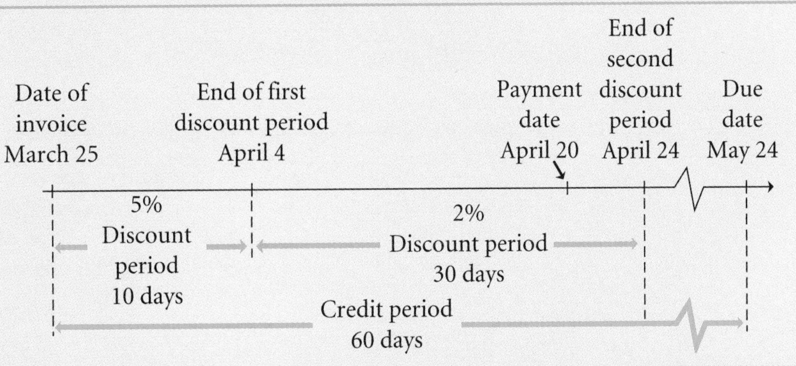

The 5% cash discount is *not* allowed; payment on April 20 is after the end of the discount period for the 5% discount. However, the 2% discount *is* allowed since payment on April 20 is within the 30 day period for the 2% discount.

Amount paid = 0.98(752.84) = $737.78

EXAMPLE I.4D

Three invoices with terms 5/10, 3/20, n/60 are paid on November 15. The invoices are for $645.00 dated September 30, $706.00 dated October 26, and $586.00 dated November 7. What is the total amount paid?

SOLUTION

Invoice Dated	End of Discount Period For 5%	End of Discount Period For 3%	Discount Allowed		Amount Paid
Sept. 30	Oct. 10	Oct. 20	None		$ 645.00
Oct. 26	Nov. 5	Nov. 15	3%	0.97(706.00)	684.82
Nov. 7	Nov. 17	Nov. 27	5%	0.95(586.00)	556.70
				Amount paid	$1886.52

C. End-of-the-month or proximo dating

End-of-the-month dating is shown in the terms of payment by the abbreviation E.O.M. (*end of month*), such as in 2/10, n/30 E.O.M. The abbreviation E.O.M. means that the discount may be taken within the stipulated number of days following the end of the month shown in the invoice date. The abbreviation has the effect of shifting the invoice date to the last day of the month. The abbreviation "prox." (meaning "in the following month") has a similar effect.

Commonly, in end-of-the-month dating, the credit period (such as n/30) is not stated. In our example, "2/10, n/30 E.O.M." would be written "2/10 E.O.M." In this case, it is understood that the end of the credit period (the due date) is *twenty* days after the last day for taking the discount.

EXAMPLE I.4E

An invoice for $1233.95 dated July 16, terms 2/10 E.O.M., is paid on August 10. What is the amount paid?

SOLUTION

The abbreviation E.O.M. means that the invoice is to be treated as if the invoice date were July 31. Therefore, the last day for taking the discount is August 10.

Amount paid = 0.98(1233.95) = $1209.27

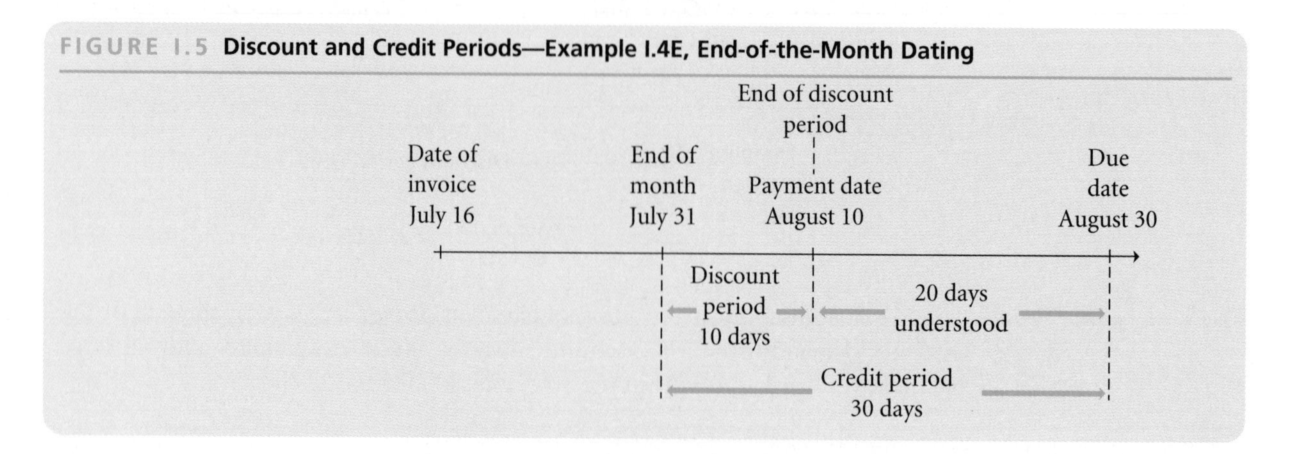

FIGURE I.5 Discount and Credit Periods—Example I.4E, End-of-the-Month Dating

D. Receipt-of-goods dating

When the abbreviation R.O.G. (*receipt of goods*) appears in the terms of payment, as in 2/10, n/30 R.O.G., the last day for taking the discount is the stipulated number of days after the date the merchandise is received rather than the invoice date. This method of offering a cash discount is used when the transportation of the goods takes a long time, as in the case of long-distance overland shipments by rail or truck, or shipments by boat.

EXAMPLE I.4F

Hansa Import Distributors has received an invoice of \$8465.00 dated May 10, terms 3/10, n/30 R.O.G., for a shipment of cuckoo clocks that arrived on July 15. What is the last day for taking the cash discount and how much is to be paid if the discount is taken?

SOLUTION

The last day for taking the discount is ten days after receipt of the shipment, that is, July 25.

Amount paid = 0.97(8465.00) = \$8211.05

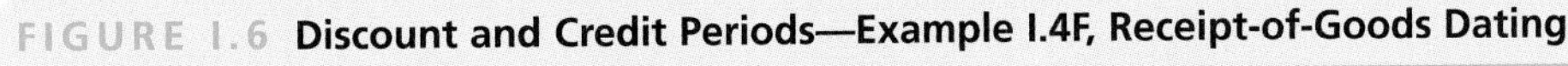

FIGURE I.6 **Discount and Credit Periods—Example I.4F, Receipt-of-Goods Dating**

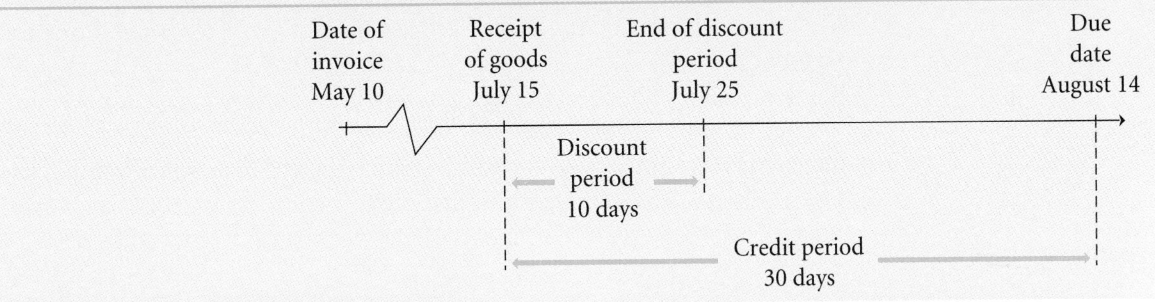

E. Partial payments and additional problems

The problem of a cash discount for a **partial payment** arises when a business pays *part* of an invoice within the discount period. In such cases, the purchaser is entitled to the cash discount on the partial amount paid.

EXAMPLE I.4G

Sheridan Service has received an invoice of \$2780.00 dated August 18, terms 2/10 E.O.M. What payment must be made on September 10 to reduce the debt

(i) by \$1000.00?
(ii) to \$1000.00?

SOLUTION

Since the terms of payment involve end-of-month dating, the last day for taking the cash discount is September 10. The discount of 2% may be taken off the partial payment.

(i) Reducing the debt by \$1000.00 requires paying \$1000.00 less the discount.
Amount paid = 1000.00(0.98) = \$980.00

(ii) Reducing the debt to \$1000.00 requires paying (\$2780.00 − 1000.00), that is, \$1780.00 less the discount.
Amount paid = 1780.00(0.98) = \$1744.40

EXAMPLE I.4H

A cheque for \$1480.22 was received on June 24 in full payment of an invoice dated June 14, terms 3/10, n/30. What was the net amount of the invoice?

SOLUTION

Since the payment was made by the last day of the discount period, the purchaser was entitled to the 3% cash discount. The payment of \$1480.22 is the amount left *after* taking 3% off the net invoice amount.

Let the net invoice amount be x.
Amount paid = NPF × Net amount of invoice

$$1480.22 = 0.97x$$

$$x = \frac{1480.22}{0.97} = \$1526.00$$

The net amount of the invoice was \$1526.00.

EXAMPLE I.4I

Applewood Supplies received a payment of \$807.50 from Sheridan Service on October 10 on an invoice of \$2231.75 dated September 15, terms 5/10 prox.

(i) For how much should Applewood credit Sheridan Service's account for the payment?
(ii) How much does Sheridan Service still owe on the invoice?

SOLUTION

Since the payment terms involve proximo dating, the payment is within the discount period. Sheridan Service is entitled to the 5% discount on the partial payment. The amount of \$807.50 represents a partial payment already reduced by 5%.

Let the credit allowed be x.
Amount paid = NPF × Credit allowed

$$807.50 = 0.95x$$

$$x = \frac{807.50}{0.95} = \$850.00$$

(i) Applewood should credit the account of Sheridan Service with \$850.00.

(ii) Sheridan Service still owes (\$2231.75 − 850.00) = \$1381.75.

EXERCISE I.4

A. Determine the amount paid to settle each of the following eight invoices on the date indicated.

	Invoice Amount	Payment Terms	Date of Invoice	Date Goods Received	Date Paid
1.	$640.00	2/10, n/30	Aug. 10	Aug. 11	Sept. 9
2.	$1520.00	3/15, n/60	Sept. 24	Sept. 27	Oct. 8
3.	$783.95	3/10, 1/20, n/60	May 18	May 20	June 5
4.	$1486.25	5/10, 2/30, n/60	June 28	June 30	July 8
5.	$1160.00	2/10 E.O.M.	Mar. 22	Mar. 29	April 10
6.	$920.00	3/15 E.O.M.	Oct. 20	Oct. 30	Nov. 12
7.	$4675.00	2/10 R.O.G.	April 15	May 28	June 5
8.	$2899.65	4/20 R.O.G.	July 17	Sept. 21	Oct. 10

B. Determine the missing values for each of the following six invoices. Assume that a partial payment was made on each of the invoices by the last day for taking the cash discount.

	Amount of Invoice	Payment Terms	Amount of Credit for Payment	Net Payment Received	Invoice Balance Due
1.	$1450.00	3/10, n/30	$600.00	?	?
2.	$3126.54	2/10 E.O.M.	$2000.00	?	?
3.	$964.50	5/20 R.O.G.	?	?	$400.00
4.	$1789.95	4/15, n/60	?	?	$789.95
5.	$1620.00	3/20 E.O.M.	?	$785.70	?
6.	$2338.36	2/10 R.O.G.	?	$1311.59	?

C. Answer the following questions.

1. Santucci Appliances received an invoice dated August 12 with terms 3/10 E.O.M. for the items listed below:

5 GE refrigerators at $980.00 each less 25%, 5%;
4 Inglis dishwashers at $696.00 each less $16\frac{2}{3}$%, 12.5%, 4%.

a. What is the last day for taking the cash discount?

b. What is the amount due if the invoice is paid on the last day for taking the discount?

c. What is the amount of the cash discount if a partial payment is made such that a balance of $2000.00 remains outstanding on the invoice?

2. Import Exclusives Ltd. received an invoice dated May 20 from Dansk Specialities of Copenhagen with terms 5/20 R.O.G. for:

100 teak trays at $34.30 each;
25 teak icebuckets at $63.60 each;
40 teak salad bowls at $54.50 each.

All items are subject to trade discounts of $33\frac{1}{3}$%, $7\frac{1}{2}$%, 5%.

(a) If the shipment was received on June 28, what is the last day of the discount period?

(b) What is the amount due if the invoice is paid in full on July 15?

(c) If a partial payment only is made on the last day of the discount period, what amount is due to reduce the outstanding balance to $2500.00?

3. What amount must be remitted if invoices dated July 25 for $929.00, August 10 for $763.00, and August 29 for $864.00, all with terms 3/15 E.O.M., are paid together on September 12?

4. The following invoices, all with terms 5/10, 2/30, n/60, were paid together on May 15. Invoice No. 234 dated March 30 is for $394.45; invoice No. 356 dated April 15 is for $595.50; and invoice No. 788 dated May 10 is for $865.20. What amount was remitted?

5. An invoice for $5275.00 dated November 12, terms 4/10 E.O.M., was received on November 14. What payment must be made on December 10 to reduce the debt to $3000.00?

6. What amount will reduce the amount due on an invoice of $1940.00 by $740.00 if the terms of the invoice are 5/10, n/30 and the payment was made during the discount period?

7. Sheridan Service received an invoice dated September 25 from Wolfedale Automotive. The invoice amount was $2540.95, and the payment terms were 3/10, 1/20, n/30. Sheridan Service made a payment on October 5 to reduce the balance due by $1200.00, made a second payment on October 15 to reduce the balance to $600.00, and paid the remaining balance on October 25.

(a) How much did Sheridan Service pay on October 5?

(b) How much did it pay on October 15?

(c) What was the amount of the final payment on October 25?

8. The Ski Shop received an invoice for $9600.00 dated August 11, terms 5/10, 2/30, n/90, for a shipment of skis. The Ski Shop made two partial payments.

(a) How much was paid on August 20 to reduce the unpaid balance to $7000.00?

(b) How much was paid on September 10 to reduce the outstanding balance by $3000.00?

(c) What is the remaining balance on September 10?

9. Jelinek Sports received a cheque for $1867.25 in partial payment of an invoice owed by The Ski Shop. The invoice was for $5325.00 with terms 3/20 E.O.M. dated September 15, and the cheque was received on October 18.

(**a**) By how much should Jelinek Sports credit the account of The Ski Shop?

(**b**) How much does The Ski Shop still owe Jelinek?

10. Darrigo Grape received an invoice for $13 780 dated September 28, terms 5/20 R.O.G., from Nappa Vineyards for a carload of grape juice received October 20. Darrigo made a partial payment of $5966.00 on November 8.

(**a**) By how much did Darrigo reduce the amount due on the invoice?

(**b**) How much does Darrigo still owe?

DID YOU KNOW?

Companies often offer credit terms of "2/10, n/30" to their customers. This means that if payment is made within 10 days of the invoice date, the customer can deduct 2% from the invoice amount. In this case, the discount period is 10 days. Did you know that companies will sometimes extend the discount period? They might do this to match competitors' discount periods or to encourage purchase of seasonal or slow-moving products. The extended discount period would be identified by adding "extra," "ex," or "X" to the regular discount period on the invoice. For example, adding 30 days to the regular discount period would appear as "2/10—30 extra" on the invoice. It would *not* appear as "2/40" because the company wants to indicate that it is granting a special discount period to a particular customer.

I.5 Markup

A. Basic concepts and calculations

The primary purpose of operating a business is to generate profits. Businesses engaged in merchandising generate profits through their buying and selling activities. The amount of profit depends on many factors, one of which is the pricing of goods. The selling price must cover

(i) the cost of buying the goods;

(ii) the operating expenses (or overhead) of the business;

(iii) the profit required by the owner to stay in business.

SELLING PRICE = COST OF BUYING + EXPENSES + PROFIT

$$S = C + E + P$$ — Formula I.8A

EXAMPLE I.5A

Sheridan Service buys a certain type of battery for $84.00 each. Operating expenses of the business are 25% of cost and the owner requires a profit of 10% of cost. For how much should Sheridan sell the batteries?

SOLUTION

Selling price = Cost of buying + Expenses + Profit
= 84.00 + 25% of 84.00 + 10% of 84.00
= 84.00 + 0.25(84.00) + 0.10(84.00)
= 84.00 + 21.00 + 8.40
= $113.40

Sheridan should sell the batteries for $113.40 to cover the cost of buying, the operating expenses, and the required profit.

Note: In Example I.5A, the selling price is $113.40 while the cost is $84.00. The difference between selling price and cost = 113.40 − 84.00 = $29.40. This difference covers operating expenses of $21.00 and a profit of $8.40 and is known as **markup**, **margin**, or **gross profit**.

MARKUP = EXPENSES + PROFIT

$M = E + P$ —— Formula I.9

Using this relationship between markup, expenses, and profit, the relationship stated in Formula I.8A becomes

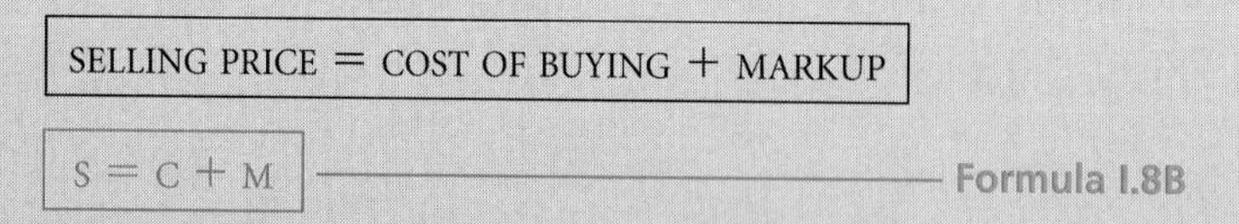

$S = C + M$ —— Formula I.8B

Figure I.7 illustrates the relationships among cost of buying (C), markup (M), operating expenses (E), profit (P), and selling price (S) established in Formulas I.8A, I.8B, and I.9.

FIGURE I.7

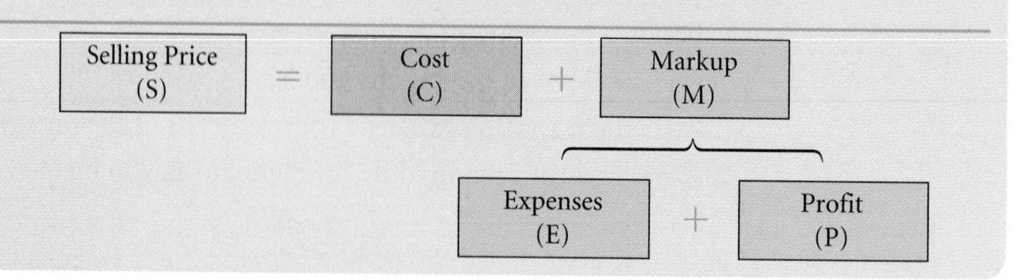

EXAMPLE I.5B

Friden Business Machines bought two types of electronic calculators for resale. Model A cost $42.00 and sells for $56.50. Model B cost $78.00 and sells for $95.00. Business overhead is 24% of cost. For each model, determine

(i) the markup (or gross profit);
(ii) the operating expenses (or overhead);
(iii) the profit.

SOLUTION

	Model A	Model B
(i)	$C + M = S$	$C + M = S$ — using Formula I.8B
	$42.00 + M = 56.50$	$78.00 + M = 95.00$
	$M = 56.50 - 42.00$	$M = 95.00 - 78.00$
	$M = 14.50$	$M = 17.00$
	The markup on Model A is \$14.50.	The markup on Model B is \$17.00.
(ii)	Expenses (or overhead)	Expenses (or overhead)
	$= 24\%$ of 42.00	$= 24\%$ of 78.00
	$= 0.24(42.00)$	$= 0.24(78.00)$
	$= 10.08$	$= 18.72$
	Overhead for Model A is \$10.08.	Overhead for Model B is \$18.72.
(iii)	$E + P = M$	$E + P = M$ — using Formula I.9
	$10.08 + P = 14.50$	$18.72 + P = 17.00$
	$P = 14.50 - 10.08$	$P = 17.00 - 18.72$
	$P = 4.42$	$P = -1.72$
	Profit on Model A is \$4.42.	Profit on Model B is −\$1.72, that is, a loss of \$1.72.

EXAMPLE I.5C

A ski shop bought 100 pairs of skis for \$105.00 per pair and sold 60 pairs for the regular selling price of \$295.00 per pair. The remaining skis were sold during a clearance sale for \$180.00 per pair. Overhead is 40% of the regular selling price. Determine

(i) the markup, the overhead, and the profit per pair of skis sold at the regular selling price;
(ii) the markup, the overhead, and the profit per pair of skis sold during the clearance sale;
(iii) the total profit realized.

SOLUTION

(i) *At regular selling price*	(ii) *At clearance price*
Markup	**Markup**
$C + M = S$	$C + M = S$
$105.00 + M = 295.00$	$105.00 + M = 180.00$
$M = \$190.00$	$M = \$75.00$
Overhead	**Overhead**
$E = 40\%$ of regular selling price	$E = 40\%$ of regular selling price
$= 0.40(295.00)$	$= 0.40(295.00)$
$= \$118.00$	$= \$118.00$
Profit	**Profit**
$E + P = M$	$E + P = M$
$118.00 + P = 190.00$	$118.00 + P = 75.00$
$P = \$72.00$	$P = -\$43.00$

(iii)

Profit from sale of 60 pairs at regular selling price = 60(72.00)	\$4320.00
Profit from sale of 40 pairs during clearance sale = 40(−43.00)	−1720.00
Total profit	\$2600.00

B. Rate of markup

A markup may be stated in one of two ways:

1. As a percent of cost.
2. As a percent of selling price.

The method used is usually determined by the way in which a business keeps its records. Since most manufacturers keep their records in terms of cost, they usually calculate markup as a percent of cost. Since most department stores and other retailers keep their records in terms of selling price, they usually calculate markup as a percent of selling price.

Computing the rate of markup involves comparing the amount of markup to a base amount. Depending on the method used, the base amount is either the cost or the selling price. Since the two methods produce different results, great care must be taken to note whether the markup is based on the cost or on the selling price.

$$\text{RATE OF MARKUP BASED ON COST} = \frac{\text{MARKUP}}{\text{COST}} = \frac{M}{C}$$ ——— **Formula I.10**

$$\text{RATE OF MARKUP BASED ON SELLING PRICE} = \frac{\text{MARKUP}}{\text{SELLING PRICE}} = \frac{M}{S}$$ ——— **Formula I.11**

EXAMPLE I.5D

Compute (a) the missing value (cost, selling price, or markup), (b) the rate of markup based on cost, and (c) the rate of markup based on selling price for each of the following:

(i) cost, \$60.00; selling price, \$75.00
(ii) cost, \$48.00; markup, \$16.00
(iii) selling price, \$88.00; markup, \$33.00
(iv) cost, \$8.00; markup, \$8.00
(v) selling price, \$24.00; markup, \$18.00

SOLUTION

	(a) Missing Value	(b) Rate of Markup Based on Cost	(c) Rate of Markup Based on Selling Price
(i)	Markup = 75.00 − 60.00 = \$15.00	$\frac{15}{60} = 0.25 = 25\%$	$\frac{15}{75} = 0.20 = 20\%$
(ii)	Selling price = 48.00 + 16.00 = \$64.00	$\frac{16}{48} = \frac{1}{3} = 33\frac{1}{3}\%$	$\frac{16}{64} = 0.25 = 25\%$
(iii)	Cost = 88.00 − 33.00 = \$55.00	$\frac{33}{55} = 0.60 = 60\%$	$\frac{33}{88} = 0.375 = 37.5\%$
(iv)	Selling price = 8.00 + 8.00 = \$16.00	$\frac{8}{8} = 1.00 = 100\%$	$\frac{8}{16} = 0.50 = 50\%$
(v)	Cost = 24.00 − 18.00 = \$6.00	$\frac{18}{6} = 3.00 = 300\%$	$\frac{18}{24} = 0.75 = 75\%$

C. Finding the cost or the selling price

When the rate of markup is given and either the cost or the selling price is known, the missing value can be found using Formula I.8B.

COST + MARKUP = SELLING PRICE	C + M = S

When using this formula, pay special attention to the base of the markup, that is, whether it is based on cost or based on selling price.

EXAMPLE I.5E

What is the selling price of an article costing \$72.00 if the markup is
(i) 40% of cost?
(ii) 40% of the selling price?

SOLUTION

(i)

$$C + M = S$$ — using Formula I.8B

$$C + 40\% \text{ of } C = S$$ — replacing M by 40% of C is the crucial step in the solution

$$72.00 + 0.40(72.00) = S$$
$$72.00 + 28.80 = S$$
$$S = 100.80$$

If the markup is 40% based on cost, the selling price is \$100.80.

(ii)

$$C + M = S$$
$$C + 40\% \text{ of } S = S$$
$$72.00 + 0.40S = S$$
$$72.00 = S - 0.40S$$
$$72.00 = 0.60S$$
$$S = \frac{72.00}{0.60}$$
$$S = 120.00$$

If the markup is 40% based on selling price, the selling price is \$120.00.

Note: In problems of this type, replace M by X% of C or X% of S before using specific numbers. This approach is used in the preceding problem and in the following worked examples.

EXAMPLE I.5F

What is the cost of an article selling for \$65.00 if the markup is
(i) 30% of selling price? (ii) 30% of cost?

SOLUTION

(i)

$$C + M = S$$
$$C + 30\% \text{ of } S = S$$ — replace M by 30% of S

$$C + 0.30(65.00) = 65.00$$
$$C + 19.50 = 65.00$$
$$C = 65.00 - 19.50$$
$$C = 45.50$$

If the markup is 30% based on selling price, the cost is \$45.50.

(ii)

$$\begin{aligned} C + M &= S \\ C + 30\% \text{ of } C &= S \quad \text{——— replace M by 30\% of C} \\ C + 0.30C &= 65.00 \\ 1.30C &= 65.00 \\ C &= \frac{65.00}{1.30} \\ C &= 50.00 \end{aligned}$$

If the markup is 30% based on cost, the cost is $50.00.

EXAMPLE I.5G Find the missing value in each of the following:

	(i)	(ii)	(iii)	(iv)	(v)
Cost	$45.00	$84.00	?	?	?
Selling price	?	?	$3.24	$23.10	$42.90
Markup based on cost	$33\frac{1}{3}\%$	?	?	37.5%	50%
Markup based on selling price	?	40%	$16\frac{2}{3}\%$	?	?

SOLUTION

(i)

$$\begin{aligned} C + M &= S \\ C + 33\tfrac{1}{3}\% \text{ of } C &= S \\ 45.00 + \tfrac{1}{3}(45.00) &= S \\ 45.00 + 15.00 &= S \\ S &= 60.00 \end{aligned}$$

$$\text{Markup based on selling price} = \frac{15}{60} = 0.25 = 25\%$$

(ii)

$$\begin{aligned} C + M &= S \\ C + 40\% \text{ of } S &= S \\ 84.00 + 0.40S &= S \\ 84.00 &= 0.60S \\ S &= \frac{84.00}{0.60} \\ S &= 140.00 \end{aligned}$$

$$\text{Markup based on cost} = \frac{140.00 - 84.00}{84.00} = \frac{56.00}{84.00} = 0.666667 = 66\tfrac{2}{3}\%$$

(iii)

$$\begin{aligned} C + M &= S \\ C + 16\tfrac{2}{3}\% \text{ of } S &= S \\ C + \tfrac{1}{6}(3.24) &= 3.24 \\ C + 0.54 &= 3.24 \\ C &= 2.70 \end{aligned}$$

$$\text{Markup based on cost} = \frac{3.24 - 2.70}{2.70} = \frac{0.54}{2.70} = 0.20 = 20\%$$

(iv)
$$C + M = S$$
$$C + 37.5\% \text{ of } C = S$$
$$C + 0.375C = 23.10$$
$$1.375C = 23.10$$
$$C = \frac{23.10}{1.375}$$
$$C = 16.80$$

$$\text{Markup based on selling price} = \frac{23.10 - 16.80}{23.10} = \frac{6.30}{23.10} = 0.272727 = 27.27\%$$

(v)
$$C + M = S$$
$$C + 50\% \text{ of } C = S$$
$$C + 0.50C = 42.90$$
$$1.50C = 42.90$$
$$C = 28.60$$

$$\text{Markup based on selling price} = \frac{42.90 - 28.60}{42.90} = \frac{14.30}{42.90} = 0.333333 = 33\tfrac{1}{3}\%$$

EXAMPLE I.5H

The Beaver Ski Shop sells ski vests for \$98.00. The markup based on cost is 75%.
(i) What did the Beaver Ski Shop pay for each vest?
(ii) What is the rate of markup based on the selling price?

SOLUTION

(i)
$$C + M = S$$
$$C + 75\% \text{ of } C = S$$
$$C + 0.75C = 98.00$$
$$1.75C = 98.00$$
$$C = 56.00$$

The Beaver Ski Shop paid \$56.00 for each vest.

(ii)
$$\text{Rate of markup based on selling price} = \frac{\text{Markup}}{\text{Selling price}}$$
$$= \frac{98.00 - 56.00}{98.00}$$
$$= \frac{42.00}{98.00} = 0.4285714 = 42.86\%$$

EXAMPLE I.5I

Sheridan Service bought four Michelin tires from a wholesaler for $343.00 and sold the tires at a markup of 30% of the selling price.

(i) For how much were the tires sold?

(ii) What is the rate of markup based on cost?

SOLUTION

(i)

$$C + M = S$$
$$C + 30\% \text{ of } S = S$$
$$343.00 + 0.30S = S$$
$$343.00 = 0.70S$$
$$S = 490.00$$

Sheridan sold the tires for $490.00.

(ii)

$$\begin{aligned} \text{Rate of markup based on cost} &= \frac{\text{Markup}}{\text{Cost}} \\ &= \frac{490.00 - 343.00}{343.00} \\ &= \frac{147.00}{343.00} = 0.428571 = 42.86\% \end{aligned}$$

EXAMPLE I.5J

The markup, or gross profit, on each of two articles is $25.80. If the rate of markup for Article A is 40% of cost while the rate of markup for Article B is 40% of the selling price, determine the cost and the selling price of each.

SOLUTION

For Article A:

$$\text{Markup (or gross profit)} = 40\% \text{ of cost}$$
$$25.80 = 0.40C$$
$$C = 64.50$$

The cost of Article A is $64.50.
The selling price is 64.50 + 25.80 = $90.30.

For Article B:

$$\text{Markup (or gross profit)} = 40\% \text{ of selling price}$$
$$25.80 = 0.40S$$
$$S = 64.50$$

The selling price of Article B is $64.50.
The cost = 64.50 − 25.80 = $38.70.

EXERCISE I.5

For each of the following six questions, determine

(a) the amount of markup;
(b) the amount of overhead;
(c) the profit or loss realized on the sale;
(d) the rate of markup based on cost;
(e) the rate of markup based on selling price.

	Cost	Selling Price	Overhead
1.	$24.00	$30.00	16% of cost
2.	$72.00	$96.00	15% of selling price
3.	$52.50	$87.50	36% of selling price
4.	$42.45	$67.92	60% of cost
5.	$27.00	$37.50	34% of selling price
6.	$36.00	$42.30	21% of cost

B. For each of the following twelve questions, compute the missing values represented by the question marks.

				Rate of Markup Based On:	
	Cost	Selling Price	Markup	Cost	Selling Price
1.	$25.00	$31.25	?	?	?
2.	$63.00	$84.00	?	?	?
3.	$64.00	?	$38.40	?	?
4.	?	$162.00	$27.00	?	?
5.	$54.25	?	?	40%	?
6.	?	$94.50	?	?	30%
7.	?	$66.36	?	50%	?
8.	?	$133.25	?	$66\frac{2}{3}\%$	?
9.	$31.24	?	?	?	60%
10.	$87.74	?	?	?	$33\frac{1}{3}\%$
11.	?	?	$22.26	?	$16\frac{2}{3}\%$
12.	?	?	$90.75	125%	?

C. Answer each of the following questions.

1. Guiseppe's buys supplies to make pizzas for $4.00. Operating expenses of the business are 110% of the cost and the profit he makes is 130% of cost. What is the regular selling price of each pizza?

2. Neptune Dive Shop sells snorkelling equipment for $50. Their cost is $25 and their operation expenses are 30% of regular selling price. How much profit will they make on each sale?

3. Mi Casa imports pottery from Mexico. Their operation expenses are 260% of cost of buying and the profit is 110% of cost of buying. They sell a vase for $14.10. What is their cost for each piece?

4. Windsor Hardware buys outdoor lights for $5.00 per dozen less 20%, 20%. The store's overhead is 45% of cost and the required profit is 15% of cost. For how much per dozen should the lights be sold?

5. A merchant buys an item listed at $96.00 less $33\frac{1}{3}$% from a distributor. Overhead is 32% of cost and profit is 27.5% of cost. For how much should the item be retailed?

6. Tennis racquets were purchased for $55.00 less 40% (for purchasing more than 100 items), and less a further 25% (for purchasing the racquets in October). They were sold for $54.45.

(a) What is the markup as a percent of cost?

(b) What is the markup as a percent of selling price?

7. A dealer bought personal computers for $1240.00 less 50%, 10%. They were sold for $1395.00.

(a) What was the markup as a percent of cost?

(b) What was the markup as a percent of selling price?

8. The Bargain Bookstore makes a gross profit of $3.42 on a textbook. The store's markup is 15% of cost.

(a) For how much did the bookstore buy the textbook?

(b) What is the selling price of the textbook?

(c) What is the rate of markup based on the selling price?

9. An appliance store sells electric kettles at a markup of 18% of the selling price. The store's margin on a particular model is $6.57.

(a) For how much does the store sell the kettles?

(b) What was the cost of the kettles to the store?

(c) What is the rate of markup based on cost?

10. The markup on an item selling for $74.55 is 40% of cost.

(a) What is the cost of the item?

(b) What is the rate of markup based on the selling price?

11. Sheridan Service sells oil at a markup of 40% of the selling price. If Sheridan paid $0.99 per litre of oil,

(a) what is the selling price per litre?

(b) what is the rate of markup based on cost?

12. The Ski Shop purchased ski poles for $12.80 per pair. The poles are marked up 60% of the selling price.

(a) For how much does The Ski Shop sell a pair of ski poles?

(b) What is the rate of markup based on cost?

13. Neal's Photographic Supplies sells a Pentax camera for $444.98. The markup is 90% of cost.

(a) How much does the store pay for this camera?

(b) What is the rate of markup based on selling price?

I.6 Markdown

A. Pricing strategies

The pricing relationship, Selling price = Cost + Expense + Profit (S = C + E + P), plays an important role in pricing. It describes how large a markup is needed to cover overhead and a reasonable profit, and how large a markdown can be tolerated. One pricing strategy is to set a selling price based on the business' "internal" factors—actual costs and expenses, and a desired profit level. However, pricing decisions must often be based on "external" market factors—competitors' prices, consumers' sensitivity to a high price, economic conditions that affect interest rates and income available for purchases, and so on. Often, selling prices must be marked down more than anticipated in response to market conditions, leading to less-than-desired profit levels. Use the pricing relationship S = C + E + P as a guide to determine the effect of markdown decisions on the operations of a business.

The cost of buying an article plus the overhead represents the **total cost**, or handling cost, of the article.

TOTAL COST = COST OF BUYING + EXPENSES

If an article is sold at a price that *equals* the total cost, the business makes no profit nor does it suffer a loss. This price is called the break-even point and is discussed in Chapter 5. Any business, of course, prefers to sell at least at a break-even price. If the price is insufficient to recover the total cost, the business will suffer an operating loss. If the price does not even cover the cost of buying, the business suffers an absolute loss. To determine the profit or loss, the following accounting relationship is used.

PROFIT = SELLING PRICE − TOTAL COST

PROFIT = REVENUE − TOTAL COST

B. Basic concepts and calculations

A **markdown** is a reduction in the price of an article sold to the consumer. Markdowns are used for a variety of purposes, such as sales promotions, meeting competitors' prices, reducing excess inventories, clearing out seasonal merchandise, and selling off discontinued items. A markdown, unlike a markup, is always stated as a percent of the price to be reduced and is computed as if it were a discount.

FIGURE I.8

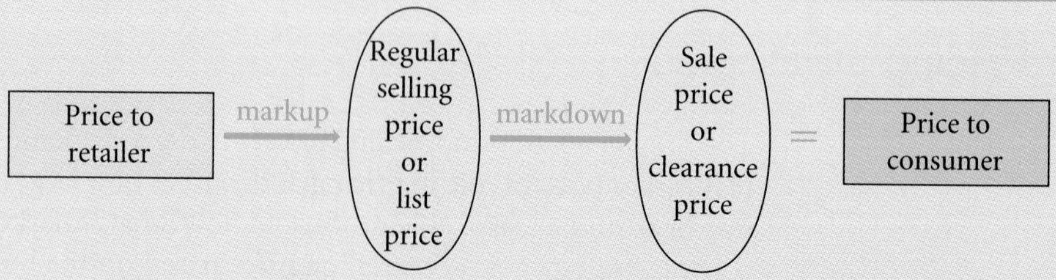

While markdowns are simple to calculate, the rather wide variety of terms used to identify both the price to be reduced (such as *regular selling price, selling price, list price, marked price, price tag*) and the reduced price (such as *sale price* or *clearance price*) introduces an element of confusion. In this text, we use **regular selling price** to describe the price to be reduced and **sale price** to describe the reduced price.

In general,

$$\text{SALE PRICE} = \text{REGULAR SELLING PRICE} - \text{MARKDOWN}$$

However, since the markdown is a percent of the regular selling price, the net price factor approach used with discounts is applicable (see Formula I.4A).

$$\text{SALE PRICE} = \text{NPF} \times \text{REGULAR SELLING PRICE}$$
$$\text{where NPF} = 100\% - \%\ \text{markdown}$$

EXAMPLE I.6A

The Cook Nook paid $115.24 for a set of dishes. Expenses are 18% of selling price and the required profit is 15% of selling price. During an inventory sale, the set of dishes was marked down 30%.

(i) What was the regular selling price?
(ii) What was the sale price?
(iii) What was the operating profit or loss?

SOLUTION

(i) Selling price = Cost + Expenses + Profit

$$S = C + 18\% \text{ of } S + 15\% \text{ of } S$$
$$S = C + 0.18S + 0.15S$$
$$S = 115.24 + 0.33S$$
$$0.67S = 115.24$$
$$S = \frac{115.24}{0.67} = \$172.00$$

The regular selling price is $172.00

(ii) Sale price = Selling price − Markdown

$$= S - 30\% \text{ of } S$$
$$= S - 0.30S$$
$$= 0.70S$$
$$= 0.70(172.00)$$
$$= \$120.40$$

The sale price is $120.40.

$$
\begin{aligned}
\text{(iii) Total cost} &= \text{Cost of buying} + \text{Expenses} \\
&= C + 18\% \text{ of } S \\
&= 115.24 + 0.18(172.00) \\
&= 115.24 + 30.96 \\
&= \$146.20
\end{aligned}
$$

$$
\begin{aligned}
\text{Profit} &= \text{Revenue} - \text{Total cost} \\
&= 120.40 - 146.20 \\
&= -\$25.80
\end{aligned}
$$

The dishes were sold at an operating loss of $25.80.

EXAMPLE I.6B

Solomon 555 ski bindings purchased for $57.75 were marked up 45% of the selling price. When the binding was discontinued, it was marked down 40%. What was the sale price of the binding?

SOLUTION

First determine the regular selling price.

$$
\begin{aligned}
C + M &= S \\
C + 45\% \text{ of } S &= S \\
57.75 + 0.45S &= S \\
57.75 &= 0.55S \\
S &= \$105.00
\end{aligned}
$$

The regular selling price is $105.00.

$$
\begin{aligned}
\text{Sale price} &= \text{Regular selling price} - \text{Markdown} \\
&= 105.00 - 40\% \text{ of } 105.00 \\
&= 105.00 - 42.00 \\
&= \$63.00
\end{aligned}
$$

Alternatively:

$$
\begin{aligned}
\text{Sale price} &= \text{NPF} \times \text{Regular selling price} \\
&= 0.60 \times 105.00 \\
&= \$63.00
\end{aligned}
$$

The sale price is $63.00.

EXAMPLE I.6C

A ski suit with a regular selling price of $280.00 is marked down 45% during Beaver Ski Shop's annual clearance sale. What is the sale price of the ski suit?

SOLUTION

$$
\begin{aligned}
\text{Sale price} &= \text{Regular selling price} - \text{Markdown} \\
&= 280.00 - 45\% \text{ of } 280.00 \\
&= 280.00 - 0.45(280.00) \\
&= 280.00 - 126.00 \\
&= \$154.00
\end{aligned}
$$

Alternatively:

$$
\begin{aligned}
\text{Sale price} &= \text{NPF} \times \text{Regular selling price} \\
&= (100\% - 45\%)(280.00) \\
&= 0.55(280.00) \\
&= \$154.00
\end{aligned}
$$

The ski suit sold for $154.00.

EXAMPLE I.6D

Lund Sporting Goods sold a bicycle regularly priced at \$195.00 for \$144.30.

(i) What is the amount of markdown?

(ii) What is the rate of markdown?

SOLUTION

(i) Markdown = Regular selling price − Sale price

$= 195.00 - 144.30$

$= \$50.70$

(ii) $\text{Rate of markdown} = \dfrac{\text{Markdown}}{\text{Regular selling price}}$

$= \dfrac{50.70}{195.00} = 0.26 = 26\%$

EXAMPLE I.6E

The Winemaker sells California concentrate for \$22.50. The store's overhead expenses are 50% of cost and the owners require a profit of 30% of cost.

(i) For how much does The Winemaker buy the concentrate?

(ii) What is the break-even price?

(iii) What is the highest rate of markdown at which the store will still break even?

(iv) What is the highest rate of discount that can be advertised without incurring an absolute loss?

SOLUTION

(i)

$$S = C + E + P$$

$$S = C + 50\%\text{ of }C + 30\%\text{ of }C$$

$$S = C + 0.50C + 0.30C$$

$$22.50 = 1.80C$$

$$C = \frac{22.50}{1.80} = \$12.50$$

The Winemaker bought the concentrate for \$12.50.

(ii) Total cost = C + 50% of C

$= 1.50C$

$= 1.50(12.50)$

$= \$18.75$

To break even, the concentrate must be sold for \$18.75.

(iii) To break even, the maximum markdown is 22.50 − 18.75 = \$3.75.

$\text{Rate of markdown} = \dfrac{3.75}{22.50} = 0.166667 = 16\tfrac{2}{3}\%$

The highest rate of markdown to break even is $16\tfrac{2}{3}\%$.

(iv) The lowest price at which the concentrate can be offered for sale without incurring an absolute loss is the cost at which the concentrate was purchased, that is, \$12.50. The maximum amount of discount is 22.50 − 12.50 = \$10.00.

$\text{Rate of discount} = \dfrac{10.00}{22.50} = 0.444444 = 44\tfrac{4}{9}\%$

The maximum rate of discount that can be advertised without incurring an absolute loss is $44\tfrac{4}{9}\%$.

C. Integrated problems

EXAMPLE I.6F

During its annual Midnight Madness Sale, The Ski Shop sold a pair of ski boots, regularly priced at \$245.00, at a discount of 40%. The boots cost \$96.00 and expenses are 26% of the regular selling price.

(i) For how much were the ski boots sold?
(ii) What was the total cost of the ski boots?
(iii) What operating profit or loss was made on the sale?

SOLUTION

(i) Sale price $= 0.60(245.00) = \$147.00$

(ii) Total cost = Cost of buying + Expenses

$$= 96.00 + 0.26(245.00)$$
$$= 96.00 + 63.70$$
$$= \$159.70$$

(iii) Profit = Revenue − Total cost

$$= 147.00 - 159.70$$
$$= -\$12.70 \text{ (a loss)}$$

Since the total cost was higher than the revenue received from the sale of the ski boots, The Ski Shop had an operating loss of \$12.70.

EXAMPLE I.6G

Big Sound Electronics bought stereo equipment at a cost of \$960.00 less 30%, 15%. Big Sound marks all merchandise at a list price that allows the store to offer a discount of 20% while still making its usual profit of 15% of regular selling price. Overhead is 25% of regular selling price. During its annual midsummer sale, the usual discount of 20% was replaced by a markdown of 45%. What operating profit or loss was made when the equipment was sold?

SOLUTION

Complex problems of this type are best solved by a systematic approach. Consider the given information step by step. The following computations are necessary to determine the profit.

(i) the cost (or purchase price) to the store;
(ii) the regular selling price required to cover cost, expenses, and the usual profit;
(iii) the list price from which the 20% discount is offered;
(iv) the midsummer sale price;
(v) the total cost (cost and expenses);
(vi) the operating profit or loss.

Step-by-Step Computations

(i) Cost = NPF × List price $= (0.70)(0.85)(960.00) = \$571.20$

(ii) Let the regular selling price be S.

$$S = C + E + P$$
$$S = C + 25\% \text{ of } S + 15\% \text{ of } S$$
$$S = C + 0.25S + 0.15S$$
$$S = 571.20 + 0.40S$$

$$0.60S = 571.20$$
$$S = \frac{571.20}{0.60} = \$952.00$$

The regular selling price is \$952.00.

(iii) List price − Discount = Regular selling price
Let the List price be L.

$$L - 20\% \text{ of } L = 952.00$$
$$L - 0.20L = 952.00$$
$$0.80L = 952.00$$
$$L = \frac{952.00}{0.80} = \$1190.00$$

The list price was \$1190.00.

(iv) Midsummer sale price = List price − Markdown

$$= 1190.00 - 45\% \text{ of } 1190.00$$
$$= 1190.00 - 0.45(1190.00)$$
$$= 1190.00 - 535.50$$
$$= \$654.50$$

(v) Total cost = Cost of buying + Expenses

$$= C + 25\% \text{ of } S$$
$$= 571.20 + 0.25(952.00)$$
$$= 571.20 + 238.00$$
$$= \$809.20$$

(vi) Profit = Sale price − Total cost

$$= 654.50 - 809.20$$
$$= -\$154.70$$

The equipment was sold at an operating loss of \$154.70.

EXAMPLE I.6H

Magder's Furniture Emporium bought a dining room suite that must be retailed for \$5250.00 to cover the cost, overhead expenses of 50% of the cost, and a normal net profit of 25% of the cost. The suite is marked at a list price so that the store can allow a 20% discount and still receive the required regular selling price.

When the suite remained unsold, the store owner decided to mark the suite down for an inventory clearance sale. To arrive at the rate of markdown, the owner decided that the store's profit would have to be no less than 10% of the normal net profit and that part of the markdown would be covered by reducing the commission paid to the salesperson. The normal commission (which accounts for 40% of the overhead) was reduced by $33\frac{1}{3}$%.

What is the maximum rate of markdown that can be advertised instead of the usual 20%?

SOLUTION

STEP 1 Determine the cost.
Let the regular selling price be S.

$$S = C + E + P$$
$$S = C + 50\% \text{ of } C + 25\% \text{ of } C$$
$$S = C + 0.50C + 0.25C$$

$$5250.00 = 1.75C$$

$$C = \frac{5250.00}{1.75} = \$3000.00$$

STEP 2 Determine the list price.

Let the list price be \$L.

List price − Discount = Regular selling price

$$L - 20\% \text{ of } L = 5250.00$$
$$L - 0.20L = 5250.00$$
$$0.80L = 5250.00$$
$$L = \frac{5250.00}{0.80} = \$6562.50$$

STEP 3 Determine the required profit.

$$\text{Normal net profit} = 25\% \text{ of cost} = 0.25(3000.00) = \$750.00$$

$$\text{Required net profit} = 10\% \text{ of normal net profit} = 0.10(750.00) = \$75.00$$

STEP 4 Determine the amount of overhead expense to be recovered.

$$\text{Normal overhead expense} = 50\% \text{ of cost} = 0.50(3000.00) = \$1500.00$$

$$\text{Normal commission} = 40\% \text{ of normal overhead expense} = 0.40(1500.00) = \$600.00$$

$$\text{Reduction in commission} = 33\tfrac{1}{3}\% \text{ of normal commission} = \tfrac{1}{3}(600.00) = \$200.00$$

$$\text{Overhead expense to be recovered} = 1500.00 - 200.00 = \$1300.00$$

STEP 5 Determine the inventory clearance price.

$$\text{Inventory clearance price} = \text{Cost} + \text{Overhead} + \text{Profit} = 3000.00 + 1300.00 + 75.00 = \$4375.00$$

STEP 6 Determine the amount of markdown.

$$\text{Markdown} = \text{List price} - \text{Inventory clearance price} = 6562.50 - 4375.00 = \$2187.50$$

STEP 7 Determine the rate of markdown.

$$\text{Rate of markdown} = \frac{\text{Amount of markdown}}{\text{List price}} = \frac{2187.50}{6562.50} = 0.333333 = 33\tfrac{1}{3}\%$$

Instead of the usual 20%, the store can advertise a markdown of $33\tfrac{1}{3}\%$.

EXERCISE I.6

A. Compute the values represented by question marks for each of the following 6 questions.

	Price	Markdown	Sale Price (S)	Cost (C)	Overhead	Total Cost	Operating Profit (Loss)
1.	$85.00	40%	?	$42.00	20% of S	?	?
2.	?	$33\frac{1}{3}\%$	$42.00	$34.44	12% of S	?	?
3.	?	35%	$62.66	?	25% of S	$54.75	?
4.	$72.80	$12\frac{1}{2}\%$	?	$54.75	20% of C	?	?
5.	?	25%	$120.00	$105.00	? of S	?	($4.20)
6.	$92.40	$16\frac{2}{3}\%$	?	?	15% of C	?	$8.46

Answer each of the following questions.

1. A cookware set that cost a dealer $440.00 less 55%, 25% is marked up to 180% of cost. For quick sale, the cookware was reduced 45%.
   (a) What is the sale price?
   (b) What rate of markup based on cost was realized?

2. A gas barbecue cost a retailer $420.00 less $33\frac{1}{3}\%$, 20%, 5%. It carries a regular selling price on its price tag at a markup of 60% of the regular selling price. During the end-of-season sale, the barbecue is marked down 45%.
   (a) What is the end-of-season sale price?
   (b) What rate of markup based on cost will be realized during the sale?

3. The Stereo Shop sold a radio regularly priced at $125.00 for $75.00. The cost of the radio was $120.00 less $33\frac{1}{3}\%$, 15%. The store's overhead expense is 12% of the regular selling price.
   (a) What was the rate of markdown at which the radio was sold?
   (b) What was the operating profit or loss?
   (c) What rate of markup based on cost was realized?
   (d) What was the rate of markup based on the sale price?

4. An automatic dishwasher cost a dealer $620.00 less $37\frac{1}{2}\%$, 4%. It is regularly priced at $558.00. The dealer's overhead expense is 15% of the regular selling price and the dishwasher was cleared out for $432.45.
   (a) What was the rate of markdown at which the dishwasher was sold?
   (b) What is the regular markup based on selling price?
   (c) What was the operating profit or loss?
   (d) What rate of markup based on cost was realized?

5. A hardware store paid $33.45 for a set of cookware. Overhead expense is 15% of the regular selling price and profit is 10% of the regular selling price. During a clearance sale, the set was sold at a markdown of 15%. What was the operating profit or loss on the sale?

6. Aldo's Shoes bought a shipment of 200 pairs of women's shoes for $42.00 per pair. The store sold 120 pairs at the regular selling price of $125.00 per pair, 60 pairs at a clearance sale at a discount of 40%, and the remaining pairs during an inventory sale at a price that equals cost plus overhead (i.e., a break-even price). The store's overhead is 50% of cost.

(a) What was the price at which the shoes were sold during the clearance sale?
(b) What was the selling price during the inventory sale?
(c) What was the total profit realized on the shipment?
(d) What was the average rate of markup based on cost that was realized on the shipment?

7. The Pottery bought 600 pans auctioned off en bloc for $4950.00. This means that each pan has the same cost. On inspection, the pans were classified as normal quality, seconds, and substandard. The 360 normal-quality pans were sold at a markup of 80% of cost, the 190 pans classified as seconds were sold at a markup of 20% of cost, and the pans classified as substandard were sold at 80% of their cost.

(a) What was the unit price at which each of the three classifications was sold?
(b) If overhead is $33\frac{1}{3}$% of cost, what was the amount of profit realized on the purchase?
(c) What was the average rate of markup based on the selling price at which the pans were sold?

8. A clothing store buys shorts for $24.00 less 40% for buying over 50 pairs, and less a further $16\frac{2}{3}$% for buying last season's style. The shorts are marked up to cover overhead expenses of 25% of cost and a profit of $33\frac{1}{3}$% of cost.

(a) What is the regular selling price of the shorts?
(b) What is the maximum amount of markdown to break even?
(c) What is the rate of markdown if the shorts are sold at the break-even price?

9. Furniture City bought chairs for $75.00 less $33\frac{1}{3}$%, 20%, 10%. The store's overhead is 75% of cost and net profit is 25% of cost.

(a) What is the regular selling price of the chairs?
(b) At what price can the chairs be put on sale so that the store incurs an operating loss of no more than $33\frac{1}{3}$% of the overhead?
(c) What is the maximum rate of markdown at which the chairs can be offered for sale in part (b)?

10. Bargain City clothing store purchased raincoats for $36.75. The store requires a gross profit of 30% of the sale price. What regular selling price should be marked on the raincoats if the store wants to offer a 25% discount without reducing its gross profit?

11. A jewellery store paid $36.40 for a watch. Store expenses are 24% of regular selling price and the normal net profit is 20% of regular selling price. During a Special Bargain Day Sale, the watch was sold at a discount of 30%. What operating profit or loss was realized on the sale?

12. The Outdoor Shop buys tents for \$264.00 less 25% for buying more than 20 tents. The store operates on a markup of 33⅓% of the sale price and advertises that all merchandise is sold at a discount of 20% of the regular selling price. What is the regular selling price of the tents?

13. The Blast bought stereo equipment listed at \$900.00 less 60%, 16⅔%. Expenses are 45% of the regular selling price and net profit is 15% of the regular selling price. The store decided to change the regular selling price so that it could advertise a 37.5% discount while still maintaining its usual markup. During the annual inventory sale, the unsold equipment was marked down 55% of the new regular selling price. What operating profit or loss was realized on the equipment sold during the sale?

14. Lund's Pro Shop purchased sets of golf clubs for \$500.00 less 40%, 16⅔%. Expenses are 20% of the regular selling price and the required profit is 17.5% of the regular selling price. The store decided to change the regular selling price so that it could offer a 36% discount without affecting its margin. At the end of the season, the unsold sets were advertised at a discount of 54% of the new regular selling price. What operating profit or loss was realized on the sets sold at the end of the season?

15. Big Boy Appliances bought self-cleaning ovens for \$900.00 less 33⅓%, 5%. Expenses are 15% of the regular selling price and profit is 9% of the regular selling price. For competitive reasons, the store marks all merchandise with a new regular selling price so that a discount of 25% can be advertised without affecting the margin. To promote sales, the ovens were marked down 40%. What operating profit or loss did the store make on the ovens sold during the sales promotion?

16. Blue Lake Marina sells a make of cruiser for \$16 800.00. This regular selling price covers overhead of 15% of cost and a normal net profit of 10% of cost. The cruisers were marked with a new regular selling price so that the marina can offer a 20% discount while still maintaining its regular gross profit. At the end of the boating season, the cruiser was marked down. The marina made 25% of its usual profit and reduced the usual commission paid to the sales personnel by 33⅓%. The normal commission accounts for 50% of the normal overhead. What was the rate of markdown?

Review Exercise

1. A toolbox is listed for $56.00 less 25%, 20%, 5%.

(a) What is the net price of the toolbox?

(b) What is the amount of discount?

(c) What is the single rate of discount that was allowed?

2. Compute the rate of discount allowed on a lawnmower that lists for $168.00 and is sold for $105.00.

3. Determine the single rate of discount equivalent to the discount series 35%, 12%, 5%.

4. A 40% discount allowed on an article amounts to $1.44. What is the net price?

5. Baton Construction Supplies has been selling wheelbarrows for $112.00 less 15%. What additional discount percent must the company offer to meet a competitor's price of $80.92?

6. A freezer was sold during a clearance sale for $387.50. If the freezer was sold at a discount of $16\frac{2}{3}$%, what was the list price?

7. The net price of a snow shovel is $20.40 after discounts of 20%, 15%. What is the list price?

8. On May 18, an invoice dated May 17 for $4000.00 less 20%, 15%, terms 5/10 E.O.M., was received by Aldo Distributors.

(a) What is the last day of the discount period?

(b) What is the amount due if the invoice is paid within the discount period?

9. Air Yukon received a shipment of plastic trays on September 2. The invoice amounting to $25 630 was dated August 15, terms 2/10, n/30 R.O.G. What is the last day for taking the cash discount and how much is to be paid if the discount is taken?

10. What amount must be remitted if the following invoices, all with terms 5/10, 2/30, n/60, are paid together on December 8?

Invoice No. 312 dated November 2 for $923.00

Invoice No. 429 dated November 14 for $784.00

Invoice No. 563 dated November 30 for $873.00

11. Delta Furnishings received an invoice dated May 10 for a shipment of goods received June 21. The invoice was for $8400.00 less $33\frac{1}{3}$%, $12\frac{1}{2}$% with terms 3/20 R.O.G. How much must Delta pay on July 9 to reduce its debt

(a) by $2000.00?

(b) to $2000.00?

12. The Peel Trading Company received an invoice dated September 20 for $16 000.00 less 25%, 20%, terms 5/10, 2/30, n/60. Peel made a payment on September 30 to reduce the debt to $5000.00 and a payment on October 20 to reduce the debt by $3000.00.

(a) What amount must Peel remit to pay the balance of the debt at the end of the credit period?

(b) What is the total amount paid by Peel?

13. Emco Ltd. received an invoice dated May 5 for $4000.00 less 15%, $7\frac{1}{2}$%, terms 3/15 E.O.M. A cheque for $1595.65 was mailed by Emco on June 15 as part payment of the invoice.

(a) By how much did Emco reduce the amount due on the invoice?

(b) How much does Emco still owe?

14. Homeward Hardware buys cat litter for $6.00 less 20% per bag. The store's overhead is 45% of cost and the owner requires a profit of 20% of cost.

(a) For how much should the bags be sold?

(b) What is the amount of markup included in the selling price?

(c) What is the rate of markup based on selling price?

(d) What is the rate of markup based on cost?

(e) What is the break-even price?

(f) What operating profit or loss is made if a bag is sold for $6.00?

15. A retail store realizes a gross profit of $31.50 if it sells an article at a margin of 35% of the selling price.

(a) What is the regular selling price?

(b) What is the cost?

(c) What is the rate of markup based on cost?

(d) If overhead expense is 28% of cost, what is the break-even price?

(e) If the article is sold at a markdown of 24%, what is the operating profit or loss?

16. Using a markup of 35% of cost, a store priced a book at $8.91.

(a) What was the cost of the book?

(b) What is the markup as a percent of selling price?

17. A bicycle helmet costing $54.25 was marked up to realize a gross profit of 30% of the regular selling price.

(a) What was the regular selling price?

(b) What was the gross profit as a percent of cost?

18. A bedroom suite that cost a dealer $1800.00 less 37.5%, 18% carries a price tag with a regular selling price at a markup of 120% of cost. For quick sale, the bedroom suite was marked down 40%.

(a) What was the sale price?

(b) What rate of markup based on cost was realized?

19. Gino's purchased men's suits for $195.00 less $33\frac{1}{3}$%. The store operates at a normal gross profit of 35% of regular selling price. The owner marks all merchandise with new regular selling prices so that the store can offer a $16\frac{2}{3}$% discount while maintaining the same gross profit. What is the new regular selling price?

20. An appliance store sold GE coffeemakers for $22.95 during a promotional sale. The store bought the coffee makers for $36.00 less 40%, 15%. Overhead is 25% of the regular selling price.

(a) If the store's markup is 40% of the regular selling price, what was the rate of markdown?

(b) What operating profit or loss was made during the sale?

(c) What rate of markup based on cost was realized?

21. Billington's buys shirts for $21.00 less 25%, 20%. The shirts are priced at a regular selling price to cover expenses of 20% of regular selling price and a profit of 17% of regular selling price. For a special weekend sale, shirts were marked down 20%.

(a) What was the operating profit or loss on the shirts sold during the weekend sale?

(b) What rate of markup was realized based on cost?

22. A jewellery store paid a unit price of $250.00 less 40%, $16\frac{2}{3}$%, 8% for a shipment of designer watches. The store's overhead is 65% of cost and the normal profit is 55% of cost.

(a) What is the regular selling price of the watches?

(b) What must the sale price be for the store to break even?

(c) What is the rate of markdown to sell the watches at the break-even price?

23. Sight and Sound bought large-screen colour TV sets for $1080.00 less $33\frac{1}{3}$%, $8\frac{1}{3}$%. Overhead is 18% of regular selling price and required profit is $15\frac{1}{3}$% of regular selling price. The TV sets were marked at a new regular selling price so that the store was able to advertise a discount of 25% while still maintaining its margin. To clear the inventory, the remaining TV sets were marked down $37\frac{1}{2}$%.

(a) What operating profit or loss is realized at the clearance price?

(b) What is the realized rate of markup based on cost?

24. Ward Machinery lists a log splitter at $1860.00 less $33\frac{1}{3}$%, 15%. To meet competition, Ward wants to reduce its net price to $922.25. What additional percent discount must Ward allow?

25. West End Appliances bought bread makers for $180.00 less 40%, $16\frac{2}{3}$%, 10%. The store's overhead is 45% of regular selling price and the profit required is $21\frac{1}{4}$% of regular selling price.

(**a**) What is the break-even price?

(**b**) What is the maximum rate of markdown that the store can offer to break even?

(**c**) What is the realized rate of markup based on cost if the bread makers are sold at the break-even price?

26. A merchant realizes a markup of $42.00 by selling an item at a markup of 37.5% of cost.

(**a**) What is the regular selling price?

(**b**) What is the rate of markup based on the regular selling price?

(**c**) If the merchant's expenses are 17.5% of the regular selling price, what is the break-even price?

(**d**) If the article is reduced for sale to $121.66, what is the rate of markdown?

27. The Knit Shoppe bought 250 sweaters for $3100.00; 50 sweaters were sold at a markup of 150% of cost and 120 sweaters at a markup of 75% of cost; 60 of the sweaters were sold during a clearance sale for $15.00 each and the remaining sweaters were disposed of at 20% below cost. Assume all sweaters had the same cost.

(**a**) What was the amount of markup realized on the purchase?

(**b**) What was the percent markup realized based on cost?

(**c**) What was the gross profit realized based on selling price?

Self-Test

1. Determine the net price of an article listed at $590.00 less 37.5%, 12.5%, $8\frac{1}{3}$%.

2. What rate of discount has been allowed if an item that lists for $270.00 is sold for $168.75?

3. Compute the single discount percent equivalent to the discount series 40%, 10%, $8\frac{1}{3}$%.

4. Discount Electronics lists an article for $1020.00 less 25% and 15%. A competitor carries the same article for $927.00 less 25%. What further discount (correct to the nearest $\frac{1}{10}$ of 1%) must the competitor allow so that its net price is the same as Discount's?

5. What amount must be remitted if the following invoices, all with terms 4/10, 2/30, n/60, are paid on May 10?

$850.00 less 20%, 10% dated March 21
$960.00 less 30%, $16\frac{2}{3}$% dated April 10
$1040.00 less $33\frac{1}{3}$%, 25%, 5% dated April 30

6. An invoice for $3200.00, dated March 20, terms 3/10 E.O.M., was received March 23. What payment must be made on April 10 to reduce the debt to $1200.00?

7. On January 15, Sheridan Service received an invoice dated January 14, terms 4/10 E.O.M., for $2592.00. On February 9, Sheridan Service mailed a cheque for $1392.00 in partial payment of the invoice. By how much did Sheridan Service reduce its debt?

8. What is the regular selling price of an item purchased for $1270.00 if the markup is 20% of the regular selling price?

9. The regular selling price of merchandise sold in a store includes a markup of 40% based on the regular selling price. During a sale, an item that cost the store $180.00 was marked down 20%. For how much was the item sold?

10. The net price of an article is $727.20 after discounts of 20% and 10% have been allowed. What was the list price?

11. An item that cost the dealer $350 less 35%, 12.5% carries a regular selling price on the tag at a markup of 150% of cost. For quick sale, the item was reduced 30%. What was the sale price?

12. Find the cost of an item sold for $1904.00 to realize a markup of 40% based on cost.

13. An article cost $900.00 and sold for $2520.00. What was the percent markup based on cost?

14. A gross profit of $90.00 is made on a sale. If the gross profit was 45% based on selling price, what was the cost?

15. An appliance shop reduces the price of an appliance for quick sale from $1560.00 to $1195.00. Compute the markdown correct to the nearest $\frac{1}{100}$ of 1%.

16. An invoice shows a net price of $552.44 after discounts of $33\frac{1}{3}$%, 20%, $8\frac{1}{3}$%. What was the list price?

17. A retailer buys an appliance for $1480.00 less 25%, 15%. The store marks the merchandise at a regular selling price to cover expenses of 40% of the regular selling price and a net profit of 10% of the regular selling price. During a clearance sale, the appliance was sold at a markdown of 45%. What was the operating profit or loss?

18. Discount Electronics buys stereos for $830.00 less 37.5%, 12.5%. Expenses are 20% of the regular selling price and the required profit is 15% of the regular selling price. All merchandise is marked with a new regular selling price so that the store can advertise a discount of 30% while still maintaining its regular markup. During the annual clearance sale, the new regular selling price of unsold items is marked down 50%. What operating profit or loss does the store make on items sold during the sale?

Challenge Problems

1. Rose Bowl Florists buys and sells roses only by the complete dozen. The owner buys 12 dozen fresh roses daily for $117. He knows that 10% of the roses will wilt before they can be sold. What price per dozen must Rose Bowl Florists charge for its saleable roses to realize a 55% markup based on selling price?
2. A merchant bought some goods at a discount of 25% of the list price. She wants to mark them at a regular selling price so that she can give a discount of 20% of the regular selling price and still make a markup of 25% of the sale price.
   a. At what percent of the list price should she mark the regular selling price of the goods?
   b. Suppose the merchant decides she must make a markup of 25% of the cost price. At what percent of the list price should she mark the regular selling price of the goods?
3. On April 13, a stereo store received a new sound system with a list price of $2500 from the manufacturer. The stereo store received a trade discount of 25%. The invoice, with terms 2/10, n/30, arrived on the same day as the sound system. The owner of the store marked up the sound system by 60% of the invoice amount (before cash discount) to cover overhead and profits. The owner paid the invoice on April 20. How much extra profit will be made on the sale, as a percent of the regular selling price, due to the early payment of the invoice?

Case Study I.1 Focusing on Prices

» Elliot's Drug Store is a small independent drugstore. It has a small but progressive camera department. Since Elliot's does not sell very many cameras in a year, it only keeps a small number in stock. Elliot's has just ordered five of the new autofocus or "idiot-proof" cameras from Kodak. Elliot's owner has been told that the cost of each camera will be $190, with terms 2/10, n/30. The Manufacturer's Suggested Retail Price (MSRP) of each camera is $425. Elliot's owner calculates that the overhead is 40% of the MSRP and that the desired profit is 15% of the MSRP.

Zellers has a large camera shop in its store in the mall in this same town. It has ordered 50 of these same cameras from Kodak. Zellers has been offered both a cash discount and a quantity discount off the list price of $190. The cash discount is 3/15, n/30, while the quantity discount is 4%. Zellers estimates its overhead is 20% of the MSRP and it would like to make a profit of 30% of the MSRP.

QUESTIONS

1. What is the cost per camera (ignoring taxes) for Elliot's Drug Store and for Zellers?

2. For each store, what is the minimum selling price required to cover cost, overhead, and desired profits?

3. If Elliot's and Zellers sell the camera at the MSRP, how much extra profit will each store make

   a. in dollars?

   b. as a percent of MSRP?

4. What rate of a markdown from MSRP can Zellers offer to cover its overhead and make its originally intended profit?

Case Study I.2 Putting a Price on the Table

Hanover Furniture Company manufactures furniture but is well known for its stylish diningroom suites. Hanover has found that there is a lot of confusion surrounding the term *list price.* There is the list price at which Hanover offers its product to the furniture stores. These furniture retailers expect to get a discount on this list price because they pay their bills early, or they order large quantities, or they offer a prestigious location for selling Hanover's excellent diningroom suites. Hanover also has a list price or Manufacturer's Suggested Retail Price (MSRP) at which it would like to see its product sold. Hanover feels that this MSRP is a fair price in comparison with competitive suites and will bring a good return to both retailer and manufacturer. Most retailers, of course, would like to advertise the list price (MSRP) less a discount so that consumers will feel that they are getting a bargain. To resolve this problem, Hanover has decided to offer its diningroom suites to retail outlets at the MSRP and offer a larger trade discount.

Putting its new policy into practice, Hanover has offered its newest dining room suite to McKay's Furniture Store for a list price (MSRP) of $1025, less a trade discount of 40%. McKay's will now advertise the suite as $1025 less 20%.

QUESTIONS

1. For how much did McKay's purchase the diningroom suite?

2. What is McKay's selling price?

3. If McKay's sells at this price, what will be the rate of markup based on cost?

4. McKay's discovers that Becker Furniture, across town, is advertising a similar diningroom suite for $779. By what additional percent must McKay's mark down its suite to match this price?

5. If McKay's marks down its suite to match the Becker Furniture advertised price, what rate of markup based on cost will McKay's make?

SUMMARY OF FORMULAS

Formula I.1A

$$\text{AMOUNT OF DISCOUNT} = \text{RATE OF DISCOUNT} \times \text{LIST PRICE}$$

Finding the amount of discount when the list price is known

Formula I.1B

$$\text{LIST PRICE} = \frac{\text{AMOUNT OF DISCOUNT}}{\text{RATE OF DISCOUNT}}$$

Finding the list price when the amount of discount is known

Formula I.1C

$$\text{RATE OF DISCOUNT} = \frac{\text{AMOUNT OF DISCOUNT}}{\text{LIST PRICE}}$$

Finding the rate of discount when the amount of discount is known

Formula I.2

$$\text{NET PRICE} = \text{LIST PRICE} - \text{AMOUNT OF DISCOUNT}$$

Finding the net amount when the amount of discount is known

Formula I.3A

$$\text{NET PRICE FACTOR (NPF)} = 100\% - \%\ \text{DISCOUNT}$$

Finding the net price factor (NPF)

Formula I.3B

$$\text{NET PRICE FACTOR (NPF)} = (1 - d)$$

where d = rate of discount in decimal form

Restatement of Formula I.3A in algebraic terms

Formula I.4A

$$\text{NET PRICE} = \text{NET PRICE FACTOR (NPF)} \times \text{LIST PRICE}$$

Finding the net amount directly without computing the amount of discount

Formula I.4B

$$N = (1 - d)L \quad \text{or} \quad N = L(1 - d)$$

Restatement of Formula I.4A in algebraic terms

Formula I.5A

$$\text{NET PRICE FACTOR (NPF) FOR THE DISCOUNT SERIES} = \text{NPF FOR THE FIRST DISCOUNT} \times \text{NPF FOR THE SECOND DISCOUNT} \times \ldots \times \text{NPF FOR THE LAST DISCOUNT}$$

Formula I.5B

$$\text{NPF FOR A DISCOUNT SERIES} = (1 - d_1)(1 - d_2)(1 - d_3) \ldots (1 - d_n)$$

Formula I.6A

$$\text{NET PRICE} = \text{NET PRICE FACTOR FOR THE DISCOUNT SERIES} \times \text{LIST PRICE}$$

Finding the net amount directly when a list price is subject to a series of discounts

Formula I.6B

$$\text{NET PRICE} = (1 - d_1)(1 - d_2)(1 - d_3)\ldots(1 - d_n)\text{L}$$

Restatement of Formula I.6A in algebraic terms

Formula I.7

SINGLE EQUIVALENT RATE OF DISCOUNT FOR A DISCOUNT SERIES
= 1 − NPF FOR THE DISCOUNT SERIES

$$= 1 - [(1 - d_1)(1 - d_2)(1 - d_3)\ldots(1 - d_n)]$$

Finding the single rate of discount that has the same effect as a given series of discounts

Formula I.8A

SELLING PRICE = COST OF BUYING + EXPENSES + PROFIT

or

$$S = C + E + P$$

Basic relationship between selling price, cost of buying, operating expenses (or overhead), and profit

Formula I.8B

SELLING PRICE = COST OF BUYING + MARKUP

or

$$S = C + M$$

Formula I.9

MARKUP = EXPENSES + PROFIT

or

$$M = E + P$$

Basic relationship between markup, cost of buying, operating expenses (or overhead), and profit

Formula I.10

$$\text{RATE OF MARKUP BASED ON COST} = \frac{\text{MARKUP}}{\text{COST}} = \frac{M}{C}$$

Finding the rate of markup as a percent of cost

Formula I.11

$$\text{RATE OF MARKUP BASED ON SELLING PRICE} = \frac{\text{MARKUP}}{\text{SELLING PRICE}} = \frac{M}{S}$$

Finding the rate of markup as a percent of selling price

GLOSSARY

Cash discount a reduction in the amount of an invoice, usually to encourage prompt payment of the invoice *(p. 483)*

Credit period the time period at the end of which an invoice has to be paid *(p. 483)*

Discount a reduction from the original price *(p. 476)*

Discount period the time period during which a cash discount applies *(p. 483)*

Discount series two or more discounts taken off a list price in succession *(p. 476)*

End-of-month dating payment terms based on the last day of the month in which the invoice is dated *(p. 483)*

Gross profit *see* **Markup**

List price price printed in a catalogue or in a list of prices *(p. 472)*

Manufacturer's suggested retail price (MSRP) catalogue or list price that is reduced by a trade discount *(p. 472)*

Margin *see* **Markup**

Markdown a reduction in the price of an article sold to the consumer *(p. 501)*

Markup the difference between the cost of merchandise and the selling price *(p. 492)*

Net factor *see* **Net price factor**

Net price the difference between a list price and the amount of discount *(p. 472)*

Net price factor (NPF) the difference between 100% and a percent discount—the net price expressed as a fraction of the list price *(p. 474)*

Ordinary dating payment terms based on the date of an invoice *(p. 483)*

Partial payment part payment of an invoice *(p. 487)*

Payment terms a statement of the conditions under which a cash discount may be taken *(p. 483)*

Proximo dating *see* **End-of-month dating**

Rate of discount a reduction in price expressed as a percent of the original price *(p. 483)*

Receipt-of-goods dating payment terms based on the date the merchandise is received *(p. 483)*

Regular selling price the price of an article sold to the consumer before any markdown is applied *(p. 502)*

Sale price the price of an article sold to the consumer after a markdown has been applied *(p. 502)*

Single equivalent rate of discount the single rate of discount that has the same effect as a specific series of discounts *(p. 479)*

Total cost the cost at which merchandise is purchased plus the overhead *(p. 501)*

Trade discount a reduction of a catalogue or list price *(p. 472)*

USEFUL INTERNET SITES

www.electronicaccountant.com

Electronic Accountant Free access to news and critical accounting industry information. This site includes Newswire, links and commentary, discussion groups, feature articles, and accounting/tax software exhibit halls.

www.ibc.ca

Insurance Bureau of Canada This site offers an overview of the industry and of recent legal and consumer-related developments and provides significant links to both business and government sites.

Instructions and Tips for Four Preprogrammed Financial Calculator Models

Different models of financial calculators vary in their operation and labelling of the function keys and face plate. This appendix provides you with instructions and tips for solving compound interest and annuity problems with these financial calculators: Texas Instruments BAII Plus and BA-35 Solar, Sharp EL-733A, and Hewlett-Packard 10B. The specific operational details for each of these calculators are given using the following framework:

A. Basic Operations

1. Turning the calculator on and off
2. Operating modes
3. Using the Second function
4. Clearing operations
5. Displaying numbers and display formats
6. Order of operations
7. Memory capacity and operations
8. Operating errors and calculator dysfunction

B. Pre-Calculation Phase (Initial Set-up)

1. Setting to the financial mode, if required
2. Adjusting the calculator's interest key to match the text presentation, if required
3. Setting to the floating-decimal-point format, if required
4. Setting up order of operations, if required

C. Calculation Phase

1. Clearing preprogrammed registers
2. Adjusting for annuities (beginning and end of period), if required
3. Entering data using cash flow sign conventions and correcting entry errors
4. Calculating the unknown variable

D. Example Calculations

1. Compound interest
2. Annuities

E. Checklist for Resolving Common Errors

Go to the section of this appendix that pertains to your calculator. You may want to flag those pages for easy future reference.

II. 1. Texas Instruments BAII Plus Advanced Business Analyst

A. Basic operations

1. Turning the calculator on and off

The calculator is turned on by pressing ON/OFF. If the calculator was turned off using this key, the calculator returns in the standard-calculator mode. If the Automatic Power Down (APD) feature turned the calculator off, the calculator will return exactly as you left it—errors and all, if that was the case. The calculator can be turned off either by pressing ON/OFF again or by not pressing any key for approximately 10 minutes, which will activate the APD feature.

2. Operating modes

The calculator has two modes: the standard-calculation mode and the prompted-worksheet mode. In the standard-calculation mode, you can perform standard math operations and all of the financial calculations presented in this text. This is the default mode for your calculator. Refer to your calculator's *Guidebook* to learn more about the worksheet mode, since it is not addressed in this appendix.

3. Using the Second function

The primary function of a key is indicated by a symbol on its face. Second functions are marked on the face plate directly above the keys. To access the Second function of a key, press 2nd ("2nd" will appear in the upper left corner of the display) and then press the key directly under the symbol on the face plate ("2nd" will then disappear from the display).

4. Clearing operations

→ clears one character at a time from the display, including decimal points.

CE/C clears an incorrect entry, an error condition, or an error message from the display.

2nd (QUIT) clears all pending operations in the standard-calculation mode and returns the display to 0.

CE/C CE/C clears any calculation you have started but not yet completed.

2nd (CLR TVM) sets the financial function registers to 0 and returns to standard-calculation mode.

5. Displaying numbers and display formats

The display shows entries and results up to 10 digits but internally stores numeric values to an accuracy of 13 digits. The default setting in the calculator is 2 decimal places. To change the number of fixed decimal places, press 2nd (Format) along with a number key for the decimal places desired. Then press ENTER to complete the installation. For a floating-decimal-point format, press 2nd (Format) 9 ENTER. Return to standard-calculation mode by pressing 2nd (QUIT).

6. Order of operations

The default for the BAII Plus is Chn. To change to AOS, which will have the calculator do all mathematical calculations in the proper order according to the rules of mathematics, press 2nd Format, arrow down four times, and with display on Chn press 2nd SET to change display to AOS, 2nd Quit to go back to the standard-calculation mode.

7. Memory capacity and operations

The calculator has ten memory addresses available, numbered 0 through 9. To store a displayed value in a memory address (0 through 9), press STO and a digit key 0 through 9. To recall a value from memory and display it, press RCL and a digit key 0 through 9. The numeric value is displayed but is also retained in that memory address.

To clear each memory address individually, store "0" in each selected memory. To clear all of the addresses at the same time, press 2nd (MEM) 2nd (CLR Work).

Memory arithmetic allows you to perform a calculation with a stored value and then store the result with a single operation. You may add, subtract, multiply, divide, or apply an exponent to the value in the memory. Use this key sequence:

(number in display) STO + (or − or × or ÷ or y^x) and a digit key 0 to 9 for the memory address.

8. Operating errors and calculator dysfunction

The calculator reports error conditions by displaying the message "Error n," where n is a number that corresponds to a particular error discussed in the calculator's *Guidebook* on pages 80–82. Errors 4, 5, 7, and 8 are the most common financial calculation errors. A list of possible solutions to calculator dysfunction is given on page 87 of the *Guidebook*. Generally, if you experience difficulties operating the calculator, press 2nd (Reset) ENTER to clear the calculator, and repeat your calculations.

B. Compound interest and annuity calculations

The BAII Plus calculator can be used for virtually all compound interest calculations using the third row of the calculator *after* the payment and interest schedules have been set up in the Second function area of the calculator. Each key represents one of the variables in the formula. The variables are:

- N—Represents time. The value is arrived at by taking the number of *years* involved in the transaction and multiplying it by the value set up in P/Y.
- I/Y—The stated or nominal yearly interest rate.
- PV—The amount of money one has at the beginning of the transaction.
- PMT—The amount of money paid on a regular basis.
- FV—The amount of money one has at the end of the transaction.

To perform compound interest or annuity calculations, the process will be to input the variables that are known and to compute the unknown variable.

For compound interest, the process will be:

1. Set up the payment and interest schedules in the Second function of the calculator. This is done by going **2nd** **P/Y** and inputting the payment and interest schedules as prompted. Since the transaction will not have any payments, simply make the payment and interest schedules the same. For example, if there are no payments and interest is compounded quarterly, the process would be **2nd** **P/Y**, 4, **ENTER**, **2nd** **Quit**. This will set up the proper schedules in both P/Y and C/Y and take you back to the calculator mode.
2. Clear out any old information with **2nd** (CLR TVM).
3. Input the variables you know.
4. Compute the variable you need to find.

For annuity calculations, the process will be:

1. Set up the calculator for either an ordinary annuity (payments made at the end) or an annuity due (payments made at the beginning). This is done by hitting **2nd** (BGN) and then setting up the display to END or BGN. **2nd** (Set) will allow you to switch between the two options. **2nd** **Quit** will take you back to the calculcator. Note if the calculator is in END mode, the display will be clear in the upper right-hand corner of the display; if it is in BGN mode, the letters BGN will appear in the upper right-hand corner.
2. Set up the payment and interest schedules in the Second function of the calculator. This is done by going **2nd** **P/Y** and inputting the payment and interest schedules as prompted. For example, if the transaction had monthly payments with quarterly compounding, the process would be **2nd** **P/Y**, 12, **ENTER**, **↓**, 4, **ENTER**, **2nd** **Quit**. This would set up monthly payments with interest compounded quarterly.
3. Clear out the old information with **2nd** (CLR TVM).
4. Input the variables you know.
5. Compute the variable you need to find.

C. Calculation phase

1, 2. The steps required to perform calculations and an example calculation appear on pages 91–92 in Chapter 3. The steps required for annuities and sample annuity calculations appear on pages 225–226 in Chapter 6.

3. Entering data using cash flow sign conventions and correcting entry errors

Data can be entered in any order, but you *must* observe the cash flow sign conventions. For compound interest calculations, always enter PV as a negative number and all other values (N, I/Y, FV) as positive numbers. An error message will be displayed when calculating I/Y or N if both FV and PV are entered using the same sign. For annuity calculations, enter either PV or PMT as negative numbers and all other values (N, I/Y, FV) as positive numbers. When PV = 0, designate PMT as the negative number and all other values (N, I/Y, FV) as positive numbers. Failure to observe this sign convention will result in either an error message in the display when calculating I/Y values or an incorrect negative number when calculating values of N.

Data entry errors can be corrected one character at a time by using → or the entry and error message can be cleared from the display by using CE/C.

4. Calculating the unknown variable

Press CPT and the financial key representing the unknown variable after all the known variable data are entered (including 0 for PV or FV if required). Successive calculations are possible because numerical values stored in the function key registers remain there until cleared or replaced. The value stored in any of the function key registers can be determined without altering its value by pressing RCL and the function key.

D. Example calculations

1. Compound interest

See page 91 in Chapter 3 for an example of a compound interest calculation using this calculator.

2. Annuities

See pages 224 and 227 in Chapter 6 for examples of annuity calculations using this calculator.

E. Checklist for resolving common errors

1. Confirm that the P/Y and C/Y are properly set.
2. Confirm that the decimal place format is set to a floating decimal point.
3. If attempting annuity calculations, check to see that the calculator is in the appropriate payment mode ("END" or "BGN").
4. Clear all function key registers before entering your data.
5. Be sure to enter a numerical value, using the cash flow sign convention, for all known variables before solving for the unknown variable, even if one of the variables is 0.

II. 2. Texas Instruments BA-35 Solar Business Analyst

A. Basic operations

1. Turning the calculator on and off

Turn on the calculator by pressing AC/ON . The calculator turns off automatically when the solar cell panel is no longer exposed to light.

2. Operating modes

The available operational modes are financial (FIN), statistical (STAT), and Profit Margin (no message). The message FIN, STAT, or no message appears in the lower left corner of the display to indicate the current mode.

Change the mode by pressing MODE until the desired mode is displayed. Changing to a new mode clears the contents from the mode registers.

3. Using the Second function

The primary function of a key is indicated by a symbol on the face of the key. Second functions are marked on the face plate directly above the keys. To access the Second function of a key, press 2nd ("2nd" will appear in the upper left corner of the display) and then press the key directly under the symbol on the face plate ("2nd" will then disappear from the display).

4. Clearing operations

AC/ON , in addition to turning the calculator on, clears the display, all pending operations, the memory, and the mode registers. Furthermore, the key also sets the calculator to floating-decimal format and to the financial mode.

CE/C clears incorrect entries, error conditions, the display, or pending operations. It does not clear the memory, the mode registers, or the display format.

2nd CE/C clears values stored in the mode registers.

→ clears the last digit entered.

5. Displaying numbers and display formats

The display shows entries and results up to 10 digits but internally stores numeric values to an accuracy of 13 digits. The calculator normally displays numbers in the floating-decimal-point format unless the result of a calculation is too large or too small to be displayed; then scientific notation is used. The floating-decimal-point format appears when the calculator is turned on. To change the format to any number of decimal places up to 9, press [2nd] (Fix) then press the appropriate digit key from [0] to [9]. The calculator can be returned to the floating-decimal-point format by pressing [2nd] (Fix) [•] or by pressing [AC/ON].

6. Memory capacity and operations

[STO] stores the displayed numeric value in the memory, replacing any value previously stored.

[SUM] or [±] [SUM] adds or subtracts the displayed numeric value to the contents of the memory.

[RCL] [2nd] (MEM) recalls and displays the number stored in memory, without affecting the memory's contents.

7. Operating errors and calculator dysfunction

"Error" appears in the display when there is an error condition. General error conditions normally occur when the function is out of range of the calculator or not defined, or a key or key sequence is pressed that cannot be performed in the current mode. In the financial mode, errors can occur when insufficient data are entered or when no solution exists.

The display can appear blank (digits do not appear) for a number of reasons. The most common are:

- The calculator has powered down due to an inadequate light source. (The remedy is to turn the calculator back on.)
- A long calculation is in progress. (The remedy is to wait for at least 20 seconds, then investigate further.)
- A function does not appear to work. (The remedy is to check that the calculator is set to the correct operating mode.)
- The number of decimal digits expected is not displayed. (The remedy is to reset the decimal format and repeat the calculation.)

For other, less common dysfunctions, refer to pages 35–36 of the *Guidebook.*

B. Pre-calculation phase

Financial mode is the default mode when the calculator is turned on. If the calculator is in one of the other modes, press MODE until "FIN" appears in the lower left corner of the display. No adjustment is required to match this calculator's performance to the text presentation. Floating-decimal-point format is the default decimal setting when the calculator is turned on.

C. Calculation phase

1. Clearing preprogrammed registers

AC/ON clears the financial modes and registers.

2. Adjusting for annuities (beginning and end of period)

The default mode for annuity calculations is "end of period." If "beginning of period" calculations are required, press 2nd (BGN). "Begin" will appear in the lower middle portion of the display. To return to "end of period" mode, press 2nd (BGN) again or press AC/ON .

3. Entering data using cash flow sign conventions and correcting entry errors

This calculator does not require data to be entered using the cash flow sign convention for simple compound interest calculations. In other words, both PV and FV can be entered as positive values. However, for problems involving annuity calculations where PV = 0, enter the PMT amount as a negative number. Otherwise, anomalies will occur in the display when calculating FV(it will appear as a negative value), N (an incorrect negative answer will appear), or *i* ("Error" appears in the display). When FV = 0 (in problems involving paying off mortgages and loans), all values can be entered as positive values. Furthermore, when calculating PMT, all other values can be entered as positive. The resulting absolute value for PMT will be correct but the sign will be positive when FV = 0 and negative when PV = 0.

Refer to "Clearing Options" above for correcting data entry errors.

4. Calculating the unknown variable

Press CPT and the financial key representing the unknown variable after all the known variable data are entered (including 0 for PV or FV if required). Successive calculations are possible because numerical values stored in the function key registers remain there until cleared or replaced. The value stored in any of the function key registers can be determined without altering its value by pressing RCL and the function key.

D. Example calculations

1. Compound interest (Example 3.2A, page 88)

Key in	*Press*	*Display shows*	
	AC/ON or 2nd (CMR)	0	clears the mode registers
6000	PV	6000	this enters the present value P (principal)
2.5	I/Y	2.5	this enters the periodic interest rate *i* as a percent
20	N	20	this enters the number of compounding periods *n*
	CPT FV	9831.698641	this computes and displays the unknown future value S

2. Annuities (Example 5.2D, page 179)

Key in	*Press*	*Display shows*	
0	PV	0	a precaution to avoid incorrect answers
10	± PMT	−10	this enters the periodic payment R
0.5	I/Y	0.5	this enters the conversion rate *i* as a percent
60	N	60	this enters the number of payments *n*
	CPT FV	697.7003047	this computes the unknown future value S_n

E. Checklist for resolving common errors

1. Check to see that your calculator is in the financial (FIN) mode.
2. Check to see that the calculator is in the appropriate payment mode ("Begin" or end mode).
3. Clear all registers before entering your variable data by pressing AC/ON or 2nd (CMR).

4. Be sure to enter values for all variables except the unknown variable, before solving for the unknown variable, even if one of variables is 0.
5. Observe the cash flow sign conventions (discussed above) when entering the data to avoid unwanted negative signs, display errors, or incorrect answers.

II. 3. Sharp EL-733A Business/Financial Calculator

A. Basic operations

1. Turning the calculator on and off

C•CE turns the calculator on. OFF turns the calculator off. To conserve battery life, the calculator will turn itself off automatically 9 to 13 minutes after the last key operation.

2. Operating modes

The available operational modes are financial (FIN), statistical (STAT), and Normal (no message). The message FIN, STAT, or no message appears in the upper right corner of the display to indicate the current mode. Change the mode by pressing 2nd F (MODE) until the desired mode is displayed.

3. Using the Second function

The primary function of a key is indicated by a symbol on the face of the key. Second functions are marked on the face plate directly above the keys. To access the Second function of a key, press 2nd F ("2ndF" will appear in the upper left corner of the display) and then press the key directly under the symbol on the face plate ("2ndF" will then disappear from the display).

4. Clearing operations

2nd F (CA) clears the numerical values and calculation commands including data for financial calculations. The contents of memory register storage are not affected.

C•CE x→M clears the memory.

C•CE clears the last entry.

C•CE C•CE clears the calculator of all data *except* the data for financial calculations.

→ clears the last digit entered.

5. Displaying numbers and display formats

The display shows entries and results up to 10 digits. The default setting in the calculator is the floating decimal. To change the number of fixed decimal places, press 2nd F (TAB) along with a number key for the decimal places desired. For a floating-decimal-point format, press 2nd F (TAB) •. The number of decimal places is retained even when the power is turned off.

Various messages can appear in the display from time to time. Refer to page 73 of the *Operation Manual and Application Manual* for a complete list.

6. Memory capacity and operations

This calculator has one memory address. To store a displayed value in memory, press [x→M].

To clear the memory of values other than zero, press [C•CE] [x→M].

To recall a value from memory and display it, press [RM].

To add a displayed amount to the value in the memory, press [M+]. To subtract a displayed amount to the value in the memory, press [±] [M+].

7. Operating errors and calculator dysfunction

Operational errors are indicated by the symbol **E** in the lower left corner of the display. See pages 74–77 of the *Operation Manual and Application Manual* for a complete description of errors and error conditions that may affect the operation and functioning of your calculator. The error symbol is cleared from the display by pressing [C•CE].

B. Pre-calculation phase

With the calculator on, set the financial mode by pressing [2nd F] (MODE) until the FIN message appears in the upper right corner of the display. The calculator requires no change to a register or mode in order to match the text presentation. To set the calculator to the floating-decimal-point format, press [2nd F] (TAB) [•].

C. Calculation phase

1. Clearing preprogrammed registers

[2nd F] (CA) clears the preprogrammed registers of numerical values and sets them to 0 for financial calculations.

2. Adjusting for annuities (beginning and end of period)

The default mode for annuity calculations is "end of period." If "beginning of period" calculations are required, press [BGN]. "BGN" will appear in the upper right corner of the display. To return to "end of period" mode, press [BGN] again. "BGN" will disappear from the display.

3. Entering data using cash flow sign conventions and correcting entry errors

Data can be entered in any order but you must observe the cash flow sign conventions to avoid operational errors and incorrect answers. For compound interest calculations, *always* designate PV as a negative number and all other values (N, *i*, FV) as positive numbers. If you do not observe this sign convention when you enter data, your answer will be the same numerical value but the opposite sign of the answer in the text. An error message will be displayed when calculating *i* or N if both FV and PV are entered using the same sign. For annuity calculations, when FV = 0, designate PV as a negative number and all other values (N, *i*, PMT) as positive numbers. When PV = 0, designate PMT as the negative number and all other values (N, *i*, FV) as positive numbers. Failure to observe this sign convention will result in either an error message in the display when calculating *i* values or an incorrect negative number when calculating values of N.

4. Calculating the unknown variable

Press [COMP] and the financial key representing the unknown variable after all the known variable data are entered (including 0 for PV or FV if required). Successive calculations are possible because numerical values stored in the function key registers remain there until cleared or replaced. The value stored in any of the function key registers can be determined without altering its value by pressing [2nd F] (RCL) and the function key.

D. Example calculations

1. Compound interest (Example 3.2A, page 88)

Key in	*Press*	*Display shows*	
	[2nd F] (CA)	no change	clears all registers
6000	[±] [PV]	−6000	this enters the present value P (principal) with the correct sign convention
2.5	[*i*]	2.5	this enters the periodic interest rate *i* as a percent
20	[N]	20	this enters the number of compounding periods *n*
	[COMP] [FV]	9831.698642	this computes and displays the unknown future value S

2. Annuities (Example 5.2D, page 179)

Key in	*Press*	*Display shows*	
0	PV	0	a precaution to avoid incorrect answers
10	± PMT	−10	this enters the periodic payment R
0.5	*i*	0.5	this enters the conversion rate *i* as a percent
60	N	60	this enters the number of payments *n*
	COMP FV	697.7003051	this computes the unknown future value S_n

E. Checklist for resolving common errors

1. Check to see that your calculator is in the financial (FIN) mode.
2. Check to see that the calculator is in the appropriate payment mode (BGN or end mode).
3. Clear all registers before entering your data by pressing 2nd F (CA).
4. Be sure to enter values for all variables except the unknown variable, before solving for the unknown variable, even if one of variables is 0.
5. Observe the cash-flow sign conventions (discussed above) when entering the data to avoid unwanted negative signs, display errors, or incorrect answers.

II. 4. Hewlett-Packard 10B Business Calculator

A. Basic operations

1. Turning the calculator on and off

Turn the calculator on by pressing C. Turn the calculator off by pressing the yellow key ■ (OFF).

To conserve energy, the calculator turns itself off automatically approximately 10 minutes after you stop using it. The calculator has a continuous memory, so turning it off does not affect the information you have stored in the memory.

2. Operating modes

You can perform all of the financial calculations presented in this text as soon as you turn on the calculator. No mode adjustment is required for financial-, statistical-, or standard-mode calculations.

3. Using the Second function

The primary function of a key is indicated by a symbol on the face of the key. Second functions are marked on the face plate directly above the keys. To access the Second function of a key, press [■] (an arrow will appear in the lower left corner of the display) and then press the key directly under the symbol on the face plate (the arrow will then disappear from the display).

4. Clearing operations

[C] [N] [Σ+], all held down at the same time, clears all memory and resets all modes.

[■] (CLEAR ALL) clears all memory, but does not reset the modes.

[←] [C] clears the message and restores the original constants.

[C] clears the entered number to 0.

[←] clears the last digit entered.

5. Displaying numbers and display formats

The display shows entries and results up to 12 digits. Brightness is controlled by pressing [C] [+] to increase brightness and pressing [C] [−] to decrease brightness. The default setting is 2 decimal places. Regardless of the display format, each number entered is stored with a signed 12-digit number and a signed 3-digit exponent. To change the number of fixed decimal places, press [■] (DISP) and a number key for the number of decimal places desired. For a floating decimal point, press [■] (DISP) [•]. To temporarily view all 12 digits, press [■] (DISP) [=].

Graphics in the display are used to indicate various settings, operating modes, error conditions, and calculator dysfunctions. Refer to the *Owner's Manual*, pages 133–135, for a complete list.

6. Memory capacity and operations

This calculator has 15 numbered registers available to store numbers, as well as a single storage register called the M register.
To store a number in the M register, press [→M].

To recall a value from the M register and display it, press [RM].

To add a displayed amount to the value in the M register, press [M+]. To subtract a displayed amount to the value in the memory, press [±] [M+].

To store a displayed value in a numbered memory register (numbered 0 to 4), press [■] (STO) and a digit key [0] through [9] and [•] [0] to [•] [4].

To recall a number from a numbered memory register, press [RCL] and the digit key for the memory register number.

7. Operating errors and calculator dysfunction

Operational errors are indicated by an error message appearing in the display. For a complete description of the error messages, refer to pages 133–135 of the *Owner's Manual*. For calculator dysfunctions, refer to pages 116–117 and page 120 of the *Owner's Manual*.

B. Pre-calculation phase

1. No additional adjustment is required to set the calculator to the financial mode. Begin calculations as soon as you turn on your calculator.

2. Adjusting the calculator's interest key to match the text presentation

The P/YR register must be set to 1 to match the calculator's performance to the text presentation. The default setting is 12. To change the value to 1, follow the key sequence below:

Key in	*Press*	*Display shows*	
	■ (CLEAR ALL)	P_Yr	clears all memory but does not reset modes. Checks the P/YR register.
1	■ (P/YR)	1.00	changes P/YR register to 1
	■ (CLEAR ALL)	0	clears the display

3. To set the decimal display format to the floating-decimal-point format, press ■ (DISP) • .

C. Calculation phase

1. Clearing preprogrammed registers

■ (CLEAR ALL) sets all key numerical registers to 0 and momentarily displays the P/YR value.

2. Adjusting for annuities (beginning and end of period)

The default mode for annuity calculations is "end of period." If "beginning of period" calculations are required, press ■ (BEG/END). "BEGIN" will appear in the lower middle portion of the display. To return to "end of period" mode, press ■ (BEG/END) again. "BEGIN" will disappear from the display.

3. Entering data using cash flow sign conventions and correcting entry errors

Data can be entered in any order using the financial function keys. To confirm the values already in the registers or to validate your data entry, press **RCL** and the desired function key.

You must observe the cash flow sign conventions to avoid errors like "no solution" or incorrect answers when calculating N. For compound interest calculations, *always* enter PV as a negative number and the other variables as positive numbers. For annuity calculations, when FV = 0, enter PV as a negative number and all other values (N, I/YR, PMT) as positive numbers. However, when PV = 0, enter PMT as a negative number and all other values (N, I/YR, FV) as positive numbers. When calculating PMT, enter FV as a negative number to match the text presentation. If you do not observe the sign convention, the numerical value you calculate will be identical to that of this text except when calculating N (your answer will be incorrect) or when calculating I/YR (an error message may appear in the display).

Data entry errors can be corrected character by character by pressing **←** or the entry can be cleared from the display by pressing **C**.

4. Calculating the unknown variable

Press the financial key representing the unknown variable after all the known variable data are entered (including 0 for PV or FV if required). Successive calculations are possible because numerical values stored in the function key registers remain there until cleared or replaced. The value stored in any of the function key registers can be determined without altering its value by pressing **RCL** and the function key.

D. Example calculations

1. Compound interest (Example 3.2A, page 88)

Key in	*Press*	*Display shows*	
	■ (CLEAR ALL)	0	clears the function key registers and confirms the value in the P/YR register
6000	**±** **PV**	−6000	this enters the present value P (principal) with the correct sign convention
2.5	**I/YR**	2.5	this enters the periodic interest rate *i* as a percent
20	**N**	20	this enters the number of compounding periods *n*
	FV	9831.69864174	this computes and displays the unknown future value S

2. Annuities (Example 5.2D, page 179)

Key in	*Press*	*Display shows*	
0	PV	0	a precaution to avoid incorrect answers
10	± PMT	−10	this enters the periodic payment R
0.5	I/YR	0.5	this enters the conversion rate *i* as a percent
60	N	60	this enters the number of payments *n*
	FV	697.700305099	this computes the unknown future value S_n

E. Checklist for resolving common errors

1. Check to determine that the P/YR register is set to 1 to match the text presentation.
2. Clear all registers before entering your data by pressing ■ (CLEAR ALL).
3. Check to see that the calculator is in the appropriate payment mode (BEGIN or end mode).
4. Be sure to enter values for all variables except the unknown variable, before solving for the unknown variable, even if one of variables is 0.
5. Observe the cash flow sign conventions (discussed above) when entering the data to avoid unwanted negative signs, display errors, or incorrect answers.

Answers to Odd-Numbered Problems, Review Exercises, and Self-Tests

CHAPTER 1

Exercise 1.1

A. **1.** 112 days
3. 166 days

B. **1.** December 1, 2007
3. April 5, 2007

Exercise 1.2

A. **1.** 0.035; 1.25
3. 0.0825; $\frac{183}{365}$

B. **1.** $1096.88
3. $95.21
5. $75.34

C. **1.** $10.87
3. $21.76

Exercise 1.3

A. **1.** $1224.00
3. 10.75%
5. 14 months
7. 144 days

B. **1.** $3296.00
3. 9.5%
5. 11 months
7. $876.00
9. $400 000
11. 9%
13. November 18, 2006
15. **(a)** $51 975.00
(b) $51 943.53
(c) 3.887%

Exercise 1.4

A. **1.** $490.13
3. $768.75
5. $849.21
7. $1298.00

B. **1.** $2542.53
3. $13 800.00
5. $26 954.84
7. $13 864.50

Exercise 1.5

A. **1.** $266.00; $13.30
3. $517.50; $547.17
5. $2025.00; 292 days

B. **1.** $1222.00
3. $1704.60
5. $1600.04
7. $9755.22
9. $9933.50

Exercise 1.6

A. **1.** $829.33
3. $617.50
5. $1103.37
7. $856.47
9. $777.81
11. $1070.39

B. **1.** $1156.80
3. $519.17
5. $1408.21
7. $1599.35

Review Exercise

1. **(a)** 172
(b) 214
3. **(a)** $1160.00
(b) $601.77
5. **(a)** $750.00
(b) $5709.97
7. $3000.17
9. 8.25%
11. 196 days
13. $1601.89
15. $3200.00
17. $1736.47
19. $2664.00
21. $3404.32
23. $1614.74
25. $961.50
27. $1587.06

Self-Test

1. $21.40
3. 6.5%
5. $6187.50
7. $4306.81
9. 359 days
11. $7432.80
13. $1163.85
15. $1799.23

CHAPTER 2

Exercise 2.1

A. **1.** December 30, 2005
3. $530.00
5. 154 days
7. $544.54

B. **1.** **(a)** March 3, 2008
(b) 155 days
(c) $21.40
(d) $861.40
3. **(a)** April 3, 2004
(b) 63 days
(c) $14.02
(d) $1264.02

Exercise 2.2

A. **1.** $631.24
3. $837.19

Exercise 2.3

A. **1.** $500.00

B. **1.** $1471.28
3. $1615.56

C. **1.** $98 627.91
3. 2.73%
5. **(a)** 2.72%
(b) $99 636.18
(c) 2.706%

Exercise 2.4

A. **1.** $37.50
3. $22.50
5. $307.56

B. **1.** $3786.27
3. $1825.63
5. $178.66

Exercise 2.5

A. **1.** **(a)** $0.16
(b) $4.05
(c) $1.24
(d) $10.00
(e) −$956.34

Exercise 2.6

A. **1.** Totals are $1233.69; $33.69; $1200.00

Review Exercise

1. **(a)** November 2
(b) $35.62
(c) $1635.62
3. $1500.00
5. $5125.75
7. $814.17
9. $1269.57
11. 6.1493%
13. $449.12
15. **(a)** $54.57; $62.49; $70.82; $69.86; $70.75
(b) −$10 623.49

Self-Test

1. $19.79
3. $1160.00
5. $1664.66

7. (a) $98 116.43
   (b) 3.680%
9. $340.26
11. Totals are $4070.41; $70.41; $4000.00

CHAPTER 3

Exercise 3.1

A. 1. 1; 0.12; 5
3. 4; 0.01375; 36
5. 2; 0.0575; 27
7. 12; $i = 0.0066667$; 150
9. 2; 0.06125; 9

B. 1. 1.7623417
3. 1.6349754
5. 4.5244954
7. 2.7092894
9. 1.7074946

C. 1. (a) 48
   (b) 2.5%
   (c) 1.025^{48}
   (d) 3.2714896

Exercise 3.2

A. 1. $713.39
3. $2233.21
5. $5468.38
7. $6639.51
9. $662.02
11. $4152.58
13. $2052.74

B. 1. $6884.47; $1884.47
3. $1441.71
5. (a) $199.26
   (b) $202.24
   (c) $203.81
   (d) $204.89
7. (a) $148.59; $48.59
   (b) $220.80; 120.80
   (c) $487.54; 387.54
9. $3712.50
11. $761.75
13. $17116.96
15. 144.65
17. (a) Bank; $6968.27; $6914.09
   (b) $54.18

C. 1. $3010.85
3. $1452.79
5. $3436.38
7. $1102.13
9. $1444.24

Exercise 3.3

A. 1. $574.37
3. $371.86
5. $409.16
7. $500.24
9. $749.91
11. $4344.21

B. 1. $1338.81; $261.19
3. $762.84
5. $3129.97
7. $1398.85

Exercise 3.4

A. 1. $1767.71; $232.29
3. $3981.87; $1018.13
5. $2642.50; $557.50
7. $1012.96; $671.00
9. $2310.82; $720.62
11. $2036.86; $540.42
13. $1332.82; $261.78

B. 1. $4589.47
3. $1345.06
5. $3800.24
7. $1561.49

C. 1. $8452.52
3. $2346.36
5. $3488.29
7. $1074.71
9. $1972.80
11. $1492.15

Exercise 3.5

A. 1. $5983.40
3. $2537.13
5. $1673.49
7. $641.36

B. 1. $3426.73
3. $2464.35
5. $1536.03
7. $987.93

C. 1. (a) $2851.94
   (b) $3265.19
   (c) $4000.00
   (d) $5610.21
3. $6805.31
5. $655.02
7. $2423.97
9. (a) $3113.90
   (b) $1847.95

Review Exercise

1. (a) $1198.28
   (b) $1221.61
   (c) $1227.05
3. (a) $2890.09
   (b) $890.09
5. (a) $6144.45; $4344.45
   (b) $3305.27; $2055.27
7. $10 681.77
9. $11 102.50
11. $4194.33
13. $9791.31
15. $2830.68; $3190.63
17. $9294.85
19. $1035.70
21. $2838.62
23. $2079.94
25. $110 440.03
27. $26 048.42
29. (a) $2742.41
   (b) $2911.55
   (c) $3281.79
31. $4857.56
33. $1820.32
35. $3574.57

Self-Test

1. $2129.97
3. 2.7118780
5. $6919.05
7. $14 711.80
9. $4504.29
11. $10 138.19
13. $2661.85
15. $848.88

CHAPTER 4

Exercise 4.1

A. 1. (a) 13.4 years
   (b) 28 quarters
   (c) 117.8 months
   (d) 8 half-years
   (e) 37.313 quarters
   (f) 37.167 half-years

B. 1. 9.329 years
3. 17.501 years
5. 5.622 years
7. November 1, 2001
9. 21 months
11. 2 years, 298 days
13. 3 years, 230 days

Exercise 4.2

A. 1. 4.5%
3. 9.6%
5. 9.778%

B. 1. 9.237%
3. (a) 10.402%
   (b) 7.585%
5. 3.5%
7. 4.771%

Exercise 4.3

A. 1. (a) 12.891%
   (b) 6.168%
   (c) 7.397%
   (d) 10.691%
3. (a) 8.9%
   (b) 6.465%
   (c) 7.383%
   (d) 4.273%

B. 1. 6.991%
3. 4.656%

5. 8.945%
7. 6.2196%
9. 7.385%
11. **(a)** $714.57
(b) $114.57
(c) 3.5567%
13. **(a)** $1831.40
(b) $631.40
(c) 4.3182%

Review Exercise

1. 5.592%
3. 1 year, 231 days
5. 22.517085 half-years (11 years, 95 days)
7. 6.03%
9. **(a)** 5.92%
(b) 14.35%
(c) 8.833%
(d) 8.24%
11. **(a)** 4.59%
(b) 5.96%
13. 8.44%
15. **(a)** 4.0%
(b) 3.7852%
(c) 3.5462%
(d) 3.2989%
17. 1.831663 years (1 year, 304 days)
19. 1.5677878 years (1 year, 208 days)
21. 6.8287904 years (6 years, 303 days–2007-09-30)

Self-Test

1. 10.50 half-years (63 months)
3. 5.535675%
5. 7.352%
7. 10.0%
9. 7.5%

CHAPTER 5

Exercise 5.1

A. **1.** **(a)** annuity certain
(b) annuity due
(c) general annuity
3. **(a)** perpetuity
(b) deferred annuity
(c) general annuity
5. **(a)** annuity certain
(b) deferred annuity due
(c) simple annuity

Exercise 5.2

A. **1.** $54 193.60
3. $59 185.19
5. $17 915.08

B. **1.** $13 045.68
3. $32 434.02
5. **(a)** $8531.12
(b) $4500.00
(c) $4031.12
7. $62 177.25

Exercise 5.3

A. **1.** $9515.19
3. $30 941.11
5. $10 544.91

B. **1.** $6897.02
3. **(a)** $9906.20
(b) $2093.80
5. **(a)** $2523.82
(b) $372.06
7. $12 710.96
9. **(a)** $13 683.13
(b) $1428.88
(c) $14 183.16
(d) $500.03
(e) $ 21.28

Exercise 5.4

A. **1.** $821.39
3. $1117.37
5. $272.73
7. $232.54
9. $1653.70

B. **1.** $207.87
3. $591.66
5. $320.61
7. $410.00
9. $290.55
11. $557.65
13. $229.33
15. $3141.41
17. $114.89
19. $1984.62

Exercise 5.5

A. **1.** 14.6023573 years (14 years, 8 months)
3. 93.0687373 months (7 years, 10 months)
5. 15.5775778 half-years (7 years, 10 months)
7. 15.395937 semi-annual periods (7 years, 9 months)
9. 71.517450 months (6 years)

B. **1.** 74.4983312 months (6 years, 3 months)
3. 69.835347 months (5 years, 10 months)
5. 8.380061 quarters (2 years, 1.14 months)
7. 40.080206 months (3 years, 4 months)
9. 26.241960 months (2 years, 3 months)

Exercise 5.6

A. **1.** 5.0%
3. 9.0%
5. 7.6%
7. 11.37%

B. **1.** 12.5%
3. 11.7%
5. 9.50%
7. 3.04%
9. 7.685%

Review Exercise

1. **(a)** $26 734.60
(b) $17 280.00
(c) $9454.60
3. $722.62
5. 12.575297 years (12 years, 7 months)
7. $411.57
9. $101 517.64
11. 10.524175 half-years (6 years)
13. 7.25%
15. $34 031.63
17. 8.5838838 half-years (9 semi-annual payments)
19. $16 102.46

Self-Test

1. $35 786.08
3. 10.9%
5. 28.06 quarters (84 months)
7. $40 385.39
9. $3268.62

CHAPTER 6

Exercise 6.1

A. **1.** $45 855.46
3. $16 317.77
5. $63 686.72
7. $32 876.06

B. **1.** $23 268.52
3. $2326.65
5. **(a)** $13 265.50
(b) $3265.50
7. $31 293.63

Exercise 6.2

A. **1.** $47 583.23
3. $42 505.51

5. $5106.97
7. $34 627.97

B. 1. $10 041.88
3. $31 736.57
5. (a) $29 829.03
(b) $5170.97
7. $80 000.02
9. $33 393.84

Exercise 6.3

A. 1. $823.60
3. $1121.26
5. $273.20
7. $1426.84
9. $271.68

B. 1. $124.41
3. $799.39
5. $330.63
7. $117.26
9. $26.63

Exercise 6.4

A. 1. 14.51 years (14 years, 6 months)
3. 92.97 months (7 years, 9 months)
5. 15.64 semi-annual periods (7 years, 10 months)
7. 15.43 half-years (7 years, 9 months)
9. 71.57 months (6 years)

B. 1. 44.76 months (3 years, 9 months)
3. 133 months
5. 13 years, 7.3 months
7. 40 deposits
9. 25 years

Exercise 6.5

A. 1. 6.01%
3. 6.86%
5. 12.41%
7. 4.40%

B. 1. 7.31%
3. 8.82%
5. 10.778%
7. 9.24%

Review Exercise

1. $13 509.62
3. $64 125.87
5. $75 962.59
7. (a) $101.81
(b) $1714.37
9. (a) 30.62 quarters (7 years, 8 months)
(b) 27.84 semi-annual periods (13 years, 11 months)
11. (a) 12.46 months (6 years, 3 months)
(b) 23.24 months (1 year, 11 months)
13. 16.33 quarters (4 years, 1 month)
15. 8.30 quarters (2 years, 1 month)
17. (a) $13 719.16
(b) $3951.16
(c) $382.25
(d) $8580.00
19. 96.82 months (8 years, 1 month)

Self-Test

1. $55 246.47
3. 17.34 quarters (4 years, 4 months)
5. $3003.81
7. 32 quarters (8 years)
9. $13 355.86; $96.29

CHAPTER 7

Exercise 7.1

A. 1. $135 334.71; $71 813.10
3. $69 033.69; $35 581.67
5. $43 738.24 $7278.61

B. 1. $204.80
3. $909.95

C. 1. n = 113 months (9 years, 5 months)
3. n = 8 years

D. 1. nominal annual rate = 7.125%
3. nominal annual rate = 12.498%

E. 1. $31 378.29
3. (a) $29 513.14
(b) $10 313.14
5. $33 338.44
7. (a) $1238.56
(b) $1575.00
(c) $336.44
9. $2150.38
11. $300.36
13. $300.74
15. $217.19
17. n = 142.48 months (11 years, 11 months)
19. n = 41 quarterly deposits
21. nominal annual rate = 8.58%

Exercise 7.2

A. 1. $39 342.74
3. $11 304.97

B. 1. $1316.98
3. $903.61

C. 1. n = 35 half-years (17.5 years)
3. n = 18 quarters (4.5 years)

D. 1. nominal annual rate = 11.28%
3. nominal annual rate = 4.89%

E. 1. $59 113.10
3. $24 111.08
5. $459.47
7. n = 37 quarters (9 years, 3 months)
9. nominal annual rate = 9.18%

Exercise 7.3

A. 1. $1403.20
3. $6427.70
5. $21 748.37
7. $10 272.72
9. $962.88

B. 1. $126 738.19
3. (a) –$147 329.91
(b) $272 000.00
(c) $124 670.09
5. $576.10
7. $2820.21
9. $2752.22
11. n = 36 months (3 years)
13. n = 26.712 months
15. $21 831.32
17. $89 397.79
19. n = 27.65 half-years
21. $147.55
23. $165.11
25. $27 246.85
27. n = 11 years
29. n = 41.52 half-years

Exercise 7.4

A. 1. $73 529.41
3. $91 027.01
5. $44 762.78
7. $96 992.34

B. 1. $214.60
3. $850.87
5. $210 240.91
7. $41 854.67
9. $166.70

Review Exercise

1. (a) S_n = $11 223.39
A_n = $5799.35
(b)
S_n(due)= $11 728.44
A_n(due)= $6060.32
3. $19 153.93
5. (a) $2638.84
(b) $3444.00
(c) $805.16
7. n = 109 months
9. n = 28 months
11. nominal annual rate = 5.51%
13. (a) $411.57
(b) $403.50
15. (a) $15 749.42
(b) $3149.42
(c) $432.91
(d) $8179.68
17. n = 51 months
19. n = 49 months
21. (a) $28 435.38
(b) $28 552.64
23. $42 092.99
25. $2736.49
27. (a) $503.87
(b) $1437.96
(c) –$275.12
(d) $1782.29
(e) $89.52
(f) $1037.88
29. n = 29 years
31. (a) n = 6.78 half-years
(b) n = 7.51 years
(c) n = 10.111884 quarters
(d) n = 8.86133 half-years
33. (a) $16 102.46
(b) $3120.21
(c) $17 860.97
(d) $120.21
35. $301.80
37. $910.15
39. n = 26 quarters (6 years, 6 months)
41. $20 964.79
43. $34 543.53
45. $8908.36
47. n = 28.605079 quarters
49. (a) $60 229.26
(b) $49 312.99
(c) $29 620.76
(d) $36 229.73
(e) $86 633.66
(f) $90 133.66
51. $318 181.82
53. $105.30
55. $28 089.24

Self-Test

1. $84 209.75
3. $1258.38
5. $357.34
7. Nominal annual rate = 8.185%
9. $43 246.07
11. $1170.69
13. $6056.04
15. $21.97
17. $122 263.83

CHAPTER 8

Exercise 8.1

A. 1. (a) $549.22
(b) $5633.77
(c) $140.84
(d) $408.38
3. (a) $1103.73
(b) $4913.61
(c) $196.54
(d) $907.19

B. 1. (a) n = 19.48 payments
(b) $2493.97
3. (a) n = 14.53
(b) $8035.94

C. 1. (a) $920.57
(b) $15 052.64
(c) $59 234.20
(d) $23 234.20
3. (a) $136.87
(b) $4199.54
(c) $31.50
(d) $105.37
5. $2054.05; totals are $14 378.41; $4378.41; $10 000.00
7. totals are $14 943.16; $5743.16; $9200.00
9. $651.11
11. $1160.09
13. (a) $3621.90
(b) $1035.27
(c) $2799.85
(d) $10 445.14; totals are $115 900.80; $30 900.80; $8500.00
15. (a) n = 14.022508
(b) $957.79
(c) $1910.53
(d) $4664.98; totals are $35 057.75; $11 057.75; $24 000.00

Exercise 8.2

A. 1. (a) $1829.69
(b) $20 286.42
(c) $819.57
(d) $1010.12
3. (a) $164.04
(b) $4563.53
(c) $22.54
(d) $141.50

B. 1. (a) n = 17.13
(b) $2685.85
3. (a) n = 14.894762
(b) $1584.96

C. 1. (a) $2790.38
(b) $34 892.23
(c) $1599.89
(d) $1190.49
3. (a) $252.15
(b) $34 200.11
(c) $7277.51
(d) $294.24
5. (a) n = 15.413529
(b) $2911.17
(c) $4311.17
7. $3212.01; totals are $22 484.09; $6484.09; $16 000.00
9. $752.68
11. (a) $318.15
(b) $3323.27
(c) $263.14
(d) $364.44

(e)

Partial Amortization Schedule

Payment number	*Amount paid*	*Interest paid*	*Principal repaid*	*Outstanding principal*
0				40 000.00
1	318.15	278.44	39.71	39 960.29
2	318.15	278.17	39.98	39 920.31
3	318.15	277.89	40.26	39 880.05
:	:	:	:	:
:	:	:	:	:
60	:	:	:	37 056.28
61	364.44	317.37	47.07	37 009.21
62	364.44	316.97	47.47	36 961.74
63	364.44	316.56	47.88	36 913.86
:	:	:	:	:

Exercise 8.3

A. **1.** $1006.24
3. $348.26
5. $306.39

B. **1.** **(a)** $n = 29.434057$
(b) $157.46
3. **(a)** $n = 47.779687$
(b) $722.00
5. $234.53
7. $301.45
9. $1152.87
11. **(a)** $n = 61.523813$
(b) $724.48
(c) $84 599.48
(d) $68 599.48

Exercise 8.4

A. **1.** **(a)** $715.83
(b) $83 375.45
(c) $641.41
3. **(a)** $503.15 rounded to $550.00; $n =$ 104.522813 months
(b) $288.05
(c) $2889.95
5. nominal annual rate compounded semi-annually = 9.50%
7. **(a)** $86 514.51
(b) $n = 165.56125$ months
(c) $54 269.19
9. $380.00; balance $38 794.01
11. Balance December 1 = $38 795.58

Review Exercise

1. **(a)** $1491.37
(b) $12 723.84
(c) $15 771.75
(d) $338.49
(e) $1247.91
(f) $4300.94; totals are $47 723.84; $12 723.84; $35 000.00
3. **(a)** $n = 24.830989$
(b) $28 940.29
(c) $426.66
(d) $1807.58
(e) $5477.40; totals are $49 664.41; $9664.41; 40 000.00
5. **(a)** $735.80
(b) $280.57
(c) $23 981.92
(d) $1000.88
(e) $15 900.11; totals are $24 843.68; $11 087.74; $13 755.94
7. **(a)** $n = 9.3198366$
(b) $1139.88
9. **(a)** $n = 11.744464$
(b) $3211.27
11. **(a)** $1091.28
(b) $125 324.69
(c) $1023.15
13. **(a)** $n = 117.33202$ months
(b) $332.79
(c) $118 274.40; $117 332.79; $941.61
15. **(a)** $161.75
(b) $1264.00
(c) $5086.52
(d) $21.41
(e) $478.06; totals are $7764.00; $1264.00; $6500.00
17. 8.38%
19. **(a)** $601.20
(b) $1651.04
(c) $25 598.05
(d) $638.94

Self-Test

1. $6027.52
3. $12 866.98
5. **(a)** $1406.95
(b) $174 506.12
(c) $1479.66
7. 10.0%

CHAPTER 9

Exercise 9.1

A. **1.** $97 780.05
3. $24 098.27
5. $52 193.61
7. $7234.24

B. **1.** $466.70
3. $14 891.87
5. $10 759.95
7. $1184.53
9. $4 278 80 1.75
11. $43 667.11
13. **(a)** $30 258.34
(b) $792.35
(c) $29 465.99
15. $5408.13

Exercise 9.2

A. **1.** **(a)** $4877.98
(b) $20 122.02
3. **(a)** $1773.12
(b) $8626.88
5. **(a)** $100.35
(b) $962.31

B. **1.** **(a)** $14 304.00; $114 304.00
(b) $6359.60; $106 359.60
3. **(a)** $–4435.32; $20 564.68
(b) $3687.99; $28 687.99
5. **(a)** $1620.28
(b) $13 902.68
(c) $13 547.25
7. $4 569 384.48
9. $109 270.75

Exercise 9.3

A. **1.** **(a)** –$77.15; $4922.85

(b)

Schedule of Accumulation of Discount

Payment interval	*coupon b = 3%*	*Interest on book i = 3.25%*	*Discount accumulated*	*Book value*	*Discount balance*
0				4922.85	77.15
1	150.00	159.99	9.99	4932.84	67.16
2	150.00	160.32	10.32	4943.16	56.84
3	150.00	160.65	10.65	4953.81	46.19
4	150.00	161.00	11.00	4964.81	35.19
5	150.00	161.36	11.36	4976.17	23.83
6	150.00	161.73	11.73	4987.90	12.10
7	150.00	162.10	12.10	5000.00	—
Total	1050.00	1127.15	77.15		

3. **(a)** $49.18;
$1079.18

(b)

Schedule of Amortization of Premium

Payment interval	*coupon b = 6%*	*Interest on book i = 5%*	*Premium amortized*	*Book value*	*Premium balance*
0				1079.18	49.18
1	60.00	53.96	6.04	1073.14	43.14
2	60.00	53.66	6.34	1066.80	36.80
3	60.00	53.34	6.66	1060.14	30.14
4	60.00	53.01	6.99	1053.15	23.15
5	60.00	52.66	7.34	1045.81	15.81
6	60.00	52.29	7.71	1038.10	8.10
7	60.00	51.90	8.10	1030.00	—
Total	420.00	370.82	49.18		

B. **1.** $976.84
3. –$950.88
5. $387.55

Exercise 9.4

A. **1.** 5.868%
3. 8.14%
5. 9.26%

Exercise 9.5

A. **1.** **(a)** $558.24
(b) $6399.60
3. **(a)** $22.03
(b) $3104.14

B. **1.** **(a)** $500.00
(b) $265.25
(c) $765.25
(d) $10 868.38
3. **(a)** $62.50
(b) $143.33
(c) $205.83
(d) $2246.16

C. **1.** **(a)** $2699.00
(b) $64 776.00
(c) $10 224.00
3. $2419.29; totals are $16 935.03; $3064.99; $20 000.02
5. $2840.83
7. **(a)** $495.27
(b) $9598.44
9. **(a)** $4275.00
(b) $1124.00
(c) $5399.00
(d) $36 976.00
11. **(a)** $302.01
(b) $21 903.91
(c) $257.62
(d) $764.19
(e) $97 252.61; totals are $54 361.80; $45 637.43; $99 999.23
13. **(a)** $24 750.00
(b) $8604.00
(c) $33 354.00
(d) $13 931.00
(e) $107 196.00
(f) $232 277.00; totals are $172 073.00; $127 929.00; $300 000.00

Review Exercise

1. **(a)** $5336.73
(b) $4550.35
3. $26 795.83
5. **(a)** $1129.61;
$21 129.61
(b) $655.17;
$22 055.17
7. **(a)** $15 620.41
(b) $85 897.02
(c) $1284.53;
$84 612.49
9. $4527.64
11. **(a)** $3469.30
(b) $24 625.42
(c) $23 330.78
13. $914.16; totals are $510.00; $595.84; $85.84
15. $21 576.64; totals are $5700.00; $5123.64; $576.64
17. $1645.34
19. 7.73%
21. **(a)** $44 248.79
(b) $48 017.12
(c) $1795.38
23. 15.24%
25. **(a)** $9268.75
(b) $28 862.02

(c) $1873.21
(d) Totals are
$92 687.50;
$17 312.54;
$110 000.04

27. (a) $13 750.00
(b) $8279.90
(c) $22 029.90
(d) $72 194.23
(e) $5989.29
(f) Totals are
$66 239.23;
$33 760.77;
$100 000.00

29. (a) $2469.40
(b) $10 612.00
(c) $21 194.78
(d) $788.92

31. (a) $n = 38.468063$
(b) $8318.12

Self-Test

1. $9269.44
3. $304.02 (premium)
5. $4507.68
7. 12.46%
9. 12.92%
11. $240.24
13. $603 102.20

CHAPTER 10

Exercise 10.1

A. **1.** Alternative 2; PV $44 634; $53 448
3. Alternative 1; PV $22 205; $23 384
5. Alternative 1 PV $2000; $2393

B. **1.** Alternative 2; PV $62 363; $73 601
3. Offer A; PV $30 026; $29 769
5. Buy; Cost of buying = $83 564; leasing = $106 036

Exercise 10.2

A. **1.** Reject; NPV is −$5367
3. Alternative 1; NPV $234; $203
5. Alternative 1; NPV $1666; $352

B. **1.** Project B; NPV $2568; $1787
3. No; NPV is −$8561
5. Yes; NPV $5696
7. Yes; NPV $5142

Exercise 10.3

A. **1.** 25.3%
3. 20.7%

B. **1.** at 18%, NPV = $6102; at 20%, NPV = −$292; R.O.I. = 19.9%
3. at 18%, NPV = $1286; at 20%, NPV = −$104; R.O.I. = 19.9%
5. at 22%, NPV = $148; at 24%, NPV = −$1538; R.O.I. = 22.2%

Review Exercise

1. Alternative B; PV $51 624; $52 161
3. Alternative 1; NPV $3916; $3597
5. −$22 226
7. at 26%, NPV = $370; at 28%, NPV = −$1939; R.O.I. = 26.3%
9. at 16%, NPV = $47 272; at 18%, NPV = −$22 226; R.O.I = 17.4%
11. at 24%, NVP = $2035; at 26%, NVP = −$184; R.O.I = 25.8%
13. Project B; PV $22 256, $22 909
15. −$1215
17. Yes; NPV = $28 940

Self-Test

1. Alternative A; PV $13 552; $10 856
3. at 16%, NPV = $3466; at 18%, NPV =−$4386; R.O.I. = 16.9%
5. Proposal B; NPV $701; $1427

APPENDIX I

Exercise I.2

A. **1.** $11.07; $13.53
3. $33\frac{1}{3}$%; $25.65
5. 15%; $252.60
7. $133.36; $50.01
9. $134.96; $50.61

B. **1.** 15.9%
3. $2436.00
5. $74.10
7. $426.00

Exercise I.3

A. **1.** $30.24; 32.5%
3. $137.89; 48.55%
5. $1583.33; 61%

B. **1.** (a) $127.68
(b) $112.32
(c) $46.8%
3. (a) 38.75%
(b) $48.2\dot{6}$%
5. 15%
7. $180.00
9. 15%

Exercise I.4

A. **1.** $640.00
3. $776.11
5. $1136.80
7. $4581.50

B. **1.** $582.00; $850.00
3. $564.50; $536.28
5. $810.00; $810.00

C. **1.** (a) Sept. 10
(b) $5276.85
(c) $103.20
3. $2507.19
5. $2184.00
7. (a) $1164.00
(b) $733.54
(c) $600.00
9. (a) $1925.00
(b) $3400.00

Exercise I.5

A. **1.** (a) $6.00
(b) $3.84
(c) $2.16
(d) 25%
(e) 20%
3. (a) $35.00
(b) $31.50
(c) $3.50
(d) $66\frac{2}{3}$%
(e) 40%
5. (a) $10.50
(b) $12.75
(c) ($2.25)

(d) $38\frac{8}{9}\%$
(e) 28%

B. 1. $6.25; 25%; 20%
3. $102.40; 60%; 37.5%
5. $75.95; $21.70; 28.6%
7. 44.24; $22.12; $33\frac{1}{3}\%$
9. $78.10; $46.86; 150%
11. $111.30; $133.56; 20%

C. 1. $13.60
3. $3.00
5. $102.08
7. (a) 150%
(b) 60%
9. (a) $36.50
(b) $29.93
(c) 21.95%
11. (a) $1.65
(b) $66\frac{2}{3}\%$

13. (a) $234.20
(b) 47.37%

Exercise I.6

A. 1. $51.00; $59.00; −$8.00
3. $96.40; $30.65; $7.91
5. $160.00; $19.20; 12%

B. 1. (a) $228.69
(b) 54%
3. (a) 40%
(b) −$8.00
(c) 10.3%
(d) $9\frac{1}{3}\%$
5. −$2.23
7. (a) $14.85; $9.90; $6.60
(b) $957.00
(c) 34.5%
9. (a) $72.00
(b) $54.00
(c) 25%

11. −$6.50
13. −$97.50
15. −$82.50

Review Exercise

1. (a) $31.92
(b) $24.08
(c) $43%
3. 45.66%
5. 15%
7. $30.00
9. September 12; $25 117.40
11. (a) $1940.00
(b) $2813.00
13. (a) $1645.00
(b) $1500.00
15. (a) $90.00
(b) $58.50
(c) 53.85%
(d) $74.88
(e) −$6.48
17. (a) $77.50
(b) 42.86%
19. $240.00
21. (a) −$0.60
(b) 26.98%
23. (a) − $13.20
(b) 25%
25. (a) $189.00
(b) 21.25%
(c) $133\frac{1}{3}\%$
27. (a) $2152.40
(b) 69.43%
(c) 40.98%

Self-Test

1. $295.77
3. 50.5%
5. $1635.04
7. $1450.00
9. $240.00
11. $348.36
13. 180%
15. 23.40%
17. −$660.45

Index

M

N

O

P